Advances in

INORGANIC CHEMISTRY

Volume 51

ADVISORY BOARD

Advances in INORGANIC CHEMISTRY

Heme-Fe Proteins

EDITED BY

A. G. Sykes

Department of Chemistry
The University of Newcastle
Newcastle upon Tyne
United Kingdom

CO-EDITED BY

Grant Mauk

Faculty of Medicine
Department of Biochemistry and Molecular Biology
The University of British Columbia, Vancouver,
British Columbia, Canada

VOLUME 51

ACADEMIC PRESS

A Harcourt Science and Technology Company

San Diego San Francisco New York
Boston London Sydney Tokyo

This book is printed on acid-free paper. 

Academic Press
A Harcourt Science and Technology Company
525 B Street, Suite 1900, San Diego, California 92101-4495, U.S.A.
http://www.apnet.com

Academic Press
24-28 Oval Road, London NW1 7DX, UK
http://www.hbuk.co.uk/ap/

International Standard Book Number: 0-12-023651-6

PRINTED IN THE UNITED STATES OF AMERICA
99 00 01 02 03 04 QW 9 8 7 6 5 4 3 2 1

CONTENTS

Clinical Reactivity of the Active Site of Myoglobin

EMMA LLOYD RAVEN AND A. GRANT MAUK

Enzymology and Structure of Catalases

PETER NICHOLLS, IGNACIO FITA, AND PETER C. LOEWEN

Horseradish Peroxidase

NIGEL C. VEITCH AND ANDREW T. SMITH

Structure and Enzymology of Diheme Enzymes: Cytochrome cd_1 Nitrate and Cytochrome c Peroxidase

VILMOS FÜLÖP, NICHOLAS J. WATMOUGH, AND STUART J. FERGUSON

Binding and Transport of Iron-Porphyrins by Hemopexin

WILLIAM T. MORGAN AND ANN SMITH

Structures of Gas-Generating Heme Enzymes: Nitric Oxide Synthase and Heme Oxygenase

THOMAS L.POULOS, HUIYING LI, C. S. RAMAN, AND DAVID J. SCHULLER

The Nitric Oxide-Releasing Heme Proteins From the Saliva of the Blood-Sucking Insect *Rhodnius prolixus*

F. Ann Walker and William R. Montfort

Heme Oxygenase Structure and Mechanism

Paul R. Ortiz de Montellano and Angela Wilks

De Novo Design and Synthesis of Heme Proteins

Brian R. Gibney and P. Leslie Dutton

ADVANCES IN INORGANIC CHEMISTRY, VOL. 51

CHEMICAL REACTIVITY OF THE ACTIVE SITE OF MYOGLOBIN

EMMA LLOYD RAVEN* and A. GRANT MAUK†

*Department of Chemistry, University of Leicester, University Road, Leicester, LE1 7RH, United Kingdom, and †Department of Biochemistry and Molecular Biology, 2146 Health Sciences Mall, University of British Columbia, Vancouver, BC, V6T 1Z3 Canada

I. Introduction

The discovery of myoglobin in skeletal muscle occurred during the last half of the 19th century though exact attribution is complicated by uncertainty at the time regarding the possible contamination of preparations by hemoglobin and by the difficulty in spectroscopic differentiation of hemoglobin from myoglobin (*1*). The first preparation of myoglobin that was spectroscopically distinguished from hemoglobin

0898-8838/01 $35.00

was obtained from dog muscle by Mörner in 1897 (*2*). Although Mörner referred to this material as myochrome, Günther subsequently proposed the term myoglobin to emphasize the similarity of the protein to hemoglobin (*1, 3*).

Myoglobin is now known to be a protein of about 150 amino acid residues that binds a single protoheme IX prosthetic group through coordination of the heme iron by a histidyl residue and a variety of noncovalent interactions between the protein and the heme. The principal functional role of myoglobin is based on its ability to bind dioxygen reversibly at the heme iron, a property of the heme that results from the environment provided for it by the protein. This fundamental property of myoglobin combined with comparative studies of a wide range of animal species led logically to the view that myoglobin serves a role in oxygen storage (*4, 5*) that is particularly important in diving mammals. Subsequent work led to the suggestion that another role of myoglobin that may be more important in other species is to facilitate the diffusion of dioxygen through muscle tissue to mitochondria (*6–9*). Theoretical (*10, 11*) and experimental (*12*) studies have challenged this view. Surprisingly, a mutant mouse devoid of myoglobin has subsequently been shown to be viable (*13*) owing to the effectiveness of a variety of compensatory physiological processes (*14*). Ironically, therefore, the true biological role of one of the most thoroughly studied proteins remains a matter of continuing investigation.

Because the apparent functional role of myoglobin is reversible binding of dioxygen, a considerable literature has developed that concerns the binding of gaseous ligands to ferromyoglobin and the binding of anionic ligands to ferrimyoglobin (metMb). With the advent of site-directed mutagenesis as a means of manipulating the structure of proteins, increasing attention has been directed toward understanding the chemical reactivity of myoglobin with a particular emphasis on understanding the chemistry of the active site of the protein. In part, this growing activity has arisen from renewed interest in understanding the way in which the protein environment of the heme dictates the chemical reactivity of this prosthetic group.

Myoglobin in many respects is the prototypical example of the larger family of heme containing proteins and enzymes that vary in function from the relatively simple process of reversible binding of an electron to the activation of dioxygen for substrate hydroxylation. The relationship between members of this family of proteins is not based simply on structural similarities but on similarities in chemical reactivity as well. As the structure of myoglobin is relatively simple compared to other heme proteins and as it was the first for which the three-dimensional structure was determined, myoglobin serves as a model for understanding the manner in which the protein environment determines reactivity of the heme.

Interest in this relationship has led to demonstration that myoglobin can participate in a number of reactions that are related to those catalyzed with far greater efficiency by heme proteins that are true enzymes. In the context of the current volume, this chemical reactivity of the active site of myoglobin is our focus of attention. Studies of the ligand binding properties of myoglobin have been reviewed elsewhere (e.g., *15–17*) and are not considered here. In view of the essential role of molecular genetics in the development of understanding of the chemical reactivity of myoglobin and the anticipation that this role will expand in the future, we begin this survey by considering advances in the engineering of myoglobin structure before considering the chemical reactivity of heme at the active site of myoglobin. In subsequent discussion of the chemistry of the Mb active site, emphasis is given in many cases to consideration of studies involving variants of Mb and references are provided to reviews of related literature that does not involve variants.

II. Cloning and Expression of Recombinant Myoglobin

Prior to the development of recombinant DNA technology, most functional studies of myoglobin concerned protein isolated from sperm whale and horse heart muscle because these sources were abundant and they were available from commercial sources. With the cloning of cDNAs for human (*18–21*), porcine (*22*), mouse (*23*), grey seal (*24, 25*), bovine (*26*), icefish (*Chionodraco rastrospinosus, 27*), and mollusk (*Aplysia limacina* (*28*) and *Biomphalaria glabrata* (*29*)) myoglobins,[1] expression systems have been developed for production of the human, porcine, and *A. limacina* proteins in *Escherichia coli* and for bovine myoglobin in yeast (*Saccharomyces cerevisiae*) (*19, 26, 28, 35, 36*). In cases for which a cDNA has not been isolated, synthetic genes have been prepared and overexpressed. This approach has been used to prepare recombinant sperm whale (*37, 38*) and horse (*39*) Mbs.[2] Amino acid sequences for the proteins that have been studied most extensively in recombinant form and for which the greatest number of variants has

[1] cDNAs for a new class of Mb with close similarities to the heme enzyme indoleamine dioxygenase have also recently appeared (*30–34*).

[2] For sperm whale Mb, an error (corresponding to a D122N mutation) was incorporated into the synthetic gene (*37*) and, in this case, the recombinant wild-type sequence (*37*) is not identical to that of the authentic wild type.

FIG. 1. Alignment of amino acid sequences for human (*18, 19, 21*), porcine (*22, 36*), horse heart (*39*), and sperm whale (*37, 38*) myoglobins.

been produced (human, porcine, horse, and sperm whale) are aligned in Fig. 1.

Compared with recombinant expression of most proteins, expression of heme proteins presents the additional challenge that arises from the need to coordinate synthesis of the prosthetic group with synthesis of the protein and from the chemical reactivity of heme. For example, significant quantities of sulfmyoglobin (*vide infra*) are generated during expression of horse heart and sperm whale Mbs in *E. coli* (*42, 43*). Other anomalous spectroscopic features have also been noted for recombinant human hemoglobin (*44*) and for recombinant cytochrome *c* peroxidase (*45*). These observations illustrate the necessity of rigorous characterization of recombinant heme proteins to ensure that subsequent functional studies are not influenced by the presence of unanticipated protein derivatives resulting from recombinant expression or from mutations of the protein.

III. Active Site Variants of Myoglobin

With the development of bacterial expression systems for Mb described earlier and with the known diversity of metal coordination

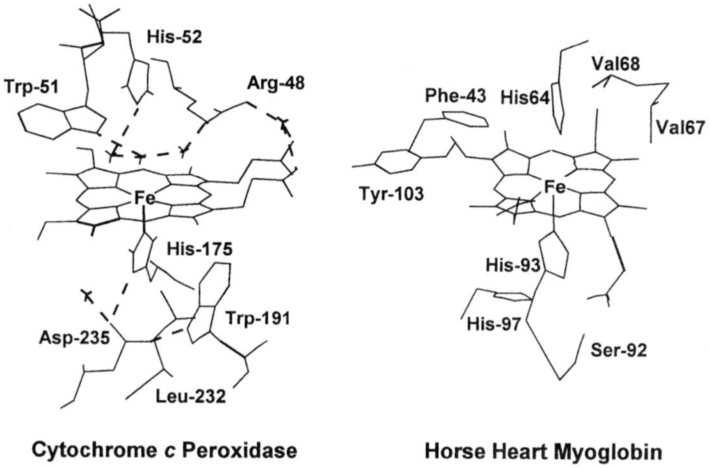

Cytochrome *c* Peroxidase Horse Heart Myoglobin

FIG. 2. Comparison of the heme environments at the active sites of horse heart myoglobin (*40*) and yeast cytochrome *c* peroxidase (*41*).

environments among heme proteins (*46–48*), considerable attention has been focused on examining the role of the axial ligands to the heme iron in determining the spectroscopic and functional properties of the protein. Although it is clear that the active site coordination environments of cytochromes P450, peroxidases, catalases, globins, and other heme proteins differ significantly from each other, the extent to which this difference contributes to the functional diversity of this family of proteins relative to the contribution provided by other structural attributes of the active sites of these proteins is not fully understood (*47–49*). For the purpose of comparison, critical residues in the heme binding pocket of horse heart Mb and yeast cytochrome *c* peroxidase are illustrated in Fig. 2. As a result, considerable interest has been directed toward changing the coordination environment of Mb in an attempt to mimic the diverse range of spectroscopic and chemical reactivities exhibited by other heme proteins. Although it is undeniably naïve to expect that identity of axial ligands alone is sufficient to reproduce the spectroscopic and functional properties of one heme protein through mutagenesis of another, such work inevitably produces new, interesting, and informative variants that provide a logical starting point for more subtle and sophisticated protein engineering efforts. For these reasons and others, a significant fraction of the myoglobin variants that have been reported at present involve modification of the heme binding pocket. As the chemical reactivity of the myoglobin active site can respond significantly to amino acid substitutions in this region of the protein, it is useful to begin by surveying the active site variants for the species of Mb that have received the greatest attention (sperm whale, horse heart, and human Mb) and by considering some of the consequences of these structural modifications.

A. Mutagenesis of the Distal Heme Binding Pocket

The distal H64 residue of Mb has been investigated extensively by site-directed mutagenesis. The emphasis of most of this effort has concerned the participation of this residue in ligand binding reactions. These studies have been reviewed in detail elsewhere (*16*) and are beyond the scope of the current discussion. The simplest mechanism by which replacement of H64 influences axial ligation results when residues incapable of H-bonding interactions are introduced at this position. Evidently, the hydrogen bond normally formed by H64 and the water molecule coordinated as the sixth ligand is sufficiently important to the stability of water coordination that such variants typically exhibit five-coordinate, high-spin metMb derivatives (*50*) with altered ligand binding properties (*16*). In a few cases, on the other hand, substitution of H64 with appropriate residues may result in coordination of the new residue to the heme iron. Variants of this type are generally less reactive in the reactions considered in this review, but they provide useful spectroscopic species for comparative studies with other types of heme proteins.

Distal pocket substitutions of various species of Mb in which a new distal ligand is provided by the protein include V68H, H64V/V68H (*51–56*), H64Y (*57–62*), V68E (*57, 63–65*), and possibly the H64R variant of sperm whale Mb (*60*). For the ferric H64Y variants of sperm whale and horse heart Mbs, electronic (*57, 61*), EPR (*57, 61, 62*), and resonance Raman (*57, 60*) spectroscopy, XANES (*61*), and X-ray crystallographic results (*62*) are all consistent with a six-coordinate species in which direct coordination of the distal tyrosyl residue occurs; however, this ligand is not coordinated following reduction of the heme iron (*57*). In this regard, the H64Y variants reproduce the behavior of HbM Saskatoon (*66, 67*) in which the distal histidyl residue of human hemoglobin is replaced with a tyrosyl residue. Demonstration of carboxylate coordination to the heme iron was provided independently by electronic (*57, 63, 64*), EPR (*57, 65*), and NMR (*64, 65*) spectroscopy of the V68E variants of sperm whale (*57*) and human (*63, 64*) Mbs. These results correspond to behavior of HbM Milwaukee, in which a valyl residue in the distal pocket is replaced with a glutamyl residue. MetHbM Milwaukee is known (*68*) to possess a high-spin, six-coordinate heme iron with a distal glutamyl residue providing the sixth ligand. As is the case for HbM Saskatoon, this variant ligand dissociates from the iron upon reduction. Coordination of an argininyl residue has been suggested from the low-spin characteristics in the resonance Raman spectrum of the H64R variant of sperm whale metMb (*60*). However, as there is no unambiguous precedent for a heme protein with a coordinated argininyl residue, the spectroscopic criteria for establishing this coordination environment are insufficient to provide a compelling argument.

B. Mutagenesis of the Proximal Heme Binding Pocket

Preparation of Mb variants in which the proximal H93 ligand is substituted is intrinsically more difficult than preparation of distal ligand variants. In all cases of which we are aware, the H93 variants are expressed as the apo-protein and require reconstitution with exogenous heme during purification. The first successful preparation of a proximal ligand variant of Mb was reported in 1990 (*57*) for the H93Y variant of sperm whale Mb. Subsequently, H93Y variants of both human (*69–71*) and horse (*72*) Mbs were reported. The spectroscopic properties of the ferric derivatives of these variants are uniformly consistent. The electronic spectra (sperm whale (*57*), human (*70*), horse (*72*)), EPR spectra (sperm whale (*57*), human (*70*), horse (*72*)), and resonance Raman spectra (sperm whale (*57*) and human (*70*)) are in agreement and are consistent with the presence of a high-spin, five-coordinate iron with tyrosine providing the fifth axial ligand. The tyrosine coordination and absence of a distally coordinate water molecule were confirmed by X-ray diffraction analysis of the H93Y variant of horse heart Mb (*72*). Interestingly, NMR spectra for this variant were consistent with rotation of the phenolate ligand on the time scale of the NMR experiment, so the crystal structure, therefore, probably represents one of a number of possible conformational orientations. NMR spectra of the corresponding variant of human Mb, however, are not consistent with phenolate rotation. This species-specific difference in behavior has been attributed (*72*) to a difference in the residue present at position 142 and emphasizes the subtle distinctions that can exist between closely related proteins.

Introduction of thiolate ligation at position 93 has also been reported (*69, 70, 73*). Again subtle species-specific differences in sequence on the proximal side of the heme binding pocket result in significant differences in behavior between closely related proteins. In this case, the spectroscopic properties of the human H93C variant (*69, 70*) were consistent with the presence of five-coordinate, cysteine-coordinated high-spin Fe(III). On the other hand, similar studies of the corresponding horse heart metMb variant (*73*) failed to indicate thiolate coordination under any conditions. In this case, the principal spectroscopic component probably comprises a five-coordinate, heme iron with a coordinated water molecule. Proximal cysteinyl coordination in the horse heart protein could be achieved, however, through concurrent replacement of the distal H64 residue with apolar valyl and isoleucyl residues (*73*) that are known (*50*) to disfavor distal coordination of a water molecule. In these double variants, the heme iron presumably lacks a distal ligand and is displaced toward the proximal C93 residue. Unfortunately, none of the H93C variants of either human or horse Mb retained the proximal ligand upon reduction of the heme iron, so the complex formed with carbon monoxide fails to exhibit a Soret maximum at 450 nm that is characteristic of cytochromes P450. In fact, spectroscopic evidence suggests that upon reduction of the H93C (and H93Y) variants of the human protein, not only does the proximal cysteinyl ligand dissociate from the heme iron, but the distal histidyl residue becomes the fifth ligand (*69, 70*).

An alternative strategy for introduction of alternative proximal ligands involves replacement of H93 with the glycyl residue to produce a so-called proximal "cavity" variant (H93G) (*74–80*) in which the proximal ligand may consist of a hydroxyl group (*78, 79*). The advantage of this variant is that it can be functionalized conveniently by the introduction of other small molecules (**X**) that can coordinate to the iron (*74*). In addition to imine ligands (**X** = imidazole, pyridine), this exchange strategy has been used to introduce both phenolate (**X** = phenol) and thiolate (**X** = ethanethiol) groups that closely resemble the spectroscopic features of the ferric catalases and cytochromes P450, respectively.

IV. Electron Transfer Reactions of Myoglobin

A. Electrochemistry

Perhaps the most fundamental functional property of a heme prosthetic group at the active site of a heme protein is the relative stability of the reduced and oxidized states of the heme iron. A number of structural characteristics of the heme binding environment provided by the apo-protein have been identified as contributing to the regulation of this equilibrium and have been reviewed elsewhere (*82–84*). Although a comprehensive discussion of these factors is not possible in the space available here, they can be summarized briefly. The two most significant influences of the reduction potential of the heme iron appear to be the dielectric constant of the heme environment (*81, 83*) and the chemical identity of the axial ligands of the heme iron (*81*). These characteristics are further modified by the orientation and hydrogen bonding interactions of the axial ligands (*85–89*) and by the orientation of the heme group in the heme binding pocket (*90*). In addition, the orientation of the heme vinyl groups and the hydrogen bonding interactions of the heme propionate groups have also been implicated as modulating influences (*91–93*). Although semiquantitative correlation of these structural features with reduction potentials is possible in some cases, comprehensive, quantitative understanding of all contributions has been elusive. Most notably, unambiguous methods for quantification of the dielectric of the heme pocket are not established. The primary experimental problem, however, has been the difficulty in varying any one of these characteristics individually without simultaneously changing another. For this reason, it is imperative that any attempt to address this issue entail a rigorous characterization of the variant or chemically modified protein to ensure that some unanticipated structural consequence of the modification will not be overlooked.

1. Methods

The first electrochemical studies of Mb were reported for the horse heart protein in 1942 (*94*) and subsequently for sperm whale Mb (e.g., *95*) through use of potentiometric titrations employing a mediator to achieve efficient equilibriation of the protein with the electrode (*96*). More recently, spectroelectrochemical measurements have also been employed (*97, 98*). The alternative methods of direct electrochemistry (*99–102*) that are used widely for other heme proteins (e.g., cytochrome *c*, cytochrome b_5) have not been as readily applied to the study of myoglobin because coupling the oxidation–reduction equilibrium of this protein to a modified working electrode surface has been more difficult to achieve. As a result, most published electrochemical studies of wild-type and variant myoglobins have involved measurements at equilibrium rather than dynamic techniques.

Recent work has resolved some of the issues that complicate direct electrochemistry of myoglobin, and, in fact, it has been demonstrated that Mb can interact effectively with a suitable electrode surface (*103–113*). This achievement has permitted the investigation of more complex aspects of Mb oxidation–reduction behavior (e.g., *106*). In general, it appears that the primary difficulty in performing direct electrochemistry of myoglobin results from the change in coordination number that accompanies conversion of metMb (six-coordinate) to reduced (deoxy) Mb (five-coordinate) and the concomitant dissociation of the water molecule (or hydroxide at alkaline pH) that provides the distal ligand to the heme iron of metMb.

A summary of reduction potentials reported for the Fe(III)/Fe(II) couple for wild-type and variant forms of horse, human and sperm whale Mbs is given in Table I. As indicated earlier, the structural characteristics of a heme protein that exert the greatest influence on the reduction potential are the axial ligands provided to the metal ion and the dielectric constant of the heme environment. For this reason, we first consider the results presented in Table I for variants involving changes in the distal and proximal ligand and then consider results for variants in which the electrostatic properties of the heme pocket have been changed without replacement of the axial ligands. Emphasis on axial ligation in this survey reflects the interest in comparison of such variants with the behavior of heme proteins that possess axial ligands different from those of wild-type myoglobin and reflects the dominance of electrochemical information for such variants in the literature.

2. Electrochemistry of Variants with an Altered Distal Ligand

The H64 distal ligand of wild-type myoglobin does not coordinate to the heme iron in either the reduced or the oxidized form of the native protein but stabilizes the coordination of a distally bound water molecule of metMb. Replacement of H64 with other amino acid residues can, therefore, change the coordination environment of the heme iron in two ways. Such variants either may possess a distal residue that is able to coordinate to the heme iron or may possess a distal residue that is incapable of either coordinating to the iron or of forming a hydrogen bond with a coordinated water molecule.

The H64Y variant of Mb is an example of the former situation in that the tyrosyl side chain coordinates to the heme iron of the oxidized variant. As expected for a variant with an anionic phenolate ligand, the reduction potential of this variant is ~40 mV lower than that of the wild-type protein (Table I). Although this change is consistent with stabilization of the oxidized form of the protein, the fact that the tyrosyl ligand is not coordinated in the reduced protein complicates quantitative interpretation of this shift in potential.

On the other hand, the H64L, H64V, and H64F variants constitute a group of proteins in which H64 is replaced by residues with side chains that are incapable of coordinating to the heme iron atom and that do not stabilize a distally coordinated water molecule through hydrogen bonding interactions. As a result, both the metMb and deoxyMb derivatives of these variants are five-coordinate. Interestingly, the reduction potentials of these variants are all somewhat greater than that for the wild-type protein and are within ~30 mV of each other ($E^{\circ} = 76$–109 mV) in both oxidation states. Interestingly, the H64G variant exhibits a coordination chemistry and midpoint potential that are virtually identical to those of the wild-type protein. A common characteristic of all the H64 variants is that their electron transfer activity as estimated from direct electrochemical measurements is enhanced relative to wild-type Mb. As observed by Van Dyke *et al.* (*106*), Mb electron transfer at the electrode surface appears to be controlled not only by oxidation-state-dependent changes in coordination number (from six- to five-coordinate) but also by the ease with which these changes are communicated to the surface of the molecule through the hydrogen bond network that links His64 to the bulk solvent (*106*). For wild-type protein, the change in oxidation state is linked through the hydrogen-bond network to the reorganization of bulk solvent; for the variants, this is not the case, and electron transfer is more facile.

In some cases, axial ligation may be modified by substitution of residues other than those that provide the axial ligands in the wild-type protein. For example, replacement of V68 with a histidyl residue results in coordination of H68 to the heme iron in both oxidation states of the protein, as is the case for cytochrome b_5. This behavior has been established both for the single variant (V68H) of the horse heart protein (*51*) and for the double variant H64V/V68H of the human and porcine proteins (*52*). Although the reduction potentials for the human H64V/V68H and horse V68H variants are similar to each other (−128 and −110 mV, respectively; Table I), they are 170 to 190 mV lower than the values of corresponding wild-type proteins. Notably, these potentials are much lower than that of microsomal cytochrome b_5 (+5 mV (*122*)) even though the heme group of the cytochrome is more exposed to bulk solvent at the surface of the protein. Furthermore, EPR data (*51*) indicate that the planes of the axial imidazole ligands in the variant Mb are presumably oriented perpendicular to each other (*123–132*), which should increase the potential as much as 50 mV relative to the parallel orientation present in cytochrome b_5.

3. *Electrochemistry of Variants with an Altered Proximal Ligand*

Evaluation of the contribution made by the proximal ligand to the oxidation-reduction equilibrium of Mb (H93 in the wild-type protein (Fig. 2)) has been more difficult because substitution of the proximal residue results in expression of apo-Mb without heme incorporation. All of the available data for these variants (Table I), therefore, derive from proteins prepared by reconstitution of purified recombinant apoprotein with exogenous heme (*69, 70, 72*). In those cases where trace quantities of native Mb are produced (*73*), heme extraction followed by reconstitution was undertaken to eliminate complications from sulfMb

TABLE I

REDUCTION POTENTIALS FOR SITE-DIRECTED VARIANTS OF HORSE, HUMAN, AND SPERM WHALE MBS

Derivative	Source	Variant	E°[a]	Axial ligation[b] Oxidized	Axial ligation[b] Reduced	Methods[c]	Conditions	Ref.
MetMb[d]	Horse	T39I/K45D/F46L/I107F	24	His/H_2O	His/–	S	Sodium phosphate, $I = 0.1$ M, pH 6.0, 25.0°C	*116*
		H64Y	20	His/Tyr[e]	His/–	S	Sodium phosphate, $I = 0.1$ M, pH 7.0, 25.0°C	*117*
		H64V	87	His/–	n.d.	S	Sodium phosphate (46 mM), pH 7.0, 25.0°C	*73*
		H64I	95	His/–	n.d.	S	Sodium phosphate (46 mM), pH 6.0, 25.0°C	*73*
		H64V/H93C	–217	Cys/–	n.d.	P	Sodium phosphate (46 mM), EDTA (10 mM), pH 8.0, 25.0°C	*73*
		H64I/H93C	–219	Cys/–	n.d.	P	Sodium phosphate (46 mM), EDTA (10 mM), pH 8.0, 25.0°C	*73*
		V67R	106	His/H_2O	His/–	S	Sodium phosphate, $I = 0.1$ M, pH 6.0, 25.0°C	*118*
		V67A/V68S	–23	His/H_2O[f]	His/–	S	Sodium phosphate, $I = 0.1$ M, pH 7.0, 25.0°C	*119*
		V68H	–110	His/His	His/His	S	Sodium phosphate, $I = 0.1$ M, pH 7.0, 25.0°C	*51*
		S92D	72	His/H_2O[f]	His/–	S	Sodium phosphate, $I = 0.1$ M, pH 6.0, 25.0°C	*120*
		H93Y	–208	Tyr/–[f]	n.d	P	Sodium phosphate (35 mM), EDTA (10 mM), pH 8.0, 25.0°C	*72*
MetMb[g]	Human	V68E	–137	His/Glu	His/–	S	Sodium phosphate, $I = 0.1$ M, pH 7.0, 25.0°C	*63, 64*
		V68D	–132	His/H_2O	His/–	S	Sodium phosphate, $I = 0.1$ M, pH 7.0, 25.0°C	*63, 64*
		V68N	–24	His/H_2O	His/–	S	Sodium phosphate, $I = 0.1$ M, pH 7.0, 25.0°C	*63, 64*
		H64V/V68H	–128	His/His[h]	His/His	R	Phosphate (0.1 M), pH 7, 20°C	*52*
		H93Y	–190	Tyr/–	–/His[i]	T	Not reported	*69, 70*
		H93C	–230	Cys/–	–/His[i]	T	Not reported	*69, 70*
MetMb[j]	Sperm whale	H64L	83	His/–[k]	His/–[k]	P	Phosphate (50 mM), pH 7.0, 25°C	*121*
		H64L	84	His/–[k]	His/–[k]	CV	Hepes (0.1 M), pH 7.0, 22°C	*106*
		H64L/F43H	88	His/H_2O[f]	His/–	P	Phosphate (50 mM), pH 7.0, 25°C	*121*
		H64L/I107L	60	***	n.d.	P	Phosphate (50 mM), pH 7.0, 25°C	*121*

		H64L/L29H	−22	His/H_2O[f]	His/−	P	Phosphate (50 mM), pH 7.0, 25°C	*121*
		H64G	65	His/H_2O[k]	His/−	CV	Hepes (0.1 M), pH 7.0, 22°C	*106*
		H64L	84	His/−	His/−	CV	Hepes (0.1 M), pH 7.0, 22°C	*106*
		H64V	76	His/−[k]	His/−	CV	Hepes (0.1 M), pH 7.0, 22°C	*106*
		H64M	98	His/−	His/−	CV	Hepes (0.1 M), pH 7.0, 22°C	*106*
		H64F	109	His/−	His/−	CV	Hepes (0.1 M), pH 7.0, 22°C	*106*
CyanometMb[l]	Horse	V67R	−392	His/CN^-	His/−[n]	CV	Tris/cacodylate, pH 7.0, $I = 0.1$ M, 20°C	*118*
		V68H	−257	His/CN^-	His/−[n]	CV	Tris/cacodylate, pH 7.0, $I = 0.1$ M, 20°C	*118*
		S92D	−412	His/CN^-	His/−[m]	CV	Tris/cacodylate, pH 7.0, $I = 0.1$ M, 20°C	*118*

[a] mV vs SHE.

[b] From analysis of various spectroscopic data unless otherwise stated. Given as proximal/distal.

[c] S, spectroelectrochemistry; P, photochemical reduction; T, reductive titration; CV, cyclic voltammetry; R, redox potentiometry.

[d] Values of $E°$ for wild-type horse Mb have been reported as 61 mV (pH 7.0, $I = 0.1$ M, 25.0°C), 64 mV (pH 6.0, $I = 0.1$ M, 25.0°C), and 45 mV (pH 8.0, $I = 0.1$ M, 25.0°C) (*114*); 46 mV (pH 6.95, phosphate, $I = 0.2$ M, 30°C) (*94*); 46 mV (pH 7.0, 0.1 M phosphate, 0.1 M NaCl) (*115*); 19 mV (pH 7.1, 0.05 M MOPS, 0.1 mM EDTA, 20°C) (*98*).

[e] Determined by X-ray crystallography (*62*).

[f] Determined by X-ray crystallography (*119*).

[g] Values of $E°$ for the wild-type protein have been reported as 59 mV (pH 7.0, $I = 0.1$ M, 25.0°C) (*63*); 50 mV (conditions not reported) (*69*).

[h] Determined from analysis of spectroscopic data. X-ray crystallographic information (*52*) for the corresponding H64V/V68H variant in porcine Mb confirmed bis-histidine axial ligation in the ferric derivative.

[i] Distal ligand proposed as arising from ligation of His64.

[j] Values for wild-type sperm whale Mb have been reported as 59 mV (pH 7.0, $I = 0.1$ M, 25.0°C) (*63*); 59 mV (0.1 M Hepes, pH 7.0, 22°C) (*106*); 52 mV (0.05 mM phosphate, pH 7.0, 25°C) (*121*); 47 mV (pH 7, 30°C) (*95*); 14 mV (pH 7.1, 0.05 M MOPS, 0.1 mM EDTA, 20°C) (*98*).

[k] Determined by X-ray crystallography (*50*).

[l] Values of E° for the wild type protein have been reported as −385 mV (pH 7.0, $I = 0.1$ M, Tris/cacodylate, 20°C) (*113*).

[m] Thermodynamic product. The initial product of the reaction is Fe(II)–CN^- (*113*).

[n] The nature of reduced form in this variant (His/His or His/−) is not known with certainty. However, CO is known to displace the distal histidine ligand in ferrous V68H, and it is, therefore, likely that the reduced form in these experiments is deoxy Mb (His/−).

(Section IX) formation to ensure homogeneity of the protein sample. This approach has been required, for example, in characterization of H93C and H93Y variants of horse heart and human Mb (*69, 72, 133*).

Replacement of histidine 93 with a tyrosyl residue (H93Y variant) in horse and human Mb leads to a dramatic decrease in reduction potential (−208 and −190 mV respectively; Table I). This finding can be understood in terms of stabilization of the oxidized protein by the anionic (electron-donating) properties of the phenolate ligand. The corresponding cysteine variant (H93C) of human Mb has a reduction potential (−230 mV) that is, again, consistent with increased electron density on the metal. This value is intermediate between that of the high-spin, five-coordinate derivative of cytochrome $P450_{cam}$ (−170 mV (*134*)) and the low-spin, six-coordinate derivative of the enzyme (−270 mV (*134*)). Similarly, the H64V/H93C and H64I/H93C variants of horse heart Mb, in which coordination of cysteine to the iron has been established (*73*), exhibit reduction potentials of −217 and −219 mV, respectively.

4. *Electrochemistry of Variants with an Electrostatically Altered Heme Binding Pocket*

Electrostatic interactions on the surface and the interior of the protein can modulate the electrochemical properties of the heme center, although the magnitude of the effect of surface electrostatic changes is more variable (*83*). Nevertheless, several attempts to rationalize these effects in terms of Coulombic interactions in other proteins have been reported (e.g., cytochrome b_5 (*135*) and high potential iron protein (*136, 137*)). However, the effects of surface electrostatic charges are not explained so simply for other proteins (e.g., cytochrome *c* (*83*) and rubrerythrin (*138*)). The first report to consider electrostatic modulation of the oxidation–reduction equilibrium of myoglobin involved investigation of a human Mb variant in which the hydrophobic V68 residue (Fig. 2) was replaced with anionic (V68E and V68D variants) and neutral (V68N variant) residues (*63, 64*). In this work, the V68D and V68E variants exhibited reduction potentials 200 mV lower than that of the wild-type protein, and the V68N variant exhibited a potential ~80 mV lower than wild-type Mb (Table I). Although the magnitude of these changes is difficult to explain quantitatively, the direction of the change for the variants with an acidic residue at this position can be understood qualitatively in terms of destabilization of the reduced protein. The behavior of the V68E variant is, however, complicated by coordination of the glutamate side chain to the heme iron in the ferric form of the protein. This is not the case, however, for the V68D variant, suggesting that the observed decrease in potential is purely a reflection of the electrostatic nature of this residue.

Seemingly complementary experiments involving horse heart metMb, on the other hand, are not readily explained (*120*). In this case, replacement of the proximal S92 residue with an aspartyl residue (S92D) increases the potential by 8 mV relative to the potential of the wild-type protein. This is a surprising result, particularly insofar as the S92 side chain is in contact with the proximal H93 heme iron ligand in the wild-type protein (Fig. 2). An extensive spectroscopic analysis of the possible basis for the increase in potential revealed that a number of small but significant and unanticipated secondary alterations are induced by this substitution. These changes include alteration in solvent accessibility of the heme, the pK_a of H97, the orientation of the axial ligands, and the hydrogen bonding properties of the proximal ligand. Evidently, these changes in structure exert mutually compensating influences on the oxidation–reduction equilibrium of the protein to result in a relatively small increase in potential. This example emphasizes the importance of assessing all the functional and spectroscopic properties of a new variant and not simply those of immediate interest.

The effect of electrostatic modifications and the role of charge compensation on the electrochemical properties of cyanmetMb (Table I) have been addressed through analysis of a series of horse heart Mb variants in which both proximal (S92D variant) and distal (V68H and V67R variants) amino acids were replaced with titratable residues (*118*). The midpoint potential of the S92D variant is 8 mV higher than that of the wild-type protein, while the potential of the V67R variant is 42 mV higher. On the other hand, the cyanide complexes of the S92D and V67R variants exhibit potentials that are 27 and 7 mV lower than that of wild-type cyanometMb, respectively. These results have been interpreted in terms of a thermodynamic driving force for electroneutrality that helps to compensate for the additional charge within the active site introduced by mutagenesis. Unfortunately, analysis of this type was not possible for the V68H variant owing to the more profound alteration in coordination environment of this variant in the oxidized form.

5. *Electrochemistry of Higher Oxidation States of Myoglobin*

Although electrochemical studies of the Fe(III)/Fe(II) couple are of considerable interest in understanding many electron transfer reactions of heme proteins, the catalytic activities of these proteins involve higher oxidation states of these proteins. Rigorous understanding of the thermodynamics of these reactions requires knowledge of the potentials for compound II/Fe(III) (i.e., Fe(IV)=O/Fe(II)) and compound I/compound II (i.e., Fe(IV)=O$^{\cdot}$/Fe(IV)=O) equilibria. The reduction potentials for interconversion of these forms of heme proteins are experimentally challenging because they are generally quite high (0.8–1.0 V) and because they are usually quite similar in value to each other. At present, these values have been reported only for wild-type sperm whale Mb (0.887 V (20°C) (*139, 140*); 0.896 V (pH 7.0, 15°C) (*141*)). Similar studies of selected variants would contribute significantly to our understanding of the catalytic activities exhibited by these variants and are discussed further in the discussion that follows.

B. Electron Transfer Kinetics

The kinetics of myoglobin oxidation and reduction have been studied by a variety of experimental techniques that include stopped-flow kinetics, pulse radiolysis, and flash photolysis. In considering this work, attention is directed first at studies of the wild-type protein and then at experiments involving variants of Mb.

1. *Stopped-Flow Kinetics*

Stopped-flow kinetics studies of metMb reduction have investigated reduction of the protein by a number of inorganic and organic reducing agents. For example, Fleischer and co-workers studied the anaerobic reduction of metMb by $[Cr(H_2O)_6]^{2-}$ (*142*) while others used dithionite (*143–145*), $[Fe(CDTA)]^-$ (*146*) and $[Fe(EDTA)]^{2-}$ (*147, 148*), $[Fe(bpy)]^{2+}$ and $[Fe(NTA)]^{2+}$ (*149*), and ascorbate (*150*). Each of these studies revealed different aspects of the reaction. For example, dithionite reduction of various ligand-bound complexes of metMb at alkaline pH was found to require dissociation of the bound ligand prior to reduction of the iron by SO^{2-} radical, except for the imidazole and cyanide complexes, which were reduced prior to ligand dissociation (*144*). On the other hand, at acidic pH, only the cyanide complex was found to undergo reduction prior to ligand dissociation (*143*). The similarity in reduction potentials of Mb and $Fe(CDTA)^-$ permitted use of this reagent to study the reduction of the cyanide and nitric oxide complexes of metMb and the oxidation kinetics of MbO_2 (*146*). The kinetics by which $Fe(EDTA)^{2-}$ reduces metMb were analyzed by Marcus theory to demonstrate the relative inefficiency of Mb in electron transfer relative to cytochromes (*147*) and to study the thermodynamics and pH dependence of reaction (*148*).

2. *Pulse Radiolysis*

Pulse radiolysis has also been used to study the reduction of various Mb derivatives by hydrated electrons (*151–154*). With this technique, it was possible to study reduction of ligand-bound forms of metMb at cryogenic temperature and thereby identify reduced, ligand-bound forms of the protein (*152*) and to reduce oxyMb to produce ferryl (Fe(IV)=O)

Mb. Subsequent application of this technique to reduction of various derivatives of reduced and oxidized myoglobin led to the observation that the rate of reduction by hydrated electrons depends primarily on the net charge of the protein and the dissociation constant for formation of ligand bound derivatives of metMb.

3. *Flash Photolysis*

Flash photolysis has been used in a variety of ways to study several types of electron transfer reactions in which myoglobin can participate. Winkler, Gray, and colleagues have studied intramolecular electron transfer of myoglobin derivatives in which ruthenium complexes of various structure have been appended to create a second electron transfer center at histidyl residues placed at various positions on the surface of the protein. Through a variety of photochemical strategies, this group has used these modified forms of myoglobin to study intramolecular oxidation and reduction of the heme iron in an effort to define the dependence of the rate of intramolecular electron transfer on the nature of the protein structure located between the electron donor and acceptor centers. Detailed discussion of these issues is beyond the scope of this chapter, but a comprehensive review of this work is available (*155*).

An alternative application of flash photolysis to study myoglobin electron transfer kinetics has been employed by Hoffman and co-workers (*156*). In this approach, the photoactive zinc-substituted derivative of Mb is mixed with an equivalent amount of ferricytochrome b_5 to form an electrostatically stabilized binary complex. Upon transient irradiation, the strongly reducing ^{3}Zn–Mb intermediate is formed, and the kinetics of ferricytochrome b_5 reduction within the preformed complex can be monitored spectrophotometrically. The resulting kinetics represents a mixed-order process consistent with electron transfer both within the electrostatically stabilized complex and between the dissociated components of the complex.

An alternative application of ruthenium-modified myoglobin has been reported by Fenwick *et al.* in experiments concerning the photoinitiated intramolecular reduction of ferrylMb (*157*). Through use of two different ruthenium complexes for modification of Mb, the effect of driving force on this reaction could be considered, and a reorganization energy for the reaction was estimated to be 1.8–2.1 eV, depending on the reduction potential of the ruthenium complex used for protein modification. This relatively unreactive character of the ferryl center to electron transfer was rationalized by these authors as resulting from the absence of a strong H-bond donor to the oxene ligand to the iron in the distal heme pocket of Mb and the lack of a strong H-bond to the proximal ligand to the heme iron.

4. Electron Transfer Kinetics of Myoglobin Variants

Following their initial studies of the oxidation of myoglobin by the Fe(III)NTA complexes discussed above (*149*), Saltman and co-workers found that oxidation of myoglobin by Cu(II)NTA exhibits biphasic kinetics in stopped-flow studies (*158*). This behavior was interpreted as arising from bimolecular reduction of the copper complex at the partially exposed edge of the heme prosthetic group and from reduction of the copper following binding of the complex at a specific site on the surface of the protein. NMR spectroscopy was used to implicate histidyl residues in copper binding. As a metal binding site involving histidyl residues had been identified on the surface of sperm whale myoglobin in early X-ray diffraction studies, the involvement of this site in electron transfer reaction of Mb with Cu(II)NTA was subsequently evaluated through kinetic studies of variants (159). This work implicated H48 in the binding of the complex ~13 Å from the heme iron and provided a structural basis for interpretation of the electron transfer kinetics in terms of two alternative mechanisms.

The kinetics by which $Fe(CN)_6^{3-}$ oxidizes pig deoxyMb was studied by Zhang *et al.* in experiments that examined the possible involvement of K45 through investigation of a family of variants in which this residue was replaced by serine, histidine, glutamate, and arginine (*160*). Despite the fact that this residue normally forms an H-bond with one of the heme propionate groups and that its replacement with at least some of these residues should have significant consequences on the electrostatic character of the surface of the protein adjacent to the partially exposed edge of the heme group, substitution of this residue had little effect on the kinetics of reaction with this oxidant. Subsequently, Dunn *et al.* studied the same reaction for a series of myoglobin variants in which the distal H64 residue was replaced with residues that alter the coordination environment of the heme iron (*161*). This work led to the finding that variants that are five coordinate in both oxidation states are 10- to 15-fold more reactive than is the wild-type protein.

As emphasized by this latter study, it has been relatively straightforward to identify myoglobin variants that are five-coordinate in both oxidation states, but it has been far more difficult to identify variants or derivatives that are six-coordinate in both oxidation states. Myoglobins with this characteristic would have the potential to provide considerable insight into the role of various types of axial ligand in regulating the electron transfer reactivity of cytochromes and other types of heme proteins. At present, progress toward this goal has been limited. For example, independent studies of porcine, human,and horse Mb variants have identified a variant that exhibits bis-histidine axial ligation in *both* oxidized and reduced states of Mb. In these studies, a variety of spectroscopic data for the H64V/V68H double variants of human and porcine (*52, 54*) and for the V68H single variant of horse Mb (*51*) provided clear and mutually consistent spectroscopic evidence for bis-histidine coordination in both oxidation states of these variants. These conclusions were confirmed by crystallographic analysis of the porcine double variant, which demonstrated that H68 is coordinated to the heme iron with the plane of the imidazole ring oriented perpendicular to the imidazole ring of the proximal H93 ligand (Fig. 2). In fact, EPR spectra of the H64V/V68H and V68H variants (*51, 52*) provided accurate prediction of the perpendicular orientation of the proximal and distal ligands in these variants fully consistent with the crystallographic structure. The effect of axial ligand orientation on the electronic, spectroscopic, and functional properties of model heme complexes and their implications for the corresponding properties of heme proteins have been studied extensively by Walker, Scheidt, Strouse, and their colleagues (*123–132*). At present, however, the electron transfer kinetics of these variants have not yet been studied.

C. Autoxidation of Oxymyoglobin

The autoxidation of oxymyoglobin is the process by which oxymyoglobin spontaneously forms metmyoglobin (*162*). The rate of this process varies with pH, dioxygen concentration, anion concentration, and species of myoglobin, and these dependences have led to a variety of proposals for the mechanism of this reaction. Early studies of autoxidation proposed that the reaction proceeds through oxidation of a deoxyMb intermediate by dioxygen. Although this mechanism explains the increased rate of autoxidation at lower concentrations of dioxygen, it is less clear why dioxygen should bind to the iron atom in some encounters and function as an oxidant in others. Caughey and co-workers suggested that chloride binding in or near the active site of the deoxygenated protein could mediate outer-sphere electron transfer from the reduced iron atom to dioxygen (*163*). In the first use of variants to study this problem, Springer *et al.* reported that nine variants of sperm whale Mb in which the distal H64 residue was replaced by other residues exhibited 40- to 350-fold increase in autoxidation that correlated with a decrease in affinity for binding dioxygen (*164*). On the other hand, similar studies of a family of sperm whale oxyMb variants, in which Leu29, a residue in the distal heme binding pocket, was replaced with a series of nonpolar residues of varying size, led to the observation that residues of larger volume at this position stabilize the protein to autoxidation (*165*).

These studies were subsequently expanded by kinetic analysis of autoxidation by about 25 variants of sperm whale and pig myoglobin that led to the conclusion that autoxidation of most of these proteins exhibits a combination of two mechanisms. At high concentration of dioxygen, the dominant mechanism involves dissociation of the neutral superoxide radical from oxyMb, a process that is promoted as pH is lowered; at low concentration of dioxygen, the authors propose a bimolecular mechanism in which dioxygen oxidizes deoxyMb in which a water molecule is weakly coordinated to the iron atom (*166*). For the wild-type protein, the distal histidyl residue stabilizes the coordination of dioxygen to the heme iron through hydrogen bond formation, so superoxide radical dissociation is disfavored. In variants lacking a distal residue capable of hydrogen bonding interactions, the second mechanism is favored because dissociation of superoxide radical is not hindered in these variants and the weak coordination of a water molecule to the reduced iron atom of deoxyMb is not facilitated by possible hydrogen bond formation.

Although these reports demonstrate the contribution that can be made by use of variant forms of Mb in the study of long-recognized but incompletely understood behavior of the protein, they represent only part of the extensive literature concerning the pH dependence, dioxygen dependence, and species dependence of autoxidation kinetics. A detailed discussion of all the relevant mechanistic issues related to autoxidation of oxyMb is beyond the scope of the current chapter, but a thorough survey of this subject has been provided by Shikama (*162*).

V. Peroxidase Activity

A. The Peroxidase Catalytic Mechanism

The heme peroxidase superfamily of enzymes catalyzes the H_2O_2-dependent oxidation of a wide variety of substrates:

$$\text{Peroxidase} + H_2O_2 \rightarrow \text{Compound I} + H_2O \qquad (1)$$

$$\text{Compound I} + S_{red} \rightarrow \text{Compound II} + S_{ox} \qquad (2)$$

$$\text{Compound II} + S_{red} \rightarrow \text{Peroxidase} + S_{ox} + H_2O \qquad (3)$$

In this representation, compounds I and II represent two-equivalent and one-equivalent oxidized forms of the enzyme, respectively, and S_{red} and S_{ox} represent reduced and oxidized substrate, respectively. Direct comparison of the active sites of cytochrome *c* peroxidase (C*c*P), the first peroxidase for which a structure was determined, and Mb demonstrates some of the structural similarities of and differences between the two proteins (Fig. 2). In both proteins, distal and proximal histidyl residues are present as axial ligands, and the proximal histidyl residue is involved in hydrogen bonding interactions with other amino acid residues through the δ-nitrogen. However, the detailed hydrogen bonding interactions of the proximal ligands of the two proteins are different (*vide infra*). On the other hand, while the distal heme pocket of Mb is relatively hydrophobic and favors reversible binding of dioxygen to the heme iron, the distal heme pocket of C*c*P includes a number of hydrophilic residues that are believed to favor the cleavage of the O–O bond during catalysis. Indeed, the intimate mechanism of peroxidase catalysis and the roles of various active site residues, first proposed in 1980 by Poulos and Kraut (*167*), has been largely confirmed by an extensive series of experiments with various site-directed variants of C*c*P and, subsequently, HRP (reviewed in *168–173*). Although the mechanism proposed for this reaction (Fig. 3) and the evidence on which it is based are not discussed in detail here, a brief overview is useful.

In essence, peroxide bond cleavage by C*c*P is favored by the presence of a distal base catalyst (H52) and a positively charged argininyl residue (R48). The reaction of the enzyme with hydrogen peroxide leads, ultimately, to the formation of a form of the enzyme that is oxidized by two equivalents relative to the "resting" enzyme to form an intermediate known as compound I through the heterolytic cleavage of a peroxo-bound intermediate. The rate of compound I formation is rapid ($\sim 10^7$ $M^{-1}s^1$ (*170, 173*)). In most peroxidases, compound I exhibits an oxyferryl heme iron and a porphyrin π-cation radical species. C*c*P is unusual in that the second oxidizing equivalent resides as a stable π-cation radical at W191 (*174, 175*) adjacent to the proximal H175 residue. The proposed (*41, 176*) role of the negatively charged D235 residue on the proximal side of the heme is to stabilize compound I by means of a strong hydrogen bond to H175 that increases the imidazolate character of this heme iron ligand. This proposal led to development of the "push–pull" mechanism for peroxidase catalysis (*176*). In this mechanism, the hydrogen bond formed by D235 and the proximal H175 ligand effectively "pushes" electrons into the proximal ligand and endows it with greater imidazolate character than is the case for Mb to stabilize higher oxidation states of the heme iron. At the same time, the distal H52 and R48 residues "pull" electrons toward the distal side of the heme in a

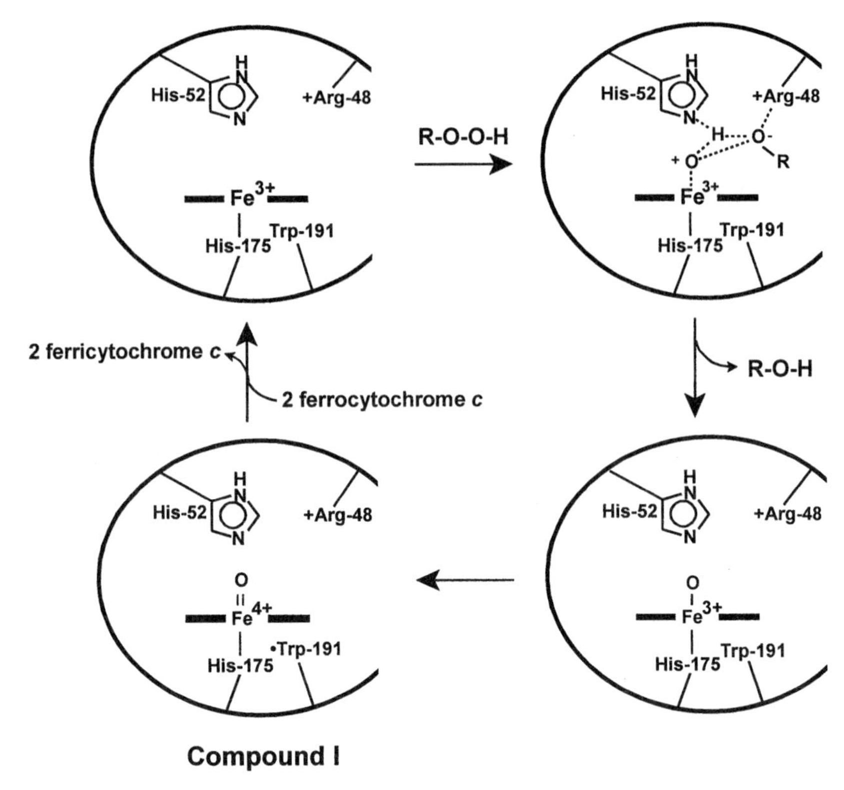

FIG. 3. Catalytic mechanism proposed for cytochrome *c* peroxidase (*167*).

process that involves acceptance of a proton from the hydrogen peroxide by H52 and transfer of this proton to the other oxygen atom of the peroxide to generate a good leaving group. R48 stabilizes the incipient negative charge on the outer oxygen atom of the peroxide during O–O bond cleavage through electrostatic interaction.

B. Reaction of Myoglobin with Hydrogen Peroxide

1. Modification of metMb by Hydrogen Peroxide

As described above, efficient peroxidase catalysis requires rapid reaction of the enzyme with H_2O_2 coupled with the formation of a discrete compound I species. As initially observed by George and Irvine in 1952 (*177*), the reaction of metMb with H_2O_2 is much slower than the corresponding reaction of peroxidases. The myoglobin derivative produced by this reaction was referred to by these authors as ferryl myoglobin

(Fe(IV)=O), a form that is oxidized by one equivalent relative to metMb. As hydrogen peroxide presents two oxidizing equivalents to the protein, the initial product of this reaction should have been oxidized by one equivalent more than expected for the ferryl derivative (which corresponds to peroxidase compound II). Although the resolution to this question evaded elucidation for some time, an early hint was provided by subsequent observation of a free radical following reaction of metMb with peroxide (*178–180*). The nature of this radical remained speculative until it was recognized that reaction of metMb with hydrogen peroxide leads to formation of a family of oxidized Mb species in which one tyrosyl residue or another is oxidized to a radical center. In considering the events that follow reaction of myglobin with peroxides, we begin by considering the reaction at the iron center.

The reaction of the classical peroxidases with peroxides results exclusively in heterolytic cleavage (reviewed in Ref. *173*) of the dioxygen bond to produce water from hydrogen peroxide and a ferryl center coupled to a porphyrin or protein radical in a reaction that is promoted by the "push–pull" mechanism (*176*) described above. In view of the facts that Mb lacks a polar, cationic group in the distal heme binding pocket that is analogous to R48 of C*c*P and that the proximal H93 ligand of Mb is not influenced by a strong hydrogen bonding interaction as is H175 of C*c*P, the behavior of Mb in this reaction is not necessarily similar to that of a peroxidase. Conceivably, Mb could promote homolytic O–O bond cleavage in which a hydroxyl radical is formed from the peroxide with subsequent abstraction of an electron from the protein to create a protein-centered radical. To address this issue, Allentoff *et al.* (*181*) studied the reaction of metMb with an organic peroxide (4-hydroperoxy-4-methyl-2,6-di-*tert*-butylcyclohexa-2,5-dien-1-one (BHTOOH)) to evaluate the mechanism by which the heme iron center of Mb reacts with peroxides. Their results indicated that for this substrate, Mb exhibits both heterolytic and homolytic peroxide bond cleavage with about equal preference for the two pathways (Fig. 4). Interestingly, selected active site variants exhibited similar behavior.

In defining the reaction of metMb with hydrogen peroxide further, Ortiz de Montellano and co-workers took advantage of the fact that myoglobin from various species vary in tyrosine content. Specifically, they noted that sperm whale Mb possesses three tyrosyl residues (Y103, Y146, and Y151) and horse heart Mb has two tyrosyl residues (Y103 and Y146), whereas red kangaroo has just one tyrosyl residue (Y146). They also noted that upon reaction with hydrogen peroxide, sperm whale Mb is unique among these three species in that it forms a protein dimer and the other two species do not. By isolating the dimeric

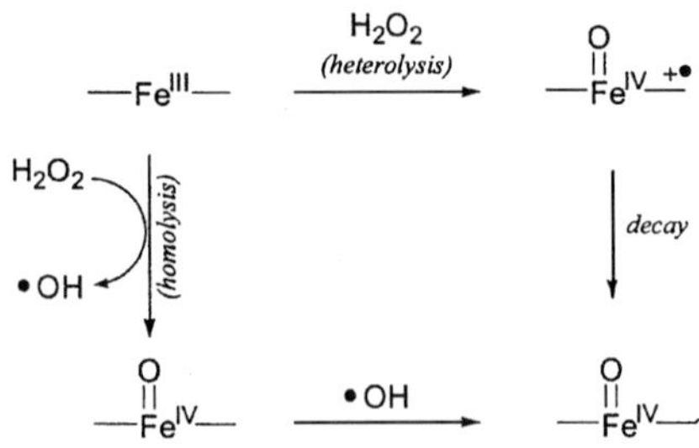

FIG. 4. Heterolytic and hemolytic pathways of peroxide cleavage by myoglobin. (modified from Ref. *198*).

Mb product and characterizing its structure with HPLC peptide mapping, this group demonstrated that sperm whale myoglobin forms a dityrosine crosslink between Y103 on the surface of one molecule of Mb and Y151 on the surface of another (*182*). Additional studies determined that zinc-substituted sperm whale myoglobin, which does not form an intermediate analogous to the ferryl derivative, does not form a dimeric product in this reaction (*182*). This observation was interpreted to mean that formation of the radical on Y151 results from intramolecular electron transfer from this tyrosyl residue to the heme iron rather than from bimolecular reaction with the ferryl group of another Mb molecule. In a subsequent study, this group demonstrated that one of the products (no more than 18% of the total) formed by horse heart metMb upon reaction with hydrogen peroxide is a modified form of the protein in which the heme group is covalently bound to the protein (*183*). Again, structural characterization was achieved through HPLC peptide mapping to implicate crosslinking of the heme through a *meso*-carbon to Y103. These results were subsequently substantiated through related studies of site-directed variants of sperm whale myoglobin in which tyrosyl residues were replaced individually and in pairs (*184*). Interestingly, replacement of all of the tyrosyl residues with phenylalanine did not eliminate formation of a radical product (*184*), but further modification of the tyrosine-deficient variant by replacement of H64 with a valyl residue eliminated formation of the protein-centered radical (*185, 186*).

The sequence of events following the reaction of metMb with hydrogen peroxide has also been investigated through the use of spin trapping agents. Initial studies of this type with 5,5-dimethylpyrroline *N*-oxide (DMPO) led to the identification of Y103 as the primary site of DMPO adduct formation, a modification that was blocked by specific iodination of this residue (*187*). The identity of the radical trapped in this reaction was initially concluded to be a tyrosine peroxyl radical (*187*), but mass spectrometry and isotopic labeling studies with ^{17}O demonstrated the trapping of a tyrosine phenoxyl radical (*188*). In the absence of Y103, only a weak EPR signal of oxidized DMPO could be detected, and this observation was attributed to oxidation of the compound by the peroxidase activity of the protein. Y151 was subsequently shown to form an adduct with trapping agent 2-methyl-2-nitrosopropane (MNP) (*189*).

The production of a peroxyl radical following addition of hydrogen peroxide to metMb was subsequently observed by Kelman *et al.* under different reaction conditions (*190*). In this work, hydrogen peroxide was added to the protein under aerobic and anaerobic conditions in the presence and absence of DMPO, and the reaction mixture then frozen in liquid nitrogen within 10 sec. The EPR spectra of the samples produced in this fashion were then recorded. Although the site of the radical could not be adduced from these experiments, perturbation of the EPR spectrum of the reaction mixtures reacted aerobically in the presence of ^{17}O-enriched dioxygen provided strong confirmation for the formation of a peroxyl radical. Subsequent spin-labeling experiments with 3,5-dibromo-4-nitrosobenzensulfonic acid (DB-NBS) followed by EPR analysis of spin-labeled peptides generated by proteolytic digestion of spinlabeled metMb generated following addition of hydrogen peroxide to metMb led to the identification of the peroxyl radical as residing on a tryptophanyl residue (*191*). Localization of this peroxyl radical to W14 was ultimately achieved through related studies of sperm whale Mb variants (*192*). At present, reaction of hydrogen peroxide with metMb is believed to result in oxidation of the heme iron to the ferryl (Fe(IV)=O) derivative and transiently produce an H64 radical immediately prior to intramolecular electron transfer to produce the peroxyl-W14 radical. This species may then oxidize small molecule substrates or other proteins (*193*), or it may undergo intramolecular electron transfer to produce a phenoxy-radical center at either Y103 or Y151, which may then proceed to form dimeric Mb or the adduct with heme covalently coupled to the protein.

2. *Kinetics of metMb Oxidation by Hydrogen Peroxide*

The kinetics of the reaction between metMb and peroxides has attracted attention in studies that have focused on the role of the distal H64 ligand and other amino acid residues present in the distal heme pocket. Brittain *et al.* reported stopped-flow studies of the reaction of hydrogen peroxide with seven variants in which the distal H64 residue was replaced with a series of residues of varying polarity (*194*). Although the H64Y variant was unreactive toward peroxide, the other variants exhibit increased reactivity with increasing polarity of the distal heme pocket and water coordination to the heme iron. In addition, these latter variants exhibit electronic and EPR spectra consistent with the accumulation of a bound peroxide species with a low-spin heme iron center that could be observed through freeze-quench EPR experiments. This reactive intermediate yields a species with a bleached Soret band and an absorbance at 700 nm that is consistent with reaction with a second equivalent of peroxide that results in heme degradation in a process that is reminiscent of coupled oxidation (*vide infra*). The observation of nonheme iron in the EPR spectrum of the reaction mixture at the completion of reaction is also consistent with this possibility (*194*). Similar observations were reported by Khan *et al.* for the H64G variant in a report that also examined the pH and temperature dependence of the reaction (*195*).

Related studies with a number of variants of sperm whale Mb in which H64 was replaced (*196–198*) also noted that the distal residue governs the rate of peroxide bond cleavage. Specifically, this group found that replacement of H64 decreases the second-order rate constant for formation of the oxidized species and increases the stability of the oxidized product substantially. For a number of these variants, the decrease in the rate constant for O–O bond cleavage was so great that the precursor [Mb–H_2O_2] complex (*197*) could be detected; moreover, in the reaction with *m*-chloroperbenzoic acid (*194*), spectra consistent with the formation of a compound I species were observed for the first time for the H64A, H64S, and H64L variants. Later, direct observation of compound I was reported for the reaction of the H64D variant of sperm whale Mb with H_2O_2 (*198*). In this case, the ability to detect compound I formation derives from an increased rate constant for reaction with peroxide and a decreased rate of decay.

The combined results of these studies clearly demonstrated that creation of a five-coordinate heme iron center in metMb through simple replacement of H64 does not capture the mechanism of peroxide reaction with heme iron observed with true peroxidases. This finding is not altogether surprising insofar as peroxidases, in fact, possess a distal histidyl residue. However, as noted by Watanabe and Phillips, the distance between the N^{ε} of the distal histidine and the heme iron is much smaller for globins (4.3 Å for sperm whale Mb) than for the peroxidases (6.0 Å in cytochrome *c* peroxidase) (*197*). This observation led these investigators to design a variant in which the distal residue is moved to an alternative position, farther from the iron, by changing the residues at positions 64 and 43 to create the F43H/H64L variant

(*197*). In this case, formation of compound I was also observed in the reaction with *m*-chloroperbenzoic acid (*199*) but not in the reaction with H_2O_2 (*197*), which the authors attributed to increased catalatic activity of this variant. Similarly, the second-order rate constant for reaction of this variant with H_2O_2 is increased 10-fold (*197*), consistent with the notion that the distal histidine-to-heme iron distance is a critical factor in active site regulation of heme reactivity. Although the effects of these changes on the steady-state peroxidase activity as assessed by oxidation of guaiacol and ABTS were marginal (*197*), such activity was later (*198*) improved substantially by incorporation of a carboxylate group at the distal position (*vide infra*). Notably, however, the F43H/H64L variant was found to have 50-fold greater catalase activity than the wild-type protein (*198*).

3. *Peroxidase Activity of Myoglobin*

The ability of Mb to catalyze oxidation of substrates by hydrogen peroxide was first reported in the 1950s (*200, 201*). As implied by studies discussed in the previous section, the ability to modify protein function through the techniques of molecular genetics has renewed interest in this catalytic capability of myoglobin. In part, this interest arises from an interest in defining more clearly the structural features of a heme binding site that engender peroxidase activity. One of the more important features of the active sites of these proteins and one of the more challenging structural features to manipulate through mutagenesis concerns the hydrogen bonding interactions of the proximal and distal residues and of intermediates bound transiently to the heme iron. As mentioned earlier, the proximal heme binding pocket of Mb is characterized by a number of hydrogen bonds. Specifically, the N^{δ} of H93 forms hydrogen bonds with the main-chain carbonyl oxygen of L89 and the O^{γ} atom of S92. This hydrogen bonding network extends to involve H97 and the heme 7-propionate. The proximal heme binding pocket of peroxidases, on the other hand, features a much stronger hydrogen bond between the N^{δ} of the proximal ligand and an aspartyl residue that is itself hydrogen bonded to a proximal tryptophanyl residue (Fig. 2). This triad of hydrogen bonds is believed to impart increased electron density on the proximal ligand and thereby strengthen the Fe–histidine bond to provide additional stabilization of the highly oxidized compound I intermediate. Indeed, site-directed mutagenesis studies of CcP have supported this hypothesis (*202–211*).

Efforts to examine the role of proximal hydrogen-bonding interactions in Mb by removal of S92 have been reported for the porcine (S92A,

S92V, and S92L variants) (*212*) and human (S92A variant) proteins (*213*), and attempts to incorporate a peroxidase-like histidine–aspartate hydrogen-bonding interaction have also been reported for human (*213*) and horse heart (*120*) Mbs (S92D variant in both cases). Careful examination of spectroscopic and crystallographic (the porcine S92L variant and the horse S92D variant) data for these variants has established that S92 helps to tether the proximal ligand in the correct orientation with respect to the heme (other proximal mutations involving L89 and H97 also affect, *inter alia*, the orientation of H93 (*214*)). Indeed, substitution of S92 generally leads to increased overall conformational heterogeneity on the proximal side, which, for the S92D variants, makes the formation of new hydrogen bonds with the aspartate group less likely (increased proximal disorder has also been reported following substitution of L89 (*215*)). In fact, although mutagenic substitution of S92 does eliminate potential hydrogen bonding interactions, the formation of new hydrogen bonds does not appear to occur with either the human or horse S92D variants. On the other hand, Sinclair *et al.* have reported EXAFS studies of the L89D variant of sperm whale Mb that led them to conclude that the aspartyl residue introduced at position 89 in this variant forms a strong hydrogen bond with the proximal H93 ligand (*215*). However, this conclusion has not yet been evaluated by other methods, and the peroxidase activity of this variant was not reported.

Most of the myoglobin variants discussed above possess amino acid substitutions that reproduce features of the peroxidase active site in at least a limited sense. Almost all exhibit little or no increase in peroxidase activity relative to that of the wild-type protein, and some exhibit less activity (E. L. Raven, D. P. Hildebrandt, C. L. Hunter, H.-L. Tan, M. Smith, and A. G. Mauk, unpublished). However, a 50- to 70-fold increase in peroxidase activity has been reported for a variant of sperm whale Mb in which the distal histidyl residue was replaced with an aspartyl residue (*198*). In this case, the variant was designed to mimic the active site of chloroperoxidase (*216*), in which the distal ligand is a glutamyl residue and which is known to possess both peroxidase and peroxygenase activities. An alternative means of identifying Mb variants with greater peroxidase activity is provided by the approach referred to as *in vitro* (or directed) evolution. In this strategy, the gene encoding Mb is subjected to random mutagenesis, and bacteria transformed with the mutant genes are screened for expression of Mb variants with peroxidase activity by plate assays involving oxidation of colored dyes. Initial work using this strategy led to the identification of a quadruple variant of horse heart Mb (T39I/K45D/F46L/I107F) with ~25-fold greater peroxidase activity than the wild-type protein (*217*). Kinetic analysis of this variant demonstrated that this improved activity results from an increase in the rate constant for the reaction with hydrogen peroxide (Eq. (1)) (*116*). It seems likely, however, that further use of the general approach of *in vitro* evolution, perhaps through use of DNA or gene shuffling (e.g., Ref. *218*), has the potential to produce a variant with far greater activity.

Nongenetic methods of increasing the peroxidase activity of Mb may also merit consideration. For example, horse heart myoglobin reconstituted with a chemically modified heme carrying eight carboxylate groups bound to the terminal propionates has been reported to exhibit 13-fold greater specificity than native Mb in the oxidation of guaiacol (*219*). Evaluation of the combined use of modified hemes with variant forms of Mb in modification of peroxidase activity also merits consideration. As part of such work, metal-substituted heme derivatives may also be of interest in view of the initial work of Mondal *et al.* regarding the reactivity of Mn-substituted Mb with hydrogen peroxide (*220, 221*).

VI. Lipoxygenase Activity

The H_2O_2-dependent catalytic oxidation of unsaturated fatty acids by myoglobin is well established (*223–236*) and has been implicated in myocardial reperfusion injury (*223, 224, 236*). The most detailed mechanistic information for this activity of myoglobin has been provided for oxidation of linoleic acid (*234*) although mechanisms have also been proposed for oxidation of more complex esterified lipids (*225, 227, 228*). For aerobic oxidation of linoleic acid by hydrogen peroxide catalyzed by sperm whale myoglobin, the mechanism proposed by Rao *et al.* (*234*) involves formation of a compound I–like species (i.e., Mb with a Fe(IV)=O heme and a protein-centered radical), binding of the substrate near the heme, and subsequent formation of 9-hydroxyperoxy10(*E*),12(*Z*)-octadecadienoic acid (*225, 234*). Interestingly, this product is formed in an 84:16 (9*S*):(9*R*) enantiomeric mixture (*234*). Both singlet oxygen and hydroxyl radicals have been eliminated as reactive species responsible for initiating oxidation of linoleic acid through use of scavenging agents, stereochemical arguments, and use of variant forms of sperm whale Mb (*234*). An oxidative mechanism mediated by a protein-centered radical, for example, was eliminated by the finding that reaction of the H64V/K102Q/Y103F/Y146F/Y151F variant, which does not form a protein radical on reaction with H_2O_2 (*vide supra*), is fully active in linoleic acid oxidation. From this observation, it appears that the reactive species in this reaction is ferryl heme itself (*234*).

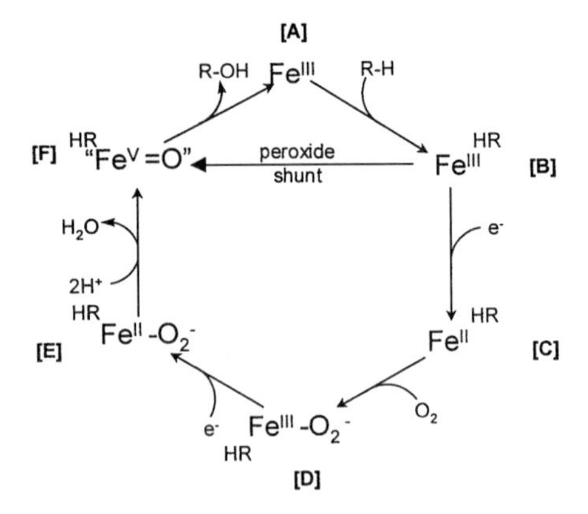

FIG. 5. Catalytic cycle of cytochrome P450. The substrate HR binds to the "resting" enzyme A to form intermediate B, which is reduced by one electron to form C and then reacts with dioxygen. The resulting ferric-peroxo intermediate D is reduced by one equivalent to form the transient oxyferrous intermediate E, which proceeds quickly to intermediate F with release of a molecule of water. F is designated "Fe(V)=O" to indicate that it is oxidized by two equivalents greater than A and not to imply anything about the true oxidation state of the iron. Intermediate F then transfers an oxygen atom to the substrate to regenerate the "resting" enzyme. The "peroxide shunt" refers to the reaction of B with hydrogen peroxide to produce the intermediate F, which can then proceed to product formation.

In subsequent work, Hamberg has reported that anerobic oxidation of lineolic acid by cumene hydroperoxide catalyzed by sperm whale Mb results in formation of five products, the two major products being 11(*R*,*S*)-hydroylinoleic acid (29% yield) and (±)*cis*-9,10-epoxy-(12*Z*)-octadecenoic acid (16%) (*235*). In this work, it was proposed that the second oxidizing equivalent required for substrate hydroxylation was probably provided by a protein-centered radical.

VII. Monooxygenase Activity

Monooxygenases catalyze introduction of one atom of a dioxygen molecule into an organic substrate concomitant with reduction of the other oxygen atom to water (*237*). The orchestration of dioxygen activation with specific binding of organic substrate and appropriately timed delivery of reducing equivalents to produce an enantiomerically pure product by heme enzymes such as cytochromes P450 (Fig. 5) represents a significantly more sophisticated chemical accomplishment than the relatively simple process of reversible binding of dioxygen by myoglobin. Nevertheless, as may be anticipated from the discussion in previous sections of this chapter, some elements of this process can be elicited from myoglobin because the critical requirements are minimal: the ability to generate a transient oxidizing species capable of oxygen atom transfer, and adequate stereochemical access of the substrate to the ferryl intermediate. A critical distinction between such assays of Mb and corresponding assays involving cytochromes P450 is that the oxidizing species of Mb must be generated by reaction with hydrogen peroxide rather than with dioxygen and a reductase. However, cytochromes P450 are also known to exhibit a mechanism referred to as the peroxide shunt in which the ferric enzyme with substrate bound reacts with peroxide to generate the ferryl intermediate directly (*237*) (Fig. 5).

In addition to the activity of the protein in substrate processing, stereospecificity of substrate oxidation is of equal concern. As a result, studies of Mb monooxygenase activity are frequently complemented by determination of the enantiomeric ratio of products (enantiomeric excess) and analysis of the fraction of peroxide oxygen transferred to product. As epoxidation and sulfoxidation reactions catalyzed by Mb have received particular attention, the following discussion considers the progress in understanding these activities of both wild-type and variant forms of the protein.

1. Epoxidation

Initial experiments with wild-type sperm whale Mb indicated that the epoxidation of styrene occurs by two concurrent mechanisms (*238*). In one of these mechanisms, the ferryl oxygen atom is transferred directly to the substrate with retention of stereochemistry and incorporation of an atom of oxygen from H_2O_2 into the epoxide product in a process analogous to that observed for cytochrome P450. In the alternative mechanism, reaction of the protein with hydrogen peroxide generates a protein-centered radical that reacts with dioxygen to produce a protein–peroxy radical that oxidizes styrene with overall loss of stereochemistry and incorporation of an atom of oxygen from dioxygen into the epoxide. Attempts to improve the stereoselectivity of the reaction by identifying the location of the protein radical initially focused (*185*) on one of several tyrosyl residues that were believed to account at least partially for the radical spin distribution (*vide supra*). However, as already discussed, it was later demonstrated that none of these residues is essential for protein radical formation. In fact, site-specific removal of all tyrosine residues (tyrosines 103, 146, and 151) from sperm whale Mb (*239*) revealed that although these residues influence the rate of epoxidation, the stereochemical specificity, and the source of epoxide oxygen, they are not critical to the protein-mediated epoxidation pathway.

Instead, a role for the distal H64 residue was identified through observation that the H64V variant exhibits improved stereoselectivity in styrene epoxidation either through improved steric access of the substrate to the heme or through increased stability of the compound I–like intermediate species (*198*). However, exclusive formation of a single enantiomeric product and complete transfer of peroxide-based ^{18}O into the product, as expected for a mechanism involving only ferryl oxygen transfer, was observed only for reactions catalyzed by the tyrosine-depleted H64V/Y103F/Y146F/Y151F variant (*239*). Subsequently, one of the two tryptophans in sperm whale Mb (W14) was also implicated in the protein-mediated pathway (*186*).

These findings have been extended in experiments by Watanabe and co-workers with the sperm whale Mb variants discussed previously in which the distal histidyl residue is relocated to a position more comparable to the position of the distal histidyl residue of cytochrome *c* peroxidase. In this work, the L29H/H64V (*240*) and F43H/H64V (*199*) variants exhibited high enantiomeric excesses and increased rates of styrene epoxidation (e.g., more than 20-fold for the F43H/H64V variant). The F43Y variant has also been reported to exhibit catalytic activity and stereospecificity in this reaction relative to the wild-type protein (*241*). For the F43H/H64V (*199, 242*), H64A, and H64S variants (*196*), a spectroscopically distinct two-equivalent oxidized species analogous to a peroxidase compound I intermediate was identified. This species and improved steric access to the heme in these variants has been proposed to account, at least partially, for the increased enantiomeric excesses of these variants.[3]

Alteration of the proximal ligand in Mb to mimic the cysteine thiolate in P450 (H93C variant) has little effect on the rate of styrene epoxidation and no effect on the enantiomeric ratio of products (*70*). Although disappointing, these results are consistent with the expectation that a minimal degree of access of substrate to the distal heme pocket is required for catalysis to occur. Unfortunately, attempts to verify this by simultaneous removal of the distal histidine (H64V/H93C and H64G/H93C variants) resulted in very poor monooxygenase activities that were believed to arise from destablization of the Mb-H_2O_2 complex prior to O–O bond cleavage (*71*).

[3] Similar strategies also have been applied successfully to HRP (*243, 244*). Hence, replacement of the distal histidine and relocation to either position 41 (F41H/H42A variant) or position 38 (R38H/H42V variant) improves the catalytic efficiency (k_{cat}/K_m) for the epoxidation and sulphoxidation reactions by ≈3000-fold, and ≈7-fold, respectively (H42A/F41H), and by ≈20-fold and ≈1400-fold, respectively (R38H/H42V).

2. *Sulfoxidation*

For many of the myoglobin variants for which the rate or the stereospecificity of alkene epoxidation have been reported, the sulfoxidation activity of the variant proteins (*71, 196, 199, 240*) (or sometimes *N*-demethylation (*70*)) has frequently been examined and, on the whole, is similarly enhanced. For the sulfoxidation reaction, the L29H/H64V and F43H/H64V variants have been shown to be the most efficient, and it has been clearly demonstrated that the mechanism for these variants involves formation of a two-electron oxidized intermediate that undergoes subsequent reduction in a single (two-electron) step (*199, 242*). In fact, it is very likely that H64 itself destabilizes "compound I" formation in Mb and that removal of this residue favors monooxygenase activity because the oxidized intermediate is more stable than in the wild type protein (*196*).

VIII. Coupled Oxidation

The reaction of oxymyoglobin (MbO_2) with ascorbate has been studied for many decades (early work has been reviewed in Refs. *245, 246*). This relatively complex reaction results in oxidation of the heme prosthetic group in a process referred to as coupled oxidation. The product of the reaction is a form of myoglobin in which the heme prosthetic group has been converted to Fe(III)-biliverdin. Prior to the discovery of the enzyme heme oxygenase (HO), coupled oxidation was generally regarded as the primary mechanism by which physiological conversion of heme to bilirubin occurs. Although the physiological significance of coupled oxidation is now known to be minimal, the similarity of the intermediate degradation products of heme formed during coupled oxidation of myoglobin to intermediates believed to participate in the catalytic cycle of heme oxygenase has resulted in continued interest in understanding the mechanism of coupled oxidation. In the discussion below, the mechanism of heme oxygenase (HO) is considered briefly and the relationship of this reaction to the coupled oxidation of myoglobin is then discussed.

A. Mechanism of Heme Oxygenase

Heme oxygenase catalyzes the regiospecific oxidation of heme to biliverdin (Fig. 6) and has been isolated and purified from several sources (*247–251*). In mammals, the enzyme is membrane-bound, a complication

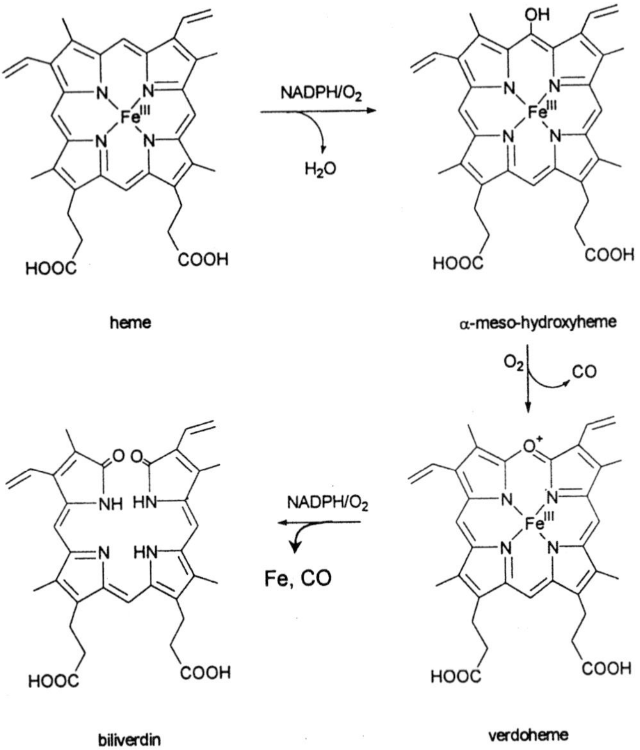

FIG. 6. Reaction catalyzed by heme oxygenase.

that has hindered detailed mechanistic studies until recently. However, the recent expression of soluble, truncated forms of HO that lack the membrane binding domain while retaining full activity (*252–254*),[4] has greatly facilitated detailed mechanistic studies of the enzyme. Heme oxygenase is an unusual enzyme insofar as it uses heme both as a prosthetic group and as a substrate. Prior to determination of the structure of the human enzyme (*260, 261*) (preliminary crystallographic information is also available for the soluble form of rat HO (*262*)), H25 had been shown conclusively to provide the fifth (proximal) axial ligand to the heme iron and to be required for catalytic activity (*263–270*).

The mechanism of HO has been reviewed comprehensively elsewhere (*271*) and is summarized briefly here. As indicated in Fig. 6, the first

[4] Expression of a bacterial HO enzyme, which is not membrane-bound (*255–257*) and of full-length HO enzymes (*258, 259*) has also been reported.

committed step in the mechanism is the formation of α-*meso*-hydroxyheme, which occurs though the formation and subsequent reduction of an oxyferrous derivative to generate what is formally a protonated ferric peroxide intermediate. Formation of this species is probably followed by electrophilic addition involving the porphyrin and the bound peroxide (*271*). The presence of a neutral imidazole ligand similar to that found in Mb has been proposed as a probable stabilizing influence for the bound peroxide species (*264, 267, 272*). (Notably, cytochrome P450 and peroxidases differ in that they have strongly electron-donating ligands that are believed to activate the peroxide bond through heterolytic cleavage.) One critical issue in the formation of this intermediate and in formation of the corresponding intermediate during coupled oxidation of myoglobin is the structural basis for the regiospecificity that results in observation of only α-hydroxyheme (nonenzymatic oxidation of heme leads to an approximately equal mixture of all four regioisomers (*246*)). Early suggestions that this specificity might be controlled by steric constraints imposed upon the heme–O_2 complex by the protein have been shown by crystallography (*261*) to be largely correct. The distal helix passes across the entire face of the heme and physically restricts access to all the *meso* positions except for the α-*meso* carbon. Indeed, spectroscopic measurements have also shown that the Fe–O–O bond is bent, possibly in the direction of the α-*meso* carbon (*268, 273*). While steric effects are clearly important, a series of experiments with various modified heme derivatives has indicated that electronic effects may also have considerable influence although currently the balance of steric *versus* electronic effects is not clear (*261, 271*).

The exact stoichiometric requirements involved in the formation of ferric verdoheme from α-*meso*-hydroxyheme have been controversial (*274–276*). Although there is general agreement that this process is oxygen-dependent, the suggestion that additional reducing equivalents are also required (*274, 277*) has been questioned (*275, 276*). The conversion of verdoheme to biliverdin is the least well-characterized step of the overall reaction, although a mechanism has been proposed (*271*).

B. Reaction of Oxymyoglobin and Ascorbate

In the presence of ascorbate and oxygen, oxyMb and other heme proteins undergo a series of reactions that resemble the catalytic cycle of HO, albeit with less efficiency (*278–281*). Although the spectroscopic similarities of Mb and corresponding derivatives of HO are remarkable (*264, 267, 272*), the mechanism of the coupled oxidation reaction is only poorly defined and differs from that described above for HO in at least three respects. First, catalytic turnover of HO requires an NADPH-dependent P450 reductase as a source of reducing equivalents, whereas coupled oxidation requires just ascorbate. Second, HO binds heme with a far lower affinity and with greater solvent exposure than does apo-Mb and other heme proteins. Third, the HO catalyzed oxidation of heme generates α-biliverdin and not Fe(III)–α-biliverdin as does coupled oxidation of Mb.[5] Nevertheless, coupled oxidation of Mb generates the α-*meso*-hydroxyheme derivative exclusively (*283*). This specificity has been attributed to the known orientation of the bound dioxygen and to steric crowding around the heme.

The initial report of a myoglobin variant that exhibits unusually efficient coupled oxidation appeared prior to the availability of the three-dimensional structure of HO and arose from a chance observation (*119*). More recently, two additional studies have used Mb variants to explore the structural basis for regiospecificity in opening of the heme ring during coupled oxidation (*121, 284*). For wild-type horse heart Mb (*119*), biphasic kinetics have been observed that appear to involve formation of oxyMb and its subsequent decay to Fe(III)–biliverdin without the formation of detectable intermediates (either α-*meso*-hydroxyheme or verdoheme). For the V67A/V68S double variant, more efficient production of Fe(III)–biliverdin was accompanied by more pronounced spectroscopic changes during the reaction that probably arise from formation and decay of verdoheme, but no change in overall specificity. Other variants (H64L, L29H/H64L, F43H/H64L, I107H/H64L, F43W/H64L, F43W, and H64D) of sperm whale Mb have since been observed to exhibit reaction kinetics and regiospecificity that differ from those of both wild-type Mb and the horse heart variant (*121*). At present, there is little consensus as to which residues are mechanistically important in Mb, which residues control regiochemistry, and how this could be altered in a systematic manner. The recent determination of the three-dimensional structure of HO may, however, precipitate further efforts to elucidate the structural issues in coupled oxidation of Mb. Already, the presence of a serine group near the heme in HO (*261*) and in the V67A/V68S double variant of horse Mb (*119*) suggests that the two processes may exhibit some instructive structural correlations.

The reaction catalyzed by HO and involved in coupled oxidation of Mb results in release of 1 mol of CO during conversion of α-*meso*-hydroxyheme to verdoheme. In a number of Mb variants, the formation

[5] Formation of biliverdin or Fe(III)–biliverdin may be dependent on the nature of the reductase, since replacement of the reductase with ascorbate in the HO-catalyzed reaction leads to formation of Fe(III)–biliverdin (*282*).

of a CO-bound species has been detected during coupled oxidation (*121*), which, by necessity, inhibits the catalytic cycle through (irreversible) binding of CO to the heme and poisoning of the catalyst. For HO, inhibition of this kind has not been observed: Indeed, it has been shown (*285*) that the verdoheme–HO complex has a much lower affinity for CO than either the heme–HO or the α-*meso*-hydroxyheme–HO complex, thereby preventing product inhibition by discriminating against CO binding after verdoheme formation (even the heme–HO complex has a much more favorable $CO{:}O_2$ binding ratio ($K_{CO}/K_{O2} = 5.4$) than does Mb ($K_{CO}/K_{O2} = 41$)). The means by which HO achieves such effective discrimination against CO binding is unclear at present but is likely to depend on a number of currently poorly defined variables (*261*) that include the polarity of the active site, specific hydrogen bond interactions (*286*), and steric effects (*287*).

IX. Sulfmyoglobin

Although Hoppe-Seyler discovered sulfhemoglobin in 1866 (*288*) and although the formation of sulfhemoglobin in response to the toxic effects of exposure to sulfides or reducing agents has been recognized for a century (*289*), the history of sulfmyoglobin begins with the work of Nicholls in 1961 (*290*). At that time, Nicholls demonstrated that the reaction of ferrylMb with a sulfide forms a green derivative of Mb in which the prosthetic group is modified. Although models for the structure of the modified heme were proposed (*291*), definitive structural characterization was elusive. In a series of papers (*292–294*), Berzofsky, Peisach, Blumberg, and colleagues reported an improved method for preparation of sulfmyoglobin and provided fundamental spectroscopic and functional characterization of the resulting product. Among their observations was the notable finding that preparation of sulfMb with $^{35}S\text{-}Na_2S$ led to quantitative incorporation of a single atom of ^{35}S in the modified prosthetic group (*294*). In addition, their analysis of the electronic and EPR spectra of ferri- and ferrosulfMb led them to propose that the sulfheme group is, in fact, a chlorin (i.e., a protoheme group in which one of the pyrrole rings is reduced). The chlorin character of sulfheme present in sulfMb was subsequently supported by resonance Raman spectroscopy (*295*).

Despite the unusual spectroscopic and structural features of sulfmyoglobin, functional characterization of this derivative has been limited. Notably, however, Berzofsky *et al.* (*292*) demonstrated that ferrosulfMb binds dioxygen reversibly but with an affinity 2.5 orders of magnitude lower than that of native deoxyMb. On the other hand, the rate constant for reduction of metsulfMb with $Fe(EDTA)^{2-}$ was found to be 2- to 7-fold greater than that of the native protein (*296*). Correction of this observation for the relatively high reduction potential for sulfMb proposed by Nicholls (*290*), however, suggested that the native protein has an intrinsic reactivity that is two- to threefold greater than that of sulfMb.

Ultimately, structural characterization of sulfMb required the application of high-field NMR spectroscopy. Through the efforts of La Mar and colleagues, the reaction of ferrylMb with sulfides was first demonstrated to produce a family of structurally related (*297*), interconverting (*298*) products rather than a single species. This added complexity combined with the variation in distribution of sulfMb isomers dependent on solution conditions combined with the inability of electronic spectroscopy to discriminate between all of the sulfMb isoforms led to a far more complex picture than had been anticipated. Ultimately, it became apparent that reduction of ferrylMb with sulfides initially produces a single, metastable sulfMb isoform (sulfMbA) that at basic pH can either proceed to a second isoform (sulfMbB) or revert to native metMb. On the other hand, at acidic pH sulfMbA can proceed to form the thermodynamically stable sulfMbC isoform. The structures of the modified sulfheme prosthetic groups present in each of these sulfMb isoforms that have been proposed on the basis of NMR studies of La Mar and co-workers are shown in Fig. 7 (*298, 299*).

Subsequently, the three-dimensional structure of the stable isoform sulfMbC was determined by X-ray crystallography (*300*). This structure confirmed the identity of the prosthetic group in this derivative and demonstrated that the structure of the protein in sulfMbC is essentially unchanged from that of native metMb. Although manipulation of the sulfheme derivatives generated in the preparation of sulfMb has been achieved through use of heme-substituted derivatives of Mb (*301, 302*), sulfMb has not been prepared from metal-substituted forms of Mb or from variants that might be useful in evaluating the role of the protein environment in determining the relative stabilities of the various sulfMb isoforms and in determining the stereochemistry of the modified prosthetic group.

X. Other Reactions of Myoglobin

The present survey has emphasized those reactions of Mb that have been studied in greatest chemical detail and, where possible, with the use of variant proteins. As a result, a number of additional peroxide-

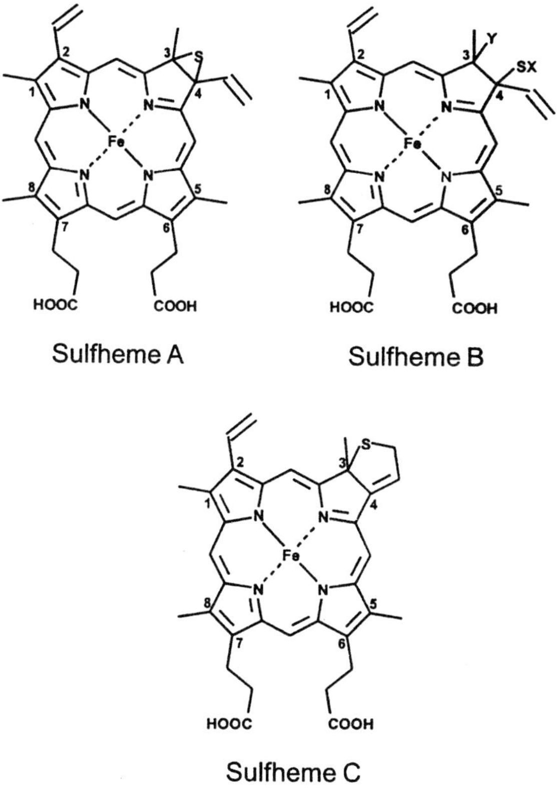
S
Fe
HOOC
COOH
Sulfheme A
Y
SX
Fe
HOOC
COOH
Sulfheme B
S
Fe
HOOC
COOH
Sulfheme C

FIG. 7. Structures of modified heme prosthetic groups identified in three isoforms of sulfmyoglobin.

dependent activities of Mb have not been considered that may merit further investigation. Some of these reactions have been reviewed (*303*), and they include oxidation of myosin (*304–306*), ethanol (*307*), ascorbic acid (*308–310*), hydroxylamine (*311*), thiols (*312*), and oxidase activity (*313*). In view of the similarity of the chemistry of myoglobin to that of hemoglobin, it is also likely that many of the reactions that are catalyzed by hemoglobin and that are reviewed elsewhere (*314–316*) are also manifested by myoglobin.

XI. Concluding Remarks

Interest in the chemistry of myoglobin originated many decades ago and has led to recognition that the environment provided to the heme prosthetic group by apoMb affords a rich and varied chemical profile with a variety of physiological and evolutionary implications. The present survey has attempted to highlight those aspects of this chemistry that have attracted the greatest attention but that have rarely been considered in juxtaposition. In view of the extensive literature related to these topics, it has not been possible to provide a comprehensive compilation of all pertinent published work. It is our hope that from the introduction provided here that the entry of the interested reader into this historically and chemically complex and intriguing area will be encouraged and facilitated.

Acknowledgments

We thank Dr. Paul Witting for helpful comments and Adam Trigg for helpful comments and asistance with preparation of figures. Work in one of our laboratories related to this topic has been supported by the MRC of Canada (Grant TM-7182 to A. G. M.).

References

1. Kagen, L. J. "Myoglobin. Biochemical, Physiological, and Clinical Aspects"; Columbia Univ. Press, 1973, p. 151.
2. Mörner, K. A. H. *Nord. Med. Arkiv.* **1897,** *30,* 1.
3. Günther, H. *Virch. Arch.* **1921,** *230,* 146.
4. Theorell, H. *Biochem. Z.* **1934,** *268,* 73.
5. Millikan, G. A. *Proc. R. Soc. London* **1937,** *B123,* 218.
6. Wittenberg, J. B. *Physiol. Rev.* **1970,** *50,* 559.
7. Wittenberg, B. A.; Wittenberg, J. B.; Caldwell, P. R. *J. Biol. Chem.* **1975,** *250,* 9038.
8. Wittenberg, B. A.; Wittenberg, J. B. *Proc. Natl. Acad. Sci. USA* **1987,** *84,* 7503.
9. Conley, K. E.; Jones, C. *Am. J. Physiol.* **1996,** *271,* C2027.
10. Federspiel, W. J. *Biophys. J.* **1986,** *49,* 857.
11. Gardner, J. D.; Schubert, R. W. *Adv. Exp. Med. Biol.* **1998,** *454,* 509.
12. Jurgens, K. L.; Peters, T.; Gros, G. *Proc. Natl. Acad. Sci. USA* **1994,** *91,* 3829.
13. Garry, D. J.; Ordway, G. A.; Lorenz, J. N.; Radford, N. B.; Chin, E. R.; Grange, R. W.; Bassel-Duby, R.; Williams, R. S. *Nature* **1998,** *395,* 905.
14. Godecke, A.; Flogel, U.; Zanger, K.; Ding, Z.; Hirchenhain, J.; Decking, U. K.; Schrader, J. *Proc. Natl. Acad. Sci. USA* **1999,** *96,* 10495.
15. Antonini, E.; Brunori, M. "Hemoglobin and Myoglobin and Their Reactions with Ligands". North-Holland/Elsevier, 1971, p. 436.
16. Sligar, S. G.; Olson, J. S.; Phillips, G. N. *Chem. Rev.* **1994,** *94,* 699.
17. Vojtechovsky, J.; Chu, K.; Berendzen, J.; Sweet, R. M.; Schlichting, I.; *Biophys. J.* **1999,** *77,* 2153.
18. Akaboshi, E. *Gene* **1985,** *33,* 241.
19. Varadarajan, R.; Boxer, S. G.; Szabo, A. *Proc. Natl. Acad. Sci. USA* **1985,** *82,* 5681.
20. Varadarajan, R.; Boxer, S. G. *Biophys. J.* **1985,** *47,* A85.

21. Weller, P.; Jeffreys, A. J.; Wilson, V.; Blanchetot, A. *EMBO J.* **1984,** *3,* 439.
22. Akaboshi, E. *Gene* **1985,** *40,* 137.
23. Blanchetot, A.; Price, M.; Jeffreys, A. J. *Eur. J. Biochem.* **1986,** *159,* 469.
24. Blanchetot, A.; Wilson, V.; Wood, D.; Jeffreys, A. J. *Nature* **1983,** *301,* 732.
25. Wood, D.; Blanchetot, A.; Jeffreys, A. J. *Nucl. Acids. Res.* **1982,** *10,* 7133.
26. Shimada, H.; Fukasawa, T.; Ishimura, Y. *J. Biochem.* **1989,** *105,* 417.
27. Sidell, B. D.; Vayda, M. E.; Small, D. J.; Moylan, T. J.; Londraville, R. L.; Yuan, M.-L.; Rosnick, K. J.; Eppley, Z. A.; Costello, L. *Proc. Natl. Acad. Sci. USA* **1997,** *94,* 3420.
28. Cutruzzola, F.; Travaglini Allocatelli, C.; Brancaccio, A.; Brunori, M. *Biochem. J.* **1996,** *314,* 83.
29. Dewilde, S.; Winnepenninckx, B.; Arndt, M. H. L.; Nascimento, D. G.; Santoro, M. M.; Knight, M.; Miller, A. N.; Kerlavaga, A. R.; Geohagen, N.; Van Marck, E.; Liu, L. X.; Weber, R. E.; Moens, L. *J. Biol. Chem.* **1998,** *273,* 13583.
30. Suzuki, T.; Takagi, T. *J. Mol. Biol.* **1992,** *228,* 698.
31. Suzuki, T. *J. Prot. Chem.* **1994,** *14,* 9.
32. Suzuki, T.; Yuasa, H.; Imai, K. *Biochim. Biophys. Acta* **1996,** *1308,* 41.
33. Suzuki, T.; Kawamichi, H.; Imai, K. *J. Prot. Chem.* **1998,** *17,* 817.
34. Kawamichi, H.; Suzuki, T. *J. Prot. Chem.* **1998,** *17,* 651.
35. Varadarajan, R.; Lambright, D.; Boxer, S. G. *Biophys. J.* **1986,** *49,* 241a.
36. Dodson, G. G.; Hubbard, R. E.; Oldfield, T. J.; Wilkinson, A. J. *Prot. Eng.* **1988,** *2,* 233.
37. Springer, B. A.; Sligar, S. G. *Proc. Natl. Acad. Sci. USA* **1987,** *84,* 8961.
38. Wilks, A.; Ortiz de Montellano, P. R. *J. Biol. Chem.* **1992,** *267,* 8827.
39. Guillemette, J. G.; Matsushima-Hibiya, Y.; Atkinson, T.; Smith, M. *Prot. Eng.* **1991,** *4,* 585.
40. Evans, S. V.; Brayer, G. D. *J. Biol. Chem.* **1988,** *263,* 4263.
41. Finzel, B.C.; Poulos, T. L.; Kraut, J. *J. Biol. Chem.* **1984,** *259,* 13027.
42. Lloyd, E.; Mauk, A. G. *FEBS Lett.* **1994,** *340,* 281.
43. Gibson, Q. H.; Regan, R.; Olson, J. S.; Carver, T. E.; Dixon, B.; Pohajdak, B.; Sharma, P. K.; Vinogradov, S. N. *J. Biol. Chem.* **1993,** *268,* 16993.
44. Shen, T. J.; Ho, N. T.; Simplaceanu, V.; Zou, M.; Green, B.; Tam, M. F.; Ho, C. *Proc. Natl. Acad. Sci. USA* **1993,** *90,* 8108.
45. Ferrer, J. C.; Ring, M.; Mauk, A. G. *Biochem. Biophys. Res. Commun.* **1991,** *176,* 1469.
46. Chapman, S. K.; Daff, S.; Munro, A. W. *Struct. Bonding* **1997,** *88,* 39.
47. Ortiz de Montellano, P. R. *Acc. Chem. Res.* **1987,** *20,* 289.
48. Dawson, J. H. *Science* **1988,** *240,* 433.
49. Rietjens, I. M. C. M.; Osman, A. M.; Veeger, C.; Zakharieva, O.; Antony, J.; Grodzicki, M.; Trautwein, A. X. *J. Biol. Inorg. Chem.* **1996,** *1,* 372.
50. Quillen, M. L.; Arduini, R. M.; Olson, J. S.; Phillips, G. N. *J. Mol. Biol.* **1993,** *234,* 140.
51. Lloyd, E.; Hildebrand, D. P.; Tu, K. M.; Mauk, A. G. *J. Am. Chem. Soc.* **1995,** *117,* 6434.
52. Dou, Y.; Admiraal, S. J.; Ikeda-Saito, M.; Krzywda, S.; Wilkinson, A. J.; Li, T.; Olson, J. S.; Prince, R. C.; Pickering, I. J.; George, G. N. *J. Biol. Chem.* **1995,** *270,* 15993.
53. Qin, J.; La Mar, G. N.; Dou, Y.; Admiraal, S. J.; Ikeda-Saito, M. *J. Biol. Chem.* **1994,** *269,* 1083.
54. Anderton, C. L.; Hester, R. E.; Moore, J. N. *Biochim. Biophys. Acta* **1995,** *1253,* 1.
55. Cortes, R.; Ascone, I.; Pin, S.; Alpert, B.; Chiu, M. L.; Sligar, S. G. *Jap. J. App. Phys.* **1993,** *32,* 544.

56. Dellalonga, S.; Bianconi, A.; Brancaccio, A.; Brunori, M.; Castellano, A. C.; Cutruzzola, F.; Hazemann, J. L.; Missori, M.; Travaglini Allocatelli, C. *Physica* **1995,** *B209,* 743.
57. Egeberg, K. D.; Springer, B. A.; Martinis, S. A.; Sligar, S. G.; Morikis, D.; Champion, P. M. *Biochemistry* **1990,** *29,* 9783.
58. Christian, J. F.; Unno, M.; Sage, J. T.; Champion, P. M.; Chien, E.; Sligar, S. G. *Biochemistry* **1997,** *36,* 11198.
59. Hargrove, M. S.; Singleton, E. W.; Quillin, M. L.; Ortiz, L. A.; Phillips, G. N.; Olson, J. S.; Mathews, A. J. *J. Biol. Chem.* **1994,** *269,* 4207.
60. Morikis, D.; Champion, P. M.; Springer, B. A.; Egeberg, K. D.; Sligar, S. G. *J. Biol. Chem.* **1990,** *265,* 12143.
61. Pin, S.; Alpert, B.; Cortes, R.; Ascone, I.; Chiu, M. L.; Sligar, S. G. *Biochemistry* **1994,** *33,* 11618.
62. Maurus, R.; Bogumil, R.; Luo, Y., Tang, H.-L. Smith, M. Mauk, A. G.; Brayer, G. D. *J. Biol. Chem.* **1994,** *269,* 12606.
63. Varadarajan, R.; Zewert, T. E.; Gray, H. B.; Boxer, S. G. *Science* **1989,** *243,* 69.
64. Varadarajan, R.; Lambright, D. G.; Boxer, S. G. *Biochemistry* **1989,** *28,* 3771.
65. Zewert, T. E.; Gray, H. B.; Bertini, I. *J. Am. Chem. Soc.* **1994,** *116,* 1169.
66. Nagai, K.; Yoneyama, Y.; Kitagawa, T. *Biochemistry* **1989,** *28,* 2418.
67. Nagai, K.; Kagimoto, A.; Hayashi, A.; Taketa, F.; Kitagawa, T. *Biochemistry* **1983,** *22,* 1305.
68. Perutz, M. F.; Pulsinelli, P. D.; Ranney, H. M. *Nature* **1972,** *237,* 259.
69. Adachi, S.; Nagano, S.; Watanabe, Y.; Ishimori, K.; Morishima, I. *Biochem. Biophys. Res. Comm.* **1991,** *180,* 138.
70. Adachi, S.; Nagano, S.; Ishimori, K.; Watanabe, Y.; Morishima, I. *Biochemistry* **1993,** *32,* 241.
71. Matsui, T.; Nagano, S.; Ishimori, K.; Watanabe, Y.; Morishima, I. *Biochemistry* **1996,** *35,* 13118.
72. Hildebrand, D. P.; Burk, D. L.; Maurus, R.; Ferrer, J. C.; Brayer, G. D.; Mauk, A. G. *Biochemistry* **1995,** *34,* 1997.
73. Hildebrand, D. P.; Ferrer, J. C.; Tang, H.-L.; Smith, M.; Mauk, A. G. *Biochemistry* **1995,** *34,* 11598.
74. DePhillis, G. D.; Decateur, S. M.; Barrick, D.; Boxer, S. G. *J. Am. Chem. Soc.* **1994,** *116,* 6981.
75. Barrick, D. *Biochemistry* **1994,** *33,* 6546.
76. Decatur, S. M.; Franzen, S.; DePhillis, G. D.; Dyer, R. B.; Woodruff, W. H.; Boxer, S. G. *Biochemistry* **1996,** *35,* 4939.
77. Rector, K. D.; Engholm, J. R.; Hill, J. R.; Myers, D. J.; Hu, R.; Boxer, S. G.; Dlott, D. D.; Fayer, M. D. *J. Phys. Chem. B* **1998,** *102,* 331.
78. Das, T. K.; Franzen, S.; Pond, A.; Dawson, J. H.; Rousseau, D. L. *Inorg. Chem.* **1999,** *38,* 1952.
79. Pond, A. E.; Roach, M. P.; Sono, M.; Rux, A. H.; Franzen, S.; Hu, R.; Thomas, M. R.; Wilks, A.; Dou, Y.; Ikeda-Saito, M.; Ortiz de Montellano, P. R.; Woodruff, W. H.; Boxer, S. G.; Dawson, J. H. *Biochemistry* **1999,** *38,* 7601.
80. Decatur, S. M.; Belcher, K. L.; Rickert, P. K.; Franzen, S.; Boxer, S. G. *Biochemistry* **1999,** *38,* 11086.
81. Moore, G. R.; Williams, R. J. P. *FEBS Lett.* **1976,** *79,* 229.
82. Moore, G. R.; Pettigrew, G.; Rogers, N. L. *Proc. Natl. Acad. Sci. USA* **1986,** *83,* 4998.
83. Mauk, A. G.; Moore, G. R. *J. Biol. Inorg. Chem.* **1997,** *2,* 119.
84. Kassner, R. J. *J. Am. Chem. Soc.* **1973,** *95,* 2674.

85. Quinn, R.; Nappa, M.; Valentine, J. S. *J. Am. Chem. Soc.* **1982,** *104,* 2588.
86. Valentine, J. S.; Sheridan, R. P.; Allen, L. C.; Kahn, P. C. *Proc. Natl. Acad. Sci. USA* **1979,** *76,* 1009.
87. Doeff, M. M.; Sweigart, D. A.; O'Brien, P. *Inorg. Chem.* **1983,** *22,* 851.
88. O'Brien, P.; Sweigart, D. A. *Inorg. Chem.* **1985,** *24,* 1405.
89. Mincey, T.; Traylor, T. G. *J. Am. Chem. Soc.* **1979,** *101,* 765.
90. Walker, F. A.; Emrick, D.; Rivera, J. E.; Hanquet, B. J.; Buttlaire, D. H. *J. Am. Chem. Soc.* **1988,** *110,* 6234.
91. Reid, L. S.; Lim, A. R.; Mauk, A. G. *J. Am. Chem. Soc.* **1986,** *108,* 8917.
92. Lee, K.-B.; Jun, E.; La Mar, G. N.; Rezzano, I. N.; Pandey, R. K.; Smith, K. M.; Walker, F. A.; Buttlaire, D. H. *J. Am. Chem. Soc.* **1991,** *113,* 3576.
93. Reid, L. S.; Mauk, M. R.; Mauk, A. G. *J. Am. Chem. Soc.* **1984,** *106,* 2182.
94. Taylor, J. F.; Morgan, V. E. *J. Biol. Chem.* **1942,** *144,* 15.
95. Brunori, M.; Saggese, U.; Rotilio, G. C.; Antonini, E.; Wyman, J. *Biochemistry* **1971,** *10,* 1604.
96. Taylor, J. F. *Methods Ezymol.* **1981,** *76,* 577.
97. Dong, S.; Niu, J.; Cotton, T. M. *Methods Enzymol.* **1995,** *246,* 701.
98. Taboy, C. H.; Bonaventura, C.; Crumbliss, A. L. *Bioelectrochem. Bioenerg.* **1999,** *48,* 79.
99. Armstrong, F. A. *Struct. Bonding* **1991,** *72,* 135.
100. Armstrong, F. A.; Hill, H. A. O.; Walton, N. J. *Quart. Rev. Biophys.* **1985,** *18,* 261.
101. Armstrong, F. A.; Hill, H. A. O.; Walton, N. J. *Acc. Chem. Res.* **1988,** *21,* 407.
102. Bowden, E. F.; Hawkridge, F. M.; Blount, H. N. in "Comprehensive Treatise of Electrochemistry"; Srinivassan, S.; Chizmaszhev, Y. A.; Bolkris, J. O. M.; Conway, B. E.; Yeager, E., Eds.; Plenum Press: New York, 1985; p. 297.
103. Taniguchi, I.; Mie, Y.; Nishiyama, K.; Brabec, V.; Novakova, O.; Neya, S.; Funasaki, N. *J. Electroanal. Chem.* **1997,** *420,* 5.
104. Tominaga, M.; Kumagai, T.; Takita, S.; Taniguchi, I. *Chem. Lett.* **1993,** 1771.
105. Taniguchi, I.; Watanabe, K.; Tominaga, M.; Hawkridge, F. M. *J. Electroanal. Chem.* **1992,** *333,* 331.
106. Van Dyke, B. R.; Saltman, P.; Armstrong, F. A. *J. Am. Chem. Soc.* **1996,** *118,* 3490.
107. Li, G.; Chen, L.; Zhu, J.; Zhu, D.; Untereker, D. F. *Electroanalysis* **1999,** *11,* 139.
108. Rusling, J. F.; Nassar, A.-E. F. *J. Am. Chem. Soc.* **1993,** *115,* 11891.
109. Nassar, A.-E. F.; Willis, W. S.; Rusling, J. F. *Anal. Chem.* **1995,** *67,* 2386.
110. Duah-Williams, L. D.; Hawkridge, F. M. J. *Electroanal. Chem.* **1999,** *466,* 177.
111. Cohen, D. J.; King, B. C.; Hawkridge, F. M. *J. Electroanal. Chem.* **1998,** *447,* 53.
112. King, B. C.; Hawkridge, F. M. *J. Electroanal. Chem. Interfac. Electrochem.* **1987,** *237,* 81.
113. King, B. C.; Hawkridge, F. M.; Hoffman, B. M. *J. Am. Chem. Soc.* **1992,** *114,* 10603.
114. Lim, A. R. Ph.D. Dissertation, Univ. British Columbia, 1989.
115. Heineman, W. R.; Meckstroth, M. L.; Norris, B. J.; Su, C.-H. *Bioelectrochem. Bioenerg.* **1979,** *5,* 577.
116. Hildebrand, D. P.; Lim, K-T.; Rosell, F. I.; Twitchett, M. B.; Wan, L.; Mauk, A. G. *J. Inorg. Biochem.* **1998,** *70,* 11.
117. Tang, H.-L.; B, C.; Mauk, A. G.; Powers, L. S.; Reddy, K. S.; Smith, M. *Biochim. Biophys. Acta* **1994,** *1206,* 90.
118. Lloyd, E.; King, B. C.; Hawkridge, F. M.; Mauk, A. G. *Inorg. Chem.* **1998,** *37,* 2888.
119. Hildebrand, D. P.; Tang, H.-L.; Luo, Y.; Hunter, C. L.; Smith, M.; Brayer, G. D.; Mauk, A. G. *J. Am. Chem. Soc.* **1996,** *118,* 12909.

120. Lloyd, E.; Burk, D. L.; Ferrer, J. C.; Maurus, R.; Doran, J.; Carey, P. R.; Brayer, G. D.; Mauk, A. G. *Biochemistry* **1996,** *35,* 11901.
121. Murakami, T.; Morishima, I.; Matsui, T.; Ozaki, S.; Hara, I.; Yang, H.-J.; Watanabe, Y. *J. Am. Chem. Soc.* **1999,** *121,* 2007.
122. Reid, L. S.; Taniguchi, V. T.; Gray, H. B.; Mauk, A. G. *J. Am. Chem. Soc.* **1982,** *104,* 7516.
123. Walker, F. A.; Reis, D.; Balke, V. L. *J. Am. Chem. Soc.* **1984,** *106,* 6888.
124. Walker, F.; Huynh, B. H.; Scheidt, W. R.; Osvath, S. R. *J. Am. Chem. Soc.* **1986,** *108,* 5288.
125. Scheidt, W. R.; Kirner, J. L.; Hoard, J. L.; Reed, C. A. *J. Am. Chem. Soc.* **1987,** *109,* 1963.
126. Hatano, K.; Safo, M. K.; Walker, F. A.; Scheidt, W. R. *Inorg. Chem.* **1991,** *30,* 1643.
127. Safo, M. K.; Gupta, G. P.; Watson, C. T.; Simonis, U.; Walker, F. A.; Scheidt, W. R. *J. Am. Chem. Soc.* **1992,** *114,* 7066.
128. Safo, M. K.; Walker, F. A.; Raitsimring, A. M.; Walters, W. P.; Dolata, D. P.; Debrunner, P. G.; Scheidt, W. R. *J. Am. Chem. Soc.* **1994,** *116,* 7760.
129. Quinn, R.; Valentine, J. S.; Byrn, M. P.; Strouse, C. E. *J. Am. Chem. Soc.* **1987,** *109,* 3301.
130. Soltis, S. M.; Strouse, C. E. J. *Am. Chem. Soc.* **1988,** *110,* 2824.
131. Inniss, D.; Soltis, S. M.; Strouse, C. E. *J. Am. Chem. Soc.* **1988,** *110,* 5644.
132. Basu, P.; Raitsimring, A. M.; Enemark, J. H.; Walker, F. A. *Inorg. Chem.* **1997,** *36,* 1088.
133. Adachi, S.; Morishima, I. *Biochemistry* **1992,** *31,* 8613.
134. Sligar, S. G. *Biochemistry* **1976,** *15,* 5399.
135. Rodgers, K. K.; Sligar, S. G. *J. Am. Chem. Soc.* **1991,** *113,* 9419.
136. Banci, L.; Bertini, I.; Gori-Savellini, G.; Luchinat, C. *Inorg. Chem.* **1996,** *35,* 4248.
137. Bertini, I.; Gori-Savellini, G.; Lucinat, C. *J. Biol. Inorg. Chem.* **1997,** *2,* 114.
138. Eidsness, M. K.; Burden, A. E.; Richie, K. A.; Kurtz, D. M., Jr.; Scott, R. A.; Smith, Eugene, T.; Ichiye, T.; Beard, B.; Min, TongPil; Kang, ChulHee. *Biochemistry*, **1999,** *38,* 14803.
139. George, P.; Irvine, D. H. *Biochem. J.* **1954,** *58,* 188.
140. George, P.; Irvine, D. H. *Biochem. J.* **1955,** *60,* 596.
141. He, B.; Sinclair, R.; Copeland, B. R.; Makino, R.; Powers, L. S. *Biochemistry* **1996,** *35,* 2413.
142. Huth, S. W.; Kimberly, K. E.; Piszkiewicz, D.; Fleischer, E. B. *J. Am. Chem. Soc.* **1976,** *98,* 8467.
143. Cox, R. P.; Hollaway, M. R. *Eur. J. Biochem.* **1977,** *74,* 575.
144. Olivas, E.; De Waal, D. J.; Wilkins, R. G. *J. Biol. Chem.* **1977,** *252,* 4038.
145. Cox, R. P. *Biochem. J.* **1977,** *167,* 493.
146. Cassatt, J. C.; Marini, C. P.; Bender, J.W. *Biochemistry* **1975,** *14,* 5470.
147. Mauk, A. G.; Gray, H. B. *Biochem. Biophys. Res. Commun.* **1979,** *86,* 206.
148. Lim, A. R.; Mauk, A. G. *Biochem. J.* **1985,** *229,* 765.
149. Eguchi, L. A.; Saltman, P. *J. Biol. Chem.* **1984,** *259,* 14337.
150. Tsukahara, K.; Yamamoto, Y. *J. Biochem. (Tokyo)* **1983,** *93,* 15.
151. Ilan, Y.; Rabani, J.; Czapski, G. *Biochim. Biophys. Acta* **1976,** *446,* 277.
152. Gasyna, Z. *Biochim. Biophys. Acta* **1979,** *577,* 207.
153. Kobayashi, K.; Hayashi, K. *J. Biol. Chem.* **1981,** *256,* 12350.
154. Pin, S.; Hickel, B.; Alpert, B.; Ferradini, C. *Biochim. Biophys. Acta,* **1989,** *994,* 47.
155. Gray, H. B.; Winkler, J. R. *Annu. Rev. Biochem.* **1996,** *65,* 537.

156. Nocek, J. M.; Sishta, B. P.; Cameron, J. C.; Mauk, A. G.; Hoffman, B. M. *J. Am. Chem. Soc.* **1997,** *119,* 2146.
157. Fenwick, C. W.; English, A. M.; Wishart, J. F. *J. Am. Chem. Soc.* **1997,** *119,* 4758.
158. Hegetschweiler, K.; Saltman, P.l; Dalvit, C.; Wright, P.E. *Biochim. Biophys. Acta* **1987,** *912,* 384.
159. Van Dyke, B. R.; Bakan, D. A.; Glover, K. A.; Hegenauer, J. C.; Saltman, P.; Springer, B. A.; Sligar, S. G. *Proc. Natl. Acad. Sci. USA* **1992,** *89,* 8016.
160. Zhang, B. J.; Smerdon, S. J.; Wilkinson, A. J.; Sykes, A. G. *J. Inorg. Biochem.* **1992,** *48,* 79.
161. Dunn, C. J.; Rohlfs, R. J.; Fee, J. A.; Saltman, P. *J. Inorg. Biochem.* **1999,** *75,* 241.
162. Shikama, K. *Chem. Rev.* **1998,** *98,* 1357.
163. Wallace, W. J.; Houtchens, R. A.; Maxwell, J. C.; Caughey, W. S. *J. Biol. Chem.* **1982,** *257,* 4966.
164. Springer, B. A.; Egeberg, K. D.; Sligar, S. G.; Rohlfs, R. J.; Mathews, A. J.; Olson, J. S. *J. Biol. Chem.* **1989,** *25,* 3057.
165. Carver, T. E.; Brantley, R. E., Jr.; Singleton, E. W.; Arduini, R. M.; Quillin, M. L.; Phillips, G. N., Jr.; Olson, J. S. *J. Biol. Chem.* **1992,** *267,* 14443.
166. Brantley, R. E., Jr. Smerdon, S. J.; Wilkinson, A. J.; Singleton, E. W.; Olson, J. S. *J. Biol. Chem.* **1993,** *268,* 6995.
167. Poulos, T. L.; Kraut, J. *J. Biol. Chem.* **1980,** *255,* 8199.
168. Smulevich, G. in "Biomolecular Spectroscopy"; Clark, R. J. H.; Hester, R. E., Eds.; John Wiley and Sons: Chichester, 1993; p. 163.
169. English, A. M.; Tsaprailis, G. *Adv. Inorg. Chem.* **1995,** *43,* 79.
170. Erman, J. E. *J. Biochem. Mol. Biol.* **1998,** *31,* 307.
171. Smith, A. T.; Veitch, N. C. *Curr. Opin. Chem. Biol.* **1998,** *2,* 269.
172. Bosshard, H. R.; Anni, H.; Yonetani, T. in "Peroxidases in Chemistry and Biology"; Everse, J.; Everse, K. E.; Grisham, M. B., Eds.; CRC Press: Boca Raton, 1991; p. 51.
173. Dunford, H. B. "Heme Peroxidases"; John Wiley: Chichester, 1999.
174. Sivaraja, M.; Goodin, D. B.; Smith, M.; Hoffman, B. M. *Science* **1989,** *245,* 738.
175. Scholes, C. P.; Liu, Y.; Fishel, L. A.; Farnum, M. F.; Mauro, J. M.; Kraut, J. *Israel J. Chem.* **1989,** *29,* 85.
176. Poulos, T. L. *Adv. Inorg. Biochem.* **1988,** *7,* 1.
177. George, P.; Irvine, D. H. *Biochem. J.* **1952,** *52,* 511.
178. King, N. K.; Winfield, M. E. *J. Biol. Chem.* **1963,** *238,* 1520.
179. Yonetani, T.; Schleyer, H. *J. Biol. Chem.* **1967,** *242,* 1974.
180. King, N. K.; Looney, F. D.; Winfield, M. E. *Biochim. Biophys. Acta* **1967,** *133,* 65.
181. Allentoff, A. J.; Bolton, J. L.; Wilks, A.; Thompson, J. A.; Ortiz de Demontellano, P. R. *J. Am. Chem. Soc.* **1992,** *114,* 9744.
182. Tew, D.; Ortiz de Montellano, P. R. *J. Biol. Chem.* **1988,** *263,* 17880.
183. Catalano, C. E.; Ortiz de Montellano, P. R. *J. Biol. Chem.* **1989,** *264,* 10534.
184. Wilks, A.; Ortiz de Montellano, P. R. O. *J. Biol. Chem.* **1992,** *267,* 8827.
185. Rao, S. I.; Wilks, A.; Ortiz de Montellano, P. R. *J. Biol. Chem.* **1993,** *268,* 803.
186. Tschirret-Guth, R. A.; Ortiz de Montellano, P. R. *Arch. Biochem. Biophys.* **1996,** *335,* 93.
187. Davies, M. J. *Biochim. Biophys. Acta* **1991,** *1077,* 86.
188. Kelman, D. J.; Mason, R. P. *Free Radic. Res. Commun.* **1992,** *16,* 27.
189. Gunther, M. R.; Tschirret-Guth, R. A.; Witkowska, H. E.; Fann, Y. C.; Barr, D. P.; Ortiz de Montellano, P. R.; Mason, R. P. *Biochem. J.* **1998,** *330,* 1293.
190. Kelman, D. J.; De Gray, J. A.; Mason, R. P. *J. Biol. Chem.* **1994,** *269,* 7458.

191. Gunther, M. R.; Kelman, D. J.; Corbett, J. T.; Mason, R. P. *J. Biol. Chem.* **1995,** *270,* 16075.
192. De Gray, J. A.; Gunther, M. R.; Tschirret-Guth, R.; Ortiz de Montellano, P. R.; Mason, R. P. *J. Biol. Chem.* **1997,** *272,* 2359).
193. Irwin, J. A.; Ostdal, Henrick; Davies, M. J. *Arch. Biochem. Biohys.* **1999,** *362,* 94.
194. Brittain, T.; Baker, A. R.; Butler, C. S.; Little, R. H.; Lowe, D. J.; Greenwood, C.; Watmough, N. J. *Biochem. J.* **1997,** *326,* 109
195. Khan, K. K.; Mondal, M. S.; Padhy, L.; Mitra, S. *Eur. J. Biochem.* **1998,** *257,* 547.
196. Matsui, T.; Ozaki, S.; Watanabe, Y. *J. Biol. Chem.* **1997,** *272,* 32735.
197. Matsui, T.; Ozaki, S.; Liong, E.; Phillips, G. N.; Watanabe, Y. *J. Biol. Chem.* **1999,** *274,* 2838.
198. Matsui, T.; Ozaki, S.; Watanabe, Y. *J. Am. Chem. Soc.* **1999,** *121,* 9952.
199. Ozaki, S.-I.; Matsui, T.; Watanabe, Y. *J. Am. Chem. Soc.* **1997,** *119,* 6666.
200. Keilin, D.; Hartree, E. F. *Biochem. J.* **1955,** *60,* 310.
201. Gibson, J. F.; Ingram, D. J. E.; Nicholls, *P. Nature* **1985,** *181,* 1398.
202. Goodin, D. B.; McRee, D. E. *Biochemistry* **1993,** *32,* 3313.
203. Smulevich, G.; Mauro, J. M.; Fishel, L. A.; English, A. M.; Kraut, J.; Spiro, T. G. *Biochemistry* **1988,** *27,* 5477.
204. Smulevich, G.; Wang, Y.; Mauro, J. M.; Wang, J.; Fishel, L. A.; Kraut, J.; Spiro, T. G. *Biochemistry* **1990,** *29,* 7174.
205. Smulevich, G.; Neri, F.; Marzocchi, M. P.; Welinder, K. G. *Biochemistry* **1996,** *35,* 10576.
206. Wang, J.; Mauro, J. M.; Edwards, S. L.; Oatley, S. J.; Fishel, L. A.; Ashford, V. A.; Xuong, N.; Kraut, J. *Biochemistry* **1990,** *29,* 7160.
207. Vitello, L. B.; Erman, J. E.; Miller, M. A.; Mauro, J. M.; Kraut, J. *Biochemistry* **1992,** *31,* 11524.
208. Satterlee, J. D.; Erman, J. E.; Mauro, J. M.; Kraut, J. *Biochemistry* **1990,** *29,* 8797.
209. Spiro, T. G.; Smulevich, G.; Su, C. *Biochemistry* **1990,** *29,* 4497.
210. Goodin, D. B. *J. Biol. Inorg. Chem.* **1996,** *1,* 360.
211. Ferrer, J. C.; Turano, P.; Banci, L.; Bertini, I.; Morris, I. K.; Smith, K. M.; Smith, M.; Mauk, A. G. *Biochemistry* **1994,** *33,* 7819.
212. Smerdon, S. J.; Krzywda, S.; Wilkinson, A. J.; Brantley, R. E.; Carver, T. E.; Hargrove, M. S.; Olson, J. S. *Biochemistry* **1993,** *32,* 5132.
213. Shiro, Y.; Iizuka, T.; Marubayashi, K.; Ogura, T.; Kitagawa, T.; Balasubramanian, S.; Boxer, S. G. *Biochemistry* **1994,** *33,* 14986.
214. Peterson, E. S.; Friedman, J. M.; Chien, E. Y. T.; Sligar, S. G. *Biochemistry* **1998,** *37,* 12301.
215. Sinclair, R.; Hallam, S.; Chen, M.; Chance, B.; Powers, L. *Biochemistry* **1996,** *35,* 15120.
216. Sundaramoorthy, M.; Terner, J.; Poulos, T. L. *Structure* **1995,** *3,* 1367.
217. Wan, L.; Twitchett, M. B.; Eltis, L. D.; Mauk, A. G.; Smith, M. *Proc. Natl. Acad. Sci. USA* **1998,** *95,* 12825.
218. Cherry, J. R.; Lamsa, M. H.; Schneider, P.; Vind, J.; Svendsen, A.; Jones, A.; Pedersen, A. H. Nat. *Biotechnol.* **1999,** *17,* 379.
219. Hayashi, T.; Hitomi, Y.; Ando, T.; Mizutani, T.; Hisaeda, Y.; Kitagawa, S.; Ogoshi, H. *J. Am. Chem. Soc.* **1999,** *121,* 7747.
220. Mondal, M. S.; Mazumdar, S.; Mitra, S. *Inorg. Chem.* **1993,** *32,* 5362.
221. Mondal, M. S.; Mitra, S. *Biochim. Biophys. Acta,* **1996,** *1296,* 1174.
222. Alayash, A. I.; Ryan, B. A.; Eich, R. F.; Olson, J. S.; Cashon, R. E. *J. Biol. Chem.* **1999,** *274,* 2029.

223. Fantone, J.; Jester, S.; Loomis, T. *J. Biol. Chem.* **1989,** *264,* 9408.
224. Grisham, M. B. J. *Free Rad. Biol. Med.* **1985,** *1,* 227.
225. Galaris, D.; Sevanian, A.; Cadenas, E.; Hochstein, P. *Arch. Biochem. Biophys.* **1990,** *281,* 163.
226. Reeder, B. J.; Wilson, M. T. *Biochem. J.* **1998,** *330,* 1317.
227. Newman, E. S. R.; Rice-Evans, C. A.; Davies, M. J. *Biochem. Biophys. Res. Commun.* **1991,** *179,* 1414.
228. Hogg, N.; Rice-Evans, C. A.; Darley-Usmar, V. M.; Wilson, M. T.; Paganga, G.; Bourne, L. *Arch. Biochem. Biophys.* **1994,** *314,* 39.
229. Kanner, J.; Harel, S. *Arch. Biochem. Biophys.* **1995,** *237,* 314.
230. Dee, G.; Rice-Evans, C.; Obeyesekera, S.; Meraji, S.; Jacobs, M.; Bruckdorfer, K. R. *FEBS Lett.* **1991,** *294,* 38.
231. Stewart, J. M. *Biochem. Cell. Biol.* **1990,** *68,* 1096.
232. Trost, L. C.; Wallace, K. B. *Biochem Biophys. Res. Commun.* **1994,** *204,* 23.
233. O'Leary, V. J.; Graham, A.; Stone, D.; Darley-Usmar, V. M. *Free Rad. Biol. Med.* **1996,** *20,* 525.
234. Rao, S. I.; Wilks, A.; Hamberg, M.; Ortiz de Montellano, P. R. *J. Biol. Chem.* **1994,** *269,* 7210.
235. Hamberg, M. Arch. *Biochem. Biophys.* **1997,** *344,* 194.
236. Galaris, D.; Cadenas, E.; Hochstein, P. *Arch. Biochem. Biophys.* **1989,** *273,* 497.
237. Ortiz de Montellano, P. R. "Cytochrome P450: Structure, Mechanism, and Biochemistry"; Plenum Press: New York, 1995.
238. Ortiz de Montellano, P. R.; Catalano, C. E. *J. Biol. Chem.* **1985,** *260,* 9265.
239. Rao, S. I.; Wilks, A.; Hamberg, M.; Ortiz de Montellano, P. R. *J. Biol. Chem.* **1994,** *269,* 7210.
240. Ozaki, S.-I.; Matsui, T.; Watanabe, Y. *J. Am. Chem. Soc.* **1996,** *118,* 9784.
241. Levinger, D. C.; Stevenson, J.-A.; Wong, L.-L. *Chem. Commun.* **1995,** 2305.
242. Ozaki, S.-I.; Yang, H.-j.; Matsui, T.; Goto, Y.; Watanabe, Y. *Tetrahedron—Asymmetry* **1999,** *10,* 183.
243. Savenkova, M. I.; Newmyer, S. L.; Ortiz de Montellano, P. R. *J. Biol. Chem.* **1996,** *271,* 24598.
244. Savenkova, M. I.; Kuo, J. M.; Ortiz de Montellano, P. R. *Biochemistry* **1998,** *37,* 10828.
245. Lemberg, R. *Pure Appl. Chem.* **1956,** *6,* 1.
246. Sano, S.; Sano, T.; Morishima, I.; Shiro, Y.; Maeda, Y. *Proc. Natl. Acad. Sci. USA* **1986,** *83,* 531.
247. McCoubrey, W. K.; Huang, T. J.; Maines, M. D. *Eur. J. Biochem.* **1997,** *247,* 725.
248. Yoshida, T.; Kikuchi, G. *J. Biol. Chem.* **1978,** *253,* 4224.
249. Yoshida, T.; Kikuchi, G. *J. Biol. Chem.* **1979,** *254,* 4487.
250. Yoshinaga, T.; Sassa, S.; Kappas, A. *J. Biol. Chem.* **1982,** *257,* 7778.
251. Bonkovsky, H. L.; Healey, J. F.; Pohl, J. *Eur. J. Biochem.* **1990,** *189,* 155.
252. Wilks, A.; Ortiz de Montellano, P. R. *J. Biol. Chem.* **1993,** *268,* 22357.
253. Ishikawa, K.; Takeuchi, N.; Takahashi, S.; Matera, C. M.; Sato, M.; Shibahara, S.; Rousseau, D. L.; Ikeda-Saito, M.; Yoshida, T. *J. Biol. Chem.* **1995,** *270,* 6345.
254. Wilks, A.; Black, S. M.; Miller, W. L.; Ortiz de Montellano, P. R. *Biochemistry* **1995,** *34,* 4421.
255. Chu, G. C.; Katakura, K.; Zhang, X.; Yoshida, T.; Ikeda-Saito, M. *J. Biol. Chem.* **1999,** *274,* 21319.
256. Chu, G. C.; Tomita, T.; Sonnichsen, F. D.; Yoshida, T.; Ikeda-Saito, M. *J. Biol. Chem.* **1999,** *274,* 24490.

257. Wilks, A.; Schmitt, M. P. *J. Biol. Chem.* **1998,** *273,* 837.
258. Rotenberg, M. O.; Maines, M. D. *J. Biol. Chem.* **1990,** *265,* 7501.
259. Ishikawa, K.; Sato, M.; Yoshida, T. *Eur. J. Biochem.* **1991,** *202,* 161.
260. Schuller, D. J.; Wilks, A.; Ortiz de Montellano, P. R.; Poulos, T. L. *Prot. Science* **1998,** *7,* 1836.
261. Schuller, D. J.; Wilks, A.; Ortiz de Montellano, P. R.; Poulos, T. L. *Nature Struct. Biol.* **1999,** *6,* 860.
262. Omata, Y.; Asada, S.; Sakamoto, H.; Fukuyama, K.; Nogichi, M. *Acta Cryst.* **1998,** *D54,* 1017.
263. Sun, J.; Loehr, T. L.; Wilks, A.; Ortiz de Montellano, P. R. *Biochemistry* **1994,** *33,* 13734.
264. Sun, J.; Wilks, A.; Ortiz de Montellano, P. R.; Loehr, T. L. *Biochemistry* **1993,** *32,* 14151.
265. Ishikawa, K.; Sato, M.; Ito, M.; Yoshida, T. *Biochem. Biophys. Res. Commun.* **1992,** *182,* 981.
266. Wilks, A.; Sun, J.; Loehr, T. L.; Ortiz de Montellano, P. R. *J. Am. Chem. Soc.* **1995,** *117,* 2925.
267. Takahashi, S.; Wang, J.; Rousseau, D. L.; Ishikawa, K.; Yoshida, T.; Host, J. R.; Ikeda-Saito, M. *J. Biol. Chem.* **1994,** *269,* 1010.
268. Takahashi, S.; Wang, J.; Rousseau, D. L.; Ishikawa, K.; Yoshida, T.; Takeuchi, N.; Ikeda-Saito, M. *Biochemistry* **1994,** *33,* 5531.
269. Maki-Ito, M.; Ishikawa, K.; Matera, K. M.; Sato, M.; Ikeda-Saito, M.; Yoshida, T. *Arch. Biochem. Biophys.* **1995,** *317,* 253.
270. Liu, Y.; Moenne-Loccoz, P.; Hildebrand, D. P.; Wilks, A.; Loehr, T. M.; Mauk, A. G.; Ortiz de Montellano, P. R. *Biochemistry* **1999,** *38,* 3733.
271. Ortiz de Montellano, P. R. *Acc. Chem. Res.* **1998,** *31,* 543.
272. Takahashi, S.; Matera, C. M.; Fujii, H.; Zhou, H.; Ishikawa, K.; Ikeda-Saito, M.; Rousseau, D. L. *Biochemistry* **1997,** *36,* 1402.
273. Takahashi, S.; Ishikawa, K., Takeuchi, N.; Ikeda-Saito, M.; Yoshida, T.; Rousseau, D. L. *J. Am. Chem. Soc.* **1995,** *117,* 6002.
274. Matera, C. M.; Takahashi, S.; Fujii, H.; Zhou, H.; Ishikawa, K.; Yoshimura, T.; Rousseau, D. L.; Yoshida, T.; Ikeda-Saito, M. *J. Biol. Chem.* **1996,** *271,* 6618.
275. Liu, Y.; Moenne-Loccoz, P.; Loehr, T. M.; Ortiz de Montellano, P. R. *J. Biol. Chem.* **1997,** *272,* 6909.
276. Sakamoto, H.; Omata, Y.; Palmer, G.; Noguchi, M. *J. Biol. Chem.* **1999,** *274,* 18196.
277. Migita, C. T.; Fujii, H.; Matera, C. M.; Takahashi, S.; Zhou, H.; Yoshida, T. *Biochim. Biophys. Acta* **1999,** *1432,* 203.
278. Warburg, O.; Negelein, E. *Chem. Ber.* **1930,** *63,* 1816.
279. Lemberg, R.; Legge, J. W.; Lockwood, W. H. *Nature* **1938,** *142,* 148.
280. Lemberg, R. *Pure Appl. Chem.* **1956,** *6,* 1.
281. Kench, J. E. *Biochem. J.* **1954,** *60,* 310.
282. Yoshida, T.; Ishikawa, K.; Sato, M. *Eur. J. Biochem.* **1991,** *199,* 729.
283. O'Carra, P.; Colleran, E. *FEBS Lett.* **1969,** *5,* 295.
284. Murakami, T.; Morishima, I.; Matsui, T.; Ozaki, S.; Watanabe, Y. *Chem. Commun.* **1998,** 773.
285. Migita, C. T.; Matera, C. M.; Ikeda-Saito, M. *J. Biol. Chem.* **1998,** *273,* 945.
286. Sigfridsson, E.; Ryde, U. *J. Biol. Inorg. Chem.* **1999,** *4,* 99.
287. Kachalova, G. S.; Popov, A. N.; Bartunik, H. D. *Nature* **1999,** *284,* 473.
288. Hoppe-Seyler, F. *Zbl. Med. Wiss.* **1866,** *4,* 436.
289. Van den Bergh, A. A. H. *Deut. Arch. Klin. Med.* **1905,** *83,* 86.

290. Nicholls, P. *Biochem. J.* **1961,** *81,* 374.
291. Morell, D. B.; Chang, Y.; Clezy, P. S. *Biochim. Biophys. Acta* **1967,** *136,* 121.
292. Berzofsky, J. A.; Peisach, J.; Blumberg, W. E. *J. Biol. Chem.* **1971,** *246,* 3367.
293. Berzofsky, J. A.; Peisach, J. Blumberg, W. E. *J. Biol. Chem.* **1971,** *246,* 7366.
294. Berzofsky, J. A.; Peisach, J.; Horecker, B. L. *J. Biol. Chem.* **1972,** *247,* 3783.
295. Andersson, L. A.; Loehr, T. M.; Lim, A. R.; Mauk, A. G. *J. Biol. Chem.* **1984,** *259,* 15340.
296. Lim, A. R.; Mauk, A. G. *Biochem. J.* **1985,** *229,* 765.
297. Chatfield, M. J.; La Mar, G. N.; Balch, A. L.; Lecomte, J. T. J. *Biochem. Biophys. Res. Commun.* **1986,** *135,* 309.
298. Chatfield, M. J.; La Mar, G. N.; Kauten, R. J. *Biochemistry*, **1987,** *26,* 6939.
299. Parker, W. O., Jr.; Chatfield, M. J.; La Mar, G. N. *Biochemistry* **1989,** *28,* 1517.
300. Evans, S. V.; Sishta, B. P.; Mauk, A. G.; Brayer, G. D. *Proc. Natl. Acad. Sci. USA* **1994,** *91,* 4723.
301. Scharberg, M. A.; La Mar, G. N. *J. Am. Chem. Soc.* **1993,** *115,* 6513.
302. Scharberg, M. A.; La Mar, G. N. *J. Am. Chem. Soc.* **1993,** *115,* 6522.
303. Giulivi, C.; Cadenas, E. *Methods Enzymol.* **1994,** *233,* 189.
304. Hanan, T.; Shaklai, N. *Eur. J. Biochem.* **1995,** *233,* 930.
305. Hanan, T.; Shaklai, N. *Free Rad. Res.* **1995,** *22,* 215.
306. Bhoite-Soloman, R. S.; Kessler-Icekson, G.; Shaklai, N. *Biochem. Int.* **1992,** *26,* 181.
307. Harada, K.; Tamura, M.; Yamazaki, I. *J. Biochem.* **1986,** *100,* 499.
308. Rice-Evans, C.; Okunade, G.; Khan, R. *Free Rad. Res. Commun.* **1989,** *7,* 45.
309. Kanner, J.; Harel, S. *Lipids* **1985,** *20,* 625.
310. Giulivi, C.; Cadenas, E. *FEBS Lett* **1993,** *332,* 287.
311. Taira, J.; Misik, V.; Riesz, P. *Biochim. Biophys. Acta* **1997,** *1336,* 502.
312. Romero, F. J.; Ordonez, I.; Arduini, A.; Cadenas, E. *J. Biol. Chem.* **1992,** *267,* 1680.
313. Osawa, Y.; Korzekwa, K. *Proc. Natl. Acad. Sci. USA* **1991,** *88,* 7081.
314. Zilletti, L.; Ciuffi, M.; Franchi-Micheli, S.; Fusi, F.; Gentilini, G.; Moneti, G.; Valoti, M.; Sgaragli, G. P. *Methods Enzymol.* **1994,** *233,* 562.
315. Mieyal, J. J.; Starke, D. W. *Methods Enzymol.* **1994,** *231,* 573.
316. Belvedere, G.; Samaja, M. *Methods Enzymol.* **1994,** *231,* 598.

ENZYMOLOGY AND STRUCTURE OF CATALASES

PETER NICHOLLS,* IGNACIO FITA,† and PETER C. LOEWEN‡

*Department of Biological Sciences, Central Campus, University of Essex, Wivenhoe Park, Colchester, CO4 3SQ, United Kingdom;
†CID–CSIC, 08034 Barcelona, Spain; and
‡Department of Microbiology, University of Manitoba, Winnipeg, Manitoba R3T 2N2

0898-8838/01 $35.00

I. Introduction

As a result of their striking ability to evolve molecular oxygen, catalases have been the subject of observation and study for well over 100 years with the first report of a biochemical characterization and naming of the enzyme appearing in 1900 (*1*). This history has been well documented (*2, 3*). The overall reaction for the classical enzyme is very simple on paper, $2H_2O_2 \rightarrow 2H_2O + O_2$, but there are two distinct stages in the reaction pathway. The first stage involves oxidation of the heme iron using hydrogen peroxide as substrate to form compound I, an oxyferryl species with one oxidation equivalent located on the iron and a second oxidation equivalent delocalized in a heme cation radical (reaction (1)). The second stage, or reduction of compound I, employs a second molecule of peroxide as electron donor providing two oxidation equivalents (reaction 2).

$$\text{Enz}\,(\text{Por–Fe}^{\text{III}}) + H_2O_2 \rightarrow \text{Cpd I}\,(\text{Por}^{+\cdot}\text{–Fe}^{\text{IV}}{=}\text{O}) + H_2O \quad (1)$$

$$\text{Cpd I}\,(\text{Por}^{+\cdot}\text{–Fe}^{\text{IV}}{=}\text{O}) + H_2O_2 \rightarrow \text{Enz}\,(\text{Por–Fe}^{\text{III}}) + H_2O + O_2 \quad (2)$$

Compound I can also undergo a one electron reduction with or without a proton resulting in the formation of an inactive compound II (Reaction (3) or (3a)).

$$\text{Cpd I}\,(\text{Por}^{+\cdot}\text{–Fe}^{\text{IV}}{=}\text{O}) + e^- \rightarrow \text{Cpd II}\,(\text{Por–Fe}^{\text{IV}}{=}\text{O}) \quad (3)$$

$$\text{Cpd I}\,(\text{Por}^{+\cdot}\text{–Fe}^{\text{IV}}{=}\text{O}) + e^- + H^+ \rightarrow \text{Cpd II}\,(\text{Por–Fe}^{\text{IV}}\text{–OH}^+) \quad (3a)$$

Heme alone can reportedly elicit a catalatic reaction (the reaction mediated by catalase) but at a much reduced, almost negligible, rate compared to the catalatic proteins containing heme, and this may explain the observation of catalase activity in enzymes not normally associated with catalatic activity (*4, 5*). Other enzymes have evolved that can catalyze a similar reaction in the absence of heme, but this review will limit itself to a consideration of heme containing proteins with catalatic activity.

Catalases can also act as peroxidases (catalyzing a peroxidatic reaction) in which electron donors are oxidized via one–one electron transfers (Reactions (4) and (5)).

$$\text{Cpd I}\,(\text{Por}^{+\cdot}\text{–Fe}^{\text{IV}}{=}\text{O}) + \text{AH2} \rightarrow \text{Cpd II}\,(\text{Por–Fe}^{\text{IV}}\text{–OH}) + \text{AH.} \quad (4)$$

$$\text{Cpd II}\,(\text{Por–Fe}^{\text{IV}}\text{–OH}) + \text{AH2} \rightarrow \text{Enz}\,(\text{Por–Fe}^{\text{III}}) + \text{AH.} \quad (5)$$

Generally, the peroxidatic reaction of true catalases is weak in comparison to actual peroxidases, but can be an important reaction in the class known as catalase-peroxidases (Section II,B).

II. Categorization

The diversity among catalases, evident in the variety of subunit sizes, the number of quaternary structures, the different heme prosthetic groups, and the variety of sequence groups, enables them to be organized in four main groups: the "classic" monofunctional enzymes (type A), the catalase-peroxidases (type B), the nonheme catalases (type C), and miscellaneous proteins with minor catalatic activities (type D).

A. Type A: Monofunctional Catalases

The largest and most extensively studied group of catalases is composed of what are effectively monofunctional enzymes. The dismutation of hydrogen peroxide is their predominant activity and any peroxidatic activity is minor and restricted to small substrates. The most convenient way of subcategorizing this group is based on subunit size with an accompanying attention to heme content. This gives rise to two subgroups, one containing small subunit enzymes (55 to 69 kDa) with heme *b* associated, and one containing large subunit enzymes (75 to 84 kDa) with heme *d* associated. The monofunctional catalases characterized in greatest detail have all proved to be active as tetramers, although dimeric, heterotrimeric, and hexameric enzymes have been reported, but never conclusively characterized. Indeed, the commonality of tetrameric structures (see below), even between the small and large subunit classes of enzymes, demands the presentation of extensive and convincing evidence to confirm any structure that is purported to be other than tetrameric.

A phylogenetic analysis of 70 monofunctional catalase sequences (*6*), now extended to include 113 sequences, has revealed a subdivision into three distinct groups or *clades,* a distinct grouping of sequences arising from a phylogenetic analysis. Clade I contains the plant enzymes and one branch of bacterial catalases. Clade II contains only large subunit catalases with bacterial and fungal origins. Clade III contains a third group of bacterial enzymes as well as fungal and animal enzymes and one enzyme with an archaebacterial origin. The main groupings are supported at very high confidence levels at the main nodes as shown

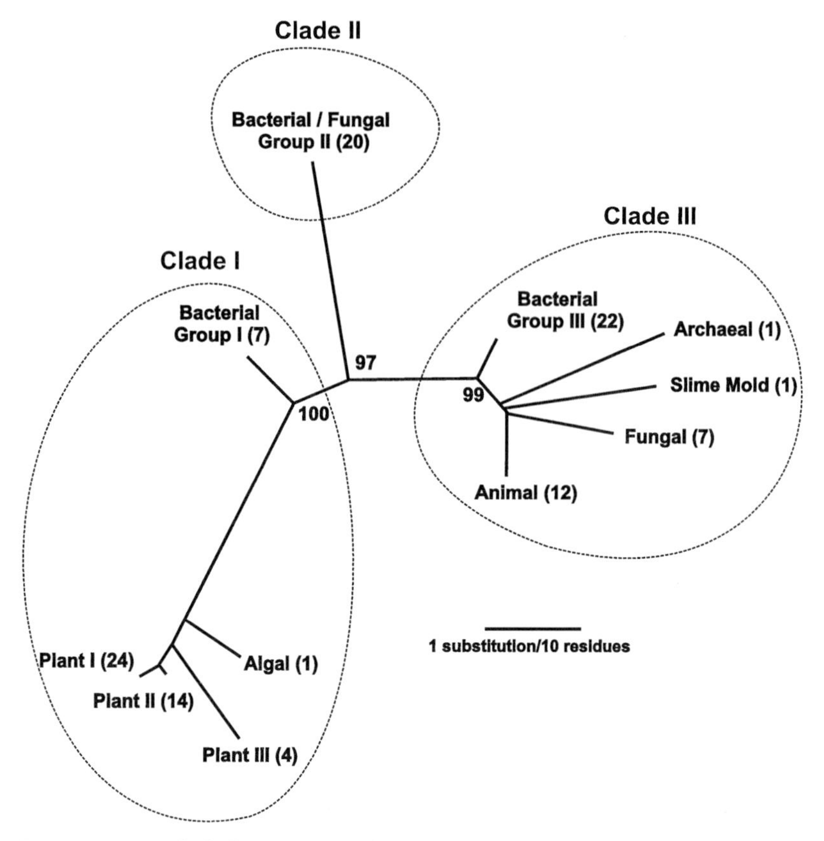

FIG. 1. Unrooted phylogenetic tree based on the core amino acid sequences of 113 catalases. The numbers at the three main nodes represent the proportion (out of 100) of bootstrap sampling that supports the topology. The three main clades are circled for clarity.

in Fig. 1. The inference from such a tree is that the main clades arose from a progenitor catalase through a minimum of at least two gene duplication events. Whether the progenitor enzyme was a large subunit or a small subunit enzyme remains the subject of discussion.

B. Type B: Catalase-Peroxidases

The next largest group of catalases are the catalase-peroxidases, so named because they exhibit a significant peroxidatic activity in addition to the catalatic activity. They have been characterized in both fungi and bacteria and resemble certain (type I) plant and fungal peroxidases in sequence. There is more uniformity in sequence within this group of catalases, which contain heme *b*, have subunits larger than 80 kDa (with a few exceptions), and are active as either dimers or tetramers. It has been hypothesized that the catalase-peroxidases may have arisen through a duplication and fusion event giving rise to two domains with similar sequences in the same subunit (*7*). One of the domains has retained activity and a greater sequence similarity to other catalase-peroxidases, while the second has evolved with greater sequence deviation into an inactive form without bound heme.

A phylogenetic analysis of the catalase-peroxidase sequences (*2*) now extended to 20 available sequences, does not reveal any major subgroupings comparable to those in the catalase family. Whether this is because of the small number of sequences or because of the homogeneity of the enzymes will become evident as further sequences come available. As a result we will, for the time being, refer to the catalase-peroxidases as a single group of enzymes.

C. Type C: Nonheme Catalases

Currently the smallest group, there are only three nonheme catalases so far characterized and an equal number sequenced, all of bacterial origin (*Lactobacillus plantarum, Thermoleophilum album,* and *Thermus thermophilus*). The active site of each of the three enzymes (8–10) contains a manganese-rich reaction center rather than a heme group, and it was this lack of a heme that led to them originally being called "pseudo-catalases." Crystal structures have been determined for the *Lactobacillus plantarum* and *Thermus thermophilus* enzymes (*11*) and have confirmed the active site as containing a bridged binuclear manganese cluster. Its mechanism of catalytic action is currently under discussion. Until more sequences are available, a phylogenetic analysis is not warranted.

D. Type D: Minor Catalases

Several heme-containing proteins, including most peroxidases (*12*), have been observed to exhibit a low level of catalatic activity, with the chloroperoxidase from *Caldariomyces fumago* exhibiting the greatest reactivity as a catalase (*13–15*). Despite the fact that there is as yet only one such example to consider, it provides an alternate mechanism for the catalatic reaction and is addressed in this review. It was first characterized for its ability to chlorinate organic substrates in the presence of chloride and hydrogen peroxide at acid pH, but was later found to have peroxidatic and catalatic properties above pH 4 in the absence of chloride ion or chloride ion and organic substrate, respectively. The enzyme has a molecular weight of 42 kDa and is active as a monomer. There are three additional classes of haloperoxidases, two of which are nonheme enzymes and are not considered here; the third, including the heme-containing bromoperoxidase from *Streptomyces violaceus,* is considered as a Type A catalase based on its sequence (*16*).

Other proteins such as methemoglobin and metmyoglobin have been observed to produce molecular oxygen in the presence of hydrogen peroxide, but at a very low rate (*5*). This may simply be a property of the heme, which can promote a low-level catalatic reaction in the absence of protein. Consequently, it is possible that all heme-containing proteins may exhibit catalatic reactions if assayed carefully, but such minor, largely nonquantifiable activities are not considered here.

III. Physiology

A. Function

What is the role of catalase in organisms? The obvious is that it protects the organism against reactive oxygen species, particularly those derived from hydrogen peroxide. The existence of so many prokaryotic catalases, as well as their occasional inducibility, suggests that a selective advantage is maintained by the ability to produce and to use catalase intermittently when any organism is liable to experience sudden increases in environmental or internally generated peroxide levels. Cultures of *Escherichia coli* subjected to long periods of aeration die off more rapidly if they lack catalase HPII (hydroperoxidase II) than if it is present (*17*), and Ma and Eaton (*18*) demonstrated a protective role for catalase (HPII or HPI) in *E. coli* cultures. In the latter report, the protection was more evident in dense than in dilute bacterial suspensions, and they speculated that a form of "group protection" against oxidative stress could have been one of the selective forces leading to the evolution of multicellular organisms. Although many populations of *E. coli* may indeed be clonal in character and thus capable of exhibiting group selection characteristics according to conventional Darwinian theory, interpretation of such results in terms of advantage to the individual cell is still possible. The individual cell, even among prokaryotes, is more likely to be at risk from internally generated than externally produced H_2O_2. Internal H_2O_2 may be dissipated either by catalatic activity or by diffusion out of the cell. The amount of catalase expressed in cells under most conditions will represent the minimum amount required to keep the maximum peroxide concentration during a pulse of production below 0.1 or 0.2 μM (*19, 20*). This can be achieved in *E. coli* by keeping about 0.1% of its total cell protein in the form of catalase. In other microorganisms much higher catalase levels may be needed to preserve a low level of peroxide under all conditions. *Rhodobacter spheroides* can reportedly synthesize up to 25% of its protein in the form of catalase (*21*).

In the case of higher eukaryotes, including humans, Nicholls and Schonbaum (*22*) had suggested that catalase might be a "fossil enzyme," present but without a functional role. This was in part based upon the finding by Aebi *et al.* (*23, 24*) of healthy acatalasics among Swiss Army recruits. One of us (P.N.) remembers the striking photograph of such a soldier, about the same age as himself, and obviously physically far stronger and fitter. Unfortunately, no longitudinal study of these acatalasics seems to have been carried out after Hugo Aebi's death. In the last 35 years we have learned much more about the roles of peroxide-generated free radicals in disease, DNA damage, and aging. P.N. therefore wonders what the Swiss acatalasics of his age look like now.

Enzymes with an intermittent role may be much more important than we thought in 1963. This was perhaps first clearly emphasized by Deisseroth and Dounce in their catalase review of 1970 (*25*). These authors also pointed out the likelihood that the specific location of most eukaryotic catalase in the peroxisomes represents a functional response to the need to decompose hydrogen peroxide generated by the aerobic oxidases present in these same organelles, including hydroxy-acid oxidases and D-amino-acid oxidases.

B. Regulation of Expression

The physiology of catalase expression and its control in bacteria, yeast, and plants has been reviewed elsewhere (*2, 26, 27*). The following precis is presented so that a summary of physiological information relevant to the detailed biochemistry is readily available.

1. *Prokaryotes*

The early work on catalase expression was carried out largely in *E. coli* and revealed two main response mechanisms. One or the other or both responses have been identified in most other bacteria expressing a catalase. The expected and most obvious response is to oxidative stress. Addition of hydrogen peroxide directly or of ascorbate, which reacts with oxygen to produce hydrogen peroxide, to the medium of exponential phase cells causes a 10- to 20-fold increase in HPI levels (*28*). This is the result of activation of OxyR, which controls the expression of eight or nine genes encoding enzymes such as HPI and alkylhydroperoxidase (*29*). As cells grow normally through exponential phase into stationary phase, the level of HPI rises about twofold and then falls slightly, a phenomenon that has been attributed, although not without some controversy, to a response to increasing levels of the alternate sigma factor RpoS in stationary phase (*30–32*). Other reagents that impose oxidative stress, such as paraquat, cause a similar response.

A less expected response is the 10- to 20-fold increase in HPII levels as cells grow into stationary phase (*28*). The explanation for this response is that the enzyme serves a protective role during periods of slow or no growth. Indeed, the mutation of *katE* results in strains that die off more rapidly during extended incubation in stationary phase (*17*). The increase in HPII is the result of increasing levels of the alternate sigma factor RpoS, which is a central control element for a generalized stress response, including starvation, acid shock, and hypertonic shift (see review in *33*). The involvement of another transcription factor controlling *katE* expression has never been demonstrated. The levels of RpoS and its influence on transcription are regulated by a complex interplay of factors working at the levels of transcrition, translation, and enzyme stability. Response to oxidative stress and response to other stresses are the two main themes found throughout the prokaryotes, with any variations presumably arising from environmental demands arising from unique habitats.

2. *Eukaryotes*

Regulation of catalase expression in eukaryotes takes place as part of a generalized response mechanism. In yeast, promoter elements of the peroxisomal catalase CTA-1 respond to glucose repression and activation by fatty acids as part of organelle synthesis. The cytosolic catalase CTT-1 responds as part of a generalized stress response to starvation, heat, high osmolarity, and H_2O_2, and there is even evidence of translational control mediated by heme availability (*26*).

Expression of the multiple catalases in plants (e.g., three in maize and four in mustard) are developmentally controlled, giving rise to complex response patterns. The picture is further complicated by overlapping responses to environmental stresses such as pathogenesis, radiation, hormones, temperature extremes, oxygen extremes, and H_2O_2 (*27*).

IV. Kinetics

A. The Catalatic Pathway

The kinetic behavior of classical catalases remains widely misunderstood, as evidenced by the frequent quoting of K_m and k_{cat}/K_m values for catalases without any rider explaining that these parameters do not have the meaning they possess for standard Michaelis–Menten enzymes, and this continues despite the fact that the matter was effectively clarified more than 50 years ago by Bonnichsen, Chance, and Theorell (*34*). As already described, catalases react with hydrogen peroxide in a two-stage process. In reaction (1), the ferric enzyme combines with hydrogen peroxide to generate water and compound I, the effective enzyme–substrate intermediate or ES, and the rate constant for this reaction is designated k_1. The reverse reaction with a rate constant of k_{-1} is negligible and will not be considered further. In reaction (2), compound I combines with a second molecule of hydrogen peroxide to regenerate the ferric enzyme, molecular oxygen, and water. The rate constant for this reaction is designated k_2, and that for the reverse reaction, which is also negligible and not considered further, k_{-2}. The combined reactions are summarized in Fig. 2A. As both reactions are peroxide-dependent, the simplest model of enzyme activity, that of Bonnichsen, Chance, and Theorell (*34*), predicts that the enzyme is never saturated with its substrate and that the turnover of substrate increases indefinitely as the peroxide concentration increases. This will be referred to as the BCT mechanism.

The velocities of reactions (1) and (2), v_1 and v_2, can be expressed in terms of the total enzyme concentration (or total heme groups) [E] and the concentration of enzyme–substrate complex [ES], as

$$v_1 = k_1[H_2O_2]([E] - [ES]) \tag{6}$$

and

$$v_2 = k_2[H_2O_2][ES]. \tag{7}$$

At steady state these two rates are equal, and we have

$$k_1[H_2O_2]([E] - [ES]) = k_2[H_2O_2][ES], \tag{8}$$

which can be simplified to

$$k_1([E] - [ES]) = k_2[ES]. \tag{9}$$

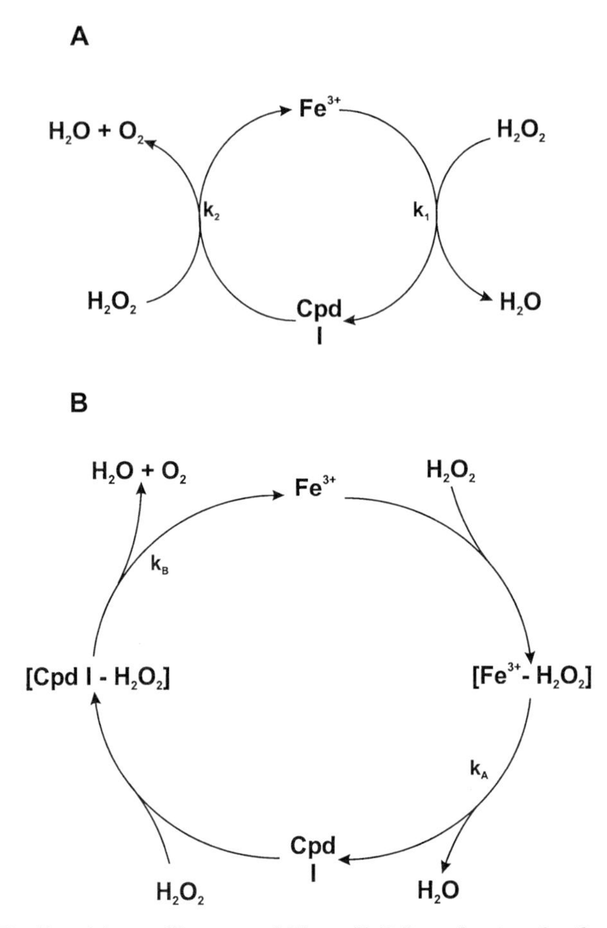

FIG. 2. The Bonnichsen, Chance, and Theorell (*34*) mechanism for the dismutation of hydrogen peroxide by catalase. (**A**) The simple ping-pong mechanism (ferric-peroxide compound cycle) involves only the successive formation and decomposition of the compound I intermediate by two successive molecules of H_2O_2. (**B**) Reversible ES (Fe^{3+}–H_2O_2) and ternary (compound I–H_2O_2]) complexes are added to the mechanism in **A.**

Solving for [ES] gives

$$[ES] = [E]k_1/(k_2 + k_1). \qquad (10)$$

It follows that the [ES]/[E] ratio is a constant:

$$[ES]/[E] = k_1/(k_2 + k_1) = a. \qquad (11)$$

Hence, the concentration of the ES complex in the steady state is independent of $[H_2O_2]$. Reaction 2 or the decomposition of compound I is the rate limiting step resulting in the overall velocity being described by

$$V = 2k_1k_2[H_2O_2][E]/(k_2 + k_1), \tag{12}$$

where the factor 2 is present because each complete cycle involves the loss of two molecules of peroxide. Equation (12) shows that the rate of peroxide decay is proportional to a constant,

$$k' = 2k_1k_2/(k_2 + k_1), \tag{13}$$

multiplied by the product of the concentrations of substrate and enzyme active sites.

Measurement of the overall rate constant, together with a measurement of the steady-state proportion of enzyme that is compound I(a), always permits calculation of the values of the two intrinsic rate constants according to

$$k_2 = k'/2a \tag{14}$$

and

$$k_1 = k'/2(1 - a). \tag{15}$$

The value of k_1 may be determined directly if a trap compound is available that reacts irreversibly and more rapidly with compound I than does a second H_2O_2.

Classical low K_m values for the mammalian enzyme that have appeared in the literature are the result of enzyme inactivation by hydrogen peroxide when measurements were carried out with peroxide levels in excess of 10 mM over time scales of 10 minutes or longer. The rapid sampling/titration method of Bonnichsen overcame the inactivation problem and permitted a satisfactory correlation of the overall catalytic measurements and Chance's observations on the intermediate complex (compound I). Eventually, the introduction of the UV detector/spectrophotometer and the consequent assay based upon the UV absorbance of peroxide (*35*) further simplified the process by eliminating the discontinuous titrimetric assay.

Obviously, there must be a limit to the turnover of any enzyme. Rates cannot theoretically go on increasing indefinitely with substrate concentration. In the case of mammalian catalases, the limits appear to lie in the range between a first order rate of $2 \times 10^6\ sec^{-1}$ and $1 \times 10^7\ sec^{-1}$ (*36*). That is, each heme active site can theoretically decompose between 2 and 10 million molecules of H_2O_2 per second. As two molecules are decomposed per turnover, that means that the lifetime of the active Michaelis–Menten complex, or compound I, lies between 0.2 and 1 microsecond.

The more complex scheme required if a true K_m is involved is shown in Fig. 2B. There are two possible steps that could provide limiting unimolecular processes, governed by k_A and k_B, the steps involved in formation of compound I and the decay of the "ternary" complex, respectively. Nicholls and Schonbaum (*22*) gave reasons the latter is the preferred limiting step for mammalian catalase. These reasons have not changed much over the past 35 years. But the increased number of catalases examined, especially the catalase-peroxidases, makes reevaluation appropriate (see below).

Several catalases, including the type B catalase-peroxidases, seem to show true substrate saturation at much lower levels of peroxide than originally observed for the mammalian enzyme (in the range of a few millimolar). This means that the limiting maximal turnover is less and the lifetime of the putative Michaelis–Menten intermediate (with the redox equivalent of two molecules of peroxide bound) is much longer. The extended scheme for catalase in Fig. 2B shows that relationships between free enzyme and compound I, and the presumed rate-limiting ternary complex with least stability or fastest decay in eukaryotic enzymes of type A and greatest stability or slowest decay in prokaryotic type B enzymes.

B. The Peroxidatic Activity of Catalases

In addition to their catalatic (peroxide dismuting) activity, catalases also use peroxides to oxidize secondary hydrogen and electron "donor" molecules. There are two major families of possible hydrogen donors: two-electron donors such as alcohols and one-electron donors such as phenols. The two modes of redox behavior are quite distinct. Keilin and Nicholls (*37*) described six donor types, classified according to their reactivities with the two catalase peroxide compounds. Figure 3A shows the basis for this classification. What has happened over the past 40 years to modify this scheme? Firstly, the number of known catalases has increased immensely, but although there are significant quantitative differences in rates of reaction with specific donor types, members of the type A family of catalases, including bovine liver catalase (BLC) and HPII, share many characteristics, including donor specificities. Less complete surveys are available for the more recently discovered catalase-peroxidases or type B enzymes. Secondly, a major development has been the discovery of the special donor role of NADPH

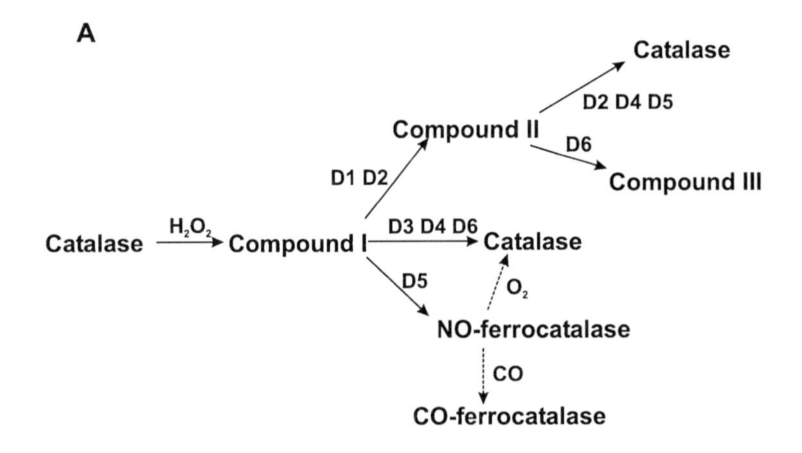

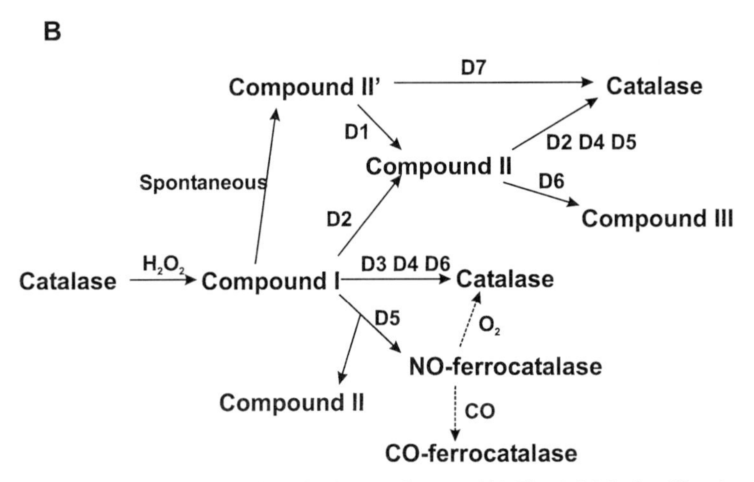

FIG. 3. Classification of catalase hydrogen donors. (**A**) The initial classification proposed by Keilin and Nicholls in 1958 (*37*). D1 donors, including ascorbate and ferrocyanide, reduce only compound I to compound II. D2 donors, including phenols and aromatic amines, reduce compound I to II and compound II to ferric enzyme. D3 donors, including alcohols and formate, reduce only compound I to ferric enzyme. D4 donors, including nitrite, reduce compound I and compound II to ferric enzyme. D5 donors, including azide and hydroxylamine, reduce compound I to ferrous enzyme and compound II to ferric enzyme. D6 donors, including H_2O_2, reduce compound I to ferric enzyme and compound II to compound III. (**B**) A revision of the scheme in part **A** that reflects the inclusion of the new donor group D7, including NADPH and NADH, which reduces a compound II precursor (compound II') to ferric enzyme. In addition, the D1-categorized ferrocyanide is now regarded as primarily a reducer of the compound II' intermediate, and the D5-categorized azide may reduce Fe via an intermediate compound II.

(listed as a seventh donor type below). Thirdly, it has been found that some type A enzymes, including both protoheme forms, such as the enzyme from *Aspergillus*, and chlorin heme forms, such as *E. coli* HPII, appear not to form compound II at all. A fourth modification of the classical scheme involves the observation that certain types of peroxide, such as *t*-butyl hydroperoxide and peroxynitrous acid, may form compound II directly by a homolytic rather than heterolytic split of the O—O bond, with consequent release of a peroxidelike radical into the medium (*t*-butylO˙ or $NO_2^{\cdot}$). In addition to these overall changes, there have been more detailed changes to the 1958 pathways, as described later.

C. Compound I and the Pathways via Compound I

The visible spectra of beef liver catalase (Type A) and its two active peroxide compounds are shown in Fig. 4. The unliganded enzyme has a Soret band at 405 nm (E_{mM}/heme ≈ 120) and a characteristic visible

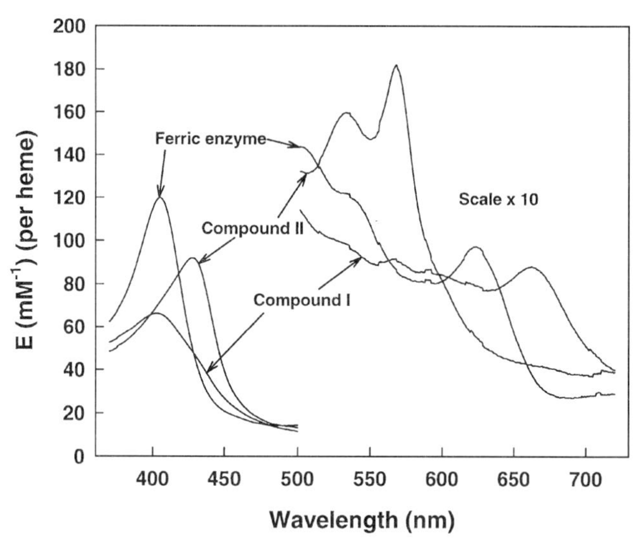

Fig. 4. Visible spectra of catalase, compound I, and compound II; 5 μM (heme) beef liver catalase (Boehringer-Mannheim) in 0.1 M potassium phosphate buffer pH 7.4, 30°C. Compound I was formed by addition of a slight excess of peroxoacetic acid. Compound II was formed from peroxoacetic acid compound I by addition of a small excess of potassium ferrocyanide. Absorbance values are converted to extinction coefficients using 120 mM^{-1} for the coefficient at 405 nm for the ferric enzyme (confirmed by alkaline pyridine hemochromogen formation). Spectra are corrected to 100% from occupancies of ≈90% compound I, 10% ferric enzyme (steady state compound I) and 88% compound II, 12% compound I (steady state compound II). The extinction coefficients for the 500 to 720 nm range have been multiplied by 10. Unpublished experiments (P.N., 1999).

spectrum with peaks at 500 and 622 nm and a shoulder at 540 nm, typical of ferric protoheme proteins, differing from the paradigmatic metmyoglobin spectrum by having a long wavelength charge transfer (CT) band at a shorter wavelength and with a higher extinction coefficient, presumably due to the proximal tyrosinate. Compound I, the porphyrin radical-ferryl state (Eq. (1)), is characterized by a much lower Soret peak (E_{mM}/heme $\approx$ 65–70) and a long wavelength band at 665–670 nm, of an intensity almost equal to that of the original CT band, indicating a disruption of the resonating π-bond system of the porphyrin ring. Beyond 700 nm there are absorbances (not shown) due to the presence of heme groups degraded to bile pigments, largely biliverdin with some retention of the heme iron to form verdohemes. Commercial beef liver catalases typically contain up to 45% bile pigment. The structure of the catalase heme groups was originally described as substantially distorted compared to those of other hemoproteins, because these verdoheme structures were averaged with the protoheme ones (*38*).

The donor types D3, D4, and D6 of Keilin and Nicholls (*37*) all reduce compound I of Type A enzymes directly to the ferric state in a two-electron process without detectable intermediates. Each of these donors is probably also able to bind in the heme pocket of the free enzyme. Alcohols (type D3) form complexes with free ferric Type A enzymes whose apparent affinities parallel the effectiveness of the same alcohols as compound I donors (*39*). Formate (type D3) reacts with mammalian ferric enzyme at a rate identical to the rate with which it reduces compound I to free enzyme (*22*). Its oxidation by compound I may thus share an initial step analogous to its complex formation with ferric enzyme. Formate also catalyzes the reduction of compound II to ferric enzyme by "endogenous" donors in the enzyme (*40, 41*). Both compound I and compound II may thus share with the free enzyme the ability to ligate formate in the heme pocket. Nitrite, which is oxidized to nitrate by a two-electron reaction with compound I (type D4), also forms a characteristic complex with free enzyme (*42*). In both cases the reaction involves the donor in its protonated (HNO_2) form.

Hydrogen peroxide itself was given a separate donor status (D6) because in addition to acting like the D3 family and reducing compound I to ferric enzyme in a single two-electron step it can also react with catalase compound II to give the "oxy" or protonated oxy species, compound III (*22*) according to

$$\text{Cpd II (Por–Fe}^{IV}\text{=OH}^+) + H_2O_2 \rightarrow \text{Cpd III (Por–Fe}^{II}\text{=O}_2\text{H}^+) + H_2O. \quad (14)$$

Compound III, like compound II, is an inactive form of catalase with respect to the normal catalatic cycle, and thus may contribute to the inactivation of the enzyme at high peroxide levels (*42*).

The classical type A enzymes show marked differences in their abilities to oxidize two-electron donors. All the enzymes initially examined, whether eukaryotic or prokaryotic in origin, were members of clade III. These are rather effective as two-electron peroxidases. But the oxidations of ethanol and formate by the paradigmatic clade II enzyme from *E. coli,* HPII, are much slower. There is little information concerning the activity of type A enzymes in clade I.

There are also substantial differences between classical type A enzymes and the type B catalase-peroxidases. The latter enzymes, although they show peroxidase activity toward donors of type D2, are inactive or only weakly active toward D3 donors such as ethanol.

D. Compound II and the Pathways via Compound II

The visible spectrum of beef liver catalase compound II is also shown in Fig. 4. The enzyme in its one-electron oxidized (ferryl) state has a Soret band at 427 nm (E_{mM}/heme $\approx$ 92) and a characteristic visible spectrum with intense peaks at 533 and 567 nm (E_{mM}/heme $\approx$ 18), indicating a low spin state, differing from the corresponding ferrylmetmyoglobin spectrum by having sharper peaks at a shorter wavelength and with higher extinction coefficients, due either to redox delocalization at the proximal tyrosinate or to protonation of the ferryl species (Eq. (3a)).

The donor types D2, D4, and D5 of Keilin and Nicholls (*37*) all reduce compound II to ferric enzyme in a one-electron process without detectable intermediates. Donors of type D2, phenols and amines, also reduce compound I to compound II. Nitrite, the only member of category D4, reduces compound I in a two-electron step as described earlier. Donors of type D1 reduce compound I to compound II, but have no appreciable effect upon compound II itself. Reactivity of the one-electron donors seems independent of heme pocket binding in the free enzyme.

The donor type D5 comprises the two species azide and hydroxylamine. These both react with the enzyme in the presence of peroxide to give rise to ferrous forms of catalase, otherwise normally inaccessible (catalase is the only common hemoprotein that is nonreducible by dithionite). The final inhibited form of catalase in the presence of azide and peroxide is NO-ferrocatalase, but not every azide molecule becomes an $NO^{\cdot}$; only in the presence of CO is there a stoichiometric inhibition of enzyme by peroxide with formation of 1 equiv of CO-ferrocatalase for every peroxide molecule added (*43*). This suggested a three-electron reduction of compound I either to give ferrocatalase, N_2, and $NO^{\cdot}$ (10–20% total) or to give ferrocatalase, $N^{\cdot}$, and N_2O (80–90% total). However, Kalyanaraman *et al.* (*45*) have demonstrated the formation of the azidyl ($N{=}N{=}N^{\cdot}$) radical in the reaction, and Lardinois and Rouxhet (*46*) proposed a role for compound II formation in catalase inhibition by azide. Nicholls and Chance (unpublished data) had in fact identified a precursor species closely similar to compound II before the appearance of ferroenzyme. The latter cannot be the result of a second reaction of azide with compound II, as the latter reaction gives rise to ferric enzyme. The simplest hypothesis involves the secondary reaction of the azidyl radical with compound II to give the ferroenzyme with either of two possible breakdown modes of the azidyl radical shown in Fig. 5. The alternative pathway in which the azidyl radical itself reacts with oxygen to give NO and N_2O (*45*) is too slow and does not seem to

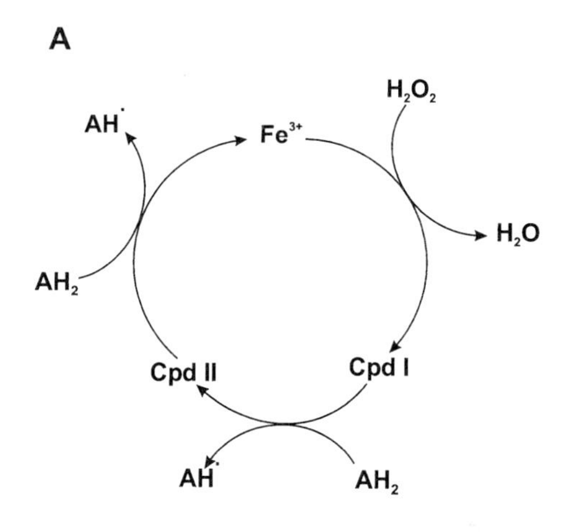

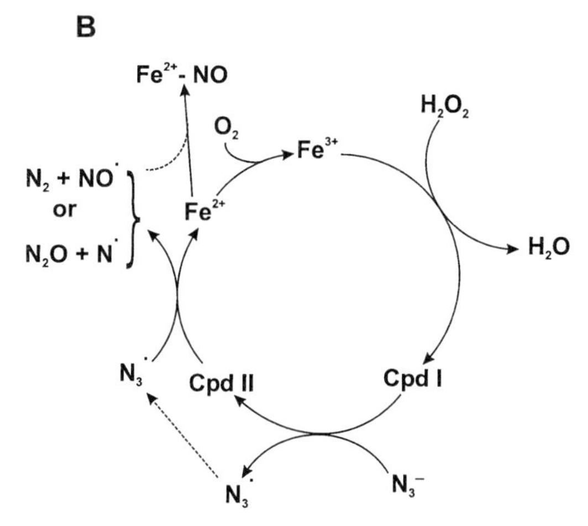

FIG. 5. (**A**) The peroxidatic reaction proceeding via compound II (the D2 family). (**B**) A catalytic cycle involving azide proceeding via compound II and ferroenzyme (the D5 family).

generate a reductant capable of reducing ferric catalase or compound II to the ferro form.

Both classical type A enzymes (clade III) and the heme d family (clade II) show a comparatively high sensitivity to azide inhibition and are reduced to ferrous forms in the presence of peroxide and azide (*47*). In contrast, the catalase-peroxidase (type B) enzymes (see below) are only weakly azide-sensitive.

E. Control by NADPH: The "Extra" Pathway

Kirkman and Gaetani (*48*) discovered the seventh catalase donor type, D7, unknown to Keilin and Nicholls (*37*). In his calculations of *in vivo* or at least "*in erythro*" rates, Nicholls (*49*) had already pondered the possibility of some unusual regeneration pathway for blood catalase, but had vaguely wondered about a role for glutathione, following his original studies on cysteine and glutathione interactions with catalase (*37*). Following the identification of tightly bound $NADP^+$/NADPH by conventional chromatographic methods, Fita and Rossmann (*50*) went back to their data and found that a piece of what previously had been thought to be disordered peptide actually fitted NADPH quite well.

Kirkman and Gaetani (*48*) were able to show not only that tightly bound $NADP^+$ or NADPH was present in mammalian catalases, but also that the presence of the reduced nucleotide decreased compound II formation. The kinetics were, however, anomalous. Of the original families of donors, two types could decrease compound II accumulation: the two-electron D3 category, exemplified by ethanol, which remove compound I before compound II can be formed spontaneously, and the one-electron category (D2), which reduce compound II itself. NADPH falls into neither category. Although it prevents compound II formation from compound I, it does not reduce compound I directly to free enzyme (that is, it is not a conventional two-electron donor), and although it prevents compound II formation, it does not reduce compound II to ferric enzyme once the former species has been produced in its absence. That is, it also cannot act as a one-electron donor. We are presented with an apparent paradox. Hillar and Nicholls (*47*) attempted to resolve the paradox by postulating an intermediate between the compound I and compound II states possessing an unique reactivity with bound NADPH, as indicated in Fig. 6. The initial (slow) step in formation of compound II from compound I lies in the migration of an oxidizing equivalent to a nearby protein residue P (as occurs rapidly and stably, for example, in cytochrome *c* peroxidase). In a second (and relatively rapid) event, the "endogenous" donor centers in the molecule (remote tyrosines, etc.)

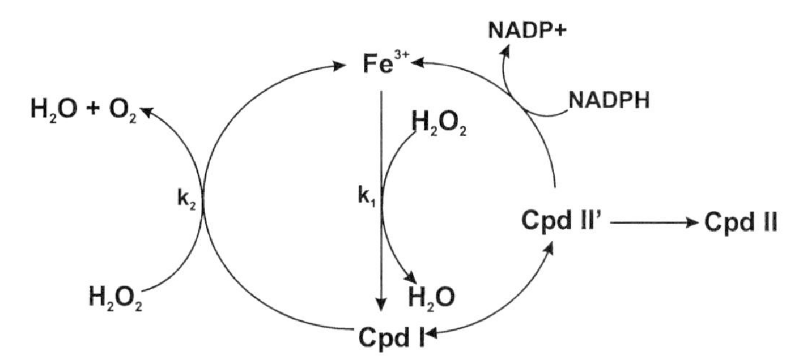

FIG. 6. The catalytic cycle involving a postulated compound II' and NADPH.

reduce the radical P˙ back to its stable PH state. What is PH/P˙? One candidate is the proximal tyrosine, oxidized to the tyrO˙ form with removal of the charged π-electron radical. A second possibility is one of the other nearby tyrosines such as Y235 or Y214 (bovine catalase numbering). A third possibility, suggested by Bicout *et al.* (*51*) (see also Ref. *3*), is that the conserved serine S196 may also act as a radical site. Little seems to be known about serine radicals in any other enzymes, and the likely thermodynamics are therefore hard to estimate.

Two current alternative views are available as to how remotely bound NADPH may work. One sees its action as involving two successive one-electron oxidations (*52, 53*). The effectiveness of NADPH in preventing compound II formation is then due to the high reactivity of the NADP˙ intermediate as reductant of the compound II generated in the first one-electron step. The other model (*47*) prefers to see NADPH as a hydride donor responsible for the almost simultaneous reduction of the ferryl iron and the protein radical species.

Kirkman and Gaetani (*54*) have reexamined the kinetics of NADPH oxidation and compound II formation by mammalian catalase. Under some experimental conditions the rate of NADPH oxidation is substantially higher than the rate of compound II formation in the absence of NADPH. This finding may be accommodated in a scheme such as that of Fig. 6 if the rate of formation of compound II' is greater than the rate of compound II formation. This can arise either if there is more than one route for compound II' decay, only one of which proceeds *via* compound II, or if the formation of compound II' from compound I is reversible, as indicated in the scheme (Fig. 6). Such reversibility is highly probable in view of the redox potentials of the species involved. Although direct measurements have not been carried out with catalases, the corresponding metmyoglobin and horse radish peroxidase peroxide compounds (*55, 56*) have potentials (E_0' values) of +880 mV (horse radish peroxidase or HRP comp. I), +900 mV (HRP comp. II), and +890 mV (ferrylmetmyoglobin). Corresponding E_0' values for the trpH/trp· and tyrOH/tyrO· pairs are +1050 and +940 mV, respectively (*57*). Although these potentials may vary in particular proteins, it seems likely that the redox gap between compound I and compound II′ (Fig. 6) will be about 60 mV in favor of compound I. Under the conditions described by Kirkman and Gaetani (*54*), the rate of formation of compound II is approximately 25% the rate of NADPH oxidation. This is consistent with the proposed mechanism provided that the rate of reaction of NADPH (k_5) is fast and the compound I ⇄ compound II′ step is an equilibrium.

The role of NADPH as "protective" donor (*58, 59*) seems to correlate well with the tendency of the different categories of catalases to form the inactive compound II. Only those classical catalases that are vulnerable to the latter inactivation step show $NADP^+$ and NADPH binding. The scheme shown in Fig. 3B summarizes the minimal revisions of the 1958 scheme (Fig. 3A) needed to accommodate the newer findings.

F. The Catalatic Activity of Catalase-Peroxidases

Members of the catalase-peroxidase family were discovered separately in different microorganisms by several groups of workers in the late 1970s and 1980s (*60–62*). It soon became evident that these enzymes are homologous to the eukaryotic type I plant peroxidases of Welinder (*63*). Although these enzymes show both catalase and peroxidase activities and thus obtained the name catalase-peroxidase (CatPx), it is not clear whether their peroxidatic activities are proportionately greater than similar activities shown by classical (type A) enzymes (*64*). What is clear is that the catalatic activity displays several unique distinguishing characteristics. Hochman and Shemesh (*65*) showed that the *Rhodopseudomonas capsulata* enzyme is characterized by instability in presence of the substrate H_2O_2, saturation kinetics ($K_m \approx 4$ mM) for the catalatic reaction, and a relative insensitivity to azide and hydroxylamine.

Follow-up experiments with the similar *Klebsiella pneumoniae* enzyme (*66, 67*) also showed that the pH profiles for CatPx are quite different from those for classical catalases. The latter's catalatic activities are essentially pH-independent from pH 5 to 10 (*68*); CatPx of *Klebsiella* showed a sharp pH optimum between pH 6 and 7 (*66*). A similar eukaryotic fungal CatPx (*69*) was also characterized by a sharp pH sensitivity and saturation kinetics ($K_m \approx 3.4$ mM) and, like the bacterial enzymes, was sensitive to incubation with peroxide but not to the classical catalase inhibitors azide (except weakly) and aminotriazole.

The low peroxidatic activities were less pH sensitive than the catalatic activity.

Most recently, the study of catalase-peroxidases has been extended to the cyanobacteria, which, unlike the eubacteria, seem often to be characterized by possession of only a CatPx (type B) and no type A catalase. Obinger *et al.* (*70*) showed that the *Anacystis nidulans* enzyme had a K_m for H_2O_2 of 4.3 mM and a maximal turnover of over 7000 sec^{-1}. At low peroxide levels the activity was about 20% that of beef liver enzyme. Peroxoacetic acid produced a typical compound I species. An analogous enzyme is found in *Synechocystis 6803* (*70, 71*). This enzyme, like the eubacterial examples, shows a narrow pH optimum between pH 6 and pH 7 (*72*) with apparent p*K* values close to 7 and between 5.5 and 6.0. The pH profiles for the peroxidatic activity were different from those for catalatic activity. Again peroxoacetate produces a characteristic compound I. However, the reaction of compound I with one-electron reductants produces a high-spin intermediate different from typical compound II. Regelsberger *et al.* (*71*) also find that the one-electron reduction product obtained by adding ascorbate to compound I is a high spin species with Soret band closely similar to that of free enzyme and visible spectrum with a peak at 626 nm compared to 631 nm for ferric enzyme.

Moreover, the steady-state spectrum in the presence of hydrogen peroxide showed no sign of the presence of compound I (identified from the peroxoacetate spectrum), and compound I produced by peroxoacetate was unaffected by addition of hydrogen peroxide. Regelsberger *et al.* (*71*) and Jakopitsch *et al.* (*72*) nevertheless interpret their results in terms of the classical "BCT" catalatic cycle (Fig. 3), attributing the differences to substantial differences in the kinetic constants for formation and decomposition of compound I, k_1 and k_2 (Eqs. (4) and (5)). An alternative analysis is represented in Fig. 7. Only the peroxidatic activity of the Type B enzymes may proceed via the usual porphyrin–cation radical compound I. The heme configuration, analogous to that of cytochrome *c* peroxidase of yeast, may permit an alternative doubly oxidized state involving a protein radical. If this is the pH-sensitive intermediate, the failure to detect appreciable amounts of compound I during the steady state and the inactivity of peroxoacetate-derived compound I toward H_2O_2 may all be explicable. In addition, the unusual spectrum of compound II of this enzyme may also indicate a more stable protein radical state than the usual ferryl state, even for this intermediate. Electron paramagnetic studies of CatPx enzymes to detect radical states are expected to be undertaken soon that may confirm, refute, or modify this scheme.

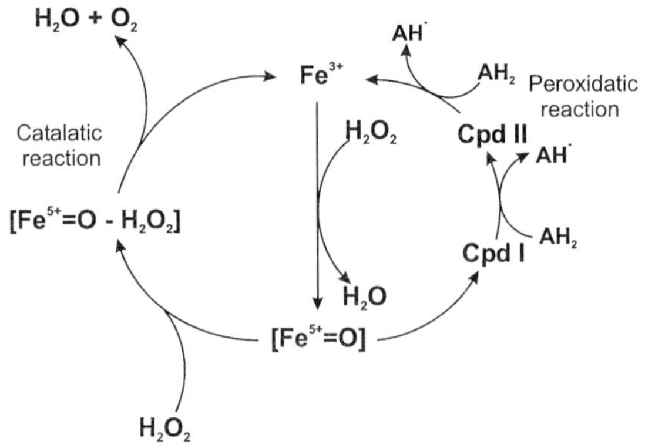

FIG. 7. Catalatic and peroxidatic reactions of type B enzymes. This represents a modification of the schemes of Figs. 2 and 5A and is proposed to account for the characteristic features of catalase-peroxidases. Compound I is drawn as $Fe^{5+}{=}O$ and can represent either a π-cation radical or alternative radical structure. The precise nature remains undefined (see Section IV,F).

V. Structure of Type A Catalases

Detailed structural information about Type A catalases is available from the crystal structures of seven monofunctional catalases that have been solved. These include representatives from small-subunit clade III enzymes of animal (bovine liver catalase or BLC (*73, 74*) and human erythrocyte (T. P. Ko, submitted to Protein Data Base as file 1qqw)), fungal (*Saccharomyces cerevisiae* SCC-A) (*75, 76*), and bacterial (*Proteus mirabilis* PMC and *Micrococcus lysodeikticus* MLC) (*77, 78*) origins, and large subunit clade II enzymes of fungal (*Penicillium vitale* PVC) (*79, 80*) and bacterial (*E. coli* HPII) (*81, 82*) origins. Despite the differences in size, all six enzymes share a number of common features that appear to be characteristic of catalases, including a homo tetrameric quaternary structure with the heme group deeply buried in a beta-barrel core structure in each subunit.

Within this common core structure, modifications have been identified that provide catalases with further unique properties. The large-subunit enzymes have extensions at both the amino and carboxyl ends, the latter having a flavodoxinlike structure, a unique His–Tyr bond, a protected cysteine, and a modified heme. NADPH binding and an oxidized methionine are found in small subunit enzymes. Identification and assignment of roles to channels providing access to and egress from the deeply buried heme have recently become the focus of study. Analysis of the structure of catalase HPII of *E. coli* has been facilitated by the construction of more than 75 mutants (Table I).

TABLE I

CHARACTERIZATION OF THE HPII MUTANT VARIANTS

Mutation	Specific activity[a] (units/mg)	Heme[b]	His392–Tyr415[c]
Wild type	14,322	*d*	y
Gln3Arg	nd (active)		
Gln3Arg/Glu21Ala	nd (active)		
Gln3Arg/Gly74Ser/Ser75Ala	nd (active)		
Δ(Gln3-Ser20)	np		
Δ(Gln3-Gly34)	np		
Δ(Gln3-Thr50)	np		
Δ(Gln3-Gly74)	np		
(Gln3-Gly74)/Val303Ser/Phe317Ser/ Phe518Ser/Leu571Stop	np		
Glu21Ala	nd (active)		
Gly34Ser/Ser35Ala	nd		
Thr50Ser	nd		
Gly74Ser/Ser75Ala	nd (active)		
His128Ala	<0.1	*b*	n
His128Asn	<0.1	*b*	n
His128Glu	np		
His128Gln	np		
His128Asn/Asn201His	<0.1	*b*	
Ser167Thr	1,100	*d* + *b*	
Ser167Ala	100	*b*	
Ser167Cys	np		
Ser167Asn	np		
Val169Ala	3,788	*d*	y
Val169Ser	3,703	*d*	y
Val169Cys	16	*b*	n
Val169Cys/Cys438Ser/Cys669Ser	10	*b*	
Asp197Ala	14,354	*d*	y
Asp197Ser	14,721	*d*	y
Asp197Ser/His395Gln	15,473	*d*	y
Asn201His	100	*b*	n
Asn201Asp	1,700	*d*	
Asn201Ala	1,300	*d*	
Asn201Gln	50	*d*	
Asn201Arg	np		
Arg260Ala	35,891	*d*	
Arg260Ala/Lys294Ala	12,880	*d*[d]	
Glu270Asp	16,137	*d* + *b*	
Glu270Asp/Glu362His	16,859	*d*	
Ile274Ser	3,416	*d*	
Ile274Ser/Pro356Leu	np		
Ile274Ser/Leu407Met	np		
Ile274Ser/Pro356Leu/Leu407Met	np		
Lys294Ala	18,625	*d* + *b*	

(continued)

TABLE I (*Continued*)

Mutation	Specific activity[a] (units/mg)	Heme[b]	His392–Tyr415[c]
Val303Ser	nd		
Phe317Ser	nd		
Pro356Leu	13,220	*d* + *b*	
Pro356Leu/Leu407Met	12,437	*d* + *b*	
Glu362His	18,325	*d* + *b*	
His392Ala	8,875	*b*	n
His392Gln	7,368	*b*	n
His392Glu	4,507	*b* + *d*[d]	n
His392Asp	2,857	*b*	n
His395Ala	9,821	*d*	y
His395Gln	9,224	*d*	y
Leu407Met	12,679	*d* + *b*	
Ser414Ala	5,511	*b* + *d*[d]	y
Tyr415Phe	np		
Tyr415His	np		
Gln419Ala	7,317	*d*[d]	y
Gln419His	11,494	*d*[d]	y
Cys438Ser	7,350	*d*	
Cys438Ala	6,100	*d*	
Cys438Ser/Cys669Ser	8,050	*d*	
Cys438Ala/Cysgg9Ala	12,870	*d*	
Phe518Ser	np		
Leu571Stop	np		
Ile593Stop	np		
Val603Stop	np		
Cys669Ser	10,840	*d* + *b*	
Cys669Ala	7,800	*d*	
Trp742Stop	2,624	*d*	
Arg744Stop	8,600	*d*	
Arg744Ala	8,786	*d*	
Arg744Lys	5,832	*d*	
Arg744Ala/Ile745Stop	6,929	*d*	
Arg744Lys/Ile745Stop	9,284	*d*	
Ile745Stop	13,753	*d*	
Pro746Stop	15,022	*d*	
Lys747Stop	15,349	*d*	
Lys750Stop	15,067	*d*	

[a] nd, not determined; np, insufficient protein accumulated in cells for purification; nd(active), near wild type levels of catalase were noted in crude extracts but enzyme was not purified.

[b] *d*, heme *d*; *b*, heme *b*. Heme composition was determined by spectral and HPLC analyses. In all cases, the enzymes contained one heme per subunit. Where no indication is given, no determination was made.

[c] Presence of the His392–Tyr415 (y, yes; n, no) covalent bond was determined by MALDI-MS analysis of trypic digest mixtures. Where no indication is given, no determination was made.

[d] There was an increased percentage of the *trans* isomer in these variants.

A. Subunit Structure

The sequences of all catalases exhibit extensive similarity in the core region defined by the β-barrel and active site to the extent that there are 15 invariant residues among 110 catalase sequences, with the variations in an additional 3 probably arising from incorrect sequencing. These include the essential histidine situated in the active site and the essential tyrosine residue that forms the fifth ligand with the heme. Many other residues are highly conserved, varying in only two or three of the catalases. Such extensive similarity speaks to a strong drive for conservation and is suggestive that the three-dimensional structure imposes limitations on the changes that are possible with retention of activity.

The sequence conservation is reflected in a highly conserved secondary and tertiary structure that is most clearly illustrated in the three-dimensional superposition of C^{α} atoms. Ignoring the C-terminal domains of PVC and HPII, the deviation of C^{α} atoms in a superposition of HPII with PVC, BLC, PMC, and MLC results in root mean square deviations of 1.1, 1.5, 1.6, and 1.5 Å for 525, 477, 471, and 465 equivalent centers, respectively (*83*). In other words, there is very little difference in the tertiary structure of the subunits over almost the complete length of the protein. The large and small subunits are shown in Fig. 8 for comparison.

The tertiary structure of small subunit enzymes can be subdivided into four distinct regions, and the C-terminal or flavodoxin domain of the large subunit enzymes becomes a fifth region. These are indicated in Fig. 8 for clarity. The first region is the amino terminal arm (Fig. 8), which extends 50 or more residues from the amino terminus almost to the essential histidine residue (to residue 53 in PMC, 60 in PVC, 73 in BLC, and 127 in HPII). There is very little structural similarity in the N-terminal region and, in the case of HPII, the structure of the terminal 27 residues is not even defined and they do not appear in the crystal structure. Within the N-terminal arm is a 20-residue helix, helix α2 in HPII, which is the first secondary structure element common to all catalases. The presence of helix α1 varies among catalases, and there is no sequence or location equivalence even when it is present.

The second region is the antiparallel β-barrel (Fig. 8) forming the core of the subunit. It includes about 250 residues from the essential histidine toward the C-terminus. The first four strands (β1–4) are contiguous and are separated from the second four strands (β5–6) by three helices (α3–5). The first four strands form the distal side of the heme pocket and portions of the second four strands participate in binding NADPH in small subunit enzymes.

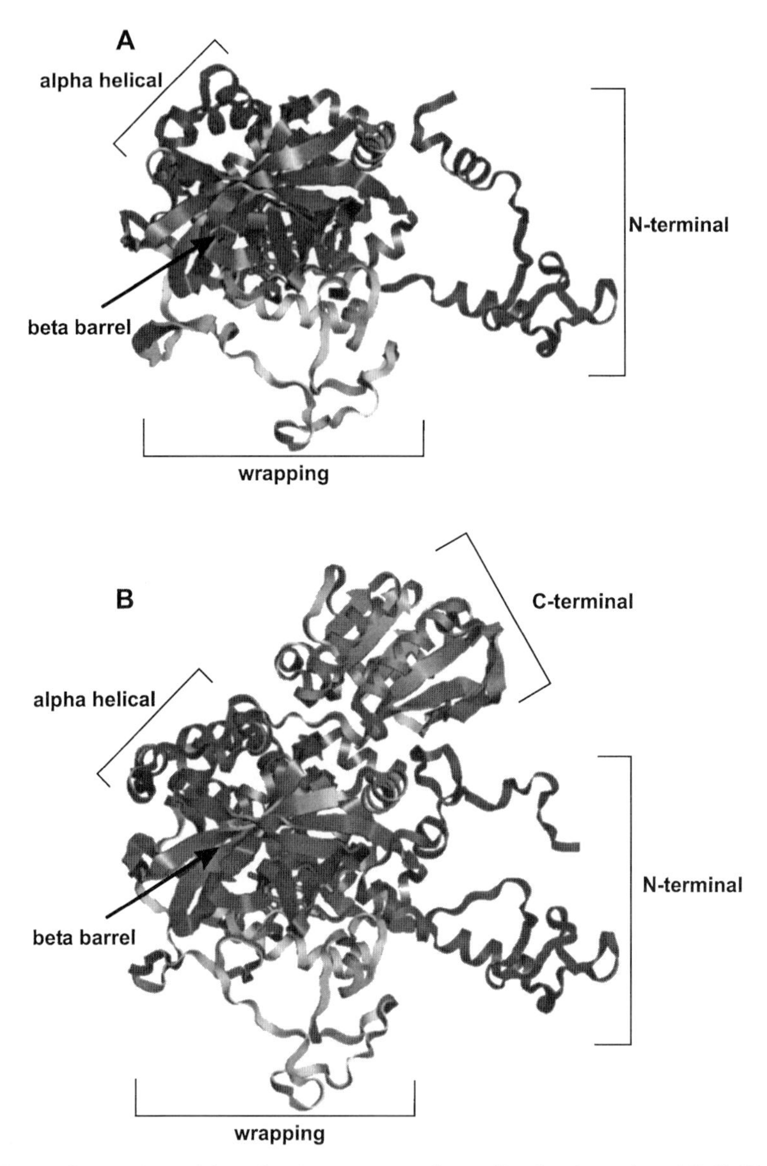

FIG. 8. Comparison of the subunit structures of a small subunit catalase (BLC) (**A**) and a large subunit catalase (HPII) (**B**). Segments including the N-terminal domain, the beta barrel core, the wrapping domain, the alpha helical domain, and the C-terminal domain (HPII) are indicated and are described in Section V,A.

The third region is the wrapping domain (Fig. 8), which includes 110 residues in an extended structure linking the β-barrel and the α-helical section. There is little secondary structure in this region, although helix α9 contains many of the residues that form the proximal side of the heme pocket, including the essential fifth ligand tyrosine. A large surface area for the formation of subunit-subunit interactions is created in this extended structure.

The fourth region is the α-helical domain (Fig. 8) containing about 60 residues organized in four contiguous α helices (α10–13) that form a close association with α-helices 3 to 5 in the β-barrel region to stabilize the structure. This region contains the C-terminus of small-subunit enzymes and, in large-subunit enzymes, may have a critical role in the folding pathway.

In large subunit enzymes (PVC and HPII), a short segment of about 30 residues links the α-helical domain to the C-terminal domain (Fig. 8). The latter segment is a conspicuous addition to the small subunit containing about 150 residues folded into a structure that resembles flavodoxin. For example, there is a root mean square deviation of 3.0 Å between flavodoxin and approximately 100 residues of the C-terminal domains of either HPII or PVC. This can be compared to the 1.8 Å root mean square deviation for 134 centers between the C-terminal domains of HPII and PVC. Unlike the N-terminal end, the final C-terminal residue Ala753 is visible in the structure of HPII. The C-terminal domain contains extensive secondary structure in the form of four α-helices (α15–18) and eight β-strands (β9–16). Despite the obvious structural similarity to flavodoxin, there is no evidence of nucleotide binding in the domain and its function remains a mystery.

Attempts to generate a "small-subunit enzyme" by removing the C-terminal domain of HPII by site-directed mutagenesis were unsuccessful (*84*). Truncation of the subunit at residues 571, 593, or 603 resulted in no protein accumulating in the cell. The lack of protein accumulation was shown to be the result of proteolytic degradation of the protein before it could fold into a protease-resistant form. Even when three hydrophobic residues that would have been exposed by removal of the domain were changed to hydrophilic residues, truncated HPII-like protein did not accumulate. Progressive truncation from the carboxy end revealed that shortening the protein past Arg744 had the same effect as removal of the complete C-terminal domain. The side chain of Arg744 extends into the C-terminal domain and is involved in a number of hydrogen bonds that are clearly important in stabilizing the structure of the domain. These results also show that a properly folded C-terminal domain is essential for the efficient folding of the whole protein into a stable, protease-resistant structure. Any interference with the folding of the domain interferes with the folding of the whole protein.

B. Quaternary Structure and Interweaving

All seven catalases, which have had their crystal structures solved, exist as homotetramers. Even HPII, which was originally characterized biochemically as a hexamer (*85*), was found to be tetrameric upon solution of the crystal structure. The reason for the biochemical determination of HPII as a hexamer was subsequently traced to the fact that the gel filtration elution volume varies with salt concentration such that elution at low salt results in a larger apparent molecular weight. The consistency in quaternary structure among both large- and small-subunit catalases, that of a tetramer, suggests that variations from the tetrameric structure among heme containing catalases will be rare.

The globular shape of the tetrameric small-subunit enzymes resembles a "dumbbell" in the *R*–*Q* orientation (Fig. 9) having dimensions of 90 by 70 by 105 Å along the *P*, *Q*, and *R* axes, respectively, and a "waist" dimension of 50 Å in the $R = 0$ plane. The large subunit enzyme has similar dimensions along the *P* and *Q* axes, although with a less obvious "waist," but is 140 Å long along the *R* axis (Fig. 9). The association of subunits gives rise to an unusual six-stranded antiparallel structure involving all four subunits. Many elements of the quaternary association have been extensively reviewed (*3*, *83*).

Perhaps the most unusual feature, and certainly the one that has the greatest implications with regard to the folding pathway and stability of the complex, is the interweaving of adjacent subunits that results in two tightly associated dimers per tetramer. The N-terminal arm of each subunit is folded underneath the wrapping domain of an adjacent subunit and the subunits are oriented such that each "*Q*-related" dimer pair contains two overlapped interactions (Fig. 10). In the case of small-subunit enzymes such as BLC, about 25 residues of the amino terminus extend beyond the overlap region, and there is very little interaction of those residues with the remainder of the subunit. The length of the overlapped segment is greater in large subunit enzymes such as HPII where 80 residues are "trapped." Furthermore, this longer section forms an extensive network of interactions with the core of the subunit. The more extensive interweaving and N-terminal interactions of HPII predict a more stable structure, and this was corroborated in a study of the denaturation patterns of HPII and BLC (*86*). HPII was activated slightly at temperatures up to 75°C and lost activity in concert with

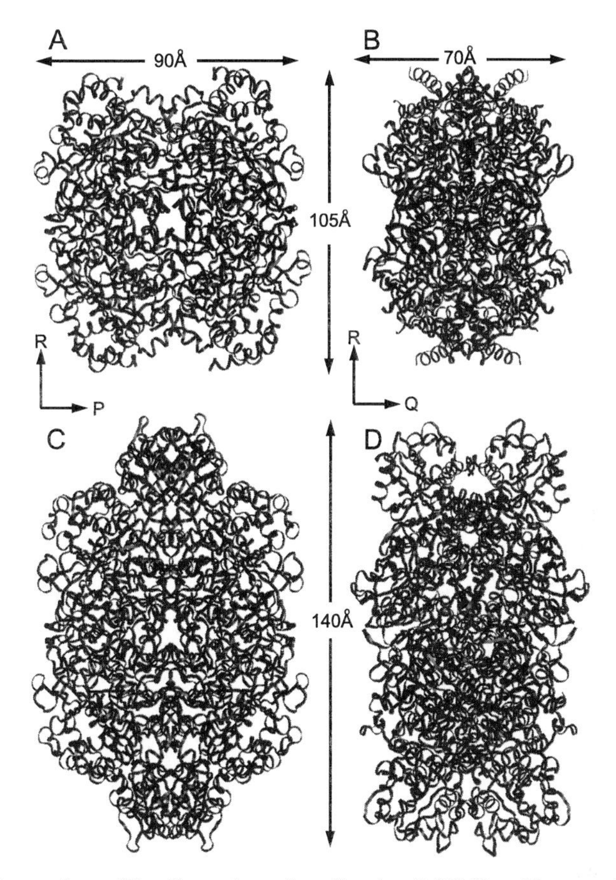

FIG. 9. Comparison of the dimensions of small-subunit (BLC) and large-subunit (HPII) catalases. BLC is shown in panels **A** and **B,** in the *R–P* and *R–Q* orientations, respectively; HPII is shown in **C** and **D,** also in the *R–P* and *R–Q* orientations, respectively.

a transition in secondary structure having a T_m of 82°C. BLC, by comparison, exhibited a similar secondary structure transition and loss of activity, but with a T_m of 56°C (*86*). The most striking illustration of the enhanced stability provided by the interwoven structure is that fact that the dimer structure does not dissociate in potassium phosphate buffer even at 95°C or in 7 M urea, 1% SDS at 60°C (Fig. 11). For comparison, BLC dimers, with the shorter overlapped segment, dissociate at room temperature in the urea–SDS solution.

Bergdoll *et al.* (*87*) have proposed that some proteins, including catalase, exhibit "arm exchange" or an interaction of one subunit with an adjacent subunit to stabilize quaternary structure, and that this

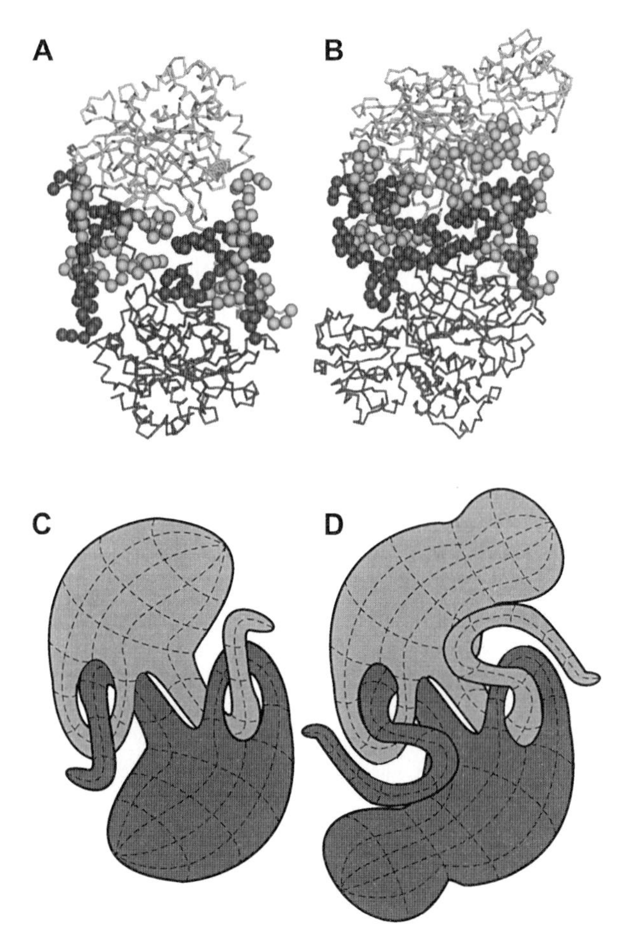

FIG. 10. Illustration of the interweaving of two subunits of the small-subunit BLC (**A**) and the large-subunit HPII (**B**). The residues in the immediate proximity of the overlaps are shown as spheres. A schematic representation of the interwoven structures are shown in **C** for BLC and **D** for HPII. Note the longer N-terminal region of HPII (80 residues in HPII as compared to 25 in BLC) that extend through the loop on the opposite subunit.

interaction is stabilized or enhanced by one or more proline residues. In the case of catalase, Pro69 of BLC, which is highly conserved among all catalases (115 of 117), was identified as the key residue in this interaction. However, the importance of prolines in catalase structure may be far more profound for two reasons. First, the interweaving of subunits in catalases is a more significant interaction than the "arm exchange" interactions or associations identified in other proteins, and, second, far more than one proline are involved when the two proline-

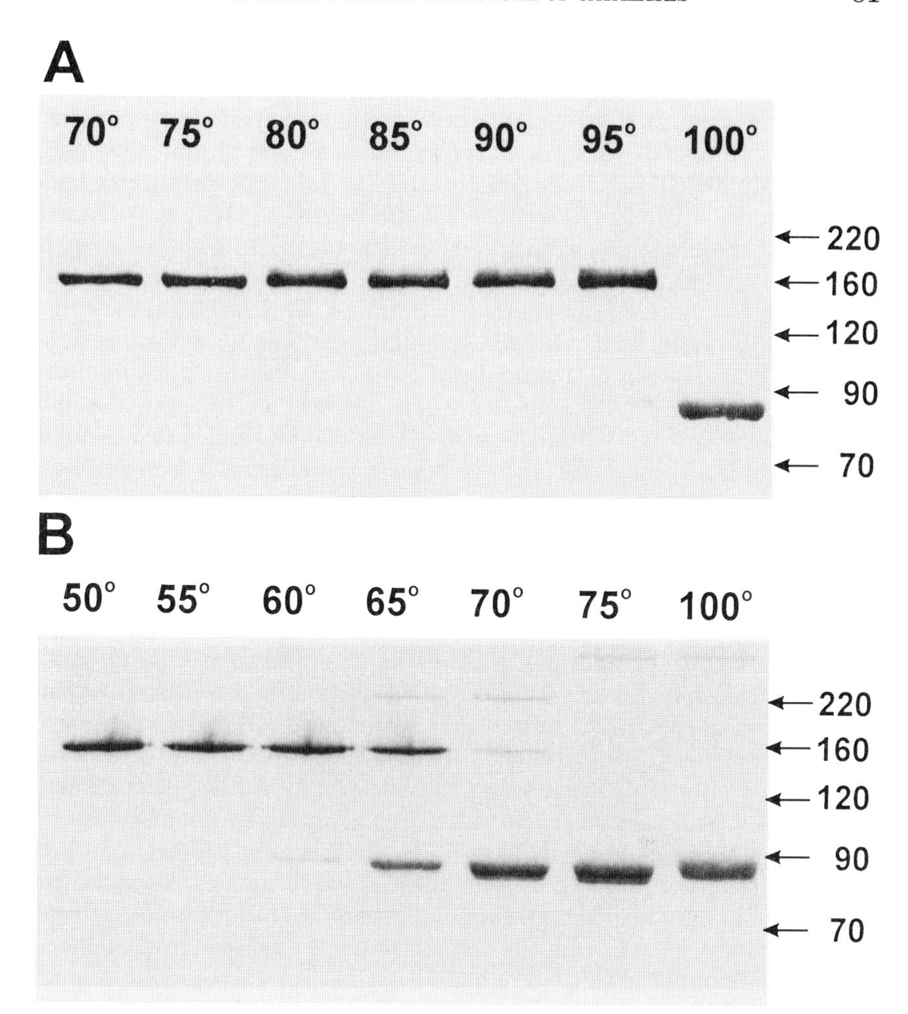

FIG. 11. Conversion of HPII dimers to monomers. In panel **A,** HPII was incubated for 10 min at 70, 75, 80, 85, 90, 95, and 100°C in 50 mM potassium phosphate buffer (pH 7). In panel **B,** HPII was incubated for 10 min at 50, 55, 60, 65, 70, 75, and 100°C in 50 mM potassium phosphate (pH 7) and 5.6 M urea. The temperatures of incubation are indicated above each lane. In both panels **A** and **B,** samples were removed, cooled to room temperature, and added to SDS-urea loading buffer. Samples were loaded and run, without further heating, on an 8% polyacrylamide gel. Reprinted with permission from Switala *et al.* (*86*). Copyright 1999 American Chemical Society.

rich regions surrounding the interwoven regions are considered. This is most evident in HPII, where 10 of 52 residues between 426 and 477, the overlapping segment of the wrapping domain, are proline, and 8 of 28 residues between 31 and 58, in the trapped N-terminal segment, are proline. There is no equivalent of the N-terminal proline-rich segment in small-subunit enzymes because they lack the N-terminal extension, and this may be another explanation for the enhanced resistance to denaturation of HPII dimers as compared to small-subunit enzymes. These proline-rich segments are in addition to the highly conserved proline identified as Pro69 in BLC (Pro123 in HPII), which may be one more component of this proline network. The prolines would contribute to enhanced rigidity in the overlapped region of the protein and thereby stabilize the quaternary structure. In addition, they may influence the folding pathway as suggested by the folding problems encountered by proteins in which the N-terminal proline-rich region up to residue 50 was deleted (*84*).

The interweaving of subunits necessitates a very complex folding process in which subunit–subunit association must occur before subunit folding is complete. In other words, the formation of secondary, tertiary, and quaternary structural elements are intermixed with quaternary interactions between partially folded subunits being formed before final tertiary and possibly even secondary structural elements can be introduced. A survey of the properties of BLC (*73, 88*), yeast catalase (*89*), and HPII (*84*) provides some insight into the folding process of catalases. For example, dimers formed by BLC have dimensions that would be expected of *R*-related dimers (*88*). Yeast accumulates large quantities of hemeless catalase monomer, and formation of the tetrameric form occurs only after heme is bound by the subunit (*89*). By contrast, HPII requires all portions of its structure to be present for correct folding into an active enzyme (*84*).

A possible mechanism based on these observations is described in Fig. 12. There is an initial folding of domains within the apocatalase, followed by association with the heme, leaving the amino terminal arm, wrapping domain, and alpha helical domain (plus carboxyl domain in large-subunit enzymes) unassociated with the core (A to B in Fig. 12). The requirement for heme early in the folding process may explain why catalases have such a high heme occupancy as compared to the catalase-peroxidases. Following complex formation with the heme, two "*R*-related" subunits will associate (B to C in Fig. 12). The next step involves the association of "*Q*-related" subunits (C to D in Fig. 12). Only one pair of "*Q*-related" subunits is shown in Fig. 12, but there would be two dimers associating simultaneously. During the "*Q*-related" subunit association, the amino terminal arms fold against the subunit, followed by the wrapping such that the α-helical domains (and C-terminal portion) fold overtop of the N-terminal segment to "trap" it (D to E in Fig. 12). In this way, each "*Q*-related" subunit pair will have two interwoven

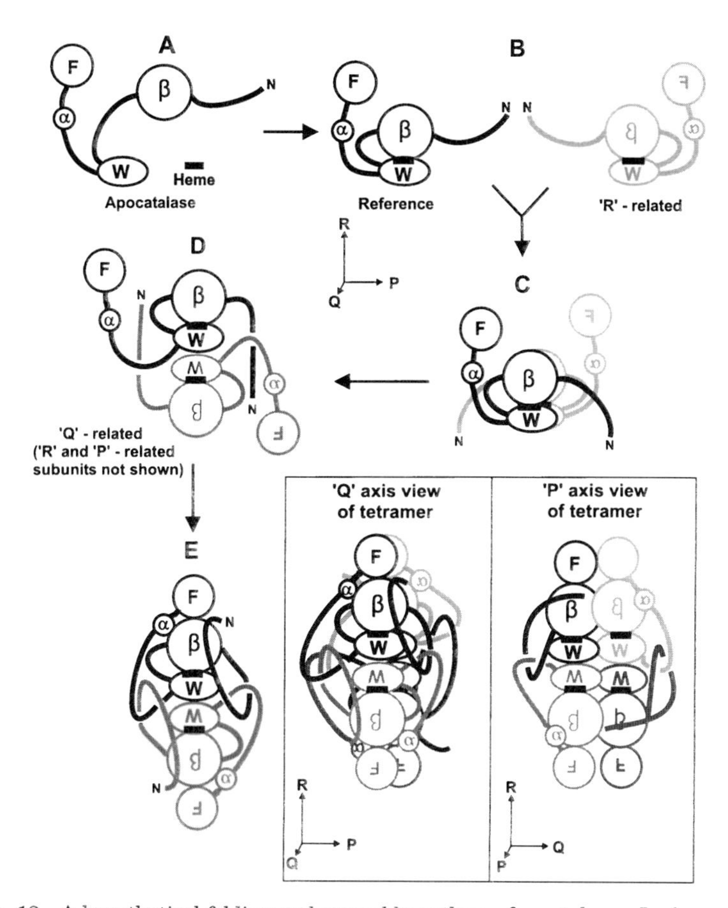

FIG. 12. A hypothetical folding and assembly pathway for catalases. In **A** secondary and tertiary folding first occurs in the individual subunits to form the β-barrel (β), wrapping domain (**W**), α-helical segment (α), and flavodoxin domain (**F**, only in HPII). In proceeding to **B,** heme is bound to each of the subunits, and this may serve as a catalyst for the rapid association of the *R*-related subunits to form the structure in **C.** In proceeding to **D,** *Q*-related subunits associate, resulting in the N-terminal arms being overlapped as the C-terminal portions fold back on themselves to form the fully folded structure shown in **E.** Only two subunits are shown in the progression from **C** to **E,** but a simultaneous folding must be occurring in the associated dimer. The fully folded tetramer is shown in two orientations.

or overlapped segments, and there will be four such interactions per tetramer. The obvious influence of the large-subunit carboxyl terminal domain on the accumulation of protein can be readily explained in such a model, as can be the importance of chaperones (*90*) in stabilizing large segments of unwound or unassociated protein. The initial formation of the *R*-related dimer shown in Fig. 6 is suggested by the dimensions of the BLC dimer, but the extreme stability of the *Q*-related dimer of HPII suggests that this alternate dimer might form. If this were the case, it would be necessary to change the order in Fig. 12 to B to D to E to tetramer. This is a relatively minor change to the pathway, however, and the main point, that the formation of tertiary and quaternary interactions are intermixed, remains the same. Cartoons of the structures of tetrameric HPII from two different orientations to further illustrate the interactions are also shown in Fig. 12.

C. Heme Composition and Location

The heme of catalases is deeply buried within the core of the catalase subunit. Protoheme IX or heme *b* is found in all small-subunit catalases so far characterized. The two large-subunit enzymes HPII and PVC have been characterized biochemically, spectrally, and structurally (*91*) as containing heme *d* in which ring III is oxidized to a *cis*-hydroxyspirolactone. Heme *b* is initially bound to both enzymes during assembly, and it is subsequently oxidized by the catalase itself during the early rounds of catalysis (*92*).

Another significant difference between the large- and small-subunit enzymes lies in the fact that the heme *d* of HPII and PVC is flipped 180° relative to the heme *b* moiety of BLC, MLC, SCC-A, and PMC (Fig. 13). This is clearly a function of the residues that form the heme pocket, although attempts to force a change in heme orientation in HPII by mutating residues that interact with the heme were unsuccessful. The heme is situated in the β-barrel and has interactions with the wrapping domain and with the amino-terminal arm of the R-related subunit. The dimensions of the pocket demand that heme bind in its final conformation and that flipping once inside the pocket not be possible.

The flipped orientation of the heme in HPII and PVC results in the oxidized ring being sufficiently well removed (7 Å) from the essential histidine (His128 in HPII) and the presumed peroxide binding site to complicate an explanation of the reaction mechanism. The explanation is further complicated by the *cis*-stereospecificity of the reaction that results in both oxygens being situated on the proximal side of the heme away from what is considered to be the normal reaction center on the distal side. This stereochemistry dictates that the hydroxyl group on the heme *d* have originated on the proximal side of the heme, and a mechanism has been proposed to explain the reaction in both PVC and HPII (*93*). The mechanism assumes that compound I is formed as a first step

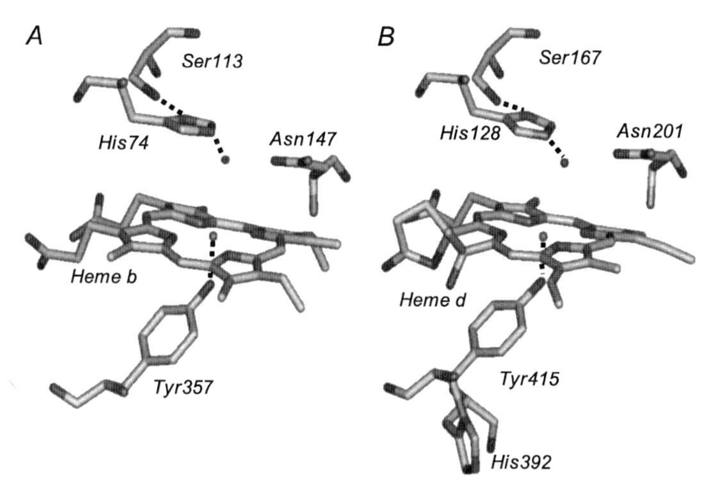

FIG. 13. Active site residues in a small-subunit catalase BLC (**A**) and a large-subunit catalase HPII (**B**). The active site residues are labeled, and hydrogen bonds are shown between the serine (113 in BLC and 167 in HPII) and the essential histidine (74 in BLC and 128 in HPII). A single water is shown hydrogen bonded to the histidine. The equivalent water in BLC is located by analogy to the position of the water in HPII. The unusual covalent bond between the N^{δ} of His392 and the C^{β} of Tyr415 in HPII is evident on the proximal side of the heme in **B.** The flipped orientations of the hemes are evident in a comparison of the two structures, as is the *cis*-hydroxyspirolactone structure of heme d in **B.**

in the heme modification, after which it reacts with a water molecule on the proximal side that acts as an electron donor to the porphyrin cation radical (Fig. 14). Cyclization to form the spirolactone follows completing the reduction of compound I. Water molecules are present on the proximal side, and a potential channel has been identified that would allow access for water. The two residues Ser414 and Gln 419 ensure the *cis* stereochemistry such that changing either residue results in more of the *trans* isomer being formed (*94*).

The mechanism in Fig. 14 applies equally well to both PVC and HPII. However, an unusual bond between the imidazole ring of His392 and the β-carbon of Tyr415 on the proximal side of the HPII heme has been identified (Fig. 13) (*93*), and subsequently its presence was correlated with heme oxidation. The apparent correlation between heme oxidation and His–Tyr bond formation suggested a mechanistic linkage between the two modifications and an alternate mechanism unique to HPII was proposed (Fig. 15). As with the first mechanism, the reaction assumes the formation of compound I that is available for reduction. Formation of the His–Tyr bond, involving a base catalyzed proton extraction from the

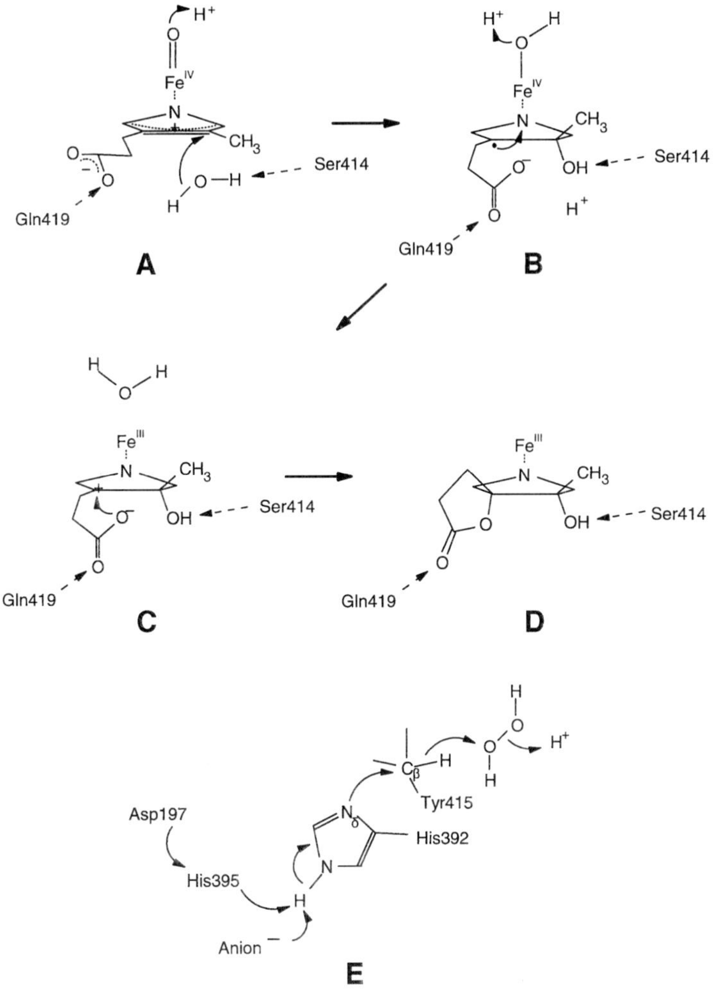

FIG. 14. A mechanism to explain heme modification in the *P. vitale* catalase and possibly *E. coli* HPII. For simplicity, the phenyl ring of Tyr415 is not shown, and only ring III of the heme and the heme iron are shown. Compound I is an oxyferryl species formed, along with water, in the reaction of one H_2O_2 with the heme. The iron is in a formal Fe^V oxidation state, but one oxidation equivalent is delocalized on the heme to create the oxo-Fe^{IV}-heme cation, shown as the starting species, compound I. A water on the proximal side of the heme is added to the heme cation species of compound I shown in **A** to generate a radical ion in **B.** The electron flow toward the oxo-iron would generate the cation shown in (**C**), leading to the spirolactone product shown in **D.** In **E,** an alternate mechanism for the His–Tyr bond formation in HPII is presented that could occur independently of the heme modification reaction. Reprinted with permission of Cambridge University Press from Bravo *et al.* (*93*).

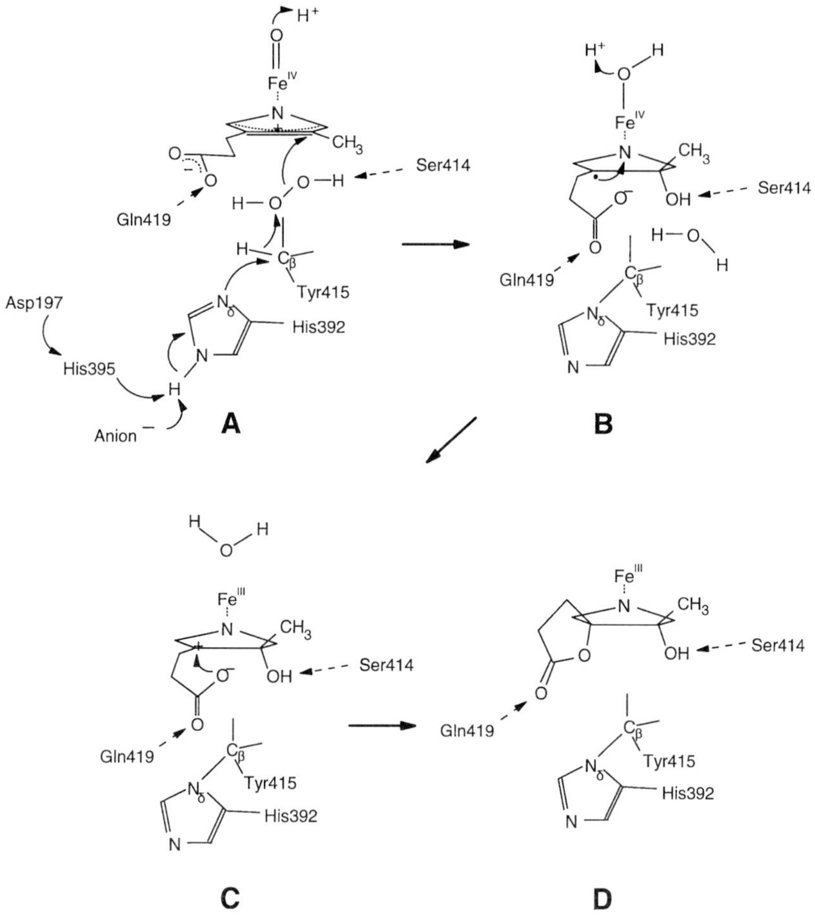

FIG. 15. A proposed mechanism coupling the formation of the His–Tyr bond to the oxidation of ring III of the heme in HPII. The mechanism begins with the formation of compound I shown in **A.** A concerted series of reactions, possibly triggered by either Asp197/His395 or by a putative anionic species bound to compound I, results in the transfer of a hydroxyl to the heme from the H_2O_2 shown in **C,** which would facilitate spirolactone cyclization to form the final product containing the His–Tyr bond and the modified heme shown in **D.** Reprinted with permission of Cambridge University Press from Bravo *et al.* (*93*).

imidazole ring of His392 by a still unidentified species, initiates a concerted reaction that results in peroxide serving as a hydroxide donor to, and consequent reducer of, the porphyrin cation radical (Fig. 15). Subsequent spirolactone formation completes the reduction of the compound I. Supporting the existence of such a mechanism is the observation that changing His392 to Gln precludes the His–Tyr bond formation and also prevents heme oxidation. Similarly, all inactive variants of HPII in which the heme is not oxidized do not contain the His–Tyr linkage. One exception to this generalization is the His392Glu variant in which some heme oxidation takes place despite the absence of His–Tyr bond formation (*93*). However, even here there appears to be an aberrant reaction pathway because a *trans* hydroxy spirolactone predominates as the oxidized heme. Surprisingly, the *P. vitale* catalase has a Gln in the similar position to the reactive His392 of HPII, but still oxidizes the heme. The conclusion seems to be that despite similar heme structures in HPII and PVC, the mechanisms leading to heme *d* may be different in the two enzymes.

One significant conclusion arising from the characterization of the His392Gln variant of HPII (*93*), which retained near-wild-type levels of activity despite containing heme *b*, is that heme *d* is not required for catalytic activity in the large-subunit enzymes. It is unreasonable to assume that such a modification would have evolved without a reason, but an unambiguous explanation still has not been found. One possibility is that the oxidized form imparts a greater resistance to heme damage and subsequent enzyme inactivation in the presence of high concentrations of hydrogen peroxide. BLC is known to have a significant population of damaged heme (*73*), and it is rapidly inactivated by peroxide concentrations above 300 mM (*95*). By comparison, HPII has nearly 100% occupancy of heme *d* and retains activity in the presence of up to 3 M hydrogen peroxide. The observation that the heme *b*–containing His392Gln variant is no more sensitive to high concentrations of peroxide than the wild-type enzyme would argue against this conclusion, but the possibility of differences in heme damage between the two enzymes after such a treatment has not been determined.

There are a number of relatively invariant residues among the catalases, but four that have been identified over the years as being essential for catalatic activity and integrity of the enzyme are a His (128 in HPII or 74 in BLC),), a Ser (167 in HPII and 113 in BLC), an Asn (201 in HPII or 147 in BLC) and a Tyr (415 in HPII or 357 in BLC), the last forming the fifth ligand with the heme iron. Changing His128 in HPII produced variants that had no detectable activity, confirming that the active site His truly is an “essential” residue for the catalatic reaction. On the other hand, replacements of either Ser167 or Asn 201 resulted in mutant variants with low levels of activity, revealing that the residues facilitated the catalatic reaction but were not “essential.” Mutations in Tyr415 abolished the accumulation of any protein, indicating that it was essential for the efficient folding of the protein into a protease resistant form. Presumably, heme binding is abolished and, in the absence of heme, correct folding of the subunit core does not occur (*92*).

D. Channels and Cavities

The cavity structure of catalases has been extensively analyzed for BLC (*38*), PMC (*77*), PVC (*80*), HPII (*82*), and SCC-A (*76*), revealing a number of large cavities that do not seem to have any role in catalysis. By contrast, smaller pockets exist on the distal side of each of the hemes deeply buried inside each of the subunits that are the active sites of the enzyme. With such deeply buried active site cavities, it is necessary to define access routes that will allow the substrate hydrogen peroxide to penetrate almost 30 Å into the protein. Furthermore, the rapid turnover rate of up to 10^6 per second strongly suggests that there must be separate inlet and outlet channels to allow the rapidly evolving oxygen to be removed without interfering with incoming substrate. The structures of all six catalases reveal two obvious channels leading from the molecular surface to the active site cavity that can fulfill this role. They have been termed major or perpendicular and minor or lateral in the various enzymes. This review will use the terminology "perpendicular" and "lateral" because they clearly describe the orientation of the channels relative to the plane of the heme (Fig. 16). Additional routes for accessing the vicinity of the active site include one leading to the molecular center from the region of the heme propionates and a second leading to the heme proximal side.

The perpendicular channel has long been considered the principal route by which substrate hydrogen peroxide accesses the active site. Two studies of the channel by site-directed mutagenesis have revealed that changes to the largely hydrophobic residues in the lower part of the channel just above the heme do affect enzyme activity in both yeast (*3, 96*) and bacterial catalases (*94*). Enlarging the channel either by changing a valine immediately above the active site histidine to alanine or by replacing phenylalanines with smaller but bulkier aliphatic groups generally causes a decrease in catalatic activity and an increase in peroxidatic activity. These results provide convincing evidence that the channel is used to access the active site. They also reveal that the size of the channel is critical for optimum catalatic activity because enlarging it, which would theoretically increase substrate accessibility, results in a reduction in catalatic activity. On the other hand, the enlarged channel was effective in increasing the peroxidatic activity.

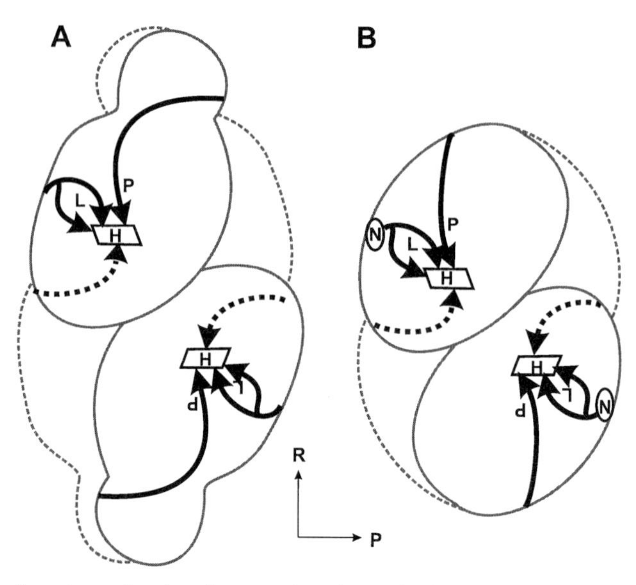

FIG. 16. A cartoon showing the putative channels that provide access to the active site of a large-subunit catalase in **A** and a small-subunit catalase in **B.** The main or perpendicular channel is labeled "**P**" and the minor or lateral channel, which is bifurcated, is labeled "**L.**" A potential channel leading to the proximal side of the heme is shown with a dashed line.

The lateral channel was referred to as the minor channel in part because it was blocked by NADPH in small-subunit enzymes, which created confusion as to whether or not it could act as an efficient channel for substrate or product movement. Furthermore, the channel is really two channels both originating at the site of NADPH binding on the surface of the enzyme. The situation is less complicated in HPII, where NADPH is not bound and the lateral channel provides clear access to the molecular surface. The location in HPII equivalent to the NADPH binding site of small subunit enzymes differs in having Arg260 forming an ionic bond with Glu270, resulting in partial occlusion of the upper branch of the lateral channel. Mutation of Arg260 to Ala significantly enlarges that portion of the channel and results in a threefold increase in specific activity and turnover rate (*95*). Furthermore, alkylated hydroxylamines and sulfhydryl reagents are more effective inhibitors of the Ala260 variant than of the wild-type HPII. It is clear that the inhibitors can access the active site through the lateral channel, but does this mean that the lateral channel also provides access for the substrate hydrogen peroxide to the active site? The answer is not clear.

The enhanced catalatic activity could arise from more facile exhaust of products just as easily as from enhanced substrate accessibility. The effect of inhibitors is a largely static process that is complete once the inhibitor has become bound in the active site. The catalatic process, on the other hand, requires a constant influx of substrate peroxide and efflux of product oxygen and water. As a result, the inlet channels for inhibitors and substrate may be different.

At this time, the proposal of additional access channels is quite conjectural. It seems likely that there is a channel or access route to the proximal side of the heme in order to provide access for the hydrogen peroxide or water needed for heme oxidation and His–Tyr bond formation. Furthermore, the electron density of compound I from PMC (*97*) reveals the presence of an anionic species that is not present in the native enzyme. However, the rapid influx–efflux rates up to 10^6 per sec needed for such a species to be a component of compound I would pose interesting constraints on a channel, and there does not seem to be a likely candidate in the region. Similarly, the potential channel leading to the cavity at the molecular center is not an ideal candidate for substrate or product movement because of its relationship to the active site residues. However, if the lateral channel is truly blocked by NADPH in small-subunit enzymes, this route may provide an alternative access or exhaust route. Both of these latter two channels require further investigation before a clear role can be ascribed to them.

E. NADPH Binding

NADPH seems to be a common component of small-subunit catalases, being present in bacterial (PMC and MLC), yeast (SCC-A and SCC-T), and mammalian (BLC) enzymes. The order of affinity is NADPH > NADH > $NADP^+$ > NAD^+, with one nucleotide bound per monomer. The nucleotide is not a compulsory cofactor and PMC has been isolated and crystallized without the nucleotide bound. This has provided an opportunity to observe that the structural adjustments required for nucleotide binding are minor. The nucleotide binds at a site about 20 Å from the heme iron in an environment that is highly conserved among the small-subunit enzymes. The cavity filled by NADPH in small-subunit enzymes is partially filled in large-subunit enzymes, such as HPII, by a segment (residues 509–595) of the α-helical and linking regions.

The role of the NADPH has not been unequivocally determined. An inactive form of small-subunit enzymes, compound II, can be formed at low peroxide concentrations as a result of a one-electron reduction of compound I. It has been proposed that NADPH serves as an electron source either to prevent the formation of compound II or to convert the inactive compound II back to the Fe^{III} state, thereby circumventing inactivation (see Section IV,B and IV,E). In the case of large-subunit enzymes, the fact that compound II has never been characterized provides an explanation for why NADPH may not be bound.

NADPH bound to most proteins has an extended conformation that presumably facilitates its role as an electron donor (*98*). However, in catalase NADPH is folded into a much more compact structure, resulting in the adenine and nicotinamide rings being only 3.8 Å apart, although not quite in a stacked conformation. Only the NADPH in flavin reductase P exists in a more compact structure with stacked adenine and nicotinamide rings 3.6 Å apart (*99*). The compact structure in catalase may allow more effective electron donation from the nicotinamide while still allowing the adenine ring to serve as an anchor for the nucleotide on the enzyme surface. The pathway of electron transfer from NADPH to the heme has been the subject of considerable conjecture, including the proposal of protein radical intermediates and of Pro to Ser or Phe to Gly to Ser channeling (*51*). The definitive experiments to characterize the transfer pathway, possibly through the mutation of residues, have yet to be carried out.

The location of NADPH in the lateral channel of small-subunit enzymes blocks the channel such that it cannot realistically be considered to have either an inlet or exhaust function in the presence of the nucleotide. For the channel to have a role in these enzymes, it is necessary to predict that NADPH binds preferentially to the enzyme in its resting state or to inactive forms of the enzyme, such that NADPH is not bound when the enzyme is actively degrading H_2O_2. Thus, the variation in NADPH occupancy among enzymes might reflect the reactive history of the enzyme immediately prior to isolation, as well as the K_a for NADPH binding and the NADPH concentration. For example, enzyme isolated from a culture with high peroxide levels may have a lower amounts of bound NADPH because the enzyme is more active.

F. Complexes

Catalases bind or react with a number of molecules that can be either substrates (hydrogen peroxide and some small alcohols) or inhibitors (cyanide, azide, etc.) (see Section IV,B). Several such intermediates have been trapped in crystalline form and subjected to structural analysis. Arguably the most interesting is that of the reaction intermediates compound I and compound II of PMC (*97*). Preformed crystals were perfused with peroxoacetic acid, which can oxidize the enzyme to form compound I but cannot reduce it back to the ground state. Subsequent reduction with dithiothreitol converted compound I to compound II.

As would be expected in an enzyme with a turnover rate that can be as high as 10^6 per second, there were no significant rearrangements in the C-alpha backbone structures of compound I and II compared to the ground state enzyme. However, there were two very significant additions of electron density to the structures. One corresponded to the expected ferryl oxygen on the distal side of the heme iron. The second corresponded to the presence of a presumed anionic species replacing a water molecule in a proximal side cavity about 18 Å from the heme iron. The presumption of negative charge was based on the basicity of the cavity and the possibility that it may serve to neutralize the positive charge on the compound I heme. This ion has been implicated as a possible catalyst in the mechanism of His–Tyr bond formation and heme oxidation in HPII (*90*). However, as noted earlier, the concept of ions flowing in and out of the enzyme with each turnover of compound I imposes the requirement of an easily accessible channel. Alternatively, the ion may remain bound while the enzyme is active and dissociate only when activity subsides. One subtle change in the catalytic center involves movement of the heme iron from 0.1 Å below the plane to 0.3 Å above the plane of the heme (*97*).

The only major difference between compounds I and II is the loss of the anion from the proximal side cavity during the formation of compound II. A subtle change in the location of the heme iron from 0.3 Å above the plane to being in the plane of the heme also occurs (*97*). Whether the loss of the putative anion is simply a reflection of the reduced positive charge on the heme in compound II or the explanation for the inactivation of compound II relative to compound I remains undetermined.

The adduct formed between PVC and the inhibitor aminotriazole has been crystallized and the structure determined (G. N. Murshudov, personal communication), providing unequivocal evidence for the mechanism of inhibition. There is a covalent linkage between the essential distal side histidine and the aminotriazole, which remains parallel to the heme forming a hydrogen bond with the active site asparagine. The formation of such a covalent linkage requires that the imidazole ring be oxidized by hydrogen peroxide either directly or indirectly, possibly involving compound I. No interaction between the heme iron and the aminotriazole is evident. By contrast, the complex of HPII with azide reveals a coordination of the azide molecule between the active site histidine and the heme iron (*83*). No other significant changes in the structure as compared to the wild-type enzyme were noted in either case.

G. Unusual Modifications

Catalases have proven to be a treasure trove of unusual modifications. The first noted modification was the oxidation of Met53 of PMC to a methionine sulfone (*77*). Met53 is situated in the distal side active site adjacent to the essential His54 in a location where oxidation by a molecule of peroxide would not be unexpected. Among the catalases whose structures have been solved, PMC is unique in having the sulfone because valine is the more common replacement in other catalases. The sulfone does not seem to have a role in the catalytic mechanism and is clearly generated as a posttranslational modification. A small number of catalases from other sources, principally bacteria, have Met in the same location as PMC, and it is a reasonable prediction that the same oxidation occurs in those enzymes as well, although this has not been demonstrated.

The subunit of HPII contains two cysteines, 438 and 669, of which Cys438 is situated in the core of the tetramer and Cys669 is situated on the surface of the C-terminal domain. Replacement of the cysteines, either individually or together, causes only small reductions in specific activity of the enzyme, indicating that neither is essential for catalysis or enzyme folding. Analysis of free sulfhydryl groups revealed that Cys448 was blocked and attempts to remove the blocking group were unsuccessful with anything but alkali (*100*). This and other analyses eliminated a number of possible modifications such as acylation, disulfide bonding, oxidation to sulfinic or sulfonic acids, and carbamoylation. Analysis of CNBr digests of HPII and its mutant variants by MALDI mass spectrometry revealed that the peptide containing Cys438 has a mass that is 43 Da larger than expected. As a control, the same peptide from the Cys438Ser variant exhibited the expected mass (allowing for the mutation). An unambiguous identification of the blocking group has not yet been achieved, and the working hypothesis at the moment is that it might involve a hemithioacetal linkage with acetaldehyde. The role of the modification also remains undetermined, although, like the heme oxidation and methionine oxidation in PMC, it may be that the modification increases the enzyme's resistance to inactivation by hydrogen peroxide. However, the remoteness of the residue from the active site may make this a questionable assertion.

The modification of heme *b* to heme *d* (*91, 92*) observed in the large-subunit enzymes HPII and PVC has already been discussed in detail. Whether all large-subunit enzymes will be found to contain such a modified heme remains to be seen, and the fact that HPII variants containing heme *b* retain activity suggests that naturally occurring large subunit catalases with heme *b* may be found.

The covalent bond linking the N^{δ} of the imidazole ring of His392 to the C^{β} of the essential Tyr415 of HPII (93) has also been described (see Section V,C and Fig. 13). It is found in HPII but not in the closely related large-subunit enzyme, PVC, where the residue equivalent to His392 is a Gln making such a bond impossible. The formation of this unusual bond seems to be linked to heme oxidation, which suggests that its role, like that of heme oxidation, may be to enhance resistance to peroxide, and is therefore another mechanism for stabilizing the enzyme. However, the absence of the covalent bond does not lead to easier denaturation, so it does not contribute measurably to thermal stability. On the other hand, the enzyme has a very rapid turnover rate at high peroxide concentrations, and the added rigidity on the proximal side may help the enzyme to maintain an active conformation in the face of high peroxide concentration or during rapid product formation.

Another unusual covalent bond has been observed in a mutant variant of HPII in which Val169 situated immediately above the essential His128 is changed to a Cys (*94*). The purified variant enzyme exhibits less than 0.1% of wild-type activity, contains heme *b*, and does not contain the His–Tyr bond. The reasons for the lack of activity were clarified only when the crystal structure was solved, revealing a covalent bond between the Cys-S and the C^{δ} of His128. The planar nature of the imidazole ring including the S is suggestive of a retention of sp^2 character or unsaturation in the imidazole ring. The imidazole ring is rotated about 30°, removing it from a being stacked over the heme, thereby interfering with its participation in the catalytic reaction. No other example of such a Cys–His bond has been reported, but the potential exists on the distal side of the heme in the KatE protein from *Xanthomonas campestris*, and this is being investigated.

VI. Structure of Type B Catalase-Peroxidases

The isolation of crystalline catalase-peroxidase has been an elusive goal that so far has not been achieved. The fact that there is no accurate view of the active sites of the enzyme has complicated a detailed study of the roles of individual residues in catalysis. Fortunately, catalase-peroxidases are similar in sequence to type I plant and fungal peroxidases, particularly in the vicinity of the heme active site. The fact that the structures of a number of peroxidases have been solved has allowed certain assumptions to be made about the identity and location of a number of highly conserved active site residues in the catalase-peroxidases based on analogy with the peroxidase structures.

Heme-containing peroxidases all contain heme *b* with a histidine imidazole as the proximal ligand. On the distal side, a histidine (#52 in cytochrome *c* peroxidase or CCP) and arginine (#48 in CCP) are highly conserved among both peroxidases and catalase-peroxidases. A third position on the distal side of the heme is also highly conserved as either a tryptophan in type I peroxidases (including yeast CCP residue #51) or as phenylalanine in all type II and III peroxidases (including horse radish peroxidase, among others). The orientation of these residues in peroxidases, using CCP as a model, is shown in Fig. 17. The relative positions of the residues in other peroxidases are virtually identical even to the extent that the phenyl and indole rings, respectively, of the phenylalanine and tryptophan options at Trp51 are coplanar when superimposed. The high conservation of these key residues, in three-dimensional orientation among the peroxidases and in sequence among the peroxidases and catalase-peroxidases, provides strong support for the existence of a similar orientation of residues in the catalase-peroxidases including KatG. Furthermore, a recent analysis of the role of these residues in *E. coli* KatG by site-directed mutagenesis of the key Arg, His, and Trp

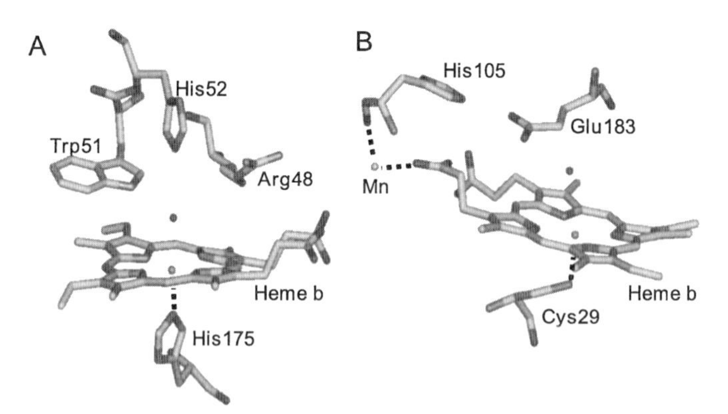

FIG. 17. Active site residues in yeast CCP (**A**) and chloroperoxidase (**B**). The active site residues are labeled, and a single water identified in the electron density situated over the heme iron is shown.

residues has confirmed that they are involved in the catalytic mechanism (*101*). The role of these residues in the catalatic mechanism is discussed later.

VII. Structure of Chloroperoxidase

The crystal structure of the monomeric chloroperoxidase from *C. fumago* has been determined (*14*), revealing a gross structure and active site that are very different from those of other catalases and peroxidases. The enzyme exists as a 42-kDa monomer containing one heme *b* group bound in an eight-helical-segment array. Within the active site and from a mechanistic standpoint, the most significant differences lie in the proximal side fifth ligand to the heme iron, which is a cysteine, rather than histidine or tryptophan, and in the presence of a single catalytic residue on the distal side of the heme, glutamic acid, rather than histidine in combination with asparagine or arginine as in catalases or catalase-peroxidases, respectively. There is a histidine on the distal side of the heme, but it is situated above the propionate side chains of the heme, seemingly too far away from the heme iron to participate directly in the formation or reduction of compound I (Fig. 17). In addition, there is a Mn^{II} associated with the heme propionates and His105, but its role is unknown.

VIII. Mechanism of the Catalatic Reaction

A. Compound I Formation

The first mechanism for the formation of compound I by either a peroxidase or a catalase was proposed in 1980 by Poulos and Kraut (*102*) based on the then newly determined structure of cytochrome *c* peroxidase. This has been reviewed recently (*103, 104*). It remains the operative mechanism for peroxidases with similar active sites and has been adapted to explain the catalatic reaction. In CCP, Arg48 and His52 on the heme distal side were identified as the catalytic determinants that bind and polarize the hydrogen peroxide molecule in concert with the heme iron. The imidazole ring of His52, which is oriented perpendicular to the plane of and about 6 Å above the heme, acts as a proton acceptor from the hydrogen peroxide, while the Arg side chain, which is situated laterally across the heme from the imidazole, stabilizes the charged intermediates. The resulting oxyferryl compound I has the iron oxidized to Fe^{IV} and has a second oxidation equivalent in the form a protein radical situated on the proximal Trp191, although in some peroxidases such as HRP, the second oxidation equivalent is found as a porphyrin cation radical. As already described, the structure of the active site of type B catalase-peroxidases is based on analogy with the active site of peroxidases. Consequently, any mechanism proposed for compound I formation will, by necessity, be similar to the mechanism proposed for CCP. That is, the His and Arg residues, 106 and 102, respectively, in HPI, participate in a polarization of the O—O bond and proton transfer to produce compound I (Fig. 18A). The electronic structure of compound I in HPI remains unclear, and there is even the potential for two different compound I species depending on the reaction pathway (see Section

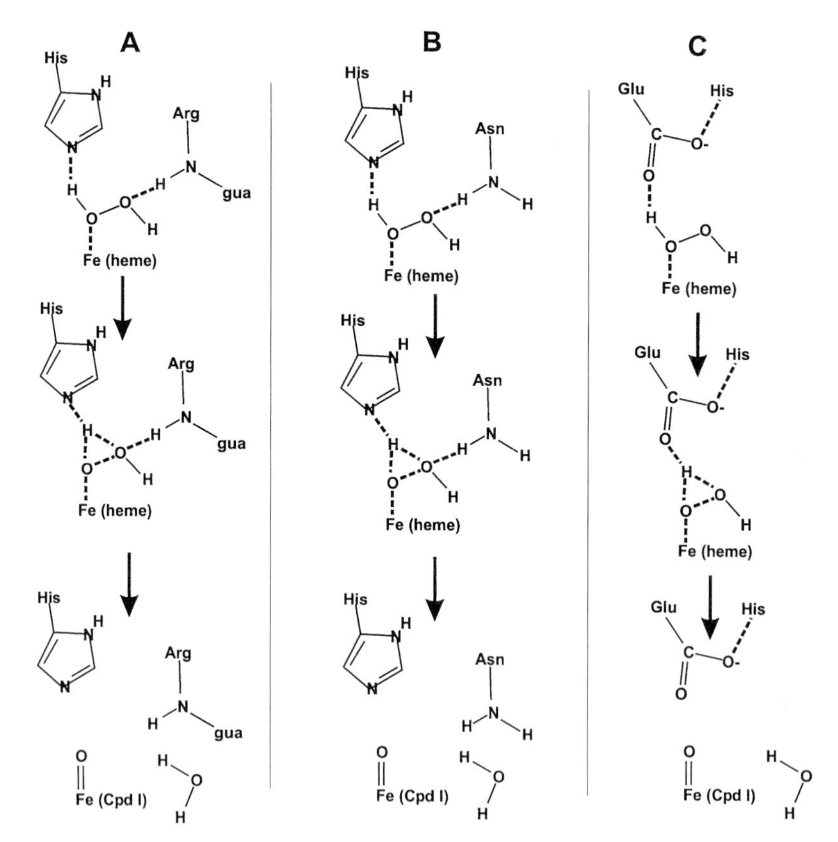

FIG. 18. Mechanisms of compound I formation in type B catalases (based on yeast CCP) (see also Section IV,F and Fig. 7) (**A**): type A catalases (**B**); and chloroperoxidase (**C**).

IV, F and Fig. 7). In catalatic mode, HPI exhibits no significant spectral change, suggesting that compound I has an oxidation equivalent in the form of a protein radical rather than a porphyrin radical. On the other hand, the W105F variant of HPI, which operates only in peroxidatic mode, has a porphyrin radical clearly evident in the absorbance and EPR spectra (*101*).

The mechanism for compound I formation in catalases is very similar to that just described for peroxidases. The active site of a type A catalase also has a histidine, which is essential for catalytic activity, and an asparagine, which, although not essential, greatly enhances catalysis (*92*). Despite the orientations of the active site residues of catalase, particularly the histidine, differing significantly from the orientations of the equivalent residues in peroxidases, the histidines have similar roles in both enzymes while the asparagine of catalase plays a role similar to the arginine of peroxidase (Fig. 18B). In contrast to its orientation in peroxidases, the imidazole of the essential histidine in catalases is coplanar with the heme, effectively stacked, about 3.5 Å above ring III of heme *b* in BLC or ring IV of heme *d* in HPII. Like the arginine in peroxidases, the asparagine residue is situated laterally across the heme from the histidine in a position where it can hydrogen bond with the hydrogen peroxide during catalysis. The presence of similar residues allowed the catalatic mechanism to be modeled on the peroxidatic mechanism (*38*). Interaction of the hydrogen peroxide with both the heme Fe^{III} and the imidazole ring of the histidine weakens and stretches the O—H bond, allowing the second oxygen of the peroxide to simultaneously form a hydrogen bond with the imidazole. The proton on the O that is interacting with the Fe is then transferred, via the imidazole, to the second oxygen, giving rise to water and compound I. The asparagine residue participates through hydrogen bonding with the peroxide and stabilizes the polarized intermediates.

For chloroperoxidase, a type D catalase, the mechanism for compound I formation must differ significantly from those just described because the only active site residue sufficiently close to the heme iron to participate directly in the reaction is glutamate183. The glutamate is hydrogen bonded with His105, but the latter is too distant to participate directly in the reaction with hydrogen peroxide. Therefore, the hydrogen peroxide must initially associate in the active site with the heme Fe^{III} and the glutamate side chain. As the reaction progresses, a second hydrogen bond may form with the highly electronegative glutamate side chain to facilitate proton transfer from the Fe—O—H to the second O, producing water and compound I (Fig. 18C). Even making allowances for possible differences arising from assays in different laboratories, the catalatic activity of the chloroperoxidase is not more than 2% that of type A or B catalases, suggesting that the active site of the chloroperoxidase is not optimized for the catalatic reaction, consistent with it being a side reaction.

The reaction of other minor or type D catalases such as methemoglobin and metmyoglobin is not treated in detail here, because they are minor activities, significantly lower than even that of chloroperoxidase. The orientation of residues on the distal side of the heme is not optimized for the catalatic reaction to the extent that there is even a sixth ligand of the heme, a histidine, that would preclude a close association of the heme with hydrogen peroxide without a significant side-chain movement. It is only after an extended treatment with H_2O_2 and oxidation of the Fe that a low level of catalatic activity becomes evident.

B. Compound I Reduction

The mechanism for the reduction of compound I by catalases must differ significantly from the mechanism in peroxidases because of the involvement of hydrogen peroxide as the two-electron donor rather than sequential organic substrates acting as one-electron donors. Based again on the peroxidatic reaction, the catalase-peroxidase or type B catalases were initially thought to employ Arg102 and His106 (HPII numbering) as catalytic residues in the reduction stage. However, recent evidence suggests that the active site tryptophan, Trp105 in HPI, plays an important role. Removal of Trp105, while not affecting compound I formation to the extent that compound I could be identified by absorbance and EPR spectrometry, significantly inhibits its subsequent reduction by hydrogen peroxide (*101*). Because it is not possible for hydrogen peroxide to bind simultaneously to all three of His106, Trp105, and Arg102, a modified second stage is proposed in which the indole ring of Trp105 has a role analogous to that of the active site Asn in type A catalases. Along with the imidazole of His106, it binds the second hydrogen peroxide and facilitates hydrogen transfer to the oxyferryl oxygen and to the imidazole ring of His106. The final transfer of the second reducing equivalent, from the imidazole of His106 to the Fe—O—H, completes the cycle, producing water and native enzyme (Fig. 19A). This mechanism is independent of the protein and porphyrin radical structure of compound I (see Section IV,F and Fig. 7).

In the type A catalases, there are only two active site residues in locations where they can influence the reaction, a histidine and an asparagine. A mechanism for compound I reduction in catalases was propsed by Fita and Rossmann (*38*) in BLC involving these two residues. The hydrogen peroxide binds to compound I, forming hydrogen bonds with the oxyferryl oxygen, the His imidazole, and the Asn side chain. This structure allows the transfer of one hydrogen to the oxyferryl oxygen and a second to the imidazole, resulting in formation of oxygen. The hydrogen and second reducing equivalent on the imidazole ring could then be transferred to the Fe^{IV}–OH to produce water and regenerate the enzyme in its native state (Fig. 19B). The role of the Asn in this stage might be to stabilize the nascent oxygen during the reaction. Direct evidence for this portion of the mechanism is lacking, although the Asn201His variant of HPII exhibits spectral changes consistent with a modification of the heme environment despite a very low catalatic activity and no oxidation of the heme (*92*).

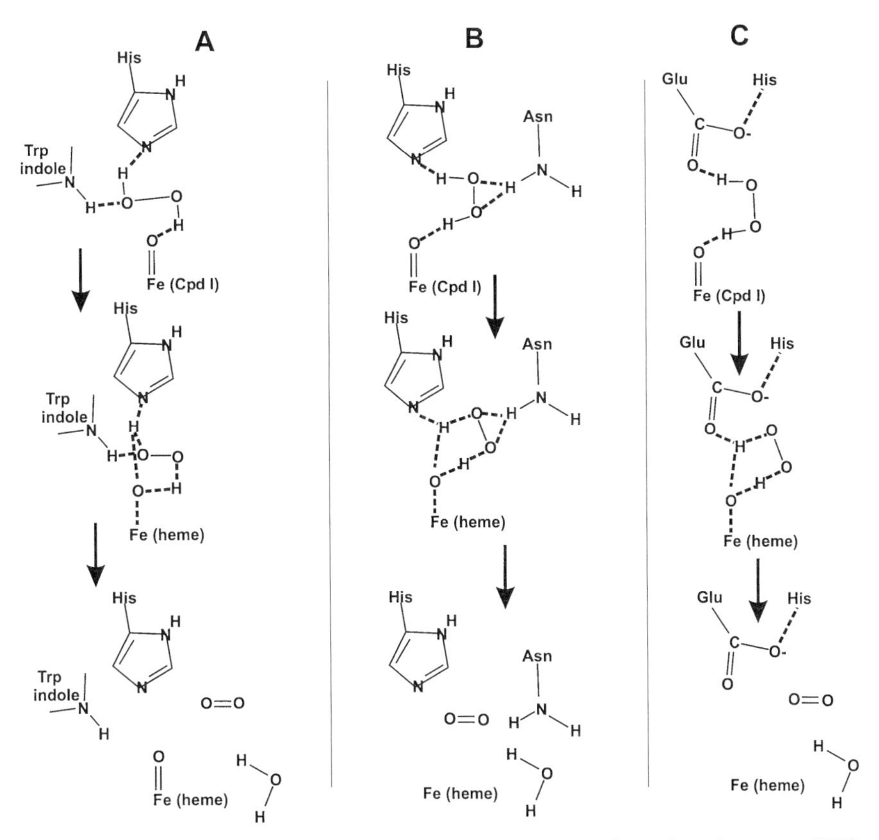

FIG. 19. Mechanisms of compound I reduction in type B catalases (based on yeast CCP) (see also Section IV,F and Fig. 7) (**A**); type A catalases (**B**); and chloroperoxidase (**C**).

The reduction of compound I in chloroperoxidase differs from both of the mechanisms in type A and B catalases because of there being only a single glutamic acid residue to catalyze the reaction. The reducing hydrogen peroxide must initially bind with the glutamate side chain and the oxyferryl oxygen. The highly electronegative glutamate will easily extract a proton from the peroxide, but it is more difficult to conceive of a hydrogen or reducing equivalent being transferred to it. Two options might be considered. In the first, the hydrogen peroxide is highly polarized by the glutamate, resulting in a hydride transfer to the oxyferryl oxygen and simultaneous proton transfer to the glutamate. The proton on the glutamate would ultimately be transferred to the Fe^{II}—O—H^- to produce water and native enzyme. In the second option, a single reducing equivalent, a hydrogen, is transferred to the oxyferryl oxygen. The second reducing equivalent and proton are transferred to the glutamate and delocalized in the intricate complex involving Glu183, His105, the propionate chains, and the Mn^{II} cation. This equivalent is then transferred back to the Fe^{IV}—O—H to complete the reaction (Fig. 19C). Obviously, the presence of glutamate alone on the distal side does not present the optimal catalytic site for reaction with hydrogen peroxide, explaining the low reactivity.

IX. Summary

Catalases continue to present a challenge and are an object of interest to the biochemist despite more than 100 years of study. More than 120 sequences, seven crystal structures, and a wealth of kinetic and physiological data are currently available, from which considerable insight into the catalytic mechanism has been gained. Indeed, even the crystal structures of some of the presumed reaction intermediates are available. This body of information continues to accumulate almost daily.

Have we exhausted catalases as a source of information about protein structure and the catalatic mechanisms? The answer is clearly no. With each structure reported comes new information, often including structural modifications seemingly unique to catalases and with roles that remain to be explained. Despite a deeply buried active site, catalases exhibit one of the fastest turnover rates determined. This presents the as yet unanswered question of how substrate can access the active site while products are simultaneously exhausted with a potential turnover rate of up to 10^6 per second. The complex folding pathway that produces the intricate interwoven arrangement of subunits also remains to be fully clarified.

The catalase-peroxidases present other challenges. More than 20 sequences are available, and interest in the enzyme arising from its involvement in the process of antibiotic sensitivity in tuberculosis-causing bacteria has resulted in a considerable body of kinetic and physiological information. Unfortunately, the determination of crystallization conditions and crystals remain an elusive goal, precluding the determination of a crystal structure. Furthermore, the presence of two possible reaction pathways, peroxidatic and catalatic, has complicated a definition of the reaction mechanisms and the identity of catalytic intermediates. There is work here to occupy biochemists for many more years.

Acknowledgments

This work was supported by operating grants from the Natural Sciences and Engineering Research Council of Canada (NSERC) to PCL and PN. The assistance of J. Switala, B. Tattrie, and M. Maj over the years is appreciated. Receipt of preprints in advance of their publication from C. Obinger is also appreciated.

References

1. Loew, O. *U.S. Dept of Agri. Repts.* **1900,** *65,* 5.
2. Loewen, P. C. In "Oxidative Stress and the Molecular Biology of Antioxidant Defenses"; Scandalios, J. G., ed.; pp. 273–308. Cold Spring Harbor Laboratory Press: Cold Spring Harbor, NY, 1997.
3. Zamocky, M.; Koller, F. *Prog. Biophys. Molec. Biol.* **1999,** *72,* 19.
4. Kremer, M. L. *Trans. Faraday Soc.* **1965,** *61,* 1453.
5. Keilin, D.; Hartree, E. F. *Nature* **1950,** *160,* 513.
6. Klotz, M. G.; Klassen, G. R.; Loewen, P. C. *Mol. Biol Evol.* **1997,** *14,* 951.
7. Welinder, K. *Curr. Opin. Struc. Biol.* **1992,** *2,* 388.
8. Kono, Y.; Fridovich, I. *J. Biol. Chem.* **1983,** *258,* 6015.
9. Allgood, G. S.; Perry, J. J. *J. Bacteriol.* **1986,** *168,* 563.
10. Waldo, G. S.; Fronko, R. M.; Penner-Hahn, J. E. *Biochemistry* **1991,** *30,* 10486.
11. Whittaker, M. M.; Barynin, V. V.; Antonyuk, S. V.; Whittaker, J. W. *Biochemistry* **1999,** *38,* 9126.
12. Arnao, M. B.; Acosta, M.; del Rio, J. A.; Varn, R.; Garcia-Canovas, F. *Biochim. Biophys. Acta* **1990,** *1041,* 43.
13. Thomas, J. A.; Morris, D. R.; Hager, L. P. *J. Biol. Chem.* **1970,** *245,* 3129.
14. Sundaramoorthy, M.; Terner, J.; Poulos, T. L. *Structure* **1995,** *3,* 1367.
15. Sun, W.; Kadima, T. A.; Pickard, M. A.; Dunford, H. B. *Biochem. Cell. Biol.* **1994,** *72,* 321.
16. Facey, S.; Gross, F.; Vining, L. C.; Yang, K.; van Pee, K. H. *Microbiology* **1996,** *142,* 657.
17. Mulvey, M. R.; Switala, J.; Borys, A.; Loewen, P. C. *J. Bacteriol.* **1990,** *172,* 6713.
18. Ma, M.; Eaton, J. W. *Proc. Nat. Acad. Sci. USA* **1992,** *89,* 7924.

19. Nicholls, P. *Biochim. Biophys. Acta*. **1965,** *99,* 286.
20. Nicholls, P. *Biochim. Biophys. Acta* **1972,** *279,* 306.
21. Clayton, R. K. *Biochim. Biophys. Acta*. **1960,** *36,* 35.
22. Nicholls, P.; Schonbaum, G. R. In "The Enzymes"; P. Boyer; H. Lardy; K. Myrback, eds.; Vol. VIII, 2nd ed., pp. 147–225; Academic Press: New York, 1963.
23. Aebi, H.; Heiniger, J.-P.; Suter, H. *Experientia* **1962,** *18,* 129.
24. Aebi, H.; Cantz, M.; Suter, H. *Experientia* **1965,** *21,* 713.
25. Deisseroth, A.; Dounce, A. L. *Physiol. Rev*. **1970,** *50,* 319.
26. Ruis, H.; Koller, F. In "Oxidative Stress and the Molecular Biology of Antioxidant Defenses"; Scandalios, J. G., ed.; pp. 309–342. Cold Spring Harbor Laboratory Press: Cold Spring Harbor, NY, 1997.
27. Scandalios, J. G., Guan, L.; Polidoros, A. N. In "Oxidative Stress and the Molecular Biology of Antioxidant Defenses"; Scandalios, J. G., ed.; pp. 343–406. Cold Spring Harbor Laboratory Press: Cold Spring Harbor, NY, 1997.
28. Loewen, P. C.; Switala, J.; Triggs-Raine, B. L. *Arch. Biochem. Biophys*. **1985,** *243,* 144.
29. Christman, M. F.; Storz, G.; Ames, B. N. *Proc. Natl. Acad. Sci. USA* **1989,** *86,* 3484.
30. Ivanova, A.; Miller, C.; Glinsky, G.; Eisenstark, A. *Molec. Microbiol*. **1994,** *12,* 571.
31. Mukhopadhyay, S.; Schellhorn, H. E. *J. Bacteriol*. **1994,** *176,* 2300.
32. Visick, J. E.; Clark, S. *J. Bacteriol*. **1997,** *179,* 3158.
33. Loewen, P. C.; Hennge-Aronis, R. *Ann. Rev. Microbiol*. **1994,** *48,* 53.
34. Bonnichsen, R. K.; Chance, B.; Theorell, H. *Acta Chem. Scand*. **1947,** *1,* 685.
35. Beers, R. F.; Sizer, I. W. *J. Biol. Chem*. **1952,** *195,* 133.
36. Ogura, Y. *Arch. Biochem. Biophys*. **1955,** *57,* 288.
37. Keilin, D.; Nicholls, P. *Biochim. Biophys. Acta* **1958,** *29,* 302.
38. Fita, I.; Rossmann, M. G. *J. Mol. Biol*. **1985,** *185,* 21.
39. Brill, A. S.; Castleman, B. W.; McKnight, M. E. *Biochemistry* **1976,** *15,* 2209.
40. Nicholls, P. *Biochem. J*. **1961,** *81,* 365.
41. Maj, M.; Loewen, P. C.; Nicholls, P. *Biochim. Biophys. Acta* **1998,** *1384,* 209.
42. Young, L. J.; Siegel, L. M. *Biochemistry* **1988,** *27,* 2790.
43. Nicholls, P. *Biochem. J.* **1964,** *90,* 331.
44. Nicholls, P. *Trans. Farad. Soc.* **1964,** *60,* 137.
45. Kalynaraman, B.; Janzen, E. G.; Mason, R. P. *J. Biol. Chem*. **1985,** *260,* 4003.
46. Lardinois, O.; Rouxhet, P. G. *Biochim. Biophys. Acta* **1996,** *1298,* 180.
47. Hillar, A.; Nicholls, P. *FEBS Lett*. **1992,** *314,* 179.
48. Kirkman, H. N.; Gaetani, G. F. *Proc. Natl. Acad. Sci. USA* **1984,** *81,* 4343.
49. Nicholls, P. *Biochim. Biophys. Acta*. **1964,** *81,* 479.
50. Fita, I.; Rossmann, M.G. *Proc. Natl. Acad. Sci. USA* **1985,** *82,* 1604.
51. Bicout, D. J.; Field, M. J.; Gouet, P.; Jouve, H.-M. *Biochim. Biophys. Acta* **1995,** *1252,* 172.
52. Olson, L. P.; Bruice, T. C. *Biochemistry* **1995,** *34,* 7335.
53. Almarsson, O.; Sinha, A.; Gopinath, E.; Bruice, T. C. *J. Amer. Chem. Soc*. **1993,** *115,* 7093.
54. Kirkman, H. N.; Rolfo, M.; Ferraris, A. M.; Gaetani, G. F. *J. Biol. Chem*. **1999,** *274,* 13908.
55. Hayashi, Y.; Yamazaki, I. *J. Biol Chem*. **1979,** *254,* 9101.
56. He, B.; Sinclair, R.; Copeland, B. R.; Makino, R.; Powers, L. S.; Yamazaki, I. *Biochemistry* **1996,** *35,* 2413.
57. DeFelippis, M. R.; Murthy, C. P.; Faraggi, M.; Klapper, M. H. *Biochemistry* **1989,** *28,* 4847.

58. Hillar, A.; Nicholls, P.; Switala, J. Loewen, P. C. *Biochem. J.* **1994,** *300,* 531.
59. Jouve, H. M.; Pelmont, J.; Gaillard, J. *Arch. Biochem. Biophys.* **1986,** *248,* 71.
60. Claiborne, A.; Malinowski, D. P.; Fridovich, I. *J. Biol. Chem.* **1979,** *254,* 11664.
61. Hochman, A.; Shemesh, A. *J. Biol. Chem.* **1987,** *262,* 6871.
62. Loewen, P. C.; Stauffer, G. V. *Mol. Gen. Genet.* **1990,** *224,* 147.
63. Welinder, K. *Biochim. Biophys. Acta* **1991,** *1080,* 215.
64. Percy, M. E. *Can. J. Biochem. Cell. Biol.* **1984,** *62,* 1006.
65. Hochman, A.; Shemesh, A. *J. Biol. Chem.* **1987,** *262,* 6871.
66. Goldberg, I.; Hochman, A. *Biochim. Biophys. Acta* **1989,** *991,* 330.
67. Hochman, A.; Goldberg, I. *Biochim. Biophys. Acta* **1991,** *1077,* 299.
68. Goldberg, I.; Hochman, A. *Arch. Biochim. Biophys.* **1989,** *268,* 124.
69. Levy, E.; Eyal, Z.; Hochman, A. *Arch. Biochim. Biophys.* **1992,** *296,* 321.
70. Obinger, C.; Regelsberger, G.; Strasser, G.; Burner, U.; Peschek, G. A. *Biochem. Biophys. Res. Commun.* **1997,** *235,* 545.
71. Regelsberger, G.; Jakopitsch, C.; Engleder, M.; Rüker, F.; Peschek, G. A.; Obinger, C. *Biochemistry* **1999,** *38,* 10480.
72. Jakopitsch, C.; Rüker, F.; Regelsberger, G.; Tockal, M.; Peschek, G.; Obinger, C. *Biol. Chem.* **1999,** *380,* in press.
73. Murthy, M. R. N.; Reid, T. J.; Sicignano, A.; Tanaka, N.; Rossmann, M. G. *J. Mol. Biol.* **1981,** *152,* 465.
74. Fita, I.; Silva, A. M.; Murthy, M. R. N.; Rossmann, M. G. *Acta Cryst. B* **1986,** *42,* 497.
75. Berthet, S.; Nykyri, L. M.; Bravo, J.; Berthet-Colominas, C.; Alzari, P.; Koller, F.; Fita, I. *Protein Sci.* **1997,** *6,* 481.
76. Mate, M. J.; Zamocky, M.; Nykyri, L. M.; Herzog, C.; Alzari, P. M.; Betzel, C.; Koller, F.; Fita, I. *J. Mol. Biol.* **1999,** *286,* 135.
77. Gouet, P.; Jouve, H. M.; Dideberg, O. *J. Mol. Biol.* **1995,** *249,* 933.
78. Murshudov, G. N.; Melik-Adamyan, W. R.; Grebenko, A. I.; Barynin, V. V.; Vagin, A. A.; Vainshtein, B.K.; Dauter, Z.; Wilson, K. *FEBS Lett.* **1992,** *312,* 127.
79. Vainshtein, B. K.; Melik-Adamyan, W. R.; Barynin, V. V.; Vagin, A. A.; Grebenko, A. I. *Nature* **1981,** *293,* 1981, 411.
80. Vainshtein, B. K.; Melik-Adamyan, W. R.; Barynin, V. V.; Vagin, A. A.; Grebenko, A. I.; Borisov, V. V.; Bartels, K. S.; Fita, I.; Rossmann, M. G. *J. Mol. Biol.* **1986,** *188,* 49.
81. Bravo, J.; Verdaguer, N.; Tormo, J.; Betzel, C.; Switala, J.; Loewen, P. C.; Fita, I. *Structure* **1995,** *3,* 491.
82. Bravo, J.; Mate, M. J.; Schneider, T.; Switala, J.; Wilson, K.; Loewen, P. C.; Fita, I. *Proteins* **1999,** *34,* 155.
83. Bravo, J.; Fita, I.; Gouet, P.; Jouve, H. M.; Melik-Adamyan, W.; Murshudov, G. N. In "Oxidative Stress and the Molecular Biology of Antioxidant Defenses"; Scandalios, J. G., ed.; pp. 407–445. Cold Spring Harbor Laboratory Press, Cold Spring Harbor, NY, 1997.
84. Sevinc, M. S.; Switala, J.; Bravo, J.; Fita, I.; Loewen, P. C. *Prot. Eng.* **1998,** *11,* 549.
85. Loewen, P. C.; Switala, J. Biochem. *Cell Biol.* **1986,** *64,* 638.
86. Switala, J.; O´Neil, J.; Loewen, P. C. *Biochemistry* **1999,** *38,* 3895.
87. Bergdoll, M.; Remy, M. H.; Cagon, C.; Masson, J. M.; Dumas, P. *Structure,* **1997,** *4,* 391.
88. Tanford, C.; Lovrien, R. *J. Am. Chem. Soc.* **1962,** *84,* 1892.
89. Ruis, H. Can. *J. Biochem.* **1979,** *57,* 1122.
90. Hook, D. W. A.; Harding, J. J. *Eur. J. Biochem.* **1997,** *247,* 380.
91. Murshudov, G. N.; Grebenko, A. I.; Barynin, V.; Dauter, Z.; Wilson, K.; Vainshtein,

B. K.; Melik-Adamyan, W.; Bravo, J.; Ferran, J. M.; Switala, J.; Loewen, P. C.; Fita, I. *J. Biol. Chem.* **1996,** *271,* 8863.
92. Loewen, P. C.; Switala, J.; von Ossowski, I.; Hillar, A.; Christie, A.; Tattrie, B.; Nicholls, P. *Biochemistry* **1993,** *32,* 10159.
93. Bravo, J.; Fita, I.; Ferrer, J. C.; Ens, W.; Hillar, A.; Switala, J.; Loewen, P. C. *Protein Sci.,* **1997,** *6,* 1016.
94. Mate, M. J.; Sevinc, M. S.; Hu, B.; Bujons, J.; Bravo, J.; Switala, J.; Ens, W.; Loewen, P. C.; Fita, I. *J. Biol. Chem.* **1999,** *274,* 27717.
95. Sevinc, M. S.; Mate, M. J.; Switala, J.; Fita, I.; Loewen; P. C. *Protein Sci.* **1999,** *8,* 490.
96. Zamocky, M.; Herzog, C.; Nykyri, L. M.; Koller, F. *FEBS Lett.* **1995,** *367,* 241.
97. Gouet, P.; Jouve, H. M.; Williams, P. A.; Andreoletti, P.; Nussaume, L.; Hadju, J. *Nature Struct. Biol.* **1996,** *3,* 951.
98. Bell, C. E.; Yeates, T. O.; Eisenberg, D. *Protein Sci.* **1997,** *6,* 2084.
99. Tanner, J. J.; Tu, S. C.; Barbour, L. J.; Barnes, C. L.; Krause, K. L. *Protein Sci.* **1999,** *8,* 1725.
100. Sevinc, S.; Ens, W.; Loewen, P. C. *Eur. J. Biochem* **1995,** *230,* 127.
101. Hillar, A. Ph.D. Thesis, 1999, University of Manitoba, Winnipeg, MB, Canada.
102. Poulos, T. L.; Kraut, J. *J. Biol. Chem.* **1980,** *255,* 8199.
103. English, A. M.; Tsaprailis, G. *Adv. Inorg. Chem.* **1996,** *43,* 79.
104. Dunford, H. B. In "Heme Peroxidases." J. Wiley & Sons: New York, 1999.

ADVANCES IN INORGANIC CHEMISTRY, VOL. 51

HORSERADISH PEROXIDASE

NIGEL C. VEITCH* and ANDREW T. SMITH†

*Jodrell Laboratory, Royal Botanic Gardens, Kew, Richmond, Surrey, TW9 3DS, UK, and
†School of Biological Sciences, University of Sussex, Falmer, Brighton, BN1 9QG, UK.

Abbreviations

APX, ascorbate peroxidase; ARP, *Arthromyces ramosus* peroxidase; BP1, barley grain peroxidase; CCP, cytochrome *c* peroxidase; CIP, *Coprinus cinereus* peroxidase; EXAFS, extended X-ray absorption fine structure; HRP, horseradish peroxidase; HRP Z (where Z = A1–A3, B1–B3, C1, C2, D, E1–E6, or N), a specific isoenzyme of horseradish peroxidase; HS, high-spin; IAA, indole-3-acetic acid; LIP, lignin peroxidase; LS, low-spin; PNP, the major cationic isoenzyme of peanut peroxidase; WT, wild-type; 5-c, five-coordinate; 6-c, six-coordinate.

0898-8838/01 $35.00

Site-directed mutants of peroxidase enzymes are represented using one-letter amino acid codes in the style X *residue number* X* *peroxidase,* which indicates that amino acid X has been substituted by amino acid X*.

I. Introduction

Few enzymes have been studied as comprehensively as horseradish peroxidase and yet continue to stimulate research across a wide range of scientific disciplines. This review describes our present state of knowledge of the enzyme, with particular emphasis on recent data obtained from both X-ray crystallography and site-directed mutagenesis. The contributions made by these techniques to our understanding of peroxidase structure and function are highlighted. Paradoxically, many questions still remain unanswered about the true *in vivo* function of horseradish peroxidase and its physiological significance, although new information has become available for peroxidases from other plant species. The enzyme from horseradish is significant commercially, with a wide variety of uses, particularly as a component of medical diagnostic kits. It seems likely that most applications will benefit from the potential to improve desirable qualities of the enzyme using both knowledge-based site-directed mutagenesis and more recent directed-evolution selection techniques. Progress in these areas will depend to some extent on the success of current studies of this important enzyme.

The source of the enzyme is the root of the horseradish, *Armoracia rusticana* P.Gaertn., B.Mey. & Scherb., a hardy perennial herb native to Europe and parts of Asia that is also cultivated in north-temperate regions of North America. This plant is a member of the *Cruciferae,* a family of considerable economic importance. Strictly speaking, horseradish peroxidase is not one enzyme, but a group of enzymes, or isoenzymes as they are more commonly referred to. The most abundant of these, the C isoenzyme (HRP C), is the subject of the majority of studies on horseradish peroxidase and also the isoenzyme used in most applications. The remaining isoenzymes are introduced in Section II. It is fortuitous that HRP C has acquired the status of a model peroxidase, probably for no other reason than its relative abundance in the plant, which assisted early attempts to purify it. The presence of peroxidase enzymes in the horseradish plant was detected as long ago as 1810 by Louis Antoine Planche, who reported that when a fresh piece of root tissue was soaked in a tincture of guaiacum (the heartwood of *Guaiacum officinale* L. and *G. Sanctum* L., which contains a number of phenolic compounds), an intense blue color appeared (*1*). The enzymes themselves were first purified by Willstätter more than a century later (*2, 3*), although material of sufficient purity for crystallization was not obtained until 1942, using a procedure designed by Theorell (*4*). Two excellent accounts have appeared that summarize the historical literature on horseradish peroxidase and describe the major developments from 1810 to 1960 (*5, 6*). These are also of interest because of the contribution made by peroxidase research to the development of ideas on biological oxidation and reduction during the early decades of the twentieth century. A monograph on peroxidase was published in 1964 by Saunders, Holmes-Siedle, and Stark, containing some 1500 references (*5*). Several reviews have appeared in the literature subsequently (*7–22*), as well as a second monograph on heme-containing peroxidases published in 1999 (*23*). It should be noted that although most of these works contain some material on horseradish peroxidase, their primary objective has been to describe peroxidases in a wider context. One exception is a review devoted solely to the structure and kinetic properties of horseradish peroxidase, published in 1991 (*24*). There are also valuable printed sources of references compiled for plant peroxidases. Two volumes have been published that cover the periods 1970–1980 and 1980–1990, with approximately 5000 references cited (*25, 26*), and a diskette of references is available that at present covers the period 1980–1998. The most recent compilations are issued in a biannual newsletter published by the International Working Group on Plant Peroxidases at the University of Geneva (*27*). This resource and the reference database can be accessed through the Internet (*28*).

II. Biochemistry and Molecular Biology

A. Isolation

The development of techniques to purify peroxidases from horseradish root has been well documented (*5, 6*). Most methods involve ammonium sulfate fractionation of the initial root extract followed by a series of column chromatography steps based on ion-exchange and size-exclusion media. One practical hazard is that volatile and intensely pungent isothiocyanates are released from the crushed root because of the hydrolytic action of the enzyme myrosinase on glucosinolates such as sinigrin (allylglucosinolate). It is perhaps fortunate that high-quality enzyme preparations have been available from commercial sources for several decades. This material is used in most investigations of the enzyme, although additional purification steps are sometimes undertaken.

A convenient measure of peroxidase purity is given by the RZ or *reinheitszahl* number, which is the ratio of the absorbances at 403 and 280 nm recorded using UV-VIS spectrophotometry. In the case of HRP C, RZ values of between 3.0 and 3.4 are optimal.

The existence of peroxidase isoenzymes in horseradish root was noted first by Theorell (*4*), and confirmed by later workers (*29, 30*). Seven isoenzymes, denoted A1, A2, A3, B, C, D, and E, were isolated and purified to homogeneity by Shannon *et al.* (*31*). The isoenzyme nomenclature reflects the variation in the value of the isoelectric point (p*I*) from acidic (A group), through neutral (B and C), to basic (D and E). Six basic E isoenzymes are now known (E1 to E6), all of which are characterized by a p*I* value of 10.6 or greater (*32*). The number of neutral isoenzymes isolated has increased to five (B1, B2, B3, C1, and C2), all with p*I* values of 5–10 (*33*). These results indicate that the horseradish root contains at least 15 peroxidase isoenzymes. Although their existence is not in dispute, the number of isoenzymes has been a matter for debate, particularly as up to 42 enzyme isoforms can be detected by isoelectric focusing (*34*). Whether these are all true gene products or in some instances artifacts of the detection method remains to be confirmed. Crosslinking with phenols and deamidation of surface residues are two modifications that could take place under some experimental conditions (*35*). Posttranslational modification of gene products by C-terminal processing may also occur, one example being the nonspecific proteolytic cleavage of HRP C to give a variant lacking one C-terminal residue (*36*). It should be stressed that there is now good evidence from the results of a systematic survey of the "model" plant *Arabidopsis thaliana* (L.) Heyhn (common name; thale cress) that a single plant species contains many true peroxidase isoenzymes (*37–40*).

An improved method for isolation of HRP C based on affinity chromatography has been developed (*41*). This exploits the differential affinity of horseradish peroxidase isoenzymes for the aromatic donor molecule benzhydroxamic acid (Section IV,F). In a single step, HRP C can be separated effectively from both the acidic A isoenzymes and other components of the root extract, a process that normally requires three or four stages of ion-exchange chromatography. A homogenous product with a RZ value of 3.25 ± 0.1 can be obtained under optimal conditions. Sepharose-bound concanavalin A has also been investigated as an affinity matrix for HRP purification (*42*), but in general is less effective. Other purification methods explored recently include the use of reversed micelles (*43, 44*) and aqueous two-phase partition (*45*). A useful comparative survey of the purity, activity, and susceptibility to inactivation of commercially available A and C isoenzymes of horseradish peroxidase has been published (*46*).

B. Sequences and Genes

The complete amino acid sequence of HRP C was determined using classical methods and published by Welinder in 1976 (*47*). There are 308 amino acid residues in a single polypeptide chain with a molecular mass of 33,890. The N-terminus is blocked by a pyroglutamate residue, and C-terminal peptides are obtained with and without the C-terminal Ser residue, as mentioned previously (*36*). Disulfide bridges occur between Cys residues 11–91, 44–49, 97–301, and 177–209. The enzyme also comprises a prosthetic group (heme) that is iron protoporphyrin IX (Section III,A), eight carbohydrate side chains (Section III,B) and two calcium ions (Section IV,D), so that the overall molecular mass is close to 44,000 (*36*). Two other HRP isoenzymes have been sequenced using similar procedures, the acidic HRP A2 (305 residues) (*15*) and the basic HRP E5 (306 residues) (*48*). The amino acid sequence identity between HRP C and these isoenzymes is 54 and 70%, respectively. HRP A1 has been sequenced partially and is reported to be approximately 80% identical to HRP A2 (*15*). In general, amino acid sequence identities among plant peroxidases range from 30 to 95% (*15, 39*).

Estimates for the number of peroxidase genes in a given plant species have increased dramatically with the knowledge that at least 28 different peroxidase genes are present in *Arabidopsis thaliana* (*37–40*). These were identified from approximately 200 peroxidase-encoding expressed sequence tags (ESTs). It is likely that this number of peroxidase genes is an underestimate of the total present. Less is known about the number of peroxidase genes in horseradish (*49*). Three cDNAs and two genomic DNAs corresponding to the mRNAs and genes for HRP C were cloned and characterized initially (*50*). The genomic clones are located in tandem on the chromosome and consist of four exons and three introns. The positions of the splice sites are identical, in agreement with a general trend observed in peroxidase genes that the positions of splice sites in mature sequences are well conserved, a feature of probable evolutionary significance (*51, 52*). The three cDNAs are remarkably homologous (91–94%) and all contain both a hydrophobic leader sequence of 30 amino acid residues and a C-terminal extension of 15 amino acid residues (*50*). The latter may be required for vacuolar targeting (*53*).

Two peroxidase-encoding genes have been found in addition to the gene family described above, and named *prxC2* and *prxC3* (*54*). The original genes designated *prxC1, prxC2,* and *prxC3* were renamed *prxC1a, prxC1b,* and *prxC1c,* respectively, owing to the high sequence homology of their cDNAs. The amino acid sequence deduced from the nucleotide sequence of *prxC1a* is identical to that of HRP C. Organ specific expression of *prxC1a, prxC1b, prxC2,* and *prxC3* in horseradish plants has been examined using Northern blot analysis (*55*). According to these studies, the family of *prxC1* genes is expressed mainly in stems (and to a lesser extent in roots), whereas *prxC2* and *prxC3* are expressed predominantly in roots. Peroxidase enzyme activity can be induced by wounding of horseradish leaves (*56*). Gene-specific probes have shown that the mRNA for *prxC2* accumulates after wounding, but not that of other peroxidases (*56*). Regulation of *prxC2* has been investigated at the genetic level in transgenic tobacco by deletion analysis of the upstream region from promoter-GUS (β-glucuronidase) gene fusions (*56*). Two *cis*-acting elements are necessary for a full response to wounding, both of which appear to recognize specific nuclear transcription factors (*56*). A cDNA encoding a *trans*-acting factor that is actively transcribed in root and stem and that can bind to one of the upstream regulatory regions of *prxC2* has been isolated (*57, 58*). The putative *trans*-acting factor protein contains a basic region, a leucine zipper, and a helix–turn–helix motif. Only one other HRP cDNA has been cloned and sequenced, corresponding to a neutral isoenzyme designated HRP N (*59*). Catalytically active recombinant HRP N has been produced in an insect-cell baculovirus system (*60*), but the plant enzyme has still to be isolated from horseradish root. Genes encoding other HRP isoenzymes have not yet been cloned, although in some cases similar counterparts have been identified in *A. thaliana* (*61, 62*). A cDNA clone corresponding to the major peroxidase of an *A. thaliana* cell suspension culture has been isolated from an *Arabidopsis* cell suspension cDNA library and sequenced (*62*). The protein encoded, ATP A2, consists of 305 amino acid residues and shows 95% amino acid sequence identity with HRP A2. The expression of the *atpA2* gene in the plant appears to be developmentally regulated and is specific to the roots (*62*).

C. Expression

The successful expression of recombinant plant peroxidases such as HRP C has been a major focus of research in a number of laboratories. Three synthetic HRP C genes based on the amino acid sequence determined by Welinder (*36, 47*) were synthesized independently in order to initiate this work (*63–65*). A number of different expression systems have been evaluated (*64, 66–73*), a summary of which is presented in Table I. Refolding of recombinant HRP C isolated from inclusion

TABLE I

SUMMARY OF EXPRESSION SYSTEMS USED TO PRODUCE RECOMBINANT HORSERADISH PEROXIDASE

Host organism	Construct and gene	Expression level	Comments and references
Aspergillus oryzae	Various fusions between the synthetic HRP C gene and CIP at both N- and C-termini.	Low	Recombinant HRP C detected on Western blots (*66*).
Baculovirus	Synthetic HRP C gene with the signal sequence of the plant enzyme.	Medium	Active glycosylated enzyme produced; yield 5–10 mg liter^{-1} (*67*).
Escherichia coli	Synthetic HRP C gene under control of the *tac* promoter.	High	*In vitro* refolding required to obtain active enzyme; yield 2–4 mg liter^{-1} (*64, 68*). Higher yields reported (*69, 70*).
Saccharomyces cerevisiae	Synthetic HRP C gene fused in-frame to pre-pro α-mating factor. Constitutive and inducible promoters tried.	Low	Low levels of hyperglycosylated enzyme secreted; yield approx. 50 μg liter^{-1} (*71, 72*).
Transgenic tobacco (*Nicotiana tabacum*)	Synthetic HRP C sequence with the signal sequence of the plant enzyme and both with and without the C-terminal extension.	Low/medium	Some active enzyme produced (*73*).

bodies of *Escherichia coli* has proved popular as a relatively low-cost method (*64, 74, 75*), which, however, is rather inefficient. The procedure requires the controlled reoxidation of reduced denatured HRP C solubilized from the inclusion bodies isolated from *E. coli*. Refolding exhibits a dependence on both Ca^{2+} and urea concentration; some of the parameters have been investigated in detail (*64*), and several protocols exist (*64, 68–70*). Highly purified recombinant HRP C (which is nonglycosylated) has been crystallized successfully (*76*) and provides good-quality diffraction data in contrast to that obtained previously from crystals of the plant enzyme (*77*). The baculovirus system leads to the production of active but highly glycosylated recombinant HRP C (*67*). Nevertheless, both expression systems have been used for production of site-directed mutants. In a recent development, directed evolution techniques have been used to identify a mutant of HRP C that exhibits some activity in

E. coli without the need for refolding (*78*). The mutant corresponds to the substitution of Asn255 by Asp at one of the *N*-linked glycosylation sites of the enzyme.

D. Relationship to Other Peroxidases

Peroxidases are distributed widely among living organisms, and many examples have been isolated and characterized. Although most incorporate a prosthetic group based on iron protoporphyrin IX (the heme-containing peroxidases), additional classes of peroxidase containing vanadium (*79, 80*) and selenium (*81*) have also been identified. One further group, known as peroxiredoxins, require no cofactors and instead generate peroxidaselike activity by means of an active site cysteine residue (*82*). Two relatively well-defined enzyme groups have been recognized among the heme-containing peroxidases, the mammalian peroxidase superfamily and the plant peroxidase superfamily. Important structural and functional differences between these two superfamilies have been reviewed previously (*19*). It should be noted that these superfamilies do not include either the di-heme cytochrome *c* peroxidases (*83*) or chloroperoxidase, an enzyme with some attributes characteristic of cytochrome P450 (*84*). The plant peroxidase superfamily comprises enzymes of bacterial, fungal, and plant origin and is divided into three classes. This arrangement was primarily the result of detailed comparisons of amino acid sequence data (*51*), but gained additional support as crystal structure data became available for representative enzymes from each class. A structure-based alignment of peroxidase amino acid sequences is shown in Fig. 1. These sequences

→

Fig. 1. A structure-based alignment of peroxidase amino acid sequences. The sequences included represent all members of the plant peroxidase superfamily for which crystal structures are available, with the exception of manganese peroxidase. Coordinate files were obtained from the Protein Data Bank at Brookhaven for cytochrome *c* peroxidase (CCP) (*85*), ascorbate peroxidase (APX) (*86*), *Arthromyces ramosus* peroxidase (ARP/CIP) (*87*), lignin peroxidase (LIP) (*88*), peanut peroxidase (PNP) (*89*), barley peroxidase (BP1) (*90*), and HRP C (*91*). The alignment was made using the programmes STAMP (*92*), AMPS (*93*), and ALSCRIPT (*94*). Key: Uppercase bold boxed regions, Cα coordinates superimpose with less than 2.5 Å rms deviation; shaded residues, those important in either the distal or proximal heme pocket; reverse font, cysteine residues; HRPC 2D line, secondary structure; "&," *N*-linked glycan attachment site; "+," calcium ligand; numbers in 2D line, disulfide pairings. Note that the amino acid sequence numbering for each peroxidase should be read from the far right-hand column. This structural alignment was kindly provided by David J. Schuller, University of California—Irvine.

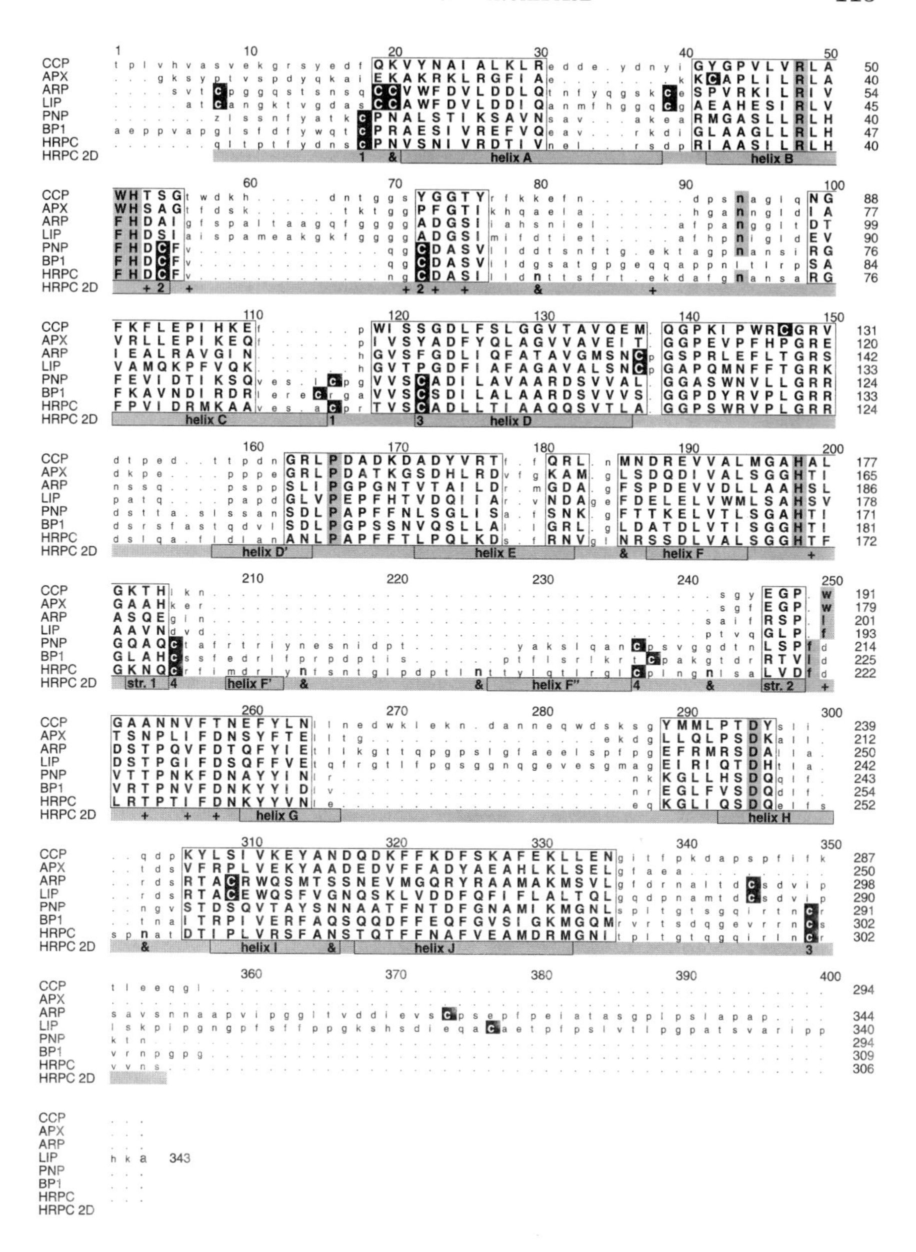

represent peroxidases of the plant peroxidase superfamily for which crystal structures have been determined (*85–91*), although manganese peroxidase has been omitted from this analysis. The overall conclusion is that structural topology is conserved more highly among the three classes than amino acid sequence identity. Yeast cytochrome *c* peroxidase, cytosol, and chloroplast ascorbate peroxidases (APX), and gene-duplicated bacterial peroxidases comprise class I, while class II contains fungal peroxidases, the most well-known examples being the inkcap peroxidase from *Coprinus cinereus* (CIP; this enzyme is essentially identical to that known as *Arthromyces ramosus* peroxidase or ARP), and lignin and manganese peroxidases from *Phanerochaete chrysosporium*. Peroxidases from plants are placed in class III ("classical" secretory plant peroxidases) or class I (ascorbate peroxidases only). One general point of difference is that class I peroxidases do not contain disulfide linkages, unlike peroxidases of classes II and III. Nevertheless, much can be learned about the role of individual amino acid residues in peroxidases of one class if careful comparison is made with the corresponding residues in the others. This approach is useful in structure:function studies of HRP C, as complementary data are often available for the enzymes CCP (class I) and CIP (class II).

III. General Features of the Enzyme

A. Heme Prosthetic Group

All HRP isoenzymes contain iron(III) protoporphyrin IX (ferriprotoporphyrin IX), the structure of which is shown in Fig. 2. The nomenclature of this prosthetic group is somewhat confusing, as two different schemes are still in current use, one introduced originally by Fischer, and the second by the IUB/IUPAC. It is the latter system that is used throughout this review. It is customary to refer to iron protoporphyrin IX simply as the "heme group." In HRP there are two axial coordination sites (the fifth and sixth positions) where ligand binding can occur. A coordinate bond between the iron atom and the Nϵ2 atom of His170 at the fifth (or proximal) coordination site links the prosthetic group to the apoenzyme. There is no additional crosslinking between the side chains of the heme group and the apoenzyme as observed for some mammalian peroxidases (*95*, and references therein).

The resting state of HRP C is high-spin Fe(III) (d^5) with a magnetic susceptibility of 5.23 Bohr magnetons (*96*). However, it cannot be described adequately by the spin quantum number $S = \frac{5}{2}$, which would correspond to a high-spin species with five unpaired d electrons and a

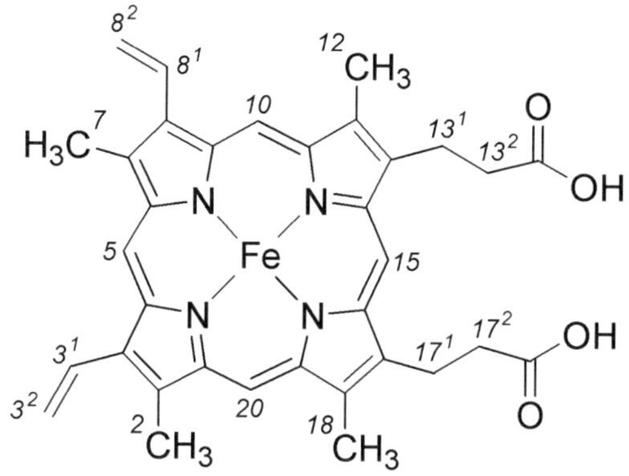

FIG. 2. Nomenclature of the heme group of horseradish peroxidase according to the IUB/IUPAC system. Proton-bearing carbon atoms are numbered on the structure and can be cross-referenced with the Fischer system as follows:

IUB/IUPAC	*2*	*3*	*5*	*7*	*8*	*10*	*12*	*13*	*15*	*17*	*18*	*20*
Fischer	*1*	*2*	α	*3*	*4*	β	*5*	*6*	γ	*7*	*8*	δ

magnetic susceptibility of 5.92 Bohr magnetons. Analysis of experimental data, including recent results from resonance Raman (*97, 98*) and ^{1}H NMR spectroscopy (*99*), favors a quantum mechanically mixed spin state arising from an admixture of high-spin ($S = \frac{5}{2}$) and intermediate-spin ($S = \frac{3}{2}$) states. This quantum-mixed spin state has also been observed spectroscopically as a component of the resting states of other class III plant peroxidases such as that from barley grain (BP1) (*100*), and in low-temperature studies of HRP A2 and C (*97, 101*). The dominant species in both solution and solid states of HRP C is five-coordinate high-spin (5-c HS), although a small amount of a 6-c HS species can be detected using resonance Raman spectroscopy. It is reasonable to maintain, however, that the resting state is essentially five-coordinate, and that the sixth (or distal) coordination site is vacant. The frequencies of the core size marker bands in resonance Raman spectra of resting state HRP C are anomalous when compared to those of other heme peroxidases (*102, 103*), and may reflect a characteristic heme distortion.

The coordination- and spin state of HRP C are dependent on pH, although little change occurs between pH 4 and pH 10. At pH 3.1, the resting state enzyme is a 5-c HS form in which His170 has been replaced as ligand by a water molecule (*104*). The scission of the Fe–His bond occurs from a 6-c intermediate with a water molecule bound at the distal site. At high pH values, a transition to an alkaline form occurs with a pK_a value of approximately 11.0 (*4, 105–107*). The controversy surrounding its identity was resolved using resonance Raman spectroscopy and isotopic labeling (*106, 107*). This revealed a 6-c LS species with hydroxide bound to heme iron at the vacant distal site. The R38G, R38L, and R38K mutants of HRP C do not show an alkaline transition, presumably because of the loss of the stabilization afforded by a critical hydrogen bond between the positively charged Arg38 side chain and the hydroxide ligand (*103*). A second important interaction is hydrogen bond donation to the side chain of distal His42. The pH dependence of two His42 mutants of the enzyme has been documented, including the interesting observation of an Fe–OH bond formed at neutral pH by H42R HRP C (*103*).

The resting states of other HRP isoenzymes have not been characterized in great detail, although some information is available for A1 and A2 (*98, 99*). These acidic isoenzymes differ from HRP C with respect to the degree of quantum mechanical mixing of high- and intermediate-spin states, which increases in the order HRP C < HRP A1 < HRP A2 (*99*). A detailed analysis of the resonance Raman spectra of resting state HRP A2 has been presented (*98*). The alkaline transition of acidic HRP isoenzymes occurs at lower pH than for the neutral isoenzymes, according to results obtained in deuterated solutions using ^{1}H NMR spectroscopy (*108*). This method gave pD values of 9.3 ± 0.1, 9.4 ± 0.1, and 9.8 ± 0.1 for HRP A1, A2, and A3, respectively, compared to 10.9 for both HRP B and C.

B. Carbohydrate

The recognition that carbohydrate residues are an integral part of horseradish peroxidase has its origin in the studies of Theorell and Åkesson on protein composition by elemental analysis (*109*). They determined that the enzyme (HRP C) comprised 18.4% carbohydrate. Once the amino acid sequence of the enzyme became available (*36, 47*), nine potential N-glycosylation sites represented by the motif Asn-X-Ser/Thr (where X can be any amino acid residue) were identified. Eight of these Asn residues have carbohydrate chains attached (the exception being the site at residues *286–288*) (*36*). The major *N*-linked glycan is the branched heptasaccharide Manα3(Manα6)(Xylβ2)Manβ4GlcNAcβ4 (Fucα3)GlcNAc (*110, 111*), which accounts for 75 to 80% of the total carbohydrate. A number of minor glycan structures have also been determined (*112, 113*), principally of the type $(\mathrm{Xyl})_x\mathrm{Man}_m(\mathrm{Fuc})_f\mathrm{GlcNAc}_2$ ($m = 2, 4, 5, 6$; $f = 0$ or 1; $x = 0$ or 1) or less commonly, $\mathrm{Man}_m\mathrm{GlcNAc}_2$

($m = 4$ to 7). The carbohydrate profile of HRP C is heterogeneous, to the extent that the type of glycan structure found at any of the eight attachment sites can vary. In addition, the total carbohydrate depends on the source of the enzyme, although percentage composition remains within the range 18 to 22% (*36, 109, 114, 115*). The high carbohydrate content and heterogeneity of HRP C are no doubt responsible for the lack of success in obtaining good-quality X-ray diffraction data for crystals of the plant enzyme (*77*). Attempts to remove the glycans by enzymatic hydrolysis have proved to be unsuccessful (*116*). Mild chemical deglycosylation using anhydrous trifluoromethanesulfonic acid in the presence of 90 mM phenol for 5 min at −5°C removes all carbohydrate except for the two GlcNAc residues at each site (*116*). Fully active, homogeneous deglycosylated enzyme can then be recovered after ion-exchange chromatography, but shows greatly reduced solubility compared to native HRP C. The glycans of HRP C are located in loop regions on the surface of the molecule (*91*) and are relatively mobile, according to ^{1}H NMR data obtained in a comparison of the wild-type plant and recombinant enzymes (the recombinant enzyme expressed in *E. coli* is nonglycosylated) (*117*).

HRP A2 has seven potential N-glycosylation sites and a carbohydrate content of approximately 20%. The carbohydrate profile is heterogeneous and the major glycan is identical to that of HRP C (*118*). An anionic peroxidase of molecular mass 55,000 isolated from cell cultures of horseradish was found to have 11 potential N-glycosylation sites and a carbohydrate content of 26% (*119*). The corresponding glycan structures are less highly processed analogs of the major glycan from HRP A2 and contain an additional GlcNAc residue. In contrast, the basic peroxidase isoenzymes of horseradish root contain significantly less carbohydrate than the acidic and neutral isoenzymes, particularly HRP E3 to E6 (*32*). Low-resolution X-ray diffraction data have been obtained for HRP E5, which has only two N-linked carbohydrate attachment sites (*48, 120*). These are located at Asn188 and Asn214 and are analogous to those of HRP C at Asn186 and Asn214, respectively. The glycosylation of plant peroxidases has been reviewed (*118*), although comparative data on glycan structures are only available for enzymes from barley (*53*), peanut (*121*), and soybean (*122*).

C. Function and Biological Roles

The majority of reactions catalyzed by horseradish peroxidase can be summarized by Eq. (1), in which AH_2 represents a reducing substrate, and AH$^{\cdot}$a free radical product:

$$H_2O_2 + 2AH_2 \longrightarrow 2AH^{\cdot} + 2H_2O \quad (1)$$

Consideration of this equation indicates that two major functions can be assigned to the enzyme: either to catalyze the conversion of hydrogen peroxide to water, or to catalyze the oxidation of a substrate molecule to a free radical product. There is little doubt that protection from peroxide is one role that can be carried out by horseradish peroxidase, but it should be noted that plants also contain ascorbate peroxidases (class I) that perform the same function. It is more difficult to state with certainty what other biological roles can be definitively associated with HRP, as most have been demonstrated only *in vitro*. Neither is it known as yet whether each isoenzyme has a specific role or roles, although this seems plausible. The following comments apply to the enzyme in general, unless reference is made to specific isoenzymes.

Several examples of crosslinking reactions catalyzed by HRP are known, such as the formation of diferulate (*123*) and dityrosine linkages (*124, 125*) from the corresponding monomers. The crosslinking of tyrosine-containing peptides has also been reported (*126*). The matrix of the plant cell wall contains structural polymers such as polysaccharides and pectins to which ferulic acid monomers are attached as acylating groups (*127*). Crosslinking reactions can therefore result in tightening of the cell wall. The alteration of the physical properties of cell walls in response to external stimuli is one process that could be mediated by peroxidases. The oxidative coupling of phenolic monomers such as *p*-coumaryl, coniferyl, and sinapyl alcohols can also be catalyzed by HRP (*128*). These reactions are essential in the biosynthesis of lignin, a complex and highly branched phenylpropanoid polymer that constitutes 20–30% of plant cell walls (*128*). Crosslinking of phenolic monomers is also important for the formation of suberin, a polymeric lignin-like material consisting of phenylpropanoid, esterified fatty acid, and alcohol components. The process of suberization is initiated when plant tissue is wounded, and provides in suberin a natural barrier that prevents moisture loss and affords some protection against subsequent pathogenic invasion. One element that is often overlooked in these important peroxidase-catalyzed reactions is the origin of the hydrogen peroxide. Experiments with the vascular tissue of spinach (*Spinacea oleracea*) hypocotyls indicate that one possible mechanism for H_2O_2 generation involves both a NAD(P)H oxidase and a CuZn superoxide dismutase, as illustrated in Eqs. (2) and (3) (*129*). A scheme similar to this was proposed previously based on results obtained with isolated cell walls of horseradish root (*130, 131*).

$$2O_2 + NAD(P)H \longrightarrow 2O_2^- + NAD(P)^+ \tag{2}$$

$$2O_2^- + 2H^+ \longrightarrow H_2O_2 + O_2 \tag{3}$$

Expression of peroxidases that catalyze crosslinking reactions can be induced by wounding or infection. Well-documented examples include the production of acidic peroxidase isoenzymes in wounded tissue of tobacco (*132*), potato (*133*), and tomato (*134*), and also in wheat plants infected with the powdery mildew fungus, *Erysiphe graminis* (*135*). It is known that peroxidase activity can be induced on wounding of horseradish leaves (*56*), but other mechanisms of defense in horseradish remain to be demonstrated.

The ability of HRP to degrade the plant hormone indole-3-acetic acid (IAA) in the absence of hydrogen peroxide was noted as early as 1955 (*136*). Plant peroxidases are now known to be of major importance in the metabolism of IAA (*137*) (note that they are often referred to as indole acetic acid oxidases in the older literature). The mechanism of IAA oxidation by HRP C is complex and has been studied experimentally in great detail by several groups (*23, 137*). Reaction products include indole-3-methanol, indole-3-aldehyde, and 3-methylene oxindole, which is probably a nonenzymatic conversion product of indole-3-methylhydroperoxide. The most important developments in this area have been reviewed (*23*).

D. Reactivity

The catalytic cycle of HRP can be represented by the sequence of reactions shown in Eqs. (4)–(7). This applies to most, but not all, reactions catalyzed by the enzyme, irrespective of the specific isoenzyme involved. Exceptions include the oxidations of iodide (*138*) and bisulfite ions (*139, 140*), both of which can involve a single two-electron reduction step.

$$E(Fe^{III}, R) + H_2O_2 \xrightarrow{k_1} \underset{\text{compound I}}{E(Fe^{IV}{=}O, R^{\bullet +})} + H_2O \tag{4}$$

$$E(Fe^{IV}{=}O, R^{\bullet +}) + AH_2 \xrightarrow{k_2} \underset{\text{compound II}}{E(Fe^{IV}{=}O, R)} + AH^{\bullet} \tag{5}$$

$$E(Fe^{IV}{=}O, R) + AH_2 \xrightarrow{k_3} E(Fe^{III}, R) + AH^{\bullet} + H_2O \tag{6}$$

$$2AH^{\bullet} \longrightarrow A_2H_2 \tag{7}$$

In this scheme, E, R, AH_2, and $AH^{\bullet}$ represent the enzyme, site of cation radical, reducing substrate, and the one-electron oxidation product of AH_2, respectively. k_1, k_2, and k_3 are the rate constants for compound I formation, compound I reduction, and compound II reduction, respectively. The first step shown in Eq. (4) involves reaction of the Fe(III) resting state of the enzyme with hydrogen peroxide to yield a high oxidation state intermediate known as compound I. In formal terms this is two oxidation state equivalents above the Fe(III) resting state. Comprehensive spectroscopic studies have shown that compound I is an oxoferryl species with a π-cation radical located on the heme group (reviewed in *24*). A short Fe–O bond length of 1.64 Å measured by extended X-ray absorption fine structure spectroscopy (EXAFS) is consistent with the assignment of compound I to an oxoferryl species (*141, 142*). One electron reduction of compound I gives a second high oxidation state intermediate known as compound II. This is also a low-spin ($S = 1$) oxoferryl species, but only one oxidation state equivalent above the Fe(III) resting state. The catalytic activity of compound II is linked to the ionization of a heme pocket group with a pK_a value of 8.6 (*143*). Above pH 8.6 the activity decreases markedly. The active protonated form of this titrating group is hydrogen bonded to the oxygen atom of the Fe(IV)=O center (*144–146*). This hydrogen-bonded interaction may explain the apparent pH dependence of the Fe–O bond length determined from EXAFS data, where a longer bond length is observed at neutral (1.90 ± 0.02 Å at pH 7) than at alkaline pH (1.72 ± 0.02 Å at pH 10) (*147*). The identity of the crucial heme pocket group remains unknown, although the distal histidine residue (His42) has been suggested (*24*). Mechanisms of compound I and II formation are discussed in Section IV,B.

HRP C is able to oxidize a wide variety of substrates, but the most common classes of compound are aromatic amines, indoles, phenols, and sulfonates. The final products of aromatic amine and phenol oxidation are often oligomers and polymers (*5*), as the primary products (for example, dimers) can also act as peroxidase substrates, thereby producing additional radical species that are able to participate in further coupling reactions (*148*). The redox potentials of the compound I/compound II and compound II/resting-state couples of HRP A2 and C have been estimated to be of the order of +1 V with reference to the standard hydrogen electrode. The pH dependence of these redox couples is complex, and there are also subtle differences between the behavior of the two isoenzymes (*149*). HRP isoenzymes are not able to oxidize every type of aromatic compound, one example being a series of substituted methoxybenzenes studied by Kersten *et al.* (*150*). These are, however, good substrates for lignin peroxidase, an enzyme that appears to have greater oxidizing power than HRP (*150, 151*). The structural reasons for the superior oxidizing power of LIP are not fully understood at present.

The site of substrate oxidation by HRP C has been explored through enzyme inactivation studies, principally with cyclopropanol hydrate (*152*), nitromethane (*153*), alkylhydrazines (*154*), phenylhydrazine (*155*), and sodium azide (*156*). The free radical species so generated react with the heme group to give adducts that can be identified; for example, incubation of HRP C with phenylhydrazine and H_2O_2 gives both C18-hydroxymethyl and C20-meso-phenyl heme derivatives (*155*). The conclusion is that substrates interact with an "exposed" heme edge comprising the C18 methyl and C20 meso protons (Fig. 2). This contrasts with results from comparable studies on hemoglobin, myoglobin, catalase, and cytochrome P450, where alkyl- and arylhydrazines modify either the iron atom or the nitrogen atoms of the heme group (*157–160*). The oxoferryl center in HRP C appears to be inaccessible to most substrates due to the presence of a so-called "protein-imposed barrier."

A number of attempts have been made to rationalize the rates of reaction of HRP C substrates with compounds I and II, including the use of Hammett correlations (*161, 162*) and frontier orbital methods (*163*). Assuming that a common heme-edge electron transfer site is accessed, the rates of reaction of phenol or indole-3-acetic acid derivatives with compound I correlate well with the reduction potentials of their cation radical products (*164*). Second-order rate constants for reduction of compounds I and II by phenols have been interpreted in the context of the Marcus theory of electron transfer (*165*). In simple terms, the lower reactivity of compound II can be ascribed to the longer electron transfer distance between the iron atom of the heme group and the substrate in this intermediate, compared to the corresponding distance from the porphyrin π-cation radical to the substrate in compound I. Similar trends have been reported for the oxidation of substituted ferrocene derivatives by compound II of HRP C (*166*). Those derivatives with bulkier cyclopentadienyl ring substituents showed lower reactivity. In these examples the effective electron transfer distance between the substrate and the iron atom of the heme group is increased as a result of steric hindrance, thus limiting access to the substrate interaction site. Substrate ionization potentials for a series of phenol derivatives have been calculated using computational methods (*167*). These were used to analyze the rates of phenol oxidation by HRP C in terms of thermodynamic driving force, and to investigate possible structure–activity relationships.

IV. Structure and Function

A. Crystal Structures and Modeling

Early attempts to crystallize plant HRP C for X-ray diffraction experiments were made by Braithwaite (*77*). The crystals obtained were tetragonal and contained approximately 32 molecules in the unit cell. Detailed analysis of the diffraction data was not pursued owing to its complexity. The inability to obtain good crystals can be attributed in part to the microheterogeneity of preparations of the plant enzyme, some of which have been analyzed by electrospray mass spectroscopy and found to contain up to 21 components, mainly consisting of glycoforms (*168*). The first X-ray crystal structure of a heme-containing peroxidase (cytochrome *c* peroxidase) was solved in 1980 (*169*), and a high-resolution refinement appeared subsequently (*85*), allowing the first comparisons of amino acid sequence and structure to be made between this enzyme and HRP C (*170*). Despite an amino acid sequence identity of only 20%, the similarity between the hydrophobicity profiles and heme-linked regions suggested that both enzymes shared a similar fold (*170, 171*). This prompted low-resolution modeling of the core α-helices and deduction of some of the helical connectivities of HRP C (*170*). Cytochrome *c* peroxidase contains no structural Ca sites, in contrast to class II fungal and class III plant peroxidases. Thus, solution of the crystal structures of the class II lignin (*88, 172*) and *Coprinus cinereus* peroxidases (*87, 173*) allowed more detailed homology modeling studies of HRP C to be undertaken. The first attempt failed to include Ca binding sites in the model (*174*), although this was later corrected (*175*). A different approach was taken by Loew and colleagues, who used structurally derived sequence alignments to identify 10 structurally conserved regions. Coordinates for these were used to construct a core peroxidase fold onto which a full protein model of HRP C was built (*176, 177*). The two Ca sites and heme-linked regions were reasonably well predicted, but a large untemplated insertion containing two additional α-helices unique to plant peroxidases was not. The publication of the first crystal structure of a class III plant peroxidase (from peanut) (*89*) led to a molecular replacement solution for recombinant HRP C (*91*), for which both crystals and heavy atom derivatives had been obtained previously (*76*). This structure has been solved to 2.15 Å resolution, and is illustrated in Fig. 3 (see color insert).

The three-dimensional structure of HRP C is characterized by 10 prominent α-helices denoted A–J. These occupy similar three-dimensional positions in all the peroxidase structures described to date for

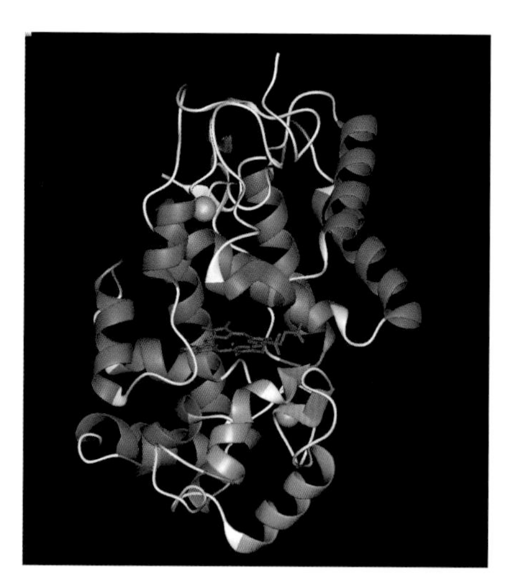

FIG. 3. Schematic representation of the X-ray crystal structure of HRP C (Brookhaven accession code 1ATJ) (*91*). Helical regions are shown in purple or blue, and β-sheet regions in yellow. The F′ and F″ α-helices, which are unique to class III plant peroxidases, are highlighted in blue. The heme group, iron(III) protoporphyrin IX, shown here in red, is sandwiched between the distal (upper) and proximal (lower) domains of the enzyme. The two structural calcium atoms are shown as blue van der Waals spheres (distal and proximal to the heme group).

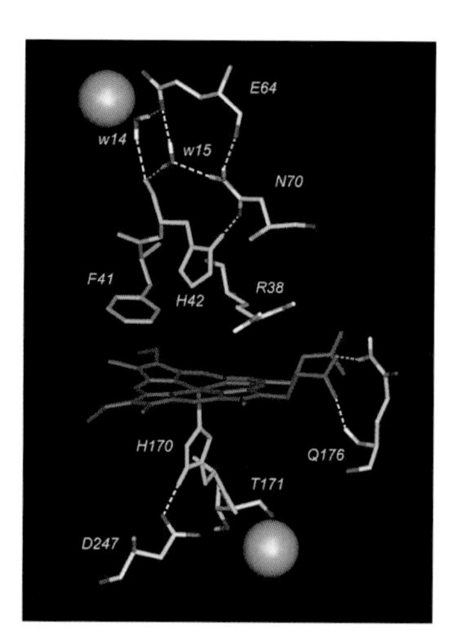

FIG. 4. Heme-linked catalytic residues and hydrogen bonding networks in HRP C (*91*). The heme group (shown in red) is coordinated to the proximal histidine residue, His170. The sixth coordination site distal to the heme group is vacant. Note the link between the distal histidine residue, His42, and the distal calcium binding site, via Asn70, Glu64, and two structural water molecules (wat14 is one of the ligands to calcium; see Fig. 7A). The residue Thr171 provides a link between His170 and the proximal calcium binding site. In addition, the proximal heme region is coupled to the heme group by hydrogen bonds between Gln176 and heme propionate C17.

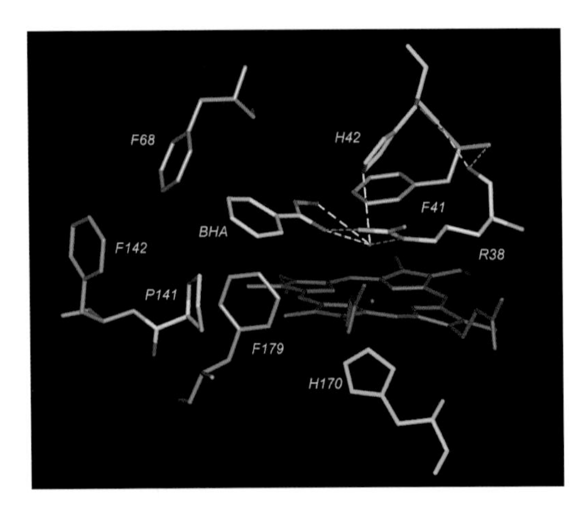

FIG. 9. Binding mode of benzhydroxamic acid (BHA) in resting state HRP C, from the X-ray crystal structure of the complex (Brookhaven accession code 2ATJ) (*196*). The side chains of three Phe residues (Phe68, Phe142, and Phe179; colored blue) in the substrate access channel contribute to aromatic donor molecule binding. Phe41 (above the heme group) is responsible in part for the closed heme architecture of the enzyme, and restricts access to the oxoferryl heme iron of compound I. The orientation of the Phe68 side chain alters on complex formation with benzhydroxamic acid; here it is seen in the so-called "lid-closed" conformation. Key hydrogen bonds between the hydroxamic acid side chain and distal residues (including a structural water molecule) are shown by dashed lines.

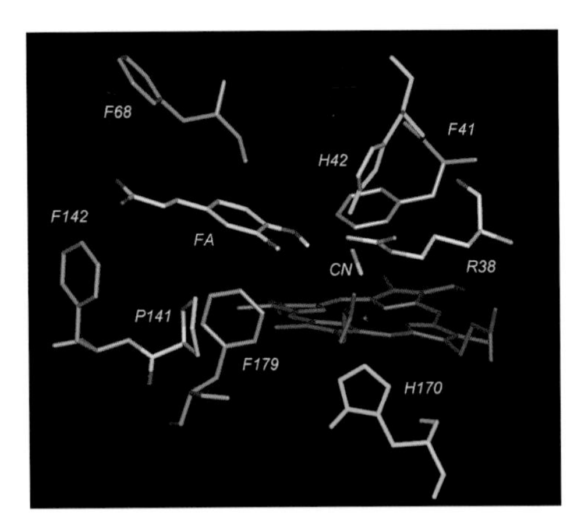

FIG. 10. Structural detail from the X-ray crystal structure of the ternary complex between ferulic acid (3-methoxy-4-hydroxycinnamic acid) and cyanide-ligated HRP C (Brookhaven accession code 7ATJ) (*195*). This view of the binding site, and the residues illustrated, are identical to Fig. 9. Ferulic acid (FA) and the cyanide ligand (CN) are shown in atom colors. Note that hydrogen bonds have been omitted from this view for clarity, although the Nε2H proton of distal His42 has been added (this is hydrogen bonded to the N atom of the cyanide ligand, which binds to the enzyme only as the protonated form). A general mechanism for substrate oxidation by HRP C has been proposed based on information from this structure (Fig. 6; Sections IV,B,2 and IV,F,4) (*195*).

members of the plant peroxidase superfamily. There are also 13 structurally conserved regions that have a root mean square deviation of 2.5 Å with respect to the Cα backbone atom positions of the other known structures. A pairwise comparison of six representative peroxidase structures shows that ascorbate peroxidase (class I) has the most average core fold (*91*). HRP C, in common with peanut peroxidase, contains three helices termed D′, F′, and F″ that are additional to the core peroxidase fold and not found within classes I and II of the superfamily. HRP C contains a long insertion between helices F and G in which the F′ and F″ helices are situated. The local structure in this region is maintained overall by a disulfide bridge between Cys177 and Cys209. The insertion is hypervariable among class III peroxidases, displaying considerable variation in both number of residues and amino acid sequence (*15*). It is still unclear how this additional structural feature relates to the function of this class of peroxidases. What is known, however, is that helix F′ provides additional contacts to the substrate access channel, some of which modulate the binding of aromatic donor molecules (*178*). As many reactions thought to be catalyzed by plant peroxidases *in vivo* involve the formation of radicals, it is tempting to speculate that this region of the structure is involved in their retention or stabilization (*91, 179*). Other aspects of the structure are examined in detail in the following sections.

B. Distal Region

1. Formation of Compound I

The distal region of HRP C (considered to be that region of the enzyme lying above the plane of the heme when viewed in the usual orientation) contains a number of amino acid residues with an essential role in catalysis. Two of the most important are the distal arginine, Arg38, and the distal histidine, His42, as highlighted in Fig. 4 (see color insert). It was Poulos and Kraut who first proposed a mechanism for the formation of compound I where the functions of these residues were defined (*180*). Their ideas developed from a detailed analysis of the three-dimensional structure of yeast cytochrome *c* peroxidase (*169*). Two critical premises are that distal histidine accepts a proton from hydrogen peroxide, and distal arginine helps to stabilize a charged intermediate (Fig. 5). It has always been supposed that a similar mechanism should operate in HRP C, given that the distal His and Arg residues are conserved in all members of the plant peroxidase superfamily (*24*). Establishing the mechanism of compound I formation in HRP C has been a major focus of

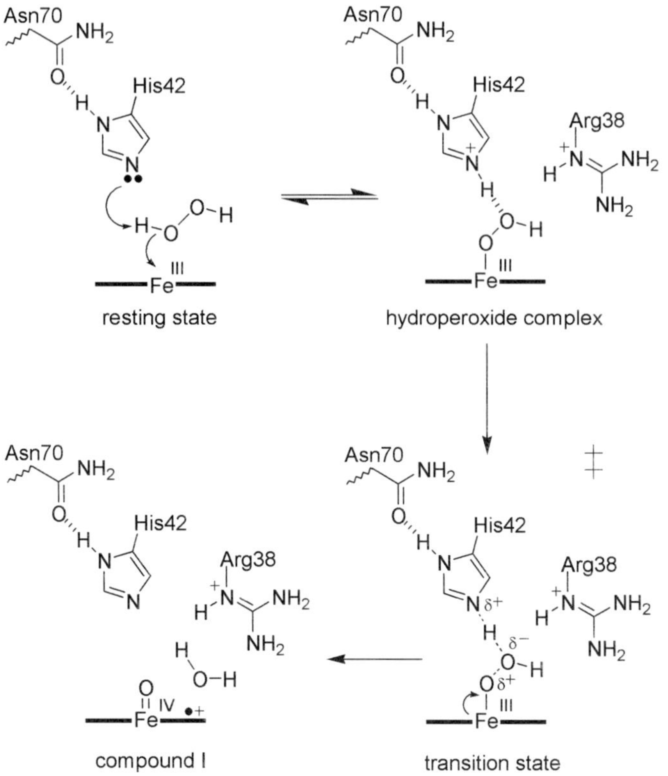

FIG. 5. The role of distal heme pocket residues of HRP C in compound I formation. A mechanism based on the original proposals of Poulos and Kraut (*180*), together with results from site-directed mutagenesis studies at Arg38, His42 and Asn70 (Section IV,B). The proximal heme ligand (His170) is omitted for clarity, but is shown in Fig. 4. There is some evidence to suggest that a neutral Fe(III)–HOOH complex may be formed prior to the Fe(III)–OOH complex shown in the scheme (*192, 194*). Note that the α-proton of H_2O_2 is abstracted by His42, and the positively charged side chain of Arg38 assists in stabilizing the transition state (‡).

site-directed mutagenesis studies. The relative contributions of heme pocket residues to the rate of compound I formation can be assessed from the data summarized in Table II.

The rate of compound I formation in HRP C is decreased by five orders of magnitude when His42 is substituted by Leu (*181*), and by six orders of magnitude when His42 is substituted by Ala (*182*). The catalytic activity can be recovered in part by addition of 2-substituted imidazoles (*182*), or by provision of a "surrogate" His residue, as with the H42A:F41H HRP C double mutant (*183*). These results support the

TABLE II

SUMMARY OF RATE CONSTANTS ASSOCIATED WITH THE FORMATION OF COMPOUND I IN WILD-TYPE HRP C AND SELECTED SITE-DIRECTED MUTANTS[a], ACCORDING TO THE FOLLOWING SCHEME:

$$E + H_2O_2 \underset{k_{-1}}{\overset{k_1}{\rightleftharpoons}} [E\text{–}(H)OOH] \xrightarrow{k_c} \text{compound I} + H_2O$$

HRP C	k_1(M^{-1} s^{-1})	k_{-1} (s^{-1})	k_c (s^{-1})	Refs.
WT	1.7×10^7	$\approx 3.0 \times 10^{-1}$	$\approx 1.0 \times 10^3$	*181*
R38A	8.0×10^4			*190*
R38H	1.3×10^6			*190*
R38K	4.0×10^4	$\approx 2.0 \times 10^{-1}$	$>5.0 \times 10^2$	*191*
R38L	1.1×10^4		$\approx 1.4 \times 10^2$	*181*
F41V	2.1×10^6			*68*
H42A	1.9×10^1			*182*
H42A:F41H	3.0×10^4			*183*
H42A:N70D	3.7×10^1			*186*
H42D	1.3×10^3			*b*
H42E	4.9×10^3			*184*
H42E:R38S	3.2×10^1			*b*
H42L	1.4×10^2	$\approx 8.0 \times 10^{-2}$	$\approx 7.0 \times 10^{-2}$	*181*
H42Q	9.6×10^1			*185*
H42V:R38H	1.6×10^3			*190*
E64G	4.3×10^5			*211*
E64P	4.4×10^5			*211*
E64S	4.3×10^5			*211*
N70D	1.5×10^6			*187*
N70V	6.0×10^5			*75*
H170A	1.6×10^1			*199*
F172Y	1.6×10^7			*203*
F221W	1.5×10^7			*204*

[a] The location of the amino acid residues is shown in Fig. 4. Single letter amino acid codes are used for the mutants.
[b] S. P. Jennings and A. T. Smith, unpublished data.

hypothesis that His42 functions by accepting a proton from hydrogen peroxide, although a precise structural location for this site relative to heme iron is not essential to achieve an intermediate rate of compound I formation. An interesting series of studies considers the effect of replacing His42 by Glu to give an enzyme that mimics in part the distal pocket of chloroperoxidase (*184, 185*). An intermediate rate of compound I formation can still be achieved with either H42E or H42D HRP C, suggesting that these residues act as alternative proton acceptors but are only weakly basic. However, introduction of a weakly basic residue at Asn70 to give H42A:N70D HRP C does not improve the rate of compound I formation in comparison to H42A HRP C (*186*). What is also fascinating is that Arg38 is essential in order to retain an intermediate rate of compound I formation in the H42E HRP C mutant, but has no obvious structural counterpart in chloroperoxidase. This is evident from the two orders of magnitude decrease in the rate of compound I formation for the H42E:R38S HRP C double mutant. The greater accessibility of the distal cavity of the H42E HRP C mutant gives rise to improved peroxygenase activity compared to wild-type enzyme (*185*). An important hydrogen bonding network extends from His42 to the distal calcium binding site through Asn70 and Glu64 (Fig. 4), a structural feature conserved in all class III peroxidase structures solved to date with the exception of barley (*89–91*). Asn70 assists in maintaining the basicity of the His42 side chain through the Asn–His couple (Fig. 5) (*75, 187–189*). If the basicity of His42 is decreased, as in N70D and N70V HRP C, this has implications not only for the rate of compound I formation, but also for the rates of reduction of compounds I and II (*187*).

Arg38 is not absolutely essential for compound I formation, but promotes proton transfer to the side chain of His42, and stabilizes the transition state for the heterolytic cleavage of the peroxide O–O bond (Fig. 5) (*181, 192*). Stopped-flow kinetic studies of the reaction between R38L HRP C and high concentrations of hydrogen peroxide at 10°C have revealed the presence of an intermediate formed prior to compound I (*192*). In this case it appears that the absence of an arginine residue at position 38 slows the rate of O–O bond cleavage sufficiently for an intermediate to accumulate. Baek and Van Wart also observed a transient intermediate (denoted compound 0) formed prior to compound I when studying the reaction between HRP C and hydrogen peroxide at low temperatures (*193*). The UV-VIS spectra of these two intermediates are not the same. Harris and Loew used theoretical calculations of electronic structure and spectra to predict that compound 0 was an Fe(III)–OOH complex, and that the intermediate obtained in the case of R38L HRP C was a neutral Fe(III)–HOOH complex (*194*).

2. *Reduction of Compounds I and II*

The mechanisms for the reduction of compounds I and II are less well understood than the mechanism of compound I formation, although a number of suggestions have been presented in the literature (*24*). It is clear from site-directed mutagenesis studies that Arg38 and His42 are also important in the reduction steps, although their precise role has been difficult to define. A detailed mechanism for substrate oxidation by plant peroxidases has been proposed, based on data from the crystal structure of a ternary complex of cyanide-ligated HRP C and ferulic acid (*195*) (Section IV,F). Here it is assumed, with some justification, that the orientation of ferulic acid in the pre-electron transfer complexes formed transiently with compounds I and II is likely to be identical to that in the ternary complex with cyanide-ligated HRP C. A second assumption is that the position of Arg38 remains the same in both compounds I and II and is unaffected by the binding of either ligands or substrates. This is supported by the fact that its position is invariant in crystal structures of HRP C (*91*), the complex of benzhydroxamic acid and HRP C (*196*), the complex of ferulic acid and HRP C (*195*), and the complex of ferulic acid with cyanide-ligated HRP C (*195*). Mechanisms for the reduction of compounds I and II are illustrated schematically in Fig. 6. One interesting feature is the participation of a Pro residue (Pro139) that is conserved throughout the plant peroxidase superfamily (see Fig. 1). The carbonyl group of Pro139 acts as a hydrogen bond acceptor for a structurally conserved water molecule, which also acts as a conduit for proton transfer from the hydroxyl group of ferulic acid to the Nε2 atom of His42. The process of proton transfer is coupled to electron transfer from ferulate to the exposed heme edge (the region comprising $C18H_3$ and C20H). A similar mechanism operates in compound II reduction, except that an additional proton is transferred to the ferryl oxygen from the protonated side chain of His42 (Fig. 6).

C. Proximal Region

The most important residue in the proximal region (considered to be that region of the enzyme lying below the plane of the heme when viewed in the usual orientation) is His170, which, as noted previously, makes a covalent bond to the heme iron atom. In most publications it is referred to as the proximal histidine residue. Other residues of interest in this region are indicated in Fig. 4, including the conserved aspartic acid residue, Asp247, the carboxylate side chain of which is hydrogen bonded to His170 Nδ1H. The strength of this interaction controls the degree of imidazolate character imparted to the proximal histidine ring, a feature that appears to be common among peroxidases of the plant peroxidase superfamily (*197*). A useful experimental parameter correlated directly to this Fe–His–Asp couple is the frequency of the Fe(II)–imidazole stretching mode observed in resonance Raman spectra of the ferrous enzyme. This is observed typically as a strong band (or bands) between 200 and 250 cm^{-1} and can be correlated with the charge status of the histidine ring from neutral imidazole to imidazolate, respectively (*198*). In Fe(II) HRP C, the ^{54}Fe isotopic shift was used to assign

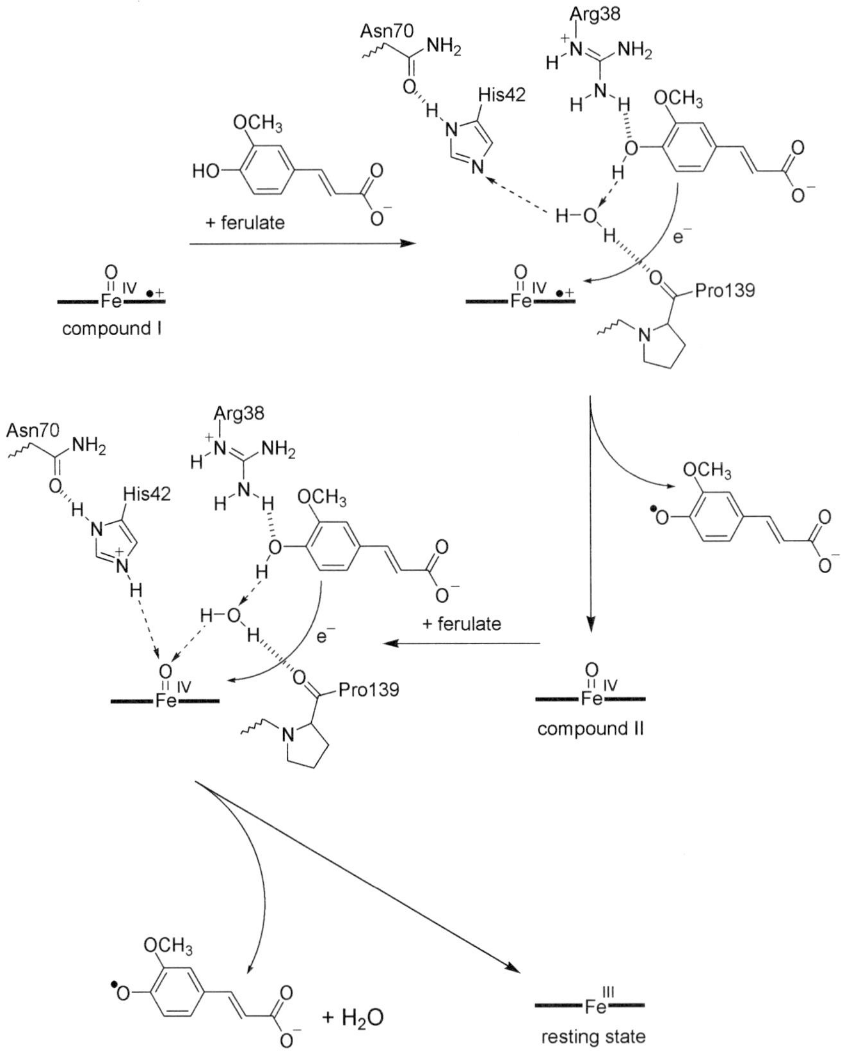

FIG. 6. Mechanisms for the reduction of compounds I and II of HRP C by ferulic acid, after Henriksen *et al.* (*195*). This scheme is based on new information from the 1.45 Å resolution crystal structure of the ternary complex of ferulic acid and cyanide-ligated HRP C (*195*). The direction of proton transfer is indicated by the dotted arrows. The mechanism is discussed in Section IV,B,2, and the crystal structure data in Section IV,F,4. Note that a distal site water molecule makes an important hydrogen bond with the backbone carbonyl group of Pro139 (a residue conserved in all members of the plant peroxidase superfamily).

an intense band at 244 cm^{-1} to the Fe(II)–imidazole stretching mode (*199*). The same mode has been assigned in a number of site-directed mutants of HRP C, including R38G (241 cm^{-1}), R38K (240 cm^{-1}), R38L (238 cm^{-1}), F41V (242 cm^{-1}), H42E (239 cm^{-1}), H42L (238 cm^{-1}), H42R (242 cm^{-1}), H42Q (239 cm^{-1}) (*103*), N70D (238 cm^{-1}), and N70V (238 cm^{-1}) (*187*), indicating that substitution of a distal residue can lead to a small decrease in the imidazolate character of the proximal His170 ring.

The reaction between H170A HRP C and H_2O_2 results in heme degradation, rather than the formation of high oxidation state intermediates (compounds I and II were not detected by spectroscopic methods) (*200*). Not surprisingly, the spin and coordination states of heme iron are altered, in this case to 6-c LS and a small amount of 5-c HS. The 6-c LS state is achieved through His42 ligation at the distal site, and aqua/hydroxo ligation at the vacant proximal site. Sufficient space exists in the cavity created by the His→Ala substitution for exogenous imidazole to bind and form a coordinate bond to heme iron. This improves the catalytic activity of the mutant 1500-fold, but does not restore it to that of the wild-type, the rate of compound I formation remaining 180-fold lower (Table II). The restoration of full catalytic activity is compromised by the inability of the H170A–imidazole complex to maintain a fully 5-c state. This indicates that one major function of His170 is to "tether" the heme iron atom and prevent formation of 6-c states (*200*). The conserved proximal His–Asp hydrogen bond is an additional structural restraint in this respect, a role that may be as significant as the modulation of the imidazolate character of the His ring. The properties of the proximal His175→Gly mutant of CCP have been described in a related study (*201*).

Relatively little has been published on other proximal site-directed mutants of HRP C, the only residues to receive attention being Phe172 (*202, 203*), Tyr185 (*202*), and Phe221 (*204*). The kinetic and spectroscopic properties of F172Y HRP C are similar to those of wild-type enzyme (*202, 203*). What is more interesting in this case is that two compound I species can be detected. The first is formed at the same rate as for wild-type enzyme, but decays rapidly to give the second, presumably by electron transfer from the porphyrin radical cation to the Tyr172 side chain. Thus, if amino acid side chains that are susceptible to oxidation are located close to the heme, then the formation of protein-based radicals may be promoted. Compound I of cytochrome *c* peroxidase has a protein radical located on Trp191 (*205, 206*). The corresponding residue in HRP C is Phe221, which has been substituted by

Trp in one investigation (*204*). On reaction of resting state F221W HRP C with hydrogen peroxide the following reaction series was observed.

$$\mathrm{Fe(III)} \xrightarrow{k_1} \mathrm{I_A} \xrightarrow{k_{ET}} \mathrm{I_B} \xrightarrow{k_2} \mathrm{I_C} \xrightarrow{k_3} \mathrm{Fe(III)} \tag{8}$$

The intermediates I_A and I_C are equivalent to compounds I and II of wild-type HRP C, whereas I_B has a compound II-type UV-VIS absorption spectrum and an EPR spectrum characteristic of a Trp radical. The rate of formation of compound I (k_1) is similar to that of wild-type enzyme, while the rate of electron transfer (k_{ET}) leading to the formation of the Trp221 radical is 65 s^{-1}. This reaction sequence is comparable to that described for F172Y HRP C (*203*). The W191F CCP mutant has also been studied and forms a transient porphyrin π-cation radical on reaction with hydrogen peroxide (*207, 208*). This undergoes an electron transfer step to yield a protein-based radical at Tyr236, with $k_{ET} = 54\ s^{-1}$. A limited study of Y185F HRP C has been reported, but no significant differences to the wild-type enzyme were observed (*202*).

D. Calcium Binding Sites

There are two calcium binding sites in HRP C, one distal and one proximal to the heme plane, as illustrated in Figs. 3 and 7 (*91*). It is significant that the spatial relationship between these sites and the heme group is conserved (Fig. 1), at least among those class II and class III peroxidases for which crystal structure data are available (note that barley peroxidase, BP1, is an exception). The donor ligands to calcium in each site are shown in detail in Fig. 7. Distal calcium is seven-coordinate with an O-donor ligand set provided by the side-chain carboxylates of Asp43 and Asp50, the side-chain hydroxyl of Ser52, the backbone carbonyl atoms of Asp43, Val46, and Gly48, and a structural water molecule, Wat15. Several of these calcium ligands originate from the loop region delineated by the disulfide bridge from Cys44 to Cys49. Proximal calcium is also seven-coordinate, with an O-donor ligand set provided by the side-chain carboxylates of Asp222 and Asp230, the side-chain hydroxyls of Thr171 and Thr225, and the backbone carbonyls of Thr171, Thr225, and Ile228. These calcium sites are identical in structure to those of peanut peroxidase (PNP), the only exception being that the residue equivalent to Ile228 of HRP C is Lys220 in PNP (*89*). The existence of two calcium binding sites may be a consequence of an early gene duplication event during the evolution of both plant and fungal peroxidases (*51, 52*).

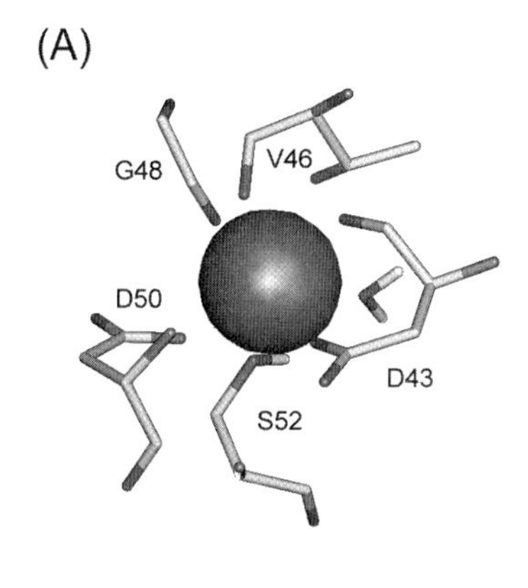

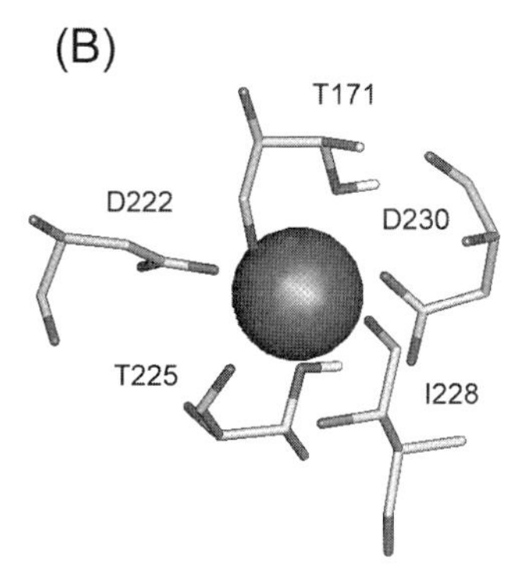

FIG. 7. The calcium binding sites of HRP C (detail). Ligands to the distal (A) and proximal (B) calcium atoms are shown. The O-donor ligand sets are listed in Section IV,D.

The presence of calcium in horseradish peroxidase was demonstrated originally by Haschke and Friedhoff, working with the C and A (unspecified, but likely to have been predominantly A2) isoenzymes (*209*). HRP C and HRP A contain 2.0 ± 0.13 and 1.4 ± 0.19 moles calcium per mole enzyme, respectively, as determined by atomic absorption spectroscopy. Incubation in 6 M guanidinium hydrochloride and 10 mM EDTA for 4 hours at neutral pH and room temperature gave calcium-depleted enzymes with specific activities decreased by 40% and 15%, respectively. The thermal stability of calcium-depleted HRP C was also reduced compared to native enzyme. Reconstitution was successful only with calcium-depleted HRP C (*209*). It remains to be established whether this reflects true structural differences between the calcium binding sites of the two isoenzymes, or is a consequence of the relatively harsh depletion protocols used. The fact that HRP C undergoes exchange with $^{45}Ca^{2+}$, and HRP A does not, adds weight to the former hypothesis (*209*). Comparison between the amino acid sequences of HRP C and HRP A2 indicates that all the residues involved in calcium binding are conserved, with the exception of Ile228, which is replaced by Ala (*15*). The calcium content of five neutral isoenzymes of horseradish peroxidase (B1, B2, B3, C1, and C2) has been determined by atomic absorption spectroscopy at 2 mol calcium per mol enzyme, as expected (*33*). Substitution of the calcium atoms of HRP C by either divalent metal cations (Ba^{2+}, Cd^{2+}, and Sr^{2+}) or lanthanides (Ln^{3+}) can be achieved readily in solution without compromising the structural and functional integrity of the enzyme (*210*). Analysis of the paramagnetically shifted hyperfine resonances in the ^{1}H NMR spectra of lanthanide-substituted HRP C indicated that tyrosyl hydroxyl, glutamate carboxyl, and aspartate carboxyl groups were involved in metal ion binding (*210*). Some evidence for structural differentiation between the two calcium binding sites has been obtained from a $^{113}Cd^{2+}$ titration of calcium-depleted HRP C monitored using ^{113}Cd NMR spectroscopy (*210*).

Removal of calcium from HRP C has a significant effect not only on enzyme activity and thermal stability, but also on the environment of the heme group. The calcium-depleted enzyme has optical, EPR, and ^{1}H NMR spectra that are different from those of the native enzyme (*211*). Temperature dependence studies indicate that the heme iron exists as a thermal admixture of high- and low-spin states. Kinetic measurements at pH 7 show that k_1, the rate constant for compound I formation, is only reduced marginally from $1.6 \pm 0.1 \times 10^7$ to 1.4×10^7 $M^{-1}s^{-1}$, whereas k_3, the rate constant for compound II reduction, is reduced from $8.1 \pm 1.6 \times 10^2$ to 3.6×10^2 $M^{-1}s^{-1}$ (reducing substrate; *p*-aminobenzoic acid), 44% of its initial value (*211*). There can be little doubt that this is the main reason for the loss of enzyme activity on calcium removal.

Crystallographic work has confirmed the existence of structural connections between the calcium binding sites and the heme group in HRP C (Fig. 4) (*91*). One of the ligands to distal calcium, Wat15, is also hydrogen bonded to Glu64 Oϵ1, which shares an additional hydrogen bond with another structural water molecule, Wat14. This water molecule and the backbone carbonyl of Glu64 are hydrogen bonded to Asp70 Nδ2H_2, whereas Asp70 Oδ1 is hydrogen bonded to Nδ1H of the distal histidine, His42. It is noteworthy that Glu64 and Asp70, residues that facilitate this connecting network of hydrogen bonds, are highly conserved among plant peroxidases (*15*). Three site-directed mutants of HRP C with Glu64 substituted by Gly (E64G), Pro (E64P), or Ser (E64S) have been studied in order to assess the importance of this residue for calcium stabilization and enzyme activity (*212*). Their calcium content is only 1 mol calcium per enzyme molecule, according to the results of inductively coupled plasma emission spectroscopy. The kinetic properties of the three mutants are significantly different from those of the wild-type enzyme, but show little variation among themselves. A 20-fold reduction in the value of V_{max} was recorded for hydroquinone oxidation under steady-state conditions. Elementary reaction rates obtained using stopped-flow techniques indicate that k_1 is decreased from $1.4 \pm 0.1 \times 10^7$ (wild-type HRP C) to $4.3 \pm 0.2 \times 10^5$ $M^{-1}s^{-1}$ (all Glu64 mutants; see Table II), a 33-fold reduction, whereas more moderate decreases were noted for k_2 (fivefold reduction) and k_3 (twofold reduction). The main contributions to the large decrease in k_1 appear to be the increase in the redox potential (Fe^{2+}/Fe^{3+}) of the mutants, and the reorientation and decrease in basicity of His42. ^1H NMR (resting and cyanide-ligated states) and resonance Raman (resting and ferrous states) spectra have been recorded for this series of mutants (*212*). The NMR spectra are remarkably similar, indicating that replacement of Glu64 results in a change to the electronic structure of the heme group that is independent of the nature of the substitution. The perturbations to the chemical shift values of the four heme methyl resonances of the resting states of the Glu64 mutants are smaller than those reported for calcium-depleted HRP C (*211*), which is consistent with the loss of only one calcium atom.

The proximal calcium binding site is coupled to the heme group by virtue of the fact that one of its ligands, Thr171, is adjacent to the proximal histidine residue, His170 (Fig. 4). The results of site-directed mutagenesis studies at this position are awaited with interest. An illustration of the importance of both calcium sites to the structure and function of HRP C is afforded by the need to incorporate calcium as a component of *in vitro* folding mixtures to obtain active recombinant enzyme from solubilized inclusion bodies (*64*).

E. Ligand Binding Sites

Interactions between small molecules and HRP are an important source of information about structure and reactivity at the heme group and its immediate environment. Several reviews are available in which ligand binding, and kinetic and spectroscopic characterization of the complexes are described (predominantly for HRP C rather than for other isoenzymes) (*8, 23, 24*). HRP ligands can be broadly classified into two groups on the basis of whether they bind to the vacant sixth coordination site of heme iron or not. Certain species such as iodide and thiocyanate bind at a site equidistant from heme methyls C2 and C18, according to ^{1}H and ^{15}N NMR studies (*213, 214*). There is no coordination to the heme iron atom and binding is relatively weak; at pH 4.0 the K_d for thiocyanate binding to HRP C is 158 $\pm$ 19 mM (*214*). The second group of ligands coordinate directly to heme iron and include species such as azide, carbon monoxide, cyanide, and fluoride. It is important to emphasize the fact that azide, cyanide, and fluoride bind to HRP C in their protonated forms (*215, 216*). Carbon monoxide (and also dioxygen) binds only to reduced HRP C, that is, to the ferrous or Fe(II) form (*217*). One of the most characteristic features of ligand binding to resting state HRP C is a change in the spin state of heme iron. The fluoride and cyanide complexes are 6-c HS ($S = \frac{5}{2}$) and 6-c LS species ($S = \frac{1}{2}$), respectively, whereas the azide complex can best be described as having an intermediate spin state. The Fe(II)–CO complex of HRP C is diamagnetic ($S = 0$).

Ligands bound at the sixth coordination site of peroxidases engage in hydrogen bonded interactions with the side chains of amino acid residues in the distal pocket. Crystal structure data obtained for the ferulic acid complex of cyanide-ligated HRP C (Section IV,F,4) indicates potential hydrogen bonds from Arg38 NϵH and His42 Nϵ2H (note that the proton from HCN is accepted by the His42 Nϵ2 atom on binding) to the cyanide nitrogen (*195*). Evidence that the protonated His42 side chain can form two types of hydrogen bond to bound cyanide has been obtained using resonance Raman spectroscopy (*218*). A crystal structure for the fluoride complex of CCP has been determined by difference Fourier techniques and partially refined to 2.5 Å (*219*). In this complex the distal Arg residue was displaced by 2.0 Å in order to optimize hydrogen-bonded interactions with the bound fluoride ligand, whereas a smaller adjustment of 0.5 Å was reported for the distal His side chain. A survey of the fluoride complexes of class I, II, and III peroxidases (and selected site-directed mutants) has been made using resonance Raman spectroscopy (*220*) and confirms the important contribution made by the distal Arg and His residues to ligand stabilization. It seems likely that the distal His residue accepts a proton from HF and establishes hydrogen-bonded contact with the fluoride anion through a bound water molecule.

The role of Arg38 and His42 in the stabilization of Fe(II)–CO complexes of HRP C has been investigated using site-directed mutagenesis (*221–223*). CO recombination rates after photolysis of the reduced enzymes increase in the order wild-type < H42R < H42L < R38L (*221*). The overall increase in CO affinity on replacing Arg38 by Leu is three orders of magnitude relative to wild-type enzyme (*221*). Recombination of CO to Fe(II) HRP A2 and HRP C isoenzymes has been compared (*224*). Two conformers of the CO complex with Fe(II) HRP C can be detected at neutral pH. These result from differences in the nature of hydrogen bonded interactions between the ligand and the distal residues Arg38 and His42 (*222, 223*). Formation of ternary complexes between aromatic donor molecules of the type Ar–CONHX (X = H, OH, NH_2, CH_3) and Fe(II)–CO HRP C removes the conformational heterogeneity of the CO unit (*225*). Arg38 and His42 impede the binding of dioxygen to Fe(II) HRP C, according to the analysis of association and dissociation rate constants obtained for wild-type enzyme and the site-directed mutants, R38G, R38K, R38L, R38S, H42L, and H42R (*226*). This oxyferrous complex is also known as compound III and may be written in formal terms as either Fe(II)–O_2 or Fe(III)–O_2^-. A transient intermediate has been detected by rapid-scan stopped-flow spectrophotometry when oxyferrous R38S HRP C decays to the ferric form (i.e., during the reaction Fe(II)–$O_2 \longrightarrow$ Fe(III) + O_2^-). The Fe(II) form of R38S HRP C shows the greatest affinity for dioxygen of all the mutants studied (*226*).

F. Aromatic Donor Molecule Binding Sites

The observation that HRP C forms stable, reversible 1 : 1 complexes with a variety of aromatic donor molecules in the absence of hydrogen peroxide (*227, 228*) has prompted an extensive research effort into the structure and location of possible binding sites (*21, 229*). Almost every spectroscopic technique has been exploited in this endeavor, and it is worth noting that the first crystal structure data for a horseradish peroxidase:aromatic donor molecule complex only became available as late as 1998 (*196*), approximately quarter of a century after the first research papers on this subject appeared in the literature (*227, 228*). An overview of the application of spectroscopic techniques to horseradish peroxidase has been published (*23*). There are many reasons for the high level of interest in these complexes, among them the fundamental insights sought into the interaction of horseradish peroxidase with substrates (whether physiological or nonphysiological), and the extensive applications of the enzyme in which aromatic donor molecules are an integral part. The advent of site-directed mutagenesis in HRP C offers the opportunity, in principle, to "design" substrate binding sites for small organic molecules, and although specific applications have yet to be realised, many useful data have been obtained for future exploitation.

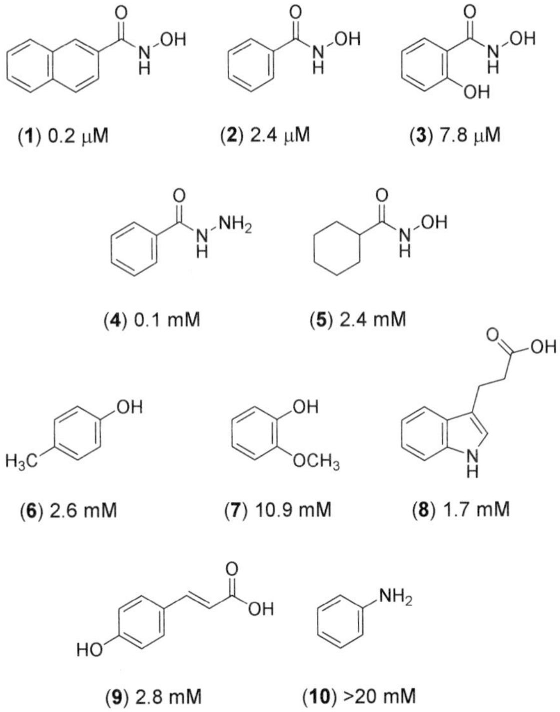

(**1**) 0.2 μM (**2**) 2.4 μM (**3**) 7.8 μM

(**4**) 0.1 mM (**5**) 2.4 mM

(**6**) 2.6 mM (**7**) 10.9 mM (**8**) 1.7 mM

(**9**) 2.8 mM (**10**) >20 mM

FIG. 8. Selected examples of aromatic donor molecules that form complexes with HRP C. The apparent dissociation constant for complex formation with the resting state plant enzyme is given (original references should be consulted for details of solution conditions and error estimations). (**1**) 2-Naphthohydroxamic acid (*228*); (**2**) benzhydroxamic acid (*228*); (**3**) 2-hydroxybenzhydroxamic acid (salicylhydroxamic acid) (*228*); (**4**) benzhydrazide (*228*); (**5**) cyclohexylhydroxamic acid (*228*); (**6**) 4-methylphenol (*p*-cresol) (*192*); (**7**) 2-methoxyphenol (guaiacol) (*192*); (**8**) indole-3-propionic acid (*241*); (**9**) *p*-coumaric acid (*238*); (10) aniline (*243*).

1. Donor Molecule Complexes of HRP C

A representative selection of aromatic donor molecules that bind to HRP C is shown in Fig. 8. The complexes can be formed with either the resting state of the enzyme, or with one of the ligand-bound forms, most commonly the cyanide-ligated state. In general little selectivity is shown by the enzyme, with most aromatic amines, indoles, phenols, and cinnamic acid derivatives exhibiting dissociation constants in the millimolar range (typically from 3 to 25 mM). However, aromatic hydroxamic acids, such as benzhydroxamic or naphthohydroxamic acids bind more strongly, with K_d values in the micromolar range (*228*).

Benzhydroxamic acid has become the donor molecule most favored by spectroscopists and crystallographers as a probe of the binding site. It has no known physiological role, although naturally occurring hydroxamic acids based on the 4-hydroxy-1,4-benzoxazin-3-one skeleton are common in the *Gramineae* (the grass family), where they are involved in plant defense (*230*), or as inhibitors of the binding of 1-naphthylacetic acid to auxin receptors (*231, 232*). Benzhydroxamic acid is also structurally similar to and isoelectronic with perbenzoic acid, an efficient oxidant of the enzyme (*233*). In a comprehensive study of the interactions between HRP C and a series of hydroxamic acids, hydrazides, and amides, Schonbaum concluded that there was no direct coordination between the hydroxamic acid side chain and heme iron, and noted that hydroxamate anions and hydrazide cations do not bind to the enzyme (*228*). His prediction that the aromatic donor molecule binding site must be "a heme-linked, extended, largely apolar region of horseradish peroxidase" has been substantiated by later work.

Structural information on aromatic donor molecule binding was obtained initially by using ^{1}H NMR relaxation measurements to give distances from the heme iron atom to protons of the bound molecule. For example, indole-3-propionic acid, a structural homologue of the plant hormone indole-3-acetic acid, was found to bind approximately 9–10 Å from the heme iron atom and at a particular angle to the heme plane (*234*). The disadvantage of this method is that the orientation with respect to the polypeptide chain cannot be defined. Other donor molecules examined include 4-methylphenol (*p*-cresol) (*235*), 3-hydroxyphenol (resorcinol), 2-methoxy-4-methylphenol and benzhydroxamic acid (*236*), methyl 2-pyridyl sulfide and methyl *p*-tolyl sulfide (*237*), and L-tyrosine and D-tyrosine (*238*). Distance constraints of between 8.4 and 12.0 Å have been reported (*235–238*). Aromatic donor proton to heme iron distances of 6 Å reported earlier for aminotriazole and 3-hydroxyphenol (resorcinol) are too short because of an inappropriate estimate of the molecular correlation time (*239*), a parameter required for the calculations. Distance information for a series of aromatic phenols and amines bound to Mn(III)-substituted HRP C has been published (*240*).

The relationship between the aromatic donor molecule binding site and the heme group has been explored further using 1D and 2D ^{1}H NMR, and enzyme inactivation studies (see Section III, D). Nuclear Overhauser enhancement data obtained in 1D NOE and 2D NOESY experiments (these provide information on proton–proton distances in a structure, with an upper limit of 5 Å) indicate that binding occurs relatively close to heme methyl $C18H_3$ (*229*), in agreement with the conclusion from inactivation work that substrates interact with the "exposed" heme edge comprising heme methyl $C18H_3$ and the heme meso proton C20H. Early models of the binding site are not in agreement, however (*236, 241*). In the case of 4-methylphenol (*p*-cresol) and indole-3-propionic acid, Morishima and Ogawa proposed a sterically restricted region on the distal side of the heme plane close to a water molecule bound at the vacant sixth coordination site of heme iron (*241*). They considered that the side chain of benzhydroxamic acid coordinated directly to heme iron in the complex formed with resting state HRP C, contrary to the earlier work of Schonbaum (*228*). Sakurada and co-workers described a site for benzhydroxamic acid on the proximal side of the heme plane, with the donor molecule bound between heme methyl $C18H_3$ and the aromatic ring of Tyr185 (*236*). The participation of aromatic amino acid side chains as components of the binding site was anticipated previously (*242*), and supported by the observation of Paul and Ohlsson that HRP C is retained by phenyl-Sepharose and not octyl-Sepharose in column chromatography (*243*). However, a tyrosine side chain in close proximity to the heme group would be susceptible to oxidation by one of the high oxidation state intermediates of HRP C. For example, Tyr103 of metmyoglobin (situated 3.3 Å from the terminal carbon of one of the heme vinyl groups) is readily oxidized by the ferryl species generated on addition of H_2O_2 to the protein (*244*). This high oxidation state intermediate is similar to compound II of plant peroxidases. It is in fact the side chains of phenylalanine, and not tyrosine residues, that contribute to aromatic donor molecule binding in HRP C, as demonstrated subsequently by ^1H NMR analysis of complexes with indole-3-propionic (*245*) and ferulic acids (*229, 246*) in which NOE connectivities between enzyme and donor molecule were recorded. The sequence-specific identification of these phenylalanine residues has been achieved using a combination of site-directed mutagenesis, ^1H NMR, and X-ray crystallography. The principal subjects for these investigations have been the complexes formed between either resting or cyanide-ligated HRP C, and benzhydroxamic acid.

2. *The Benzhydroxamic Acid Complex of Resting-State HRP C*

The X-ray crystal structure of the benzhydroxamic acid complex of resting state HRP C has been solved to 2.0 Å resolution (*196*). Important structural elements of the binding site are illustrated in Fig. 9 (see color insert). The donor molecule is located on the distal side of the heme plane and makes both hydrogen-bonded and hydrophobic interactions with the enzyme. Arg38, His42, Pro139, and a distal water molecule located 2.6 Å above heme iron contribute to an extensive hydrogen bond

TABLE III

SUMMARY OF DISSOCIATION CONSTANTS DETERMINED FOR COMPLEXES BETWEEN BENZHYDROXAMIC ACID (**2**) AND RESTING STATE HORSERADISH PEROXIDASE[a]

Enzyme	K_d(μM)	Reference
Plant HRP A2	2480	*41*
Plant HRP C	2.4	*228*
WT HRP C	2.1 ± 0.2	*251*
R38L	12100 ± 700	*181*
F41A	1.3 ± 0.3	*249*
F41V	1.2 ± 0.1	*68*
H42L	2900 ± 500	*181*
F68A	11.4 ± 0.3	*249*
N70D	970 ± 80	*189*
F142A	8.7 ± 0.2	*251*
F143A	2.8 ± 0.1	*251*
F179A	72.4 ± 2.5	*178*
F179H	13.9 ± 0.4	*178*
F179S	42.5 ± 1.1	*178*

[a] The K_d values determined for wild-type plant enzymes are given at the head of the table, followed by data obtained for wild-type recombinant HRP C and site-directed mutants of HRP C. Solution conditions typically 10–50 mM sodium phosphate or 10 mM sodium MOPS, pH 7.0, 25°C.

network with the hydroxamic acid side chain. Comparison of the amino acid sequences available for class III peroxidases indicates that Arg38, His42, and Pro139 are invariant (*15*). Substitution of Arg38 by Lys (*191*) or Leu (*181*), or His42 by Leu (*181*), leads to a dramatic decrease in the affinity of the enzyme for benzhydroxamic acid (Table III). The coordination state of heme iron in the complex has been disputed. Although the distance between heme iron and the distal water molecule in the complex (2.6–2.7 Å) is shorter than that in the enzyme alone (3.2 Å), analogous Fe–ligand distances determined from crystal structure data available for six-coordinate peroxidase complexes fall within the range of 1.8 to 2.2 Å (*247*). Resonance Raman data obtained both for single crystals and in solution demonstrates clearly that the benzhydroxamic acid complex with resting state HRP C is six-coordinate (*248*). These results disagree with the assignment of heme iron as five-coordinate based on an analysis of hyperfine-shifted resonances in ^{1}H NMR spectra of the complex (*99*). What cannot be denied, however, is the fact that the distal water molecule exerts a powerful influence on the electronic structure of heme iron. It may be that in these systems an Fe–H_2O distance of 2.6 Å effectively represents a six-coordinate state.

The aromatic ring of benzhydroxamic acid makes hydrophobic contacts with the side chains of Phe68, Gly69, Ala140, Pro141, Phe179, heme methyl C18H_3, and the heme meso proton C20H. Substitution of Phe68 by Ala leads to a relatively small decrease in affinity for benzhydroxamic acid (×5) (*249*) whereas the corresponding decrease for the F179A HRP C mutant is more significant (×35) (*178*) (Table III). One of the most intriguing features of the crystal structure data is the reorientation observed for the side chain of Phe68 in the complex compared to native enzyme, raising the possibility that it might operate as a "lid" for the donor molecule binding site (*196*). It may be that crystal packing forces are responsible in part for this phenomenon, but corroborative evidence that the binding site is a relatively mobile region exists from NMR studies (*178, 249–251*). Henriksen and co-workers noted that rotation of the Phe68 side chain would be necessary to accomodate the larger 2-naphthohydroxamic acid (Fig. 8) in the binding site (as defined crystallographically). This would allow the aromatic ring system to make additional hydrophobic contacts with Phe142 and Phe143 at the periphery of the binding site, and affords an explanation for the 10-fold greater affinity of this molecule for HRP C ($K_d = 0.2$ μM), compared to benzhydroxamic acid (*228*), assuming that the hydrogen bonded interactions made by the two hydroxamic acid side chains are the same. What is clear, however, is that both hydrogen bonded and hydrophobic interactions are necessary for "high" affinity binding of aromatic donor molecules to HRP C. Where hydrogen bonded interactions are absent or limited, as with donor molecules other than the aromatic hydroxamic acids, K_d values fall in the millimolar range (Fig. 8). In this case, substitution of Arg38 (critical to aromatic hydroxamic acid binding) has little effect; for example, the K_d values for guaiacol and *p*-cresol binding to R38L HRP C, 22.5 ± 1.0 and 7.4 ± 1.2 mM, respectively, represent no more than a twofold loss of binding affinity compared to wild-type recombinant HRP C (9.6 ± 0.1 and 2.9 ± 0.1 mM, respectively) (*192*). The 1.6 Å resolution crystal structure of the benzhydroxamic acid complex formed by resting state *Arthromyces ramosus* peroxidase affords an example of a binding site where hydrophobic contacts to the aromatic ring of the donor molecule are lacking (*252*). Although the hydroxamic acid side chain makes productive hydrogen bonds with Arg, His, and Pro residues equivalent to those of HRP C, the K_d value for the interaction is only 3.7 mM (*253*).

3. Benzhydroxamic Acid Complexes of Cyanide-ligated HRP C

Cyanide-ligated HRP C binds benzhydroxamic acid more weakly than the resting state enzyme, with K_d values of 0.15 mM (50 mM potassium phosphate, pH 6.0, 25°C) and 0.095 ± 0.005 mM (20 mM potassium phosphate, D_2O, pD 7.6, 30°C) reported for the ternary complex (*228, 250*). It has been argued that complexes formed with this six-coordinate low-spin form of the enzyme are a better model for any transient pre–electron transfer complexes formed between donor molecules and compounds I or II than complexes formed with the resting-state enzyme. The perturbation of the heme pocket due to the presence of ferryl oxygen in compounds I and II may be similar to that introduced by cyanide, which binds at the same coordination site. It must also be noted that the ternary complex is more amenable to NMR analysis, because of the superior spectral resolution and narrower linewidths in the ^1H spectrum of the low-spin cyanide-ligated enzyme (*254*). This has permitted the specific assignment of many heme and heme-linked proton resonances (*255, 256*, and references therein) that are relevant to donor molecule binding, most of which cannot be located with certainty in the corresponding spectra of the resting state enzyme (*99*). NOE connectivities can be detected in 2D NOESY spectra of the ternary complex between the aromatic protons of benzhydroxamic acid and heme methyl $C18H_3$, heme propionate $C17^{1a}H$, His42 Cϵ1H, Phe68 ArH, Ala140 CβH_3, Phe179 ArH, and Ile244 CδH_3 (*246, 250, 256*). A number of additional NOE constraints were obtained using the methyl group of a series of monomethyl-substituted benzhydroxamic acids as a probe of the binding site environment (*250*). These results agree well with crystal structure data for the benzhydroxamic acid complex of resting state HRP C (*196*), but there is no independent structure for the ternary complex available for comparison as yet. Chang and colleagues used NOE distance constraints derived from NMR experiments as verification for a theoretical model of benzhydroxamic acid bound to cyanide-ligated HRP C (*179*). This model was constructed using a computer-aided approach based on the crystal structure of the resting state enzyme and the programs, AUTODOCK and AMBER. The most favorable mode of benzhydroxamic acid binding selected using these procedures was in good agreement with the experimental NOE data. In this model, the hydroxamic acid side chain engages in hydrogen bonds with Arg38, His42, Pro139, and the nitrogen atom of the cyanide ligand.

The ternary complex of benzhydroxamic acid and cyanide-ligated HRP C is heterogenous to the extent that two binding "modes" can be detected by NMR (*250, 257*). Their relative proportions are determined by integration of the two components (termed "Y" and "Z") of the heme methyl $C18H_3$ resonance at saturation, which represent the ternary complexes formed. The dynamics of the interrelationship between the two bound complexes of benzhydroxamic acid and free benzhydroxamic acid have been analyzed in detail (*257*). The Y and Z complexes are in fast to intermediate, and slow exchange, respectively, on the NMR time scale. They interconvert at a rate of approximately 2×10^2 s^{-1} (25°C), which falls in the slow exchange regime. Measurement of benzhydroxamic acid on and off rates indicates that the Y complex is 10 times more labile than the Z complex. At 25°C, the ratio of the two complexes (Y : Z) is approximately 2 : 1, with the proportion of the Z complex increasing to approximately 50% as the temperature is decreased to 5°C. There appear to be no differences between the NOE connectivities observed from either the Y or the Z component of the $C18H_3$ resonance, indicating that the two complexes are not the result of two alternative orientations of benzhydroxamic acid when bound to the enzyme (*257*). The chemical shift profile of the $C18H_3$ resonance of the ternary complex of F142A HRP C indicates a marked decrease in the proportion of the Z component (to less than 10%) (*251*). Furthermore, this component is absent from the corresponding $C18H_3$ resonance profiles of the ternary complexes of F68A and F179A HRP C (*178, 249*). It seems likely that the heterogeneity observed for the complexes of the cyanide-ligated enzyme is due to alternative orientations of one or more of the phenylalanine side chains that participate in aromatic donor binding. This is supported by the observation from crystal structure data that Phe68 adopts an alternative orientation when benzhydroxamic acid is bound to resting-state enzyme (*196*). Furthermore, the side chain of Phe142 experienced substantial movement when the model ternary complex was subjected to an 180-ps molecular dynamics simulation (*179*). The relative contributions of Phe68, Phe142, and Phe179 to the affinity of cyanide-ligated HRP C for benzhydroxamic acid have been assessed using site-directed mutagenesis (*178*). Substitution by Ala results in a decrease in binding affinity in the order Phe179 > Phe 68 > Phe142.

4. *Ferulic Acid Complexes of HRP C*

The crystal structures of two ferulic acid complexes of HRP C have been solved, one with resting state enzyme (to 2.0 Å resolution) and the other with the cyanide-ligated enzyme (to 1.45 Å resolution) (*195*). These represent a major achievement for the crystallography of peroxidase complexes. The binary complex is heterogenous, according to the $2F_o{-}F_c$ omit difference electron density map of the active site. The disordered density observed has been interpreted in terms of three possible ferulic acid binding modes, only two of which can represent *in vivo* interaction with compounds I and II. In contrast, the ternary complex is well defined, with a single unambiguous binding mode for ferulic acid (Fig. 10; see color insert). The structure of the binding site is similar to that described for benzhydroxamic acid; thus, there are hydrogen-bonded and hydrophobic interactions between donor molecule and enzyme, but an important difference is that ferulic acid is located further from the heme group than benzhydroxamic acid. Hydrogen bonds are found between the phenolic oxygen and both Arg38 $N\eta H_2$ and an active site water molecule, as well as between the same N atom and the methoxyl oxygen of ferulic acid. The active site water molecule is also hydrogen-bonded to Pro139 CO and the N atom of the cyanide ligand. Principal hydrophobic interactions are those with Phe68, Gly69, Pro139, Ala140, Phe142, Phe179, heme methyl $C18H_3$, and heme meso proton C20H. The orientation and location of ferulic acid in the binding site is in good agreement with the results of earlier NMR studies of the ternary complex (*229, 246*).

5. *Rational Design of Substrate Binding Sites*

One of the many important applications of HRP C is in chemiluminescent reactions, the most widely used of which involves the peroxidase-catalysed oxidation of luminol to yield 3-aminophthalate and light (Section V, B). However, the luminol oxidation activity of the fungal peroxidase from *Arthromyces ramosus* (ARP, CIP) is 500-fold greater than that of HRP C, and the luminescence produced per unit of peroxidase at least 100 times as strong (*258*). Three site-directed mutants of HRP C have been constructed in order to mimic the putative binding site for luminol in ARP, namely S35K, Q176E, and the double mutant S35K:Q176E (*259*). The rationale employed here is the introduction of charged residues to maximize electrostatic interactions with the polar groups of luminol (the equivalent residues in ARP are Lys49 and Glu190). S35A and Q176A HRP C were also produced as controls. The catalytic efficiency for luminol oxidation was increased slightly by the mutations S35K ($V_{max}/K_m = 71.9$) and S35K:Q176E ($V_{max}/K_m = 105$), compared to that of wild-type HRP C ($V_{max}/K_m = 55.4$). It is clear that the introduction of these charged residues into HRP C alone is insufficient to recreate the luminol oxidation activity of ARP, for which a V_{max}/K_m value of 23,400 has been recorded. Improvement of the luminol binding affinity of HRP C by engineering additional electrostatic interactions is likely to be compromised by the negative impact of the hydrophobic side chains that constitute the donor molecule binding site. It should also be emphasized that the nature of the relationship between substrate binding affinity and activity remains unclear at present.

6. *Donor Molecule Complexes of HRP A2*

Much less is known about the structural basis for the interaction of aromatic donor molecules with HRP A2, and no crystal structure data are available as yet. In fact, HRP A2, with only 54% amino acid sequence identity to HRP C, can be considered as a distinctive enzyme with closer affinities to the acidic peroxidase isoenzyme from the tobacco plant, *Nicotiana tabacum* (*132*). The affinity of resting-state HRP A2 for benzhydroxamic acid is reduced greatly compared to HRP C (*41*), and a similar situation exists for the ternary complexes formed with the cyanide-ligated enzymes (*229, 251*). In dynamic terms, the binding of benzhydroxamic acid to cyanide-ligated HRP A2 falls in the moderately fast regime on the NMR time scale, and the complex is homogeneous (*229*). It is interesting to note that the ternary complex of F179A HRP C is similar to HRP A2 in terms of both binding affinity and dynamics (*178*). HRP A2 has a valine residue at this position (Val179), rather than the phenylalanine of HRP C (*15*). No doubt this is one of several factors that contribute to its reduced affinity for benzhydroxamic acid.

V. Applications

A. Chemical Transformations

The uses and applications of horseradish peroxidase in organic synthesis have been reviewed (*260, 261*). As a reagent it offers several advantages, including wide commercial availability, good stability under a range of conditions, and broad substrate specificity. Future developments are likely to focus on the increased use of site-directed mutants of HRP C to improve stereo- and enantiospecificity in reactions of interest. As yet these enzymes are not available commercially.

Oxidation of phenols and aromatic amines using HRP is generally of little synthetic value, as oligomers and polymers are the main products (*5, 260*). Under certain conditions oxidative coupling of phenols or naphthols to give biaryls can be achieved, but with low selectivity (*262*). In contrast, HRP can catalyze a number of useful oxidative N- and O-dealkylation reactions that are relatively difficult to carry out synthetically. This area has been described in detail by Meunier (*263*). A method for the preparation of optically active hydroperoxides using HRP C has been developed (*264*). Optically pure (*S*)-hydroperoxides and (*R*)-alcohols are obtained from reaction of racemic secondary alkyl aryl hydroperoxides with HRP C in the presence of guaiacol (2-methoxyphenol). Catalytic efficiency is reduced significantly if HRP A isoenzymes are used in place of HRP C. Enantioselective reduction of racemic hydroperoxy homoallylic alcohols to optically active hydroperoxy alcohols and unsaturated diols can also be achieved using HRP (*265*).

The catalysis of oxygen-transfer reactions by peroxidases is of interest synthetically, as high selectivity can be achieved under relatively mild conditions (*260, 261*). Site-directed mutagenesis has been used to improve the enantioselectivity of HRP-catalyzed oxidation of aryl methyl sulfides. F41L HRP C shows both increased rates of sulfoxidation and better enantioselectivity than wild-type enzyme; for example, the rate of sulfoxidation of phenyl ethyl sulfide is increased 4-fold and the enantioselectivity improved from 35 to 94 enantiomeric excess (%) (*266*). In contrast, F41T HRP C shows increased rates of sulfoxidation but lower enantioselectvity than wild-type enzyme (*266*). F41A, H42A, and H42V HRP C oxidize thioanisole at rates greater than those of the wild-type enzyme, but there is no significant change in enantioselectivity. These mutants are also able to catalyze the epoxidation of styrene, in marked contrast to wild-type enzyme, where only trace amounts of styrene oxide can be produced under the same conditions (*267*). Epoxidation of styrene can be achieved using HRP if the enzyme is incubated with H_2O_2 in the presence of a mediator such as 4-methylphenol, although styrene oxide is not the sole product of the reaction (*268*). The site-directed mutagenesis studies indicate that Phe41 and His42 hinder the approach of substrates to the oxoferryl iron of compound I. Improving access to this site by making substitutions with smaller amino acid residues offers the potential to develop oxygen transfer chemistry comparable to that achieved using cytochrome P450. Mechanisms of sulfoxidation by HRP, the H64S mutant of sperm whale myoglobin, and a synthetic model porphyrin have been compared in one communication (*269*).

Selective hydroxylation of some aromatic compounds can be achieved using HRP C in the presence of oxygen and dihydroxyfumaric acid (*270*). This process afforded L-DOPA from L-tyrosine, D-(−)-3,4-dihydroxyphenylglycine from D-(−)-4-hydroxyphenylglycine, and L-epinephrine (adrenalin) from L-(−)-phenylephrine in yields of up to 70%.

B. Biotechnology

Horseradish peroxidase is one of the most versatile biocatalysts available to the biotechnology industry, with applications that continue to be refined as our knowledge about the enzyme increases. HRP is relatively stable, shows a wide substrate specificity, and is easily coupled to other molecules such as antibodies or oligonucleotide probes. It may also be immobilized on solid media such as the surface of electrodes. Traditionally its main use has been in the analytical field, where molecules of interest can be detected and quantitated through the measurement of peroxidase activity. The literature available on HRP applications is extensive, and patent databases should not be overlooked as a source of information. Many of the most important uses have been described in a review (*271*).

Coupled enzyme assays have been developed for the determination of substances as diverse as glucose, uric acid, and cholesterol, the principal application being quantitation in biological fluids such as blood, plasma, and urine. Typical examples are illustrated by Eqs. (9)–(12).

$$\text{D-Glucose} + O_2 + H_2O \xrightarrow{\text{glucose oxidase}} \text{D-Gluconolactone} + H^+ + H_2O_2 \tag{9}$$

$$H_2O_2 + S_{red} \xrightarrow{\text{HRP}} H_2O + S_{ox} \tag{10}$$

$$\text{Uric acid} + O_2 + 2H_2O \xrightarrow{\text{uricase}} \text{Allantoin} + H_2O_2 + CO_2 \tag{11}$$

$$H_2O_2 + S_{red} \xrightarrow{\text{HRP}} H_2O + S_{ox} \tag{12}$$

HRP substrates (S_{red}) can be selected to give products that can be monitored easily by colorimetric, fluorometric or chemiluminescent methods. A popular choice in colorimetric assays is 3,3′,5,5′-tetramethylbenzidine (TMB), a colorless substance that gives a blue product (S_{ox}) on oxidation. Important considerations when choosing suitable substrates are cost, safety, sensitivity, solubility, and stability. It is sometimes necessary to use a substrate that gives an insoluble colored product, for example, in histochemical staining or membrane-bound immunoassays.

Chemiluminescent assays, which most commonly involve the oxidation of luminol or its derivatives to yield 3-aminophthalate and light, are particularly sensitive and widely used (*272*). The sensitivity can be increased further by addition of enhancers (*273*), an example of which is 4-iodophenylboronic acid (*274*). One advantage of using HRP in clinical assays is that few inhibitors of the enzyme are found in biological fluids, thus reducing the possibility of errors in quantitative work. It is important to ensure that the amount of enzyme available is sufficient to convert hydrogen peroxide stoichiometrically in the peroxidase-catalyzed reaction step. Most diagnostic test kits contain an excess of HRP, which also allows for any decrease in activity on storage (*275*). In the case of "dry" kits, which are commonly used in clinical applications, the enzymes may be immobilized on thin films or electrodes. Conjugation of HRP to antibodies, immunoglobulins, DNA probes, and other biomolecules using bifunctional crosslinking reagents is employed routinely for use in immunoassays and immunodiagnostic kits (*276–279*). In these systems it is usually convenient to detect peroxidase activity colorimetrically.

Environmental applications of HRP include immunoassays for pesticide detection and the development of methods for waste water treatment and detoxification. Examples of the latter include removal of aromatic amines and phenols from waste water (*280–282*), and phenols from coal-conversion waters (*283*). A method for the removal of chlorinated phenols from waste water using immobilised HRP has been reported (*284*). Additives such as polyethylene glycol can increase the efficiency of peroxidase-catalyzed polymerization and precipitation of substituted phenols and amines in waste or drinking water (*285*). The enzyme can also be used in biobleaching reactions, for example, in the decolorization of bleach plant effluent (*286*).

New applications for HRP are likely to develop from current research into biocatalytic reactions in nonaqueous solvents (*287*). Well-established patterns of reactivity can change dramatically outside of the aqueous environment. One example is the hydroxylation of aromatic hydrocarbons in aqueous solution, a reaction catalyzed by cytochrome P450 but not by HRP. Hydroxylation of benzene can be carried out efficiently by polyethylene-glycolated HRP if benzene is used as the solvent for the reaction (*288*). The structural reasons for this surprising change in reactivity have been examined by resonance Raman spectroscopy (*289*). A contributory factor is improved substrate access to heme iron, the result of disruption to hydrogen bonding networks in the heme pocket. The oxidation of *o*-phenylenediamine in anhydrous benzene can be catalyzed by a surfactant-HRP complex (*290*), a new application with implications for the treatment of carcinogens in waste organic solvents. A biosensor for the detection of compounds in organic solvents has also been developed by immobilization of HRP on a dimethylformamide–polyhydroxyl cellulose organohydrogel (*291*).

C. Folding and Stability

The stability of peroxidases under nonequilibrium conditions (which apply in most applications) is related in part to their affinity for calcium. It is known that addition of calcium can increase the shelf-life of stored HRP and maintain the activity of the enzyme. It is also an essential requirement for successful refolding of recombinant HRP C (*64*). The heme group contributes to thermodynamic stability and resistance to proteolysis, but is not essential for refolding of the enzyme, in contrast to other heme-containing enzymes such as cytochrome P450.

A detailed study of tryptophan fluorescence during guanidinium chloride-induced denaturation and renaturation of plant HRP C (the enzyme has a single Trp residue, Trp117) has identified a distinct equilibrium intermediate of the apoenzyme in 0.5 M guanidinium chloride (*292*). CD studies have shown that this has native-like secondary structure but poorly defined tertiary structure (*292*). The secondary structure is lost between 1.2 and 2.7 M guanidinium chloride, and a fluorescence-detectable conformational transition involving the D–D′ loop (which includes Trp117) occurs above 4 M guanidinium chloride. The half-life for unfolding of plant HRP C in 6 M guanidinium chloride is relatively slow (519 s), in contrast to CCP (14.3 s), emphasizing the remarkable kinetic stability of the plant peroxidase (*293*). Denaturation of the holoenzyme occurs in two distinct steps, unfolding of the backbone followed by loss of the heme group. Similarly, refolding of HRP involves formation of secondary structure prior to heme capture. The influence of the N-linked glycans of HRP C on folding has been investigated through studies of the chemically deglycosylated enzyme, which retains only the two terminal GlcNAc residues at each of the eight glycosylation sites (*116, 294*). Unfolding in guanidinium chloride was found to be 2–3 times faster for deglycosylated HRP C than for the fully glycosylated plant enzyme (*294*). This suggests that the presence of glycans decreases or "dampens" the dynamic fluctuations of the polypeptide chain. Thermal denaturation of HRP C can be monitored by FT-IR techniques (*295*). A slow alteration of secondary structure is observed on heating, and the polypeptide is completely unfolded at 90°C.

Chemical modification of surface residues of HRP is one method which may offer some improvement in thermal or long-term stability of the enzyme. The ϵ-amino groups of the six surface Lys residues can be modified by reaction with carboxylic anhydrides and picryl sulfonic acid (*296*). In this example the number of sites modified was found to be more significant than the chemical nature of the modification, at least as a criterion for improved stability. Other methods explored include the use of bifunctional crosslinking reagents to couple surface sites on the enzyme (*297*). Future developments are likely to be concerned with the selection of site-directed mutants of HRP C that show enhanced thermal stability. Dramatic increases in thermal stability of up to 190-fold have been reported recently for mutants of *Coprinus cinereus* peroxidase (CIP) generated using a directed evolution approach (*298*).

D. Inactivation by Peroxides

One limiting factor that must be considered in applications using HRP is the susceptibility of the enzyme to inactivation by peroxides. This process takes place when reducing substrates (S_{red}) are absent (or their source becomes exhausted), and excess peroxide reacts with compound I (*299–302*). The outcome of reactions undergone by the compound I–peroxide complex depends on the nature of the peroxide itself. In the case of hydrogen peroxide, there are three subsequent pathways to consider. The catalytic pathways comprise either a two-electron reduction step (catalase-like) that returns the complex to resting-state enzyme with the generation of molecular oxygen, or two single-electron transfer steps that result in the generation of compound II, compound III, and the superoxide radical anion. A competing inactivation pathway results in the eventual formation of an inactive verdohemeperoxidase called P-670. Compound III can also be formed by reaction of ferrous HRP with dioxygen (Section IV,E), or by reaction of the resting state ferric enzyme with superoxide ($O_2^{\cdot -}$) or the hydroperoxyl radical ($HO_2^{\cdot}$). The reaction of compound II with H_2O_2, which is part of the second catalytic pathway mentioned above, is a complex process where two mechanisms operate concurrently (*303*). The combined catalytic pathways (catalase-like and compound III generating) protect against inactivation.

An experimental model constructed for HRP inactivation by peroxides allows the calculation of a partition ratio, *r*, defined as the number of turnovers of H_2O_2 by the enzyme before inactivation (*299, 304*). The acidic isoenzyme HRP A2 ($r = 1360$) is more resistant to inactivation than HRP C ($r = 625$), whereas recombinant wild-type HRP C (the nonglycosylated enzyme produced in *E. coli*) is less resistant ($r = 335$) than plant HRP C (*302*). Substitution of Phe143 by Ala in HRP C has little effect on susceptibility to H_2O_2 inactivation ($r = 385$) (*305*), whereas site-directed mutagenesis of the distal residues Arg38 and His42 has a dramatic effect. The partition ratios for R38K, R38L and H42L HRP C are 20, 27, and 32, respectively (*302, 305*), indicating that these residues help to limit inactivation by H_2O_2, in addition to their catalytic functions (Section IV,B). A useful series of complementary studies has been carried out using *m*-chloroperbenzoic acid instead of hydrogen peroxide (*305–307*). In this case the partition ratio is reduced dramatically (typical values are 2 or 3) as the catalase-like pathway is inoperative (*305*). The extent of inactivation is determined by the reactivity of the complex formed between compound I and *m*-chloroperbenzoic acid. Competition between the subsequent catalytic and inactivation pathways gives compound II (67%) and the inactive P-670 (33%) (*307*). A detailed mechanism for this process has been developed using data from conventional and stopped-flow kinetics (*307*). Site-directed mutants of HRP C showing increased resistance to peroxide inactivation have yet to be developed. However, directed evolution techniques have been used to generate site-directed mutants of CIP with oxidative stability up to 100 times greater than that of wild-type enzyme (*298*). This is an important achievement with implications for the direction of future research on HRP C, where similar techniques are beginning to be applied (*78*).

Acknowledgments

The authors thank Michael Gajhede, University of Copenhagen, for providing the coordinates of the ferulic acid complex of HRP C, and David J. Schuller, University of California—Irvine, for the structural alignment in Fig. 1. Research in the authors' laboratories was supported in part by the European Union Biotechnology Programme "Towards Designer Peroxidases" (BIO4-CT97-2031), and the Training and Mobility of Researchers Programme "Peroxidases in Agriculture, the Environment, and Industry" (FMRX-CF98-0200).

References

1. Planche, L. A. *Bulletin de Pharmacie* **1810,** *2,* 578–580.
2. Willstätter, R.; Pollinger, A. *Liebigs Ann.* **1923,** *430,* 269–319.
3. Willstätter, R. *Ber. Deutsch. Chem. Ges.* **1926,** *59B,* 1–12.
4. Theorell, H. *Arkiv Kemi Min. Geol.* **1942,** *16A,* 1–11.
5. Saunders, B. C.; Holmes-Siedle, A. G.; Stark, B. P. "Peroxidase: The Properties and Uses of a Versatile Enzyme and some Related Catalysts"; Butterworths: London, 1964.
6. Paul, K. G. In "Molecular and Physiological Aspects of Plant Peroxidases"; Greppin, H., Penel, C., Gaspar, T., Eds.; University of Geneva: Geneva, 1986; pp. 1–14.
7. Paul, K. G. In "The Enzymes"; 2nd ed., Boyer, P. D., Lardy, H., Myrback, K., Eds.; Academic Press: New York, 1963; Vol. 8, Part B, pp. 227–274.
8. Dunford, H. B.; Stillman, J. S. *Coord. Chem. Rev.* **1976,** *19,* 187–251.
9. Williams, R. J. P.; Moore, G. R.; Wright, P. E. In "Biological Aspects of Inorganic Chemistry"; Addison, A. W., Cullen, W. R., Dolphin, D., James, B. R., Eds.; Interscience: New York, 1977; pp. 369–401.
10. Hewson, W. D.; Hager, L. P. In "The Porphyrins"; Dolphin, D., Ed.; Academic Press: New York, 1979; Vol. 7B, pp. 295–332.
11. Dunford, H. B. *Adv. Inorg. Biochem.* **1982,** *4,* 41–68.
12. Frew, J. E.; Jones, P. In "Advances in Inorganic and Bioinorganic Mechanisms"; Sykes, A. G., Ed.; Academic Press: London, 1984; Vol. 3, pp. 175–212.
13. Dawson, J. H. *Science* **1988,** *240,* 433–439.
14. Everse, J.; Everse, K. E.; Grisham, M. B. "Peroxidases in Chemistry and Biology"; CRC Press: Boca Raton, FL, 1991; Vol. I & II.

15. Welinder, K. G. In "Plant Peroxidases 1980–1990: Progress and Prospects in Biochemistry and Physiology": Penel, C., Gaspar, T., Greppin, H., Eds.; University of Geneva: Geneva, 1992; pp. 1–24.
16. Poulos, T. L. *Curr. Opin. Biotech.* **1993,** *4,* 484–489.
17. Poulos, T. L.; Fenna, R. E. In "Metal Ions in Biological Systems: Metalloenzymes Involving Amino Acid Residue and Related Radicals"; Sigel, H., Sigel, A., Eds.; Dekker: New York, 1994; Vol. 30, pp. 25–75.
18. English, A. M. In "Encyclopedia of Inorganic Chemistry: Iron: Heme Proteins, Peroxidases, & Catalases"; King, R. B., Ed.; Wiley: Chichester, 1994; Vol. 4, pp. 1682–1697.
19. English, A. M.; Tsaprailis, G. *Adv. Inorg. Chem.* **1995,** *43,* 79–125.
20. Banci, L. *J. Biotechnol.* **1997,** *53,* 253–263.
21. Smith, A. T.; Veitch, N. C. *Curr. Opin. Chem. Biol.* **1998,** *2,* 269–278.
22. Isaac, I. S.; Dawson, J. H. In "Essays in Biochemistry"; Ballou, D. P., Ed.; Portland Press: London, 1999; Vol. 34, pp. 51–69.
23. Dunford, H. B. "Heme Peroxidases"; Wiley-VCH: New York, 1999.
24. Dunford, H. B. In "Peroxidases in Chemistry and Biology"; Everse, J. Everse, K. E. Grisham, M. B., Eds.; CRC Press: Boca Raton, FL, 1991; Vol. II, pp. 1–24.
25. Gaspar, T.; Penel, C.; Thorpe, T.; Greppin, H. "Peroxidases 1970–1980. A Survey of Their Biochemical and Physiological Roles in Higher Plants"; University of Geneva: Geneva, 1982.
26. Gaspar, T.; Penel, C.; Greppin, H. "Plant Peroxidases 1980–1990. Progress and Prospects in Biochemistry and Physiology"; University of Geneva: Geneva, 1992.
27. Penel, C.; Gaspar, T.; Greppin, H. *Plant Peroxidase Newsletter*; University of Geneva: Geneva, 1993–2000, Issues 1–14.
28. International Working Group on Plant Peroxidases; University of Geneva: Geneva, http://www.unige.ch/LABPV/perox.html.
29. Jermyn, M. A.; Thomas, R. *Biochem. J.* **1954,** *56,* 631–639.
30. Paul, K. G. *Acta Chem. Scand.* **1958,** *12,* 1312–1318.
31. Shannon, L. M.; Kay, E.; Lew, J. Y. *J. Biol. Chem.* **1966,** *241,* 2166–2172.
32. Aibara, S.; Kobayashi, T.; Morita, Y. *J. Biochem.* **1981,** *90,* 489–496.
33. Aibara, S.; Yamashita, H.; Mori, E.; Kato, M.; Morita, Y. *J. Biochem.* **1982,** *92,* 531–539.
34. Hoyle, M. C. *Plant Physiol.* **1977,** *60,* 787–793.
35. Srivastava, O. P.; Van Huystee, R. B. *Bot. Gaz.* **1977,** *138,* 457–464.
36. Welinder, K. G. *Eur. J. Biochem.* **1979,** *96,* 483–502.
37. Welinder, K. G.; Jespersen, H. M.; Kjærsgård, I. V. H.; Østergaard, L.; Abelskov, A. K.; Hansen, L. N.; Rasmussen, S. K. In "Plant Peroxidases: Biochemistry and Physiology"; Obinger, C., Burner, U., Ebermann, R., Penel, C., Greppin, H., Eds.; University of Geneva: Geneva, 1996; pp. 173–178.
38. Simon, P.; Capelli, N.; Flach, J.; Overney, S.; Tognelli, M.; Penel, C.; Greppin, H. In "Plant Peroxidases: Biochemistry and Physiology"; Obinger, C., Burner, U., Ebermann, R., Penel, C., Greppin, H., Eds.; University of Geneva: Geneva, 1996; pp. 179–183.
39. Kjærsgård, I. V. H.; Jespersen, H. M.; Rasmussen, S. K.; Welinder, K. G. *Plant Mol. Biol.* **1997,** *33,* 699–708.
40. Østergaard, L.; Pedersen, A. G.; Jespersen, H. M.; Brunak, S.; Welinder, K. G. *FEBS Lett.* **1998,** *433,* 98–102.
41. Reimann, L.; Schonbaum, G. R. *Methods Enzymol.* **1978,** *52,* 514–521.
42. Brattain, M. G.; Marks, M. E.; Pretlow, T. G. *Anal. Biochem.* **1976,** *72,* 346–352.

43. Huang S.-Y.; Lee, Y.-C. *Bioseparation* **1994,** *4,* 1–5.
44. Regalado, C.; Asenjo, J. A.; Pyle, D. L. *Enzyme Microb. Technol.* **1996,** *18,* 332–339.
45. Miranda, M. V.; Lahore, H. M. F.; Cascone, O. *Appl. Biochem. Biotechnol.* **1995,** *53,* 147–154.
46. Hiner, A. N. P.; Hernández-Ruíz, J.; García-Cánovas, F.; Arnao, M. B.; Acosta, M. *Biotechnol. Bioeng.* **1996,** *50,* 655–662.
47. Welinder, K. G. *FEBS Lett.* **1976,** *72,* 19–23.
48. Morita, Y.; Mikami, B.; Yamashita, H.; Lee, J. Y.; Aibara, S.; Sato, M.; Katsube, Y.; Tanaka, N. In "Biochemical, Molecular and Physiological Aspects of Plant Peroxidases"; Lobarzewski, J., Greppin, H., Penel, C., Gaspar, T., Eds.; University of Geneva: Geneva, 1991; pp. 81–88.
49. Fujiyama, K.; Intapruk, C.; Shinmyo, A. *Biochem. Soc. Trans.* **1995,** *23,* 245–246.
50. Fujiyama, K.; Takemura, H.; Shibayama, S.; Kobayashi, K.; Choi, J.-K.; Shinmyo, A.; Takano, M.; Yamada, Y.; Okada, H. *Eur. J. Biochem.* **1988,** *173,* 681–687.
51. Welinder, K. G. *Curr. Opin. Struct. Biol.* **1992,** *2,* 388–393.
52. Welinder, K. G.; Gajhede, M. In "Plant Peroxidases: Biochemistry and Physiology"; Welinder, K. G., Rasmussen, S. K., Penel, C., Greppin, H., Eds.; University of Geneva: Geneva, 1993; pp. 35–42.
53. Johansson, A.; Rasmussen, S. K.; Harthill, J. E.; Welinder, K. G. *Plant Mol. Biol.* **1992,** *18,* 1151–1161.
54. Fujiyama, K.; Takemura, H.; Shinmyo, A.; Okada, H.; Takano, M. *Gene* **1990,** *89,* 163–169.
55. Kawaoka, A.; Sato, S.; Nakahara, K.; Matsushima, N.; Okada, N.; Sekine, M.; Shinmyo, A.; Takano, M. *Plant Cell Physiol.* **1992,** *33,* 1143–1150.
56. Kawaoka, A.; Kawamoto, T.; Ohta, H.; Sekine, M.; Takano, M.; Shinmyo, A. *Plant Cell Rep.* **1994,** *13,* 149–154.
57. Kawaoka, A.; Kawamoto, T.; Sekine, M.; Yoshida, K.; Takano, M.; Shinmyo, A. *Plant J.* **1994,** *6,* 87–97.
58. Shinmyo, A.; Yoshida, K.; Ohashi, A.; Simokawatoko, Y. In "Plant Peroxidases: Biochemistry and Physiology"; Obinger, C., Burner, U., Ebermann, R., Penel, C., Greppin, H., Eds.; University of Geneva: Geneva, 1996; pp. 313–316.
59. Bartonek-Roxå, E.; Eriksson, H.; Mattiasson, B. *Biochim. Biophys. Acta* **1991,** *1088,* 245–250.
60. Bartonek-Roxå, E.; Holm, C. *Biotechnol. Techniques* **1999,** *13,* 69–73.
61. Intapruk, C.; Higashimura, N.; Yamamoto, K.; Okada, N.; Shinmyo, A.; Takano, M. *Gene* **1991,** *98,* 237–241.
62. Østergaard, L.; Abelskov, A. K.; Mattsson, O.; Welinder, K. G. *FEBS Lett.* **1996,** *398,* 243–247.
63. Ortlepp, S. A.; Pollard-Knight, D.; Chiswell, D. J. *J. Biotechnol.* **1989,** *11,* 353–364.
64. Smith, A. T.; Santama, N.; Dacey, S.; Edwards, M.; Bray, R. C.; Thorneley, R. N. F.; Burke, J. F. *J. Biol. Chem.* **1990,** *265,* 13335–13343.
65. Jayaraman, K.; Fingar, S. A.; Shah, J.; Fyles, J. *Proc. Natl. Acad. Sci. USA* **1991,** *88,* 4084–4088.
66. Vind, J.; Dalbøge, H. In "Plant Peroxidases: Biochemistry and Physiology"; Welinder, K. G., Rasmussen, S. K., Penel, C., Greppin, H., Eds.; University of Geneva: Geneva, 1993; pp. 483–488.
67. Hartmann, C.; Ortiz de Montellano, P. R. *Arch. Biochem. Biophys.* **1992,** *297,* 61–72.
68. Smith, A. T.; Sanders, S. A.; Thorneley, R. N. F.; Burke, J. F.; Bray, R. C. *Eur. J. Biochem.* **1992,** *207,* 507–519.

69. Egorov, A. M.; Gazaryan, I. G.; Kim, B. B.; Doseeva, V. V.; Kapeliuch, J. L.; Veryovkin, A. N.; Fechina, V. A. *Annals NY Acad. Sci.* **1994,** *721,* 73–82.
70. Gazaryan, I. G.; Doseeva, V. V.; Galkin, A. G.; Tishkov, V. I.; Mareeva, E. A.; Orlova, M. A. *Russ. Chem. Bull.* **1995,** *44,* 363–366.
71. Vlamis-Gardikas, A.; Smith, A. T.; Clements, J. M.; Burke, J. F. *Biochem. Soc. Trans.* **1992,** *20,* 111S.
72. Vlamis-Gardikas, A.; Smith, A. T.; Clements, J. M.; Burke, J. F. In "Plant Peroxidases: Biochemistry and Physiology"; Welinder, K. G., Rasmussen, S. K., Penel, C., Greppin, H., Eds.; University of Geneva: Geneva, 1993; pp. 475–478.
73. Pellegrineschi, A.; Kis, M.; Dix, I.; Kavanagh, T. A.; Dix, P. J. *Biochem. Soc. Trans.* **1995,** *23,* 247–250.
74. Egorov, A. M.; Gazaryan, I. G.; Savelyev, S. V.; Fechina, V. A.; Veryovkin, A. N.; Kim, B. B. *Annals NY Acad. Sci.* **1991,** *646,* 35–40.
75. Nagano, S.; Tanaka, M.; Watanabe, Y.; Morishima, I. *Biochem. Biophys. Res. Commun.* **1995,** *207,* 417–423.
76. Henriksen, A.; Gajhede, M.; Baker, P.; Smith, A. T.; Burke, J. F. *Acta Cryst.* **1995,** *D51,* 121–123.
77. Braithwaite, A. *J. Mol. Biol.* **1976,** *106,* 229–230.
78. Lin, Z.; Thorsen, T.; Arnold, F. H. *Biotechnol. Prog.* **1999,** *15,* 467–471.
79. Butler, A. *Curr. Opin. Chem. Biol.* **1998,** *2,* 279–285.
80. Littlechild, J. *Curr. Opin. Chem. Biol.* **1999,** *3,* 28–34.
81. Stadtman, T. C. *Annu. Rev. Biochem.* **1990,** *59,* 111–127.
82. Choi, H.-J.; Kang, S. W.; Yang, C.-H.; Rhee, S. G.; Ryu, S.-E. *Nat. Struct. Biol.* **1998,** *5,* 400–406.
83. Fülöp, V.; Ridout, C. J.; Greenwood, C.; Hajdu, J. *Structure* **1995,** *3,* 1225–1233.
84. Sundaramoorthy, M.; Terner, J.; Poulos, T. L. *Structure* **1995,** *3,* 1367–1377.
85. Finzel, B. C.; Poulos, T. L.; Kraut, J. *J. Biol. Chem.* **1984,** *259,* 13027–13036.
86. Patterson, W. R.; Poulos, T. L. *Biochemistry* **1995,** *34,* 4331–4341.
87. Kunishima, N.; Fukuyama, K.; Matsubara, H.; Hatanaka, H.; Shibano, Y.; Amachi, T. *J. Mol. Biol.* **1994,** *235,* 331–344.
88. Poulos, T. L.; Edwards, S. L.; Wariishi, H.; Gold, M. H. *J. Biol. Chem.* **1993,** *268,* 4429–4440.
89. Schuller, D. J.; Ban, N.; Van Huystee, R. B.; McPherson, A.; Poulos, T. L. *Structure* **1996,** *4,* 311–321.
90. Henriksen, A.; Welinder, K. G.; Gajhede, M. *J. Biol. Chem.* **1998,** *273,* 2241–2248.
91. Gajhede, M.; Schuller, D. J.; Henriksen, A.; Smith, A. T.; Poulos, T. L. *Nat. Struct. Biol.* **1997,** *4,* 1032–1038.
92. Russell, R. B.; Barton, G. J. *Proteins* **1992,** *14,* 309–323.
93. Barton, G. J. *Methods Enzymol.* **1990,** *183,* 403–428.
94. Barton, G. J. *Protein Eng.* **1993,** *6,* 37–40.
95. Kooter, I. M.; Pierik, A. J.; Merkx, M.; Averill, B. A.; Moguilevsky, N.; Bollen, A.; Wever, R. *J. Am. Chem. Soc.* **1997,** *119,* 11542–11543.
96. Theorell, H.; Ehrenberg, A. *Arch. Biochem. Biophys.* **1952,** *41,* 442–461.
97. Smulevich, G.; English, A.M.; Mantini, A. R.; Marzocchi, M. P. *Biochemistry* **1991,** *30,* 772–779.
98. Feis, A.; Howes, B. D.; Indiani, C.; Smulevich, G. *J. Raman Spectrosc.* **1998,** *29,* 933–938.
99. De Ropp, J. S.; Mandal, P.; Brauer, S. L.; La Mar, G. N. *J. Am. Chem. Soc.* **1997,** *119,* 4732–4739.

100. Howes, B. D.; Schiødt, C. B.; Welinder, K. G.; Marzocchi, M. P.; Ma, J.-G.; Zhang, J.; Shelnutt, J. A.; Smulevich, G. *Biophys. J.* **1999,** *77,* 478–492.
101. Maltempo, M. M.; Ohlsson, P. I.; Paul, K. G.; Petersson, L.; Ehrenberg, A. *Biochemistry* **1979,** *18,* 2935–2941.
102. Smulevich, G.; Paoli, M.; Burke, J. F.; Sanders, S. A.; Thorneley, R. N. F.; Smith, A. T. *Biochemistry* **1994,** *33,* 7398–7407.
103. Howes, B. D.; Rodriguez-Lopez, J. N.; Smith, A. T.; Smulevich, G. *Biochemistry* **1997,** *36,* 1532–1543.
104. Smulevich, G.; Paoli, M.; De Sanctis, G.; Mantini, A. R.; Ascoli, F.; Coletta, M. *Biochemistry* **1997,** *36,* 640–649.
105. Foote, N.; Gadsby, P. M. A.; Berry, M. J.; Greenwood, C.; Thomson, A. J. *Biochem. J.* **1987,** *246,* 659–668.
106. Sitter, A. J.; Shifflett, J. R.; Terner, J. *J. Biol. Chem.* **1988,** *262,* 13032–13038.
107. Feis, A.; Marzocchi, M. P.; Paoli, M.; Smulevich, G. *Biochemistry* **1994,** *33,* 4577–4583.
108. Gonzalez-Vergara, E.; Meyer, M.; Goff, H. M. *Biochemistry* **1985,** *24,* 6561–6567.
109. Theorell, H.; Åkesson, Å. *Arkiv Kemi Min. Geol.* **1943,** *17B,* No. 7.
110. McManus, M. T.; McKeating, J.; Secher, D. S.; Osborne, D. J.; Ashford, D.; Dwek, R. A.; Rademacher, T. W. *Planta* **1988,** *175,* 506–512.
111. Kurosaka, A.; Yano, A.; Itoh, N.; Kuroda, Y.; Nakagawa, T.; Kawasaki, T. *J. Biol. Chem.* **1991,** *266,* 4168–4172.
112. Yang, B. Y.; Gray, J. S. S.; Montgomery, R. *Carbohydrate Res.* **1996,** *287,* 203–212.
113. Takahashi, N.; Lee, K. B.; Nakagawa, H.; Tsukamoto, Y.; Masuda, K.; Lee, Y. C. *Anal. Biochem.* **1998,** *255,* 183–187.
114. Maehly, A. C. *Methods Enzymol.* **1955,** *2,* 801–813.
115. Paul, K. G.; Stigbrand, T. *Acta Chem. Scand.* **1970,** *24,* 3607–3617.
116. Tams, J. W.; Welinder, K. G. *Anal. Biochem.* **1995,** *228,* 48–55.
117. Veitch, N. C.; Williams, R. J. P.; Bray, R. C.; Burke, J. F.; Sanders, S. A.; Thorneley, R. N. F.; Smith, A. T. *Eur. J. Biochem.* **1992,** *207,* 521–531.
118. McManus, M. T.; Ashford, D. A. *Plant Peroxidase Newsletter* **1997,** *10,* 15–23.
119. Harthill, J. E.; Ashford, D. A. *Biochem. Soc. Trans.* **1992,** *20,* 113S.
120. Morita, Y.; Funatsu, J.; Mikami, B. In "Plant Peroxidases: Biochemistry and Physiology"; Welinder, K. G., Rasmussen, S. K., Penel, C., Greppin, H., Eds.; University of Geneva: Geneva, 1993; pp. 1–4.
121. Van Huystee, R. B.; Sesto, P. A.; O'Donnell, J. P. *Plant Physiol. Biochem.* **1992,** *30,* 147–152.
122. Gray, J. S. S.; Yang, B. Y.; Hull, S. R.; Venzke, D. P.; Montgomery, R. *Glycobiology* **1996,** *6,* 23–32.
123. Markwalder, H. U.; Neukom, H. *Phytochemistry* **1976,** *15,* 836–837.
124. Gross, A. J.; Sizer, I. W. *J. Biol. Chem.* **1959,** *234,* 1611–1614.
125. Fry, S. C. *J. Exp. Bot.* **1987,** *38,* 853–862.
126. Michon, T.; Chenu, M.; Kellershon, N.; Desmadril, M.; Guéguen, J. *Biochemistry* **1997,** *36,* 8504–8513.
127. Fry, S. C. In "Molecular and Physiological Aspects of Plant Peroxidases"; Greppin, H., Penel, C., Gaspar, T., Eds.; University of Geneva: Geneva, 1986; pp. 169–182.
128. Lewis, N. G.; Davin, L. B.; Sarkanen, S. In "Comprehensive Natural Products Chemistry"; Barton, D. H. R., Nakanishi, K., Meth-Cohn, O. Eds.; Elsevier: Oxford, 1999; Vol. 3, pp. 617–745.
129. Ogawa, K.; Kanematsu, S.; Asada, K. *Plant Cell Physiol.* **1997,** *38,* 1118–1126.
130. Gross, G. G. *Phytochemistry* **1977,** *16,* 319–321.

131. Gross, G. G.; Janse, C.; Elstner, E. F. *Planta* **1977,** *136,* 271–276.
132. Lagrimini, L. M.; Burkhart, W.; Moyer, M.; Rothstein, S. *Proc. Natl. Acad. Sci. USA* **1987,** *84,* 7542–7546.
133. Roberts, E.; Kutchan, T.; Kolattukudy, P. E. *Plant Mol. Biol.* **1988,** *11,* 15–26.
134. Roberts, E.; Kolattukudy, P. E. *Molec. Gen. Genet.* **1989,** *217,* 223–232.
135. Rebmann, G.; Hertig, C.; Bull, J.; Mauch, F.; Dudler, R. *Plant Mol. Biol.* **1991,** *16,* 329–331.
136. Kenten, R. H. *Biochem. J.* **1955,** *59,* 110–121.
137. Campa, A. In "Peroxidases in Chemistry and Biology"; Everse, J. Everse, K. E. Grisham, M. B., Eds.; CRC Press: Boca Raton, FL, 1991; Vol. II, pp. 25–50.
138. Roman, R.; Dunford, H. B. *Biochemistry* **1972,** *11,* 2076–2082.
139. Roman, R.; Dunford, H. B. *Can. J. Chem.* **1973,** *51,* 588–596.
140. Araiso, T.; Miyoshi, K.; Yamazaki, I. *Biochemistry* **1976,** *15,* 3059–3063.
141. Penner-Hahn, J.; McMurry, T. J.; Renner, M.; Latos-Grazynsky, L.; Eble, K. S.; Davis, I. M.; Balch, A. L.; Groves, J. T.; Dawson, J. H.; Hodgson, K. O. *J. Biol. Chem.* **1983,** *258,* 12761–12764.
142. Penner-Hahn, J.; Eble, K. S.; McMurry, T. J.; Renner, M.; Balch, A. L.; Groves, J. T.; Dawson, J. H.; Hodgson, K. O. *J. Am. Chem. Soc.* **1986,** *108,* 7819–7825.
143. Critchlow, J. E.; Dunford, H. B. *J. Biol. Chem.* **1972,** *247,* 3703–3713.
144. Sitter, A. J.; Reczek, C. M.; Terner, J. *J. Biol. Chem.* **1985,** *260,* 7515–7522.
145. Makino, R.; Uno, T.; Nishimura, Y.; Iizuka, T.; Tsuboi, M.; Ishimura, Y. *J. Biol. Chem.* **1986,** *261,* 8376–8382.
146. Hashimoto, S.; Tatsuno, Y.; Kitagawa, T. *Proc. Natl. Acad. Sci. USA* **1986,** *83,* 2417–2421.
147. Chance, B.; Powers, L.; Ching, Y.; Poulos, T.; Schonbaum, G. R.; Yamazaki, I.; Paul, K. G. *Arch. Biochem. Biophys.* **1984,** *235,* 596–611.
148. Dunford, H. B. In "Molecular and Physiological Aspects of Plant Peroxidases"; Greppin, H., Penel, C., Gaspar, T., Eds.; University of Geneva: Geneva, 1986; pp. 15–23.
149. Hayashi, Y.; Yamazaki, I. *J. Biol. Chem.* **1979,** *254,* 9101–9106.
150. Kersten, P. J.; Kalyanaraman, B.; Hammel, K. E.; Reinhammer, B.; Kirk, T. K. *Biochem. J.* **1990,** *268,* 475–480.
151. Joshi, D. K.; Gold, M. H. *Eur. J. Biochem.* **1996,** *237,* 45–57.
152. Wiseman, J. S.; Nichols, J. S.; Kolpak, M. X. *J. Biol. Chem.* **1982,** *257,* 6328–6332.
153. Porter, D. J. T.; Bright, H. J. *J. Biol. Chem.* **1983,** *258,* 9913–9924.
154. Ator, M. A.; David, S. K.; Ortiz de Montellano, P. R. *J. Biol. Chem.* **1987,** *262,* 14954–14960.
155. Ator, M. A.; Ortiz de Montellano, P. R. *J. Biol. Chem.* **1987,** *262,* 1542–1551.
156. Ortiz de Montellano, P. R.; David, S. K.; Ator, M. A.; Tew, D. *Biochemistry* **1988,** *27,* 5470–5476.
157. Augusto, O.; Kunze, K. L.; Ortiz de Montellano, P. R. *J. Biol. Chem.* **1982,** *257,* 6231–6241.
158. Ringe, D.; Petsko, G. A.; Kerr, D. E.; Ortiz de Montellano, P. R. *Biochemistry* **1984,** *23,* 2–4.
159. Ortiz de Montellano, P. R.; Kerr, D. E. *J. Biol. Chem.* **1983,** *258,* 10558–10563.
160. Jonen, H. G.; Werringloer, J.; Prough, R. A.; Estabrook, R. W. *J. Biol. Chem.* **1982,** *257,* 4404–4411.
161. Zhang, H.; Dunford, H. B. *Can. J. Chem.* **1993,** *71,* 1990–1994.
162. Dunford, H. B.; Adeniran, A. J. *Arch. Biochem. Biophys.* **1986,** *251,* 536–542.
163. Sakurada, J.; Sekiguchi, R.; Sato, K.; Hosoya, T. *Biochemistry* **1990,** *29,* 4093–4098.
164. Candeias, L. P.; Folkes, L. K.; Wardman, P. *Biochemistry* **1997,** *36,* 7081–7085.

165. Folkes, L. K.; Candeias, L. P. *FEBS Lett.* **1997,** *412,* 305–308.
166. Ryabov, A. D.; Goral, V. N.; Ivanova, E. V.; Reshetova, M. D.; Hradsky, A.; Bildstein, B. *J. Organometallic Chem.* **1999,** *589,* 85–91.
167. Van Haandel, M. J. H.; Rietjens, I. M. C. M.; Soffers, A. E. M. F.; Veeger, C.; Vervoort, J.; Modi, S.; Mondal, M. S.; Patel, P. K.; Behere, D. V. *J. Biol. Inorg. Chem.* **1996,** *1,* 460–467.
168. Green, B. M.; Oliver, R. W. A. *Biochem. Soc. Trans.* **1991,** *19,* 929–935.
169. Poulos, T. L.; Freer, S. T.; Alden, R. A.; Edwards, S. L.; Skogland, U.; Takio, K.; Eriksson, B.; Xuong, N. H.; Yonetani, T.; Kraut, J. *J. Biol. Chem.* **1980,** *255,* 575–580.
170. Welinder, K. G. *Eur. J. Biochem.* **1985,** *151,* 497–504.
171. Henrissat, B.; Saloheimo, M.; Lavaitte, S.; Knowles, J. K. C. *Proteins* **1990,** *8,* 251–257.
172. Piontek, K.; Glumoff, T.; Winterhalter, K. *FEBS Lett.* **1993,** *315,* 119–124.
173. Petersen, J. F. W.; Kadziola, A.; Larsen, S. *FEBS Lett.* **1994,** *339,* 291–296.
174. Banci, L.; Carloni, P.; Savellini, G. G. *Biochemistry* **1994,** *33,* 12356–12366.
175. Banci, L.; Carloni, P.; Diaz, A.; Savellini, G. G. *J. Biol. Inorg. Chem.* **1996,** *1,* 264–272.
176. Loew, G. H.; Du, P.; Smith, A. T. *Biochem. Soc. Trans.* **1995,** *23,* 250–256.
177. Zhao, D.; Gilfoyle, D. J.; Smith, A. T.; Loew, G. H. *Proteins Struct. Funct. Genet.* **1996,** *26,* 204–216.
178. Veitch, N. C.; Gao, Y.; Smith, A. T.; White, C. G. *Biochemistry* **1997,** *36,* 14751–14761.
179. Chang, Y.-T.; Veitch, N. C.; Loew, G. H. *J. Am. Chem. Soc.* **1998,** *120,* 5168–5178.
180. Poulos, T. L.; Kraut, J. *J. Biol. Chem.* **1980,** *255,* 8199–8205.
181. Rodriguez-Lopez, J. N.; Smith, A. T.; Thorneley, R. N. F. *J. Biol. Inorg. Chem.* **1996,** *1,* 136–142.
182. Newmyer, S. L.; Ortiz de Montellano, P. R. *J. Biol. Chem.* **1996,** *271,* 14891–14896.
183. Savenkova, M. I.; Newmyer, S. L.; Ortiz de Montellano, P. R. *J. Biol. Chem.* **1996,** *271,* 24598–24603.
184. Tanaka, M.; Ishimori, K.; Morishima, I. *Biochem. Biophys. Res. Commun.* **1996,** *227,* 393–399.
185. Tanaka, M.; Ishimori, K.; Mukai, M.; Kitagawa, T.; Morishima, I. *Biochemistry* **1997,** *36,* 9889–9898.
186. Savenkova, M. I.; Ortiz de Montellano, P. R. *Arch. Biochem. Biophys.* **1998,** *351,* 286–293.
187. Nagano, S.; Tanaka, M.; Ishimori, K.; Watanabe, Y.; Morishima, I. *Biochemistry* **1996,** *35,* 14251–14258.
188. Mukai, M.; Nagano, S.; Tanaka, M.; Ishimori, K.; Morishima, I.; Ogura, T.; Watanabe, Y.; Kitagawa, T. *J. Am. Chem. Soc.* **1997,** *119,* 1758–1766.
189. Tanaka, M.; Nagano, S.; Ishimori, K.; Morishima, I. *Biochemistry* **1997,** *36,* 9791–9798.
190. Savenkova, M. I.; Ortiz de Montellano, P. R. *Biochemistry* **1998,** *37,* 10828–10836.
191. Smith, A. T.; Sanders, S. A.; Sampson, C.; Bray, R. C.; Burke, J. F.; Thorneley, R. N. F. In "Plant Peroxidases: Biochemistry and Physiology"; Welinder, K. G., Rasmussen, S. K., Penel, C., Greppin, H., Eds.; University of Geneva: Geneva, 1993; pp. 159–168.
192. Rodriguez-Lopez, J. N.; Smith, A. T.; Thorneley, R. N. F. *J. Biol. Chem.* **1996,** *271,* 4023–4030.
193. Baek, H. K.; Van Wart, H. E. *J. Am. Chem. Soc.* **1992,** *114,* 718–725.
194. Harris, D. L.; Loew, G. H. *J. Am. Chem. Soc.* **1996,** *118,* 10588–10594.
195. Henriksen, A.; Smith, A. T.; Gajhede, M. *J. Biol. Chem.* **1999,** *274,* 35005–35011.

196. Henriksen, A.; Schuller, D. J.; Meno, K.; Welinder, K. G.; Smith, A. T.; Gajhede, M. *Biochemistry* **1998,** *37,* 8054–8060.
197. Goodin, D. B.; McRee, D. E. *Biochemistry* **1993,** *32,* 3313–3324.
198. Smulevich, G. *Biospectroscopy* **1998,** *4,* S3–S17.
199. Teraoka, J.; Kitagawa, T. *J. Biol. Chem.* **1981,** *256,* 3969–3977.
200. Newmyer, S. L.; Sun, J.; Loehr, T. M.; Ortiz de Montellano, P. R. *Biochemistry* **1996,** *35,* 12788–12795.
201. Sun, J.; Fitzgerald, M. M.; Goodin, D. B.; Loehr, T. M. *J. Am. Chem. Soc.* **1997,** *119,* 2064–2065.
202. Ortiz de Montellano, P. R.; Harris, R. Z.; Hartmann, C.; Miller, V. P.; Newmyer, S. L.; Ozaki, S. In "Plant Peroxidases: Biochemistry and Physiology"; Welinder, K. G., Rasmussen, S. K., Penel, C., Greppin, H., Eds.; University of Geneva: Geneva, 1993; pp. 137–142.
203. Miller, V. P.; Goodin, D. B.; Friedman, A. E.; Hartmann, C.; Ortiz de Montellano, P. R. *J. Biol. Chem.* **1995,** *270,* 18413–18419.
204. Morimoto, A.; Tanaka, M.; Takahashi, S.; Ishimori, K.; Hori, H.; Morishima, I. *J. Biol. Chem.* **1998,** *273,* 14753–14760.
205. Sivaraja, M.; Goodin, D. B.; Smith, M.; Hoffman, B. M. *Science* **1989,** *245,* 738–740.
206. Huyett, J. E.; Doan, P. E.; Gurbiel, R.; Houseman, A. L. P.; Sivaraja, M.; Goodin, D. B.; Hoffman, B. M. *J. Am. Chem. Soc.* **1995,** *117,* 9033–9041.
207. Erman, J. E.; Vitello, L. B.; Mauro, J. M.; Kraut, J. *Biochemistry* **1989,** *28,* 7992–7995.
208. Musah, R. A.; Goodin, D. B. *Biochemistry* **1997,** *36,* 11665–11674.
209. Haschke, R. H.; Friedhoff, J. M. *Biochem. Biophys. Res. Commun.* **1978,** *80,* 1039–1042.
210. Morishima, I.; Kurono, M.; Shiro, Y. *J. Biol. Chem.* **1986,** *261,* 9391–9399.
211. Shiro, Y.; Kurono, M.; Morishima, I. *J. Biol. Chem.* **1986,** *261,* 9382–9390.
212. Tanaka, M.; Ishimori, K.; Morishima, I. *Biochemistry* **1998,** *37,* 2629–2638.
213. Sakurada, J.; Takahashi, S.; Hosoya, T. *J. Biol. Chem.* **1987,** *262,* 4007–4010.
214. Modi, S.; Behere, D. V.; Mitra, S. *J. Biol. Chem.* **1989,** *264,* 19677–19684.
215. Holzwarth, J. F.; Meyer, F.; Pickard, M.; Dunford, H. B. *Biochemistry* **1988,** *27,* 6628–6633.
216. Dunford, H. B.; Hewson, W. D.; Steiner, H. *Can. J. Chem.* **1978,** *56,* 2844–2852.
217. Hayashi, Y.; Yamada, H.; Yamazaki, I. *Biochim. Biophys. Acta* **1976,** *427,* 608–616.
218. Hashimoto, S.; Takeuchi, H. *J. Am. Chem. Soc.* **1998,** *120,* 11012–11013.
219. Edwards, S. L.; Poulos, T. L.; Kraut, J. *J. Biol. Chem.* **1984,** *259,* 12984–12988.
220. Neri, F.; Kok, D.; Miller, M. A.; Smulevich, G. *Biochemistry* **1997,** *36,* 8947–8953.
221. Meunier, B.; Rodriguez-Lopez, J. N.; Smith, A. T.; Thorneley, R. N. F.; Rich, P. R. *Biochemistry* **1995,** *34,* 14687–14692.
222. Rodriguez-Lopez, J. N.; George, S. J.; Thorneley, R. N. F. *J. Biol. Inorg. Chem.* **1998,** *3,* 44–52.
223. Feis, A.; Rodriguez-Lopez, J. N.; Thorneley, R. N. F.; Smulevich, G. *Biochemistry* **1998,** *37,* 13575–13581.
224. Doster, W.; Bowne, S. F.; Frauenfelder, H.; Reinisch, L.; Shyamsunder, E. *J. Mol. Biol.* **1987,** *194,* 299–312.
225. Holzbaur, I. E.; English, A. M.; Ismail, A. A. *J. Am. Chem. Soc.* **1996,** *118,* 3354–3359.
226. Rodriguez-Lopez, J. N.; Smith, A. T.; Thorneley, R. N. F. *J. Biol. Chem.* **1997,** *272,* 389–395.
227. Critchlow, J. E.; Dunford, H. B. *J. Biol. Chem.* **1972,** *247,* 3714–3725.
228. Schonbaum, G. R. *J. Biol. Chem.* **1973,** *248,* 502–511.
229. Veitch, N. C. *Biochem. Soc. Trans.* **1995,** *23,* 232–240.

230. Niemeyer, H. M. *Phytochemistry* **1988,** *27,* 3349–3358.
231. Ray, P. M.; Dohrmann, U.; Hertel, R. *Plant Physiol.* **1977,** *59,* 357–364.
232. Venis, M. A.; Watson, P. J. *Planta* **1978,** *142,* 103–107.
233. Schonbaum, G. R.; Lo, S. *J. Biol. Chem.* **1972,** *247,* 3353–3360.
234. Burns, P. S.; Williams, R. J. P.; Wright, P. E. *J. Chem. Soc. Chem. Commun.* **1975,** 795–796.
235. Banci, L.; Bertini, I.; Bini, T.; Tien, M.; Turano, P. *Biochemistry* **1993,** *32,* 5825–5831.
236. Sakurada, J.; Takahashi, S.; Hosoya, T. *J. Biol. Chem.* **1986,** *261,* 9657–9662.
237. Casella, L.; Gullotti, M.; Ghezzi, R.; Poli, S.; Beringhelli, T.; Colonna, S.; Carrea, G. *Biochemistry* **1992,** *31,* 9451–9459.
238. Casella, L.; Gullotti, M.; Poli, S.; Bonfà, M.; Ferrari, R. P.; Marchesini, A. *Biochem. J.* **1991,** *279,* 245–250.
239. Leigh, J. S.; Maltempo, M. M.; Ohlsson, P. I.; Paul, K. G. *FEBS Lett.* **1975,** *51,* 304–308.
240. Saxena, A.; Modi, S.; Behere, D. V.; Mitra, S. *Biochim. Biophys. Acta* **1990,** *1041,* 83–93.
241. Morishima, I.; Ogawa, S. *J. Biol. Chem.* **1979,** *254,* 2814–2820.
242. Williams, R. J. P.; Wright, P. E.; Mazza, G.; Ricard, J. R. *Biochim. Biophys. Acta* **1975,** *412,* 127–147.
243. Paul, K. G.; Ohlsson, P. I. *Acta Chem. Scand.* **1978,** *B32,* 395–404.
244. Davies, M. J. *Biochim. Biophys. Acta* **1991,** *1077,* 86–90.
245. Veitch, N. C.; Williams, R. J. P. *Eur. J. Biochem.* **1990,** *189,* 351–362.
246. Veitch, N. C.; Williams, R. J. P. In “Biochemical, Molecular and Physiological Aspects of Plant Peroxidases”; Lobarzewski, J., Greppin, H., Penel, C., Gaspar, T., Eds.; University of Geneva: Geneva, 1991; pp. 99–109.
247. Edwards, S. L.; Poulos, T. L. *J. Biol. Chem.* **1990,** *265,* 2588–2595.
248. Smulevich, G.; Feis, A.; Indiani, C.; Becucci, M.; Marzocchi, M. P. *J. Biol. Inorg. Chem.* **1999,** *4,* 39–47.
249. Veitch, N. C.; Gilfoyle, D. J.; White, C. G.; Smith, A. T. In “Plant Peroxidases: Biochemistry and Physiology”; Obinger, C., Burner, U., Ebermann, R., Penel, C., Greppin, H., Eds.; University of Geneva: Geneva, 1996; pp. 1–6.
250. Veitch, N. C.; Williams, R. J. P. *Eur. J. Biochem.* **1995,** *229,* 629–640.
251. Veitch, N. C.; Williams, R. J. P.; Bone, N. M.; Burke, J. F.; Smith, A. T. *Eur. J. Biochem.* **1995,** *233,* 650–658.
252. Itakura, H.; Oda, Y.; Fukuyama, K. *FEBS Lett.* **1997,** *412,* 107–110.
253. Gilfoyle, D. J.; Rodriguez-Lopez, J. N.; Smith, A. T. *Eur. J. Biochem.* **1996,** *236,* 714–722.
254. La Mar, G. N.; De Ropp, J. S. In “Biological Magnetic Resonance”; Berliner, L. J., Reuben, J., Eds.; Plenum Press: New York, 1993; Vol. 12, pp. 1–78.
255. Thanabal, V.; De Ropp, J. S.; La Mar, G. N. *J. Am. Chem. Soc.* **1987,** *109,* 265–272.
256. De Ropp, J. S.; Mandal, P. K.; La Mar, G. N. *Biochemistry* **1999,** *38,* 1077–1086.
257. La Mar, G. N.; Hernández, G.; De Ropp, J. S. *Biochemistry* **1992,** *31,* 9158–9168.
258. Akimoto, K.; Shinmen, Y.; Sumida, M.; Asami, S.; Amachi, T.; Yoshizumi, H.; Saeki, Y.; Shimizu, S.; Yamada, H. *Anal. Biochem.* **1990,** *189,* 182–185.
259. Tanaka, M.; Ishimori, K.; Morishima, I. *Biochemistry* **1999,** *38,* 10463–10473.
260. Van Deurzen, M. P. J.; Van Rantwijk, F.; Sheldon, R. A. *Tetrahedron* **1997,** *53,* 13183–13220.
261. Colonna, S.; Gaggero, N.; Richelmi, C.; Pasta, P. *Trends Biotechnol.* **1999,** *17,* 163–168.

262. Schmitt, M. M.; Schüler, E.; Braun, M.; Häring, D.; Schreier, P. *Tetrahedron Lett.* **1998,** *39,* 2945–2946.

263. Meunier, B. In "Peroxidases in Chemistry and Biology"; Everse, J. Everse, K. E. Grisham, M. B., Eds.; CRC Press: Boca Raton, FL, 1991; Vol. II, pp. 201–217.

264. Adam, W.; Hoch, U.; Lazarus, M.; Saha-Möller, C. R.; Schreier, P. *J. Am. Chem. Soc.* **1995,** *117,* 11898–11901.

265. Adam, W.; Lazarus, M.; Hoch, U.; Korb, M. N.; Saha-Möller, C. R.; Schreier, P. *J. Org. Chem.* **1998,** *63,* 6123–6127.

266. Ozaki, S.-I.; Ortiz de Montellano, P. R. *J. Am. Chem. Soc.* **1994,** *116,* 4487–4488.

267. Newmyer, S. L.; Ortiz de Montellano, P. R. *J. Biol. Chem.* **1995,** *270,* 19430–19438.

268. Ortiz de Montellano, P. R.; Grab, L. A. *Biochemistry* **1987,** *26,* 5310–5314.

269. Goto, Y.; Matsui, T.; Ozaki, S.-I.; Watanabe, Y.; Fukuzumi, S. *J. Am. Chem. Soc.* **1999,** *121,* 9497–9502.

270. Klibanov, A. M.; Berman, Z.; Alberti, B. N. *J. Am. Chem. Soc.* **1981,** *103,* 6263–6264.

271. Ryan, O.; Smyth, M. R.; Ó Fágáin, C. In "Essays in Biochemistry"; Ballou, D. P., Ed.; Portland Press: London, 1994; Vol. 28, pp. 129–146.

272. Coulet, P. R.; Blum, L. J. *Trends Anal. Chem.* **1992,** *11,* 57–61.

273. Easton, P. M.; Simmonds, A. C.; Rakishev, A.; Egorov, A. M.; Candeias, L. P. *J. Am. Chem. Soc.* **1996,** *118,* 6619–6624.

274. Kricka, L. J.; Cooper, M.; Ji, X. *Anal. Biochem.* **1996,** *240,* 119–125.

275. Krell, H.-W. In "Biochemical, Molecular and Physiological Aspects of Plant Peroxidases"; Lobarzewski, J., Greppin, H., Penel, C., Gaspar, T., Eds.; University of Geneva: Geneva, 1991; pp. 469–478.

276. Nakane, P. K.; Kawaoi, A. *J. Histochem. Cytochem.* **1974,** *22,* 1084–1091.

277. Nilsson, P.; Bergquist, N. R.; Grundy, M. S. *J. Immunol. Methods* **1981,** *41,* 81–93.

278. Tijssen, P.; Kurstak, E. *Anal. Biochem.* **1984,** *136,* 451–457.

279. Peeters, J. M.; Hazendonk, T. G.; Beuvery, E. C.; Tesser, G. I. *J. Immunol. Methods* **1989,** *120,* 133–143.

280. Klibanov, A. M.; Morris, E. D. *Enzyme Microbiol. Technol.* **1986,** *3,* 119–122.

281. Nicell, J. A.; Bewtra, J. K.; Taylor, K. E.; Biswas, N.; St. Pierre, C. C. *Wat. Sci. Technol.* **1992,** *25,* 157–164.

282. Buchanan, I. D.; Nicell, J. A. *Biotechnol. Bioeng.* **1997,** *54,* 251–261.

283. Klibanov, A. M.; Tu, T.-M.; Scott, K. P. *Science* **1983,** *221,* 259–261.

284. Tatsumi, K.; Wada, S.; Ichikawa, H. *Biotechnol. Bioeng.* **1996,** *51,* 126–130.

285. Wu, J. N.; Bewtra, J. K.; Biswas, N.; Taylor, K. E. *Can. J. Chem. Engineering* **1994,** *72,* 881–886.

286. Paice, M. G.; Jurasek, L. *Biotechnol. Bioeng.* **1984,** *26,* 477–480.

287. Khmelnitsky, Y. L.; Rich, J. O. *Curr. Opin. Chem. Biol.* **1999,** *3,* 47–53.

288. Akasaka, R.; Mashino, T.; Hirobe, M. *Bioorg. Med. Chem. Lett.* **1995,** *5,* 1861–1864.

289. Mabrouk, P. A.; Spiro, T. G. *J. Am. Chem. Soc.* **1998,** *120,* 10303–10309.

290. Kamiya, N.; Okazaki, S. Y.; Goto, M. *Biotechnol. Technique* **1997,** *11,* 375–378.

291. Guo, Y. Z.; Dong, S. J. *Anal. Chem.* **1997,** *69,* 1904–1908.

292. Pappa, H. S.; Cass, A. E. G. *Eur. J. Biochem.* **1993,** *212,* 227–235.

293. Tsaprailis, G.; Chan, D. W. S.; English, A. M. *Biochemistry* **1998,** *37,* 2004–2016.

294. Tams, J. W.; Welinder, K. G. *FEBS Lett.* **1998,** *421,* 231–236.

295. Holzbauer, I. E.; English, A. M.; Ismail, A. A. *Biochemistry* **1996,** *35,* 5488–5494.

296. Ugarova, N. N.; Rozhkova, G. D.; Berezin, I. V. *Biochim. Biophys. Acta* **1979,** *570,* 31–42.

297. Miland, E.; Smyth, M. R.; Ó Fágáin, C. *Enzyme Microb. Technol.* **1996,** *19,* 242–249.

298. Cherry, J. R.; Lamsa, M. H.; Schneider, P.; Vind, J.; Svendsen, A.; Jones, A.; Pedersen, A. H. *Nat. Biotechnol.* **1999,** *17,* 379–384.
299. Arnao, M. B.; Acosta, M.; Del Río, J. A.; Varón, R.; García-Cánovas, F. *Biochim. Biophys. Acta* **1990,** *1041,* 43–47.
300. Acosta, M.; Arnao, M. B.; Del Río, J. A.; García-Cánovas, F. In "Biochemical, Molecular and Physiological Aspects of Plant Peroxidases"; Lobarzewski, J., Greppin, H., Penel, C., Gaspar, T., Eds.; University of Geneva: Geneva, 1991; pp. 175–184.
301. Acosta, M.; Arnao, M. B.; Hernandez-Ruiz, J.; García-Cánovas, F. In "Plant Peroxidases: Biochemistry and Physiology"; Welinder, K. G., Rasmussen, S. K., Penel, C., Greppin, H., Eds.; University of Geneva: Geneva, 1993; pp. 201–205.
302. Acosta, M.; Hernandez-Ruiz, J.; García-Cánovas, F.; Rodriguez-Lopez, J. N.; Arnao, M. B. In "Plant Peroxidases: Biochemistry and Physiology"; Obinger, C., Burner, U., Ebermann, R., Penel, C., Greppin, H., Eds.; University of Geneva: Geneva, 1996; pp. 76–81.
303. Nakajima, R.; Yamazaki, I. *J. Biol. Chem.* **1987,** *262,* 2576–2581.
304. García-Cánovas, F.; Tudela, J.; Varón, R.; Vazquez, A. M. *J. Enz. Inhib.* **1989,** *3,* 81–90.
305. Hiner, A. N. P.; Hernández-Ruíz, J.; García-Cánovas, F.; Smith, A. T.; Arnao, M. B.; Acosta, M. *Eur. J. Biochem.* **1995,** *234,* 506–512.
306. Arnao, M. B.; Hernández-Ruíz, J.; Varón, R.; García-Cánovas, F.; Acosta, M. *J. Mol. Catal. A Chem.* **1995,** *104,* 179–191.
307. Rodriguez-Lopez, J. N.; Hernández-Ruiz, J.; Garcia-Cánovas, F.; Thorneley, R. N. F.; Acosta, M.; Arnao, M. B. *J. Biol. Chem.* **1997,** *272,* 5469–5476.

STRUCTURE AND ENZYMOLOGY OF TWO BACTERIAL DIHEME ENZYMES: CYTOCHROME *cd*$_1$ NITRITE REDUCTASE AND CYTOCHROME *c* PEROXIDASE

VILMOS FÜLÖP,[1] NICHOLAS J. WATMOUGH,[2] and STUART J. FERGUSON[3]

[1]Department of Biological Sciences, University of Warwick, Coventry CV4 7AL, United Kingdom; [2]Centre for Metalloprotein Spectroscopy and Biology, School of Biological Sciences, University of East Anglia, Norwich NR4 7TJ, United Kingdom; and [3]Department of Biochemistry and Oxford Centre for Molecular Sciences, University of Oxford, Oxford OX1 3QU, United Kingdom

I. Introduction

The membranes of most species of bacteria contain an electron transport system. This serves to carry electrons energetically downhill from reductants—which can include not only intracellular molecules such as NADH but also extracellular sources of electrons such as hydrogen

0898-8838/01 $35.00

gas, sulfide, or methanol (to list just three)—to an oxidant. The last can include oxygen, as in higher cells, but also a range of other molecules and ions, including nitrate, nitrite, nitric oxide, nitrous oxide, sulfate, and dimethyl sulfoxide (again to list just a small selection). The reader is referred to Richardson (*1*) for more complete lists of reductants and oxidants that can be used by bacteria. Individual species of bacteria can use subsets of these reductants and oxidants, but in all cases the purpose of the oxidoreduction reaction is to generate a proton electrochemical gradient across the membrane that in turn can drive endergonic processes such as ATP synthesis. This requirement to generate a proton electrochemical gradient means that at least some of the electron transport components must be integral membrane components. However, very often the enzymes catalyzing the final step in an electron transport process are water soluble and are not directly involved in the electron transport processes that drive proton translocation across the membrane. In this article we wish to discuss the properties of two such enzymes, a nitrite reductase and a cytochrome *c* peroxidase. These two enzymes are both found in the bacterial periplasm, each contain two heme groups per polypeptide chain, and are frequently found in the same species of bacteria.

The nitrite reductase of concern here is cytochrome cd_1, a name that recognizes the content of one *c*-type cytochrome center and one d_1 heme per polypeptide chain. The former center is defined by the covalent attachment of Fe protoporphyrin IX to the polypeptide chain as a consequence of thioether bond formation between the vinyl groups of the porphyrin and two thiols of cysteine residues in a CXXCH motif (Fig. 1) The d_1 heme is unique to this type of enzyme and is characterised by partial saturation of two of the rings and the presence of carbonyl groups (Fig. 1). As explained above, the enzyme receives electrons from the electron transport system of a bacterium; these are used to reduce nitrite to nitric oxide. The exact electron donor molecules are not known for certain, but broadly speaking are either relatively low molecular weight *c*-type cytochromes or type 1 blue Cu proteins (i.e., cupredoxins). These in turn are reduced by the cytochrome bc_1 complex (Fig. 2). This process of nitrite reduction is one step in the overall process of denitrification that entails the sequential reduction of nitrate, via nitrite, nitric oxide, and nitrous oxide to nitrogen gas. This series of reduction reactions is part of the biogeochemical nitrogen cycle and generally occurs only when oxygen is not available to the bacteria. The physiological significance of denitrification is clear.

In contrast, the physiological significance of the reaction catalyzed by bacterial cytochrome *c* peroxidase is not so clear. The reaction is

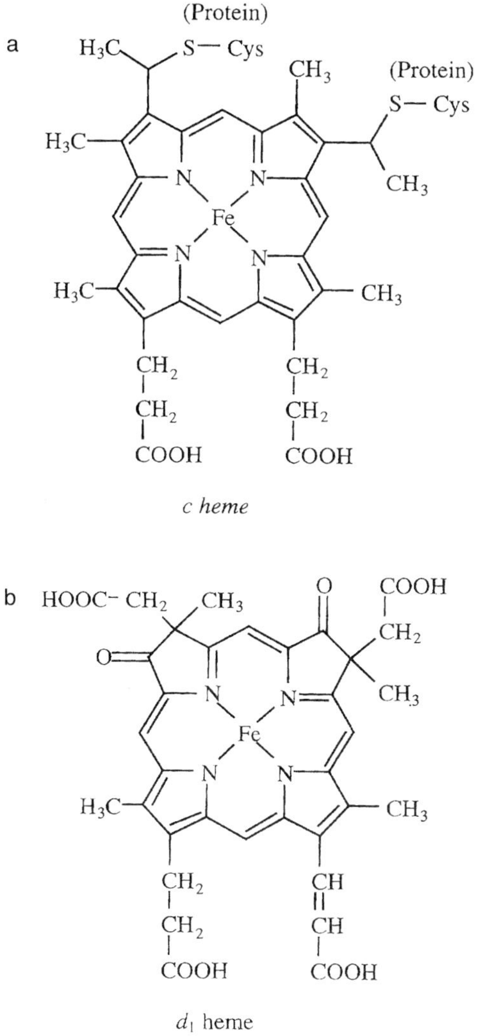

a

c heme

b

d_1 heme

FIG. 1. (a) The heme group in a *c*-type cytochrome showing the covalent attachment. (b) The d_1 heme group.

reduction of hydrogen peroxide to water. In common with the nitrite reductase discussed earlier, this peroxidase receives its electrons from the electron transfer chain via *c*-type cytochromes and cupredoxins (Fig. 2). The enzyme seems to be expressed under conditions of restricted oxygen concentration, but the source of the hydrogen peroxide is not always clear, and in contrast to the situation with nitrite reductase, there are

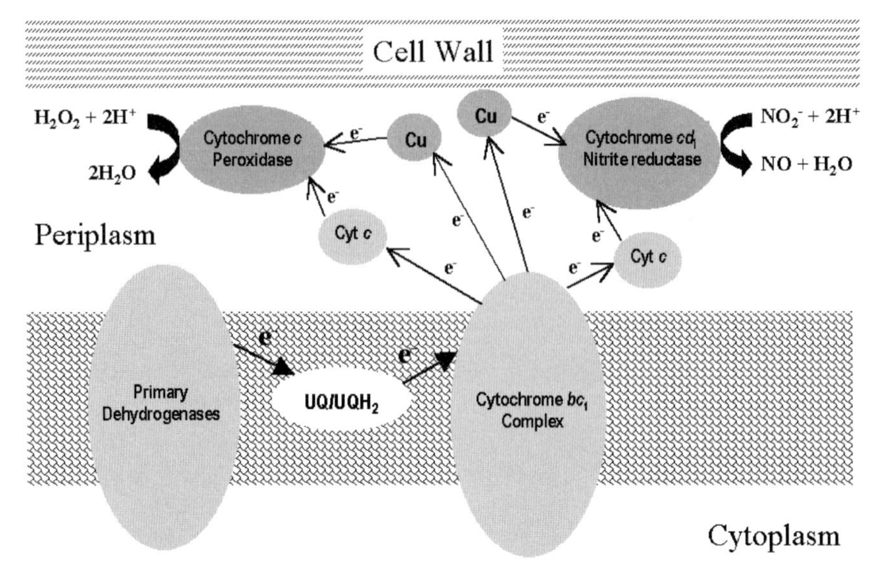

FIG. 2. The periplasmic location in Gram-negative bacteria of the two diheme enzymes, cytochrome cd_1 nitrite reductase and cytochrome *c* peroxidase. The diagram shows a grossly oversimplified version of a typical bacterial electron transport system. For example, the cytochrome oxidases that catalyze reduction of oxygen to water as an energetically favorable alternative to reduction of nitirite or peroxide are not shown. Distinct primary dehydrogenases catalyse electron abstraction from a variety of electron donors, such as NADH or succinate. Cyt *c* and Cu represent a *c*-type cytochrome and a cupredoxin respectively, each of which is currently considered to be an electron donor to the peroxidase and nitrite reductase, at least in some organisms.

few reports of the functioning of this enzyme in cells. Cytochrome *c* peroxidase is commonly found in denitrifying bacteria, but is not restricted to this type of organism. This peroxidase contains two *c*-type cytochrome centers per polypeptide chain, one of which provides the catalytic center. The other *c*-type center receives electrons from electron donors. Thus, the environment provided by the folding of the polypeptide chain drastically alters the properties of the two chemically identical heme groups. Many heme proteins have noncovalently bound heme, for example, hemoglobin. At the outset of this article we note that the advantages of covalent attachment via two thioether bonds in *c*-type cytochromes are not entirely apparent (*2*).

High-resolution crystal structures are known for both the diheme enzymes to be discussed in this article. We describe these structures and relate them to the mechanisms of the two enzymes before trying to compare the common features of these two types of enzyme. For example, they are both able to interact, at least *in vitro,* with *c*-type cytochromes and cupredoxins; they both have a receiver *c* center for electrons coming from a donor; and apparently they both have catalytically inert resting states. In addition, both enzymes provide examples of reactions that can also be catalyzed by a different type of enzyme. Thus, copper nitrite reductase and yeast cytochrome *c* peroxidase each catalyze one of two reactions discussed here.

II. Cytochromes cd_1

A. Properties of Cytochromes cd_1

The physiological reaction catalyzed by cytochrome cd_1 is the one-electron reduction of nitrite to nitric oxide. For this reaction a source of electrons, discussed in the Introduction, as well as protons is needed; water is the other reaction product. In addition, this enzyme also catalyzes the reduction of oxygen to water. The ability to catalyze this four-electron reduction is the reason why the enzyme is known as soluble cytochrome oxidase in some of the older literature. However, today this reaction is not thought to be physiologically relevant, although it is of mechanistic interest, as we discuss later. A third reaction catalyzed by a cytochrome cd_1 is the two-electron reduction of hydroxylamine to ammonia. Again, this is not thought to be physiologically significant.

The reduction of nitrite to nitric oxide poses a particular problem in that the reaction product, nitric oxide, generally has a very high affinity for heme Fe, especially when the latter is in the ferrous state. Before discussing further aspects of the enzymology of cytochrome cd_1 we will review what is known of the three-dimensional structure of the enzyme. Unusually, the structure differs depending on the particular bacterial source of the enzyme. Here, we describe first the structure of the enzyme from *Paracoccus pantotrophus* and subsequently that from *Pseudomonas aeruginosa*.

B. Structure of *Paracoccus pantotrophus* Cytochrome cd_1

The structure of *P. pantotrophus* (fomerly *Thiospharea pantotropha,* (*3*)) cytochrome cd_1, the first of this type to have its crystal structure solved (*4*), shows that the enzyme is a homodimer of 567 aminoacid residues and each subunit contains both a *c*-type cytochrome center and a d_1 heme center (Fig. 3). The d_1 heme (Fig. 1) is unique to this class of enzyme and on that basis alone might be expected to be the catalytic site. The *c*-type cytochrome centers, which are defined by the covalent attachment of the heme to the polypeptide are usually, but

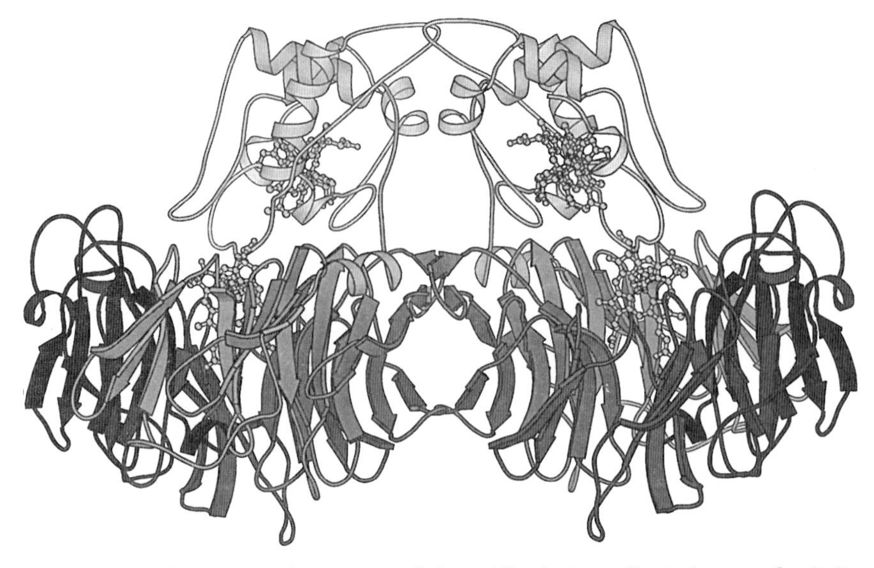

FIG. 3. The X-ray crystal structure of the oxidized state of cytochrome cd_1 nitrite reductase from *P. pantotrophus* (drawn from PDB entry 1 qks using MolScript (*98, 99*)).

not always, involved in electron transfer. Thus, if nothing functional was known about the enzyme, one might reasonably surmise that the *c*-type cytochrome center was the point of entry of electrons into the enzyme from which they would transfer to the d_1 heme. That electron transfer takes place only within a single monomer is expected, but not strictly proven, because the interheme distances across the dimer interface are on the order of 30 Å or more (Fig. 3). Such a distance is generally thought to be incompatible with biological activity, because predicted rates of electron transfer over such distances are of the order of one or two events per hour or more (*5*), scarcely compatible with enzyme turnover on the millisecond or second time scale. This conclusion leaves unanswered, of course, the question as to why the enzyme is a dimer. The crystallographic asymmetric unit contains the whole dimer, and there is some evidence that the two subunits, A and B, are not independent of one another because they always have a slightly different structure (*4, 6*). For example, in the oxidized enzyme, $N_{\delta 1}$ of His 17 is hydrogen bonded to the main-chain carbonyl of Ala 101 and to a water, whereas in the B subunit the corresponding histidine residue makes only a hydrogen bond only to a water molecule (*6*). More differences between the two subunits are described later.

The atomic resolution (1.28 Å) structure of oxidized *P. pantotrophus* cytochrome cd_1 "as isolated" revealed unexpected ligation of the heme

Fe atoms. It had been anticipated that the *c*-type cytochrome center would have His/Met coordination, but His/His is observed. The former is the more usual coordination, especially at the high potential end ($E^{o\prime} > +200$ mV) of the typical bacterial electron transfer chain to which the nitrite reductase is connected (Fig. 2) (*7*). The second curious feature is that the d_1 heme iron is also six-coordinate; thus, the enzyme does not offer a substrate-binding site at either heme. In addition to an expected axial histidine ligand there was an axial tyrosine (residue 25) ligand to the d_1 heme (Fig. 4a). Each monomer is organized into two domains,

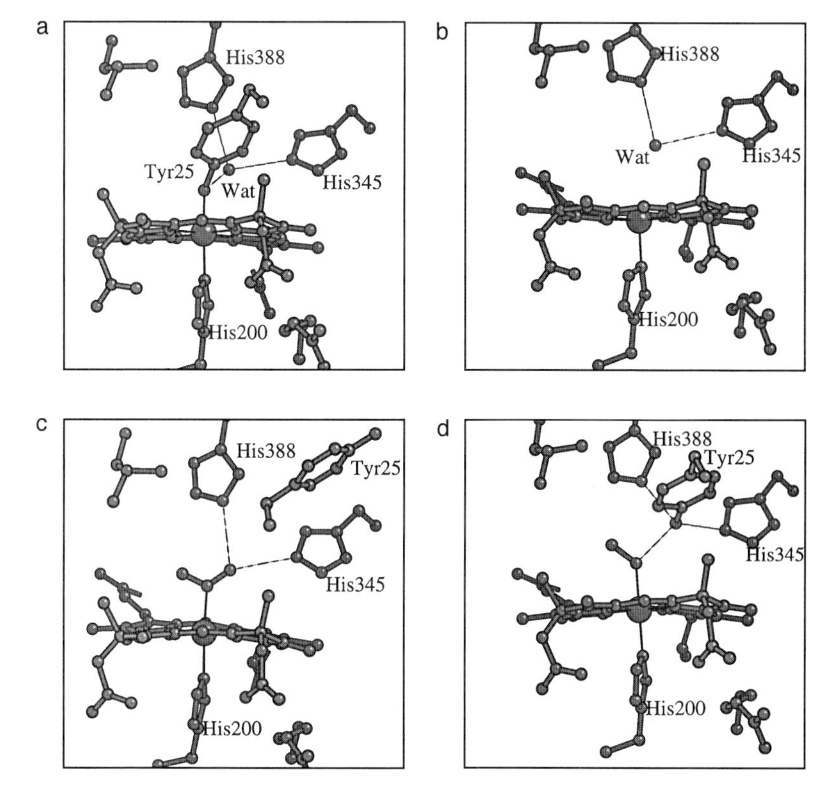

FIG. 4. The environment of the d_1 heme in various states of the *P. pantotrophus* cytochrome cd_1. (a) The oxidized enzyme. (b) The reduced enzyme. (c) The reduced enzyme to which nitrite has been added. (d) The reduced enzyme that has converted bound nitrite to nitric oxide. Tyr25 provides an axial oxygen ligand to the d_1 heme iron in the oxidized enzyme (a). This Tyr25 is displaced in the reduced enzyme (b) but can be seen returning to the d_1 heme in the structures with nitrite (c) or nitric oxide (d) bound. His 345 and His 388 are proposed proton donors to the substrate. In (b) the d_1 heme iron is five-coordinate. The water molecule (Wat) that is seen above the heme ring is not coordinated to the iron. Dashed lines represent potential hydrogen bonds. (Drawn from PDB entries 1 qks, 1 aof, 1 aom and 1 aoq.)

with the *c* heme being in a mainly α-helical domain and the d_1 heme being in an eight-bladed β-propeller domain (Fig. 3). A notable point is that the Tyr 25 ligand is provided by the α-helical cytochrome *c* domain, and only a few residues away is the His 17 ligand to the *c* heme iron. Positioned above the d_1 heme ring are two histidine residues (Fig. 4a) that can be envisaged as proton donors to one oxygen of the nitrite, thus generating water as one of the reaction products (*4*). This would be in agreement with previous mechanistic proposals that envisaged loss of water as a first step, leaving an NO bound to Fe, which is formally in the Fe(III) state (reviewed in *8*).

As mentioned earlier, the electron donor proteins to cd_1 are thought to include *c*-type cytochromes and cupredoxins. In the case of *P. pantotrophus,* two obvious molecules that could fulfil this role are cytochrome c_{550} and pseudoazurin. Although both these molecules will act as *in vitro* electron donors to the cytochrome cd_1, there is as yet no direct proof that they act in this way *in vivo*. Whole cells of *Paracoccus denitrificans* show no significant attenuation of the rate of nitrite reduction when the expression of cytochrome c_{550} is not possible as a result of disruption of its gene (*9*). Thus, either cytochrome c_{550} is not *in vivo* an electron donor to the nitrite reductase, or its role can be fulfilled by another protein(s). The latter might be pseudoazurin (*10*). The properties of mutants specifically deleted in both of cytochrome c_{550} and pseudoazurin are awaited with interest. However, the *in vitro* experiments alone raise the interesting question as to how proteins of such different structures as cytochrome c_{550} and pseudoazurin could interact with cytochrome cd_1. It has been recognized that each of these putative donor proteins has a surface hydrophobic patch that is studded with positively charged residues. In principle, these can contribute to pseudospecific docking interaction with a complementary hydrophobic patch, studded with negative residues, on the surface of the cytochrome *c* domain of cytochrome cd_1 (*11*). This proposal remains to be tested by experiment. To summarize, the structure of the oxidized enzyme does not show an obvious catalytic site and suggests that electron transfer between the *c* and d_1 hemes could be fast. It also provides a proposal as to where the donor electron transport proteins might bind or dock.

Fortunately, the structure of the fully reduced *P. pantotrophus* cytochrome cd_1 could be obtained, and this has provided important clues as to how the enzyme functions, as well as raising unexpected questions (*12*). First of all, the structure of the reduced enzyme reveals an unprecedented switch of one axial ligand at the *c* heme from His to Met (Figs. 5a,b). The second important feature revealed by this structure was that the Tyr 25 had vacated the d_1 heme iron coordination

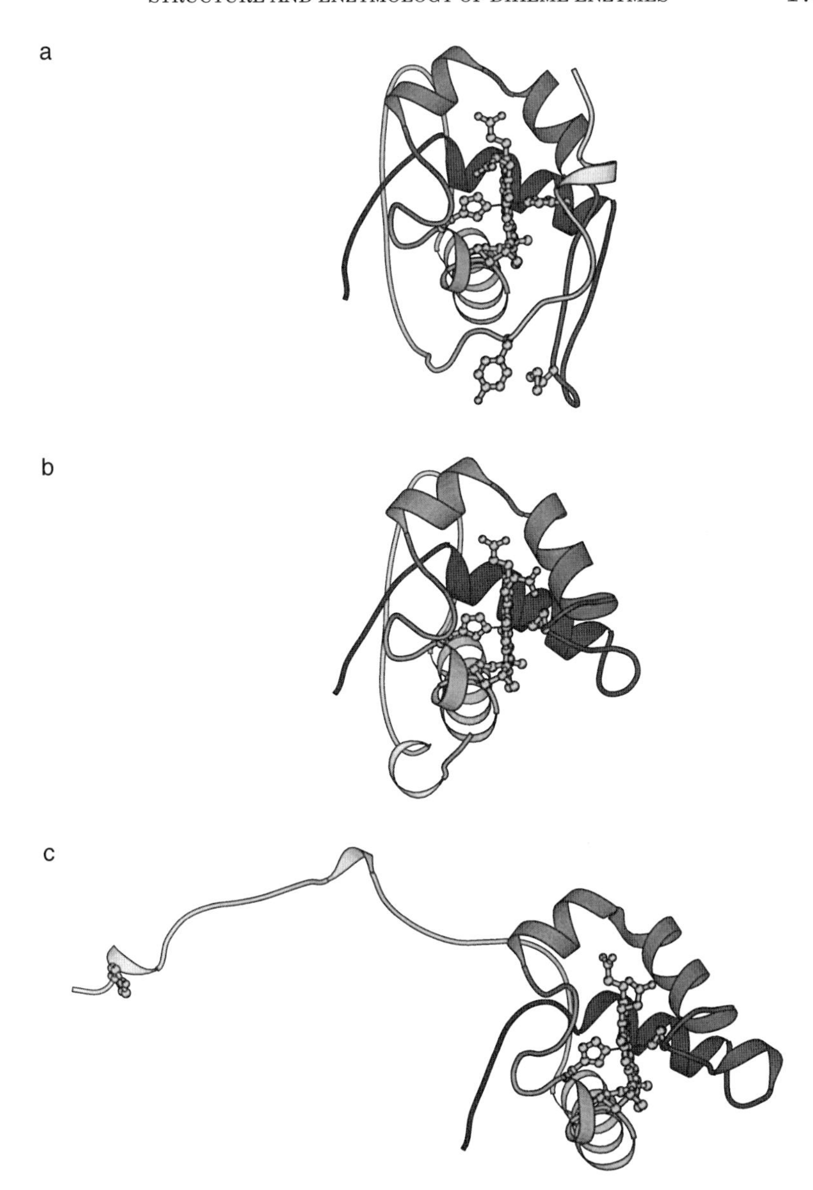

FIG. 5. The *c*-heme domain in cytochromes cd_1 from *P. pantotrophus* and *P. aeruginosa*. (a) Oxidised form of *P. pantotrophus* cytochrome cd_1 showing bis-histidine ligation of the heme iron. Tyr 25 that is a ligand to the d_1 heme can be seen at the bottom of the figure adjacent to Met 106. (b) Reduced form of *P. pantotrophus* cytochrome cd_1 showing the movement of Met106, which replaces His17 as a ligand. (c) Oxidized form of *P. aeruginosa* cytochrome cd_1 showing His/Met ligation coordination of the heme iron. To the left of the figure can be seen Tyr10, which is adjacent to the d_1 heme in the other subunit of the dimer. (Drawn from PDB entries 1 qks, 1 aof, and 1 nir.)

sphere, thus generating a five-coordinate site that could be anticipated as providing the nitrite binding site (Fig. 4b). The latter proposition was quickly supported by the observation that if nitrite was diffused into the reduced crystals, and followed by rapid freezing, then several different structures could be obtained with either nitrite or nitric oxide bound (Figs. 4c,d). In one of these structures, one of the oxygens in nitrite was clearly within hydrogen bonding distance of both histidines, already mentioned as putative proton-donor residues (Fig. 4c). The position of the Tyr 25 residue varies depending on whether the ligand bound is nitrite or nitric oxide. With the latter, the tyrosine is in a position as if poised to enter the active site and displace the bound nitric oxide (Fig. 4d). The X-ray structures naturally do not tell us if the heme centers are in the oxidized or reduced states, but we can evaluate some possibilities. In the structure with nitrite bound to the d_1 heme iron it seems very likely that the d_1 heme iron must be in the reduced state. This is because there is no evidence that nitrite will readily bind to the oxidized form of the enzyme. If the d_1 heme is in the reduced state, then the next question to be addressed is that of the oxidation state of the c heme center on the same polypeptide chain. This could be either oxidized or reduced, because the d_1 heme might already have carried out one reductive turnover of a nitrite ion to nitric oxide, received an electron from the c heme, and bound a second nitrite, in which case the c heme would be oxidized. Alternatively, the c heme might still be reduced if the structure is of an enzyme that has bound nitrite, but not brought about any catalysis at this point. When the cytochrome c domain is inspected for this nitrite-bound structure, it is found that the ligation is His/His, suggesting that it is in the oxidized state. This observation is important because it shows that the His/His liganded state does not need tyrosine anchored to the Fe of d_1 heme. The same issue, of course, pertains to the c heme domain when nitric oxide is bound to the d_1 heme iron. Again all polypeptides observed with nitric oxide bound also have His/His ligation of the c-type heme center. Also, it is not possible to deduce the oxidation state of the d_1 heme from the structure, although ruffling of the heme varies depending upon which ligand is bound.

The question of the oxidation state of the heme iron with NO bound might in principle thought to be accessible by consideration of the Fe–N distance and bond angles. However, with respect to distance it is important to realize that X-ray structures do not in themselves determine distances such as that from Fe to N to sub-angstrom resolution. The procedure is to fit electron density using restraints frequently obtained from small molecule studies. Thus, unless the resolution of a protein crystal structure is remarkably high, close to 1 Å, it is not possible to state definitively whether the Fe–N distance is in fact 1.8 or 2.0 Å. There are two reasons for this. First, the resolution of the electron density map (i.e., interpretable interatomic distances) is approximately the higher resolution limit of the diffraction data used for map calculation and/or structure refinement (*13*). Second, the electron density is subject to ripples around atomic positions due to Fourier series termination error. Not only do these ripples limit resolution, but they also affect the atomic positions in the electron density map (*13*). The principal cause of the series termination error is the incompleteness of the observed data set, and it is usually described by the low- and high-resolution limits. If the high-resolution limit is caused by the weak diffraction of the crystal, this limit would not cause significant artifacts in the map. But in practical terms, the Fourier series is always truncated, because diffraction data sets are never 100% complete. This is typically due to the limited data available using the common oscillation method around a single axis, overloaded strong reflections on the detector, and the low-resolution limit caused by the beam stop. The amplitude of the ripple around an atom is proportional to its atomic number, and the wavelength of the ripple is approximately the resolution limit of the data. For example, a 2.5 Å high-resolution limit will cause features of the electron density map to be surrounded by ripples with a wavelength of about 2.5 Å (*13*). The positions of "strong scatterers," such as atoms in an iron–sulfur cluster, are usually well determined in a typical protein structure. The problem typically arises when a relatively "weak scatterer" (NO, OH^-, etc.) is riding on the big waves around the metal center because of limited resolution effects. Partial occupancy of the bound ligand and mixed oxidation or spin states of the metal can further complicate deduction of metal to ligand bond lengths. There are statements in the literature that a metal to ligand bond is, say, 1.7 Å in length, when the diffraction data were only collected to, say, 2.4 Å. In such circumstances the assignment of such a bond length is not accurate. Thus, the temptation to deduce too much about the metal–ligand distances needs to be resisted. This limitation on low- and medium-resolution X-ray diffraction data illustrates why EXAFS can be a valuable complementary technique for probing metal-to-ligand interactions.

Nevertheless, in the case of cytochrome cd_1, the data do fit better to 2.0 Å (using 1.8 Å resolution close to complete data sets) for the Fe–N distance, and this might be taken as tentative evidence for the Fe(III) oxidation state. The bond angle for Fe–N–O cannot automatically be used as diagnostic of the oxidation state. Whereas such bond angles can be diagnostic for small molecules, other factors such as the presence of amino acid side chains can seriously affect the geometry of a heme ligand inside a protein. Thus, although small molecule studies on heme–NO complexes show that the Fe(III) derivatives have an essentially linear Fe–N–O group and a very short bond Fe–N of 1.63 to 1.65 Å in either five- or six-coordinate species (*14*), the deviation from this pattern seen in nitrite reductase cannot be taken as definite evidence against the Fe–NO species being in the (III) state. Small-molecule Fe(II)–NO complexes have an Fe–N–O bond angle of 140–150° and a short Fe–N bond distance of 1.72 to 1.74 Å (*14*). A further feature of the Fe(II)–NO state is that the bond *trans* to the NO is very long in the six-coordinate Fe(II) species; this bond lengthening is not present in the structures of cytochrome cd_1 with NO bound. In fact, notwithstanding the earlier caveats about metal–ligand bond distances, the Fe–histidine bond distance in this complex shortens slightly to 1.96 Å, compared with 1.98–2.00 Å in other ligand states, with everything restrained to 2.00 Å. Thus, although the Fe–N–O angle suggests that the NO-bound form is a ferrous state (formally at least), not all the evidence supports this view.

The observations made by X-ray crystallography would clearly be consistent with a reaction mechanism for cytochrome cd_1 nitrite reductase in which heme ligand switching occurred at the *c* heme and at the d_1 heme each time a nitric oxide molecule was produced. It is notable that microspectrophotometry of crystals has shown that, following reduction and exposure to nitrite, the crystals regain the characteristic spectrum of the fully oxidized enzyme (*12*). This means that in the crystalline state, the nitrite reductase does not get trapped in a dead-end state but rather performs rounds of catalysis before regaining the initial structural state. The observation of the Tyr 25 residue poised to return to the d_1 heme iron is also consistent with such a conclusion. This tyrosine residue could play a key role in ensuring that the nitric oxide is displaced from a Fe(III)–d_1 heme before the arrival of an electron generates the Fe(II) oxidation state.

Such an involvement of an amino acid side-chain ligand switch within each catalytic cycle was a novel proposal and as such needs to be scrutinized by a variety of experimental procedures as well as analysis in the context of information known for cytochrome cd_1 nitrite reductase from another source (see later discussion). However, it is interesting to note that something similar has been proposed for the protocatechuate 3,4-dioxygenase enzyme from *Pseudomonas putida* (*15*). On the other hand, bacterial cytochrome *c* peroxidase offers an example where ligand switching seemingly relates only to an activation phenomenon.

It is interesting, as has been proposed, that the peroxidase from *P. denitrificans* may undergo a Met/His ligand switching at center (II) (N-terminal domain) (*16, 17*). This proposal, which remains to be tested experimentally, is discussed further in Section III,E.

C. Structure of *Pseudomonas aeruginosa* Cytochrome cd_1

Over the years the cytochrome cd_1 that has received most attention is from *P. aeruginosa*. Although cytochrome cd_1 is a specialized enzyme, it is remarkable that the structure of the *P. aeruginosa* enzyme is not identical in several critical respects to the enzyme from *P. pantotrophus*, most notably in terms of the coordination of the *c* heme and d_1 heme iron centers. In the oxidized structure of the *P. aeruginosa* enzyme, one sees that the overall structure of the dimer is the same in the sense that the *c*-type cytochrome domain is helical and separated from the eight-bladed β-propeller domain that binds the d_1 heme center (*18*). However, there are intriguing differences. First, the fold of the cytochrome *c* domain of the oxidized *P. aeruginosa* (Fig. 5c) enzyme is essentially the same as in the reduced *P. pantotrophus* (Fig. 5b) enzyme, and the heme iron coordination is His/Met. The iron of the d_1 heme does not have a tyrosine ligand; in contrast to the enzyme from *P. pantotrophus* the sixth ligand is hydroxide (Fig. 6), but this in turn is bonded to Tyr 10.

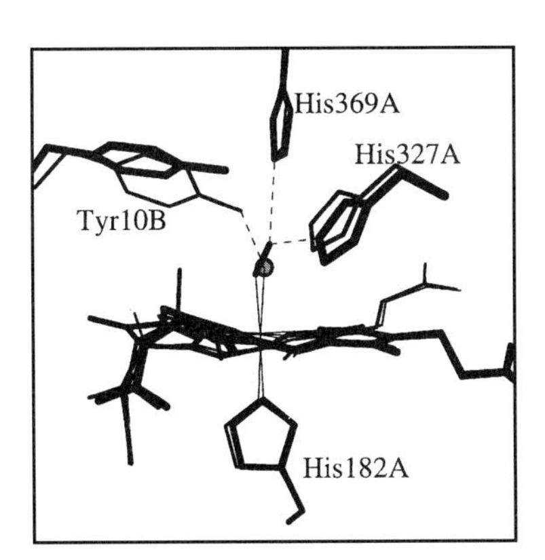

Fig. 6. Structural changes at the d_1 heme of *P. aeruginosa* cytochrome cd_1 upon reduction and binding of nitric oxide. There are no significant changes in the *c* heme domain. The residue positions shown by thin lines are those of the oxidized protein with hydroxide ion; those in bold represent the reduced enzyme with nitric oxide bound to the d_1 heme iron. Changes in the d_1 conformation can also be seen. The figure shows the d_1 heme from the A subunit; note that Tyr10 is provided from the B subunit. Dashed lines represent potential hydrogen bonds. (Drawn from PDB entries 1 nir and 1 nno.)

However, the latter residue is in no sense equivalent to the Tyr 25 of the *P. pantotrophus* enzyme. The Tyr 10, which is not an essential residue (*19*), is provided by the other subunit to that in which it is positioned close to the d_1 heme iron (Fig. 6). In other words, there is a crossing over of the domains. A reduced state structure of the *P. aeruginosa* enzyme has only been obtained with nitric oxide bound to the d_1 heme iron (*20*) (Fig. 6). As expected, the heme *c* domain is unaltered by the reduction, but the Tyr 10 has moved away from the heme d_1 iron, and clearly the hydroxide ligand to the d_1 heme has dissociated so as to allow the binding of the nitric oxide (Fig. 6). This form of the enzyme was prepared by first reducing with ascorbate and then adding nitrite.

Although the conformational changes in the *P. aeruginosa* enzyme are clearly less pronounced than those in the *P. pantotrophus* enzyme, the driving forces for the changes still require to be understood. It had been postulated (*20*) that reduction of the *c*-type cytochrome domain might lead to conformational changes leading to the release of the hydroxide from the d_1 heme Fe. However, a recent study contradicts this view. It has been possible to obtain crystals of the *P. aeruginosa* enzyme in which the *c* heme is reduced but the d_1 heme is oxidized, and this condition persists for sufficient time for the structure to be obtained. This shows that hydroxide is still bound to the d_1 heme, and that reduction of the d_1 heme, either alone or in combination with reduction of the *c* type heme, is therefore responsible for the conformational change at the d_1 heme iron (*21*). Reduction of the d_1 heme iron alone seems likely to be the essential factor, not least because no major change in the *c* heme domain conformation occurs upon reduction, and therefore it is difficult to see how there could be any effect of reduction of the *c* heme center relayed to the d_1 heme. Pulse radiolysis work with the *P. aeruginosa* enzyme has shown that the electron transfer from the *c* heme to the d_1 heme is very slow (on the order of seconds) (K. Kobayashi *et al.*, unpublished observations). This would be consistent with the idea that it is arrival of an electron at the d_1 heme that triggers the conformational change. This change could easily limit the electron transfer rate. There is clearly a parallel here with the *P. pantotrophus* enzyme, where the reduction of the d_1 heme, it has been argued, triggers the conformational change (see earlier discussion).

It has long been known that, under some conditions at least, electron transfer between the *c* and d_1 hemes of the *P. aeruginosa* enzyme is slow and requires times of the order of seconds (*22*). What does this mean? It is not necessarily related to the loss of the hydroxide ligand from the d_1 heme iron, because under some experimental conditions used, azurin (a cupredoxin) was present and the enzyme was reduced at the outset, and it is the movement of an electron from the *c* to the d_1 heme that is slow despite the presence of nitric oxide as a ligand on the d_1 heme. This raises a problem because the nitric oxide gets trapped, and thus what seems to be a dead-end ferrous d_1 heme nitric oxide complex is formed. An expected gating mechanism that might be anticipated to prevent the formation of an Fe(II)–d_1 heme nitric oxide complex appears not to operate. However, it is clear that under some conditions the rate of interheme electron transfer can be much faster. An example is provided by the work of Wilson *et al.* (*23*). They were able to trigger electron transfer by photodissociation of carbon monoxide from a mixed valence form of the enzyme. Electrons moved from the d_1 to the *c* heme at a rate of thousands per second. The dissociation of carbon monoxide and study of the kinetics of its rebinding also gave insight into the dynamics of the d_1 heme pocket. Again the meaning of these observations in the wider context of enzyme function is not clear.

It has long been assumed that azurin is an *in vivo* electron donor to cytochrome cd_1 of *P. aeruginosa*. The construction of mutants of *P. aeruginosa* in which one or both of the genes for azurin and cytochrome c_{551} have been deleted has led to the conclusion that *in vivo* cytochrome c_{551} is essential for the donation of electrons to the nitrite reductase and that azurin is ineffective (*24*). The discrepancy between *in vivo* and *in vitro* observations either could be reconciled if it is the failure of azurin to accept electrons from the cytochrome bc_1 complex or another donor that is responsible for its ineffectiveness *in vivo*.

D. Solution Spectroscopy of *P. pantotrophus* Cytochrome cd_1

One expects structures of specialized proteins such as cytochrome cd_1 to be independent of the source of the enzyme. Thus, the two different structures seen for the enzymes from *P. pantotrophus* and *P. aeruginosa* naturally raise the question as to whether crystallization has generated a structure for one of the two enzymes that does not reflect the solution structure. However, the oxidized *P. pantotrophus* protein, as isolated, has in solution the same His/His coordination of the *c*-type heme as is observed in the oxidized state of the crystallized enzyme. This was determined from magnetic circular dichroism (MCD) spectroscopy. Other solution spectroscopic measurements have shown that ligation of the d_1 heme is very likely the same as in the crystalline state. These studies also showed that the d_1 heme iron appeared to be in an unusual room temperature high/low equillibrium (*26*). Comparable studies on the enzymes from *P. aeruginosa* and *Pseudomonas stutzeri* have indicated that in solution, as in the crystal of *P. aeruginosa,* the *c* heme ligands are

His/Met and the d_1 heme ligation is probably hydroxide rather that tyrosine. Thus, solution studies, as well as analyses of primary sequences, confirm that cytochrome cd_1 from different sources can have different ligands to the heme irons.

An important question about cytochrome cd_1 is why the d_1 heme has been uniquely recruited to this enzyme. Detailed spectroscopic analysis of the d_1 heme center suggests that the relative energy levels of the d orbitals associated with the d_1 heme iron differ from the usual pattern in that for the low-spin ferric state the four-electron reduced heme d_1 macrocycle stabilizes $(d_{xz}, d_{yz})^4(d_{xy})^1$ relative to the more usual $(d_{xy})^2(d_{xz}, d_{yz})^3$ electronic configuration (*25, 26*). This "inverted" ground state represents a relocation of unpaired spin density from orbitals extending above and below the heme plane to an orbital lying in that plane, and this will have important consequences for the binding of π-unsaturated ligands such as nitric oxide, in this case the product molecule. Specifically, NO would be less strongly bound to the d_1 heme then to the usual b-type heme. This would be a desirable property of cytochrome cd_1; for further discussion the reader is referred to Watmough *et al.* (*27*), and references therein.

E. Mechanistic Studies on *P. pantotrophus* Cytochrome cd_1

1. Electron Transfer Studied by Pulse Radiolysis

Information about the kinetics of electron transfer between the hemes and of chemical events at the hemes is sparse for the enzyme from *P. pantrotrophus,* or even for that from the better-known related organism *P. denitrificans,* which has been studied for many years. The first reported kinetic study on the *P. pantotrophus* enzyme is that of Kobayashi *et al.* (*28*), in which pulse radiolysis methodology was used. In this work exceedingly rapid reduction of the c-type cytochrome center in the enzyme was followed by electron transfer to the d_1 heme on the millisecond time scale. This method involves one-electron processes only, and so the c heme, once reoxidized by the d_1 heme, remains oxidized. Interpretation of this millisecond rate of electron transfer is not straightforward because we do not know the driving force, that is, the redox potentials of the c- and d_1-type hemes. However, under the conditions of the pulse radiolysis experiment the electron transfer from the c-type center to the d_1 heme occurs essentially to completion. This implies that, under these conditions at least, the redox potential difference is on the order of at least 100 mV. A difference of 100 mV and an edge-to-edge heme distance of 11 Å would suggest that electron transfer might be faster than is observed, as judged by current theories (*5*). However, these theories suppose that no chemical bond rearrangements accompany the electron transfer event. In the case of cytochrome cd_1 from *P. pantotrophus,* at least two chemical bond rearrangements might accompany the oxidation/reduction processes. These are the ligand switching at the *c* heme and the dissociation of Tyr 25 from the d_1 heme iron. For the following reasons it is likely in the pulse radiolysis experiment that the ligands do not change at the *c*-type center, but do so at the d_1 heme:

1. The spectrum in the Soret region, indicative of the *c*-type center of the enzyme, immediately following the reduction is not identical to that obtained when the enzyme is fully reduced under steady-state/equilibrium conditions. This suggests that the reduced *c*-type cytochrome center formed under the conditions of the pulse radiolysis experiment retained the His/His coordination.
2. The speed of the reduction by the solution radical generated in the pulse radiolysis experiment is also consistent with this proposal; conformational rearrangements are highly unlikely on the microsecond time scale.
3. The midpoint redox potential of a His/His coordinated heme is likely to be less than +100 mV. Given that the d_1 heme is catalyzing a reaction with a midpoint potential of approximately +350 mV, it can be expected to have a potential considerably more positive than +100 mV. This would account for the stoichiometric transfer of electrons from the *c* heme to the d_1 heme under the conditions of the pulse radiolysis experiment.
4. The d_1 heme probably loses its Tyr 25 ligand under these conditions, because if nitrite is present during the pulse radiolysis experiment, then although the rate of electron transfer between the hemes is essentially unaltered, there is evidence for a chemical process taking place at the d_1 heme center (*28*). This suggests that, at least in the presence of nitrite, the arrival of an electron at the d_1 heme triggers the dissociation of the Tyr 25 ligand.

All these observations prompt the question as to what the redox potentials are for the *c* and d_1 hemes under equilibrium conditions. This issue is currently under study, and all that can be said here is that the enzyme does not give a straightforward redox titration. In particular, and in contrast to the observations made under pulse radiolysis conditions, the *c* and d_1 hemes titrate together, suggesting a cooperativity of behavior between them (see note added in proof). It remains to elucidate the molecular basis for this effect.

2. Mechanism of Nitrite Reduction

It is generally accepted that the reaction catalyzed by cytochrome cd_1 begins by nitrite binding through its nitrogen atom to the ferrous d_1 heme iron. As discussed earlier for *P. pantotrophus,* this is supported by crystallographic data (*12*). However, strictly speaking, it is not possible to exclude that during steady-state turnover nitrite might bind to the ferric state of the d_1 heme. When in rapid reaction studies nitrite binds to the reduced d_1 heme, there is evidence for protonation and rapid dehydration of nitrite and oxidation of the d_1 heme iron, resulting in a species that can formally be regarded as Fe^{3+}–NO or Fe^{2+}–NO^+. Under some conditions the latter species must be relatively long-lived, because it can undergo rehydration in ^{18}O-containing water to give ^{18}O-labeled nitrite, which eventually yields ^{18}O-labeled nitric oxide. The Fe^{3+}–NO species can also be trapped by nucleophiles such as azide, hydroxylamine, or aniline. This species has not so far been directly detected by a spectroscopic method during steady-state catalysis. However, it has been detected by adding nitric oxide to the oxidized form of *P. stutzeri* cytochrome cd_1 (*29*). The method used was FTIR. In due course it will be of interest to discover whether stopped-flow FTIR can detect this species during catalyzic turnover.

It is expected that the Fe^{3+}–NO species will decompose to release product nitric oxide and generate a ferric form of the d_1 heme iron center. The iron atom may then be five- or six-coordinated. Six-coordination would arise in the mechanism proposed by Fülöp *et al.* (*4*), in which the Tyr 25 ligand was predicted to become a d_1 heme iron ligand following the departure of nitric oxide. In principle, an alternative mechanism would be that the ferric d_1 heme–NO complex might become reduced before releasing nitric oxide. Such reduction might be relatively facile because an electron can be available on the *c* heme which is only approximately 11 Å away, and ferrous heme–NO complexes are usually much more stabile then their ferric counterparts, which would provide a driving force for electron transfer from the *c* heme center. However, the greater stability of ferrous heme–NO complexes suggests that formation of such a species during catalysis would lead to an inhibited enzyme from which NO release would be very slow. Thus we might expect cytochrome cd_1 to have important design features for avoiding formation of the ferrous d_1 heme–NO complex. The slow rate of interheme electron transfer has suggested such a mechanism. However, the often observed formation of the ferrous d_1 heme–NO complex suggests that experimental conditions that would promote the release of NO have not yet been identified.

In a stopped-flow study on cytochrome cd_1 from *P. aeruginosa,* the ferrous d_1 heme–NO species was formed despite electron transfer from the c to the d_1 heme being relatively slow, rate constant approximately 1 s^{-1}, for the relatively short distance between the two hemes (*32*). Such a distance would normally predict much faster electron transfer. The relatively slow interheme electron transfer rate has been observed on a number of occasions, and before the structure of the protein was known was thought to reflect the relatively large interheme separation distance and/or relative orientation of the two hemes (*30*). The crystal structures provide no evidence for either of these proposals; there is nothing unusual about the relative orientation of the c and d_1 heme groups.

In summary, there is still much to understand about the nitrite reduction reaction. The crystal structures have shown how nitrite can bind to the d_1 heme iron and protons can be provided to one of its oxygen atoms from two histidine residues. However, as yet no rapid reaction study has detected the release of product nitric oxide rather than the formation of the inhibitory dead-end ferrous d_1 heme–NO complex. It is also not clear why the rate of interheme electron transfer is so slow over 11 Å when nitrite or nitric oxide is the ligand to the d_1 heme.

3. *The Oxidase Reaction of Cytochrome* cd_1

There are several demonstrations that cytochrome cd_1 catalyzes the reduction of molecular oxygen to water. Exactly how the enzyme catalyzes this reaction is of some interest, because the crystal structure shows that the catalytic center is mononuclear and expected to handle one electron at a time. If we assume that electron transfer between subunits cannot occur, then only two of the four electrons required for reduction of one oxygen atom can obviously be stored on one subunit of the enzyme before reduction of oxygen commences. Thus, it might be anticipated that some intermediates of oxygen reduction are relatively long-lived.

The oxidase reaction was first studied some years ago for the *P. aeruginosa* enzyme, but the most recent study, made in the knowledge of the crystal structure, has been on the enzyme from *P. pantotrophus*. In addition to providing some intriguing insights as to the oxidase reaction, this study has also provided valuable new information about the general properties of cytochrome cd_1.

Studies on the reaction of *P. pantotrophus* cytochrome cd_1 with oxygen required that the enzyme be initially in the reduced state. Although the latter can be obtained using dithionite as reductant, this is best avoided because oxidation products, such as SO_2, can perturb the system. Consequently, the enzyme was reduced with ascorbate. A feature of this reaction was the slow time course, even when a mediator such as a ruthenium complex was included. Complete reduction took more than 2 hours (*31*). When the reduced enzyme was mixed with oxygen, visible absorption spectroscopy showed that oxidation at the d_1 center was substantial after 25 ms. The *c* heme center was substantially oxidized after 50 ms. Thus, interheme electron transfer took place on the millisecond time scale (*31*). This contrasts with studies on the nitrite reductase activity of the *P. aeruginosa* enzyme, when the corresponding process occurs on the seconds time scale (*32*). However, it should be noted that there is an old report of much faster interheme electron transfer in the *P. aeruginosa* enzyme, when oxygen was the substrate (*33*). For the present one must conclude that the interheme electron transfer rate is significantly affected by the nature of the chemical events occurring at the d_1 heme.

The initial oxidation of the d_1 heme by oxygen is followed by chemical events, the interpretation of which was assisted by parallel freeze-quenched EPR measurements. There was one unexpected process detected by optical spectroscopy. This was rereduction of the *c* heme center by ascorbate on the seconds time scale. This was surprising because the initial reduction by ascorbate took more than 2 hours.

Freeze-quenched EPR studies were done at various time points following mixing of ascorbate reduced enzyme with oxygen (*31*). Information about ligands and oxidation states of both the *c* and d_1 heme centers was obtained. As expected, after 25 ms a signal from the ferric *c*-type heme was observed. Importantly, this signal was not the same as that obtained from the "as prepared" oxidized enzyme used in both crystallographic and previous spectroscopic studies (*4, 26*). The latter was clearly diagnostic of bis-histidine iron coordination with a single *g*-value at 3.05. This contrasted with signals at $g = 2.93$ and $g = 2.26$, which were observed following exposure of the ascorbate-reduced enzyme to oxygen. Thus, the ferric *c*-type cytochrome center can access at least two different coordination states. There are two main possible interpretations. The first is that the $g = 2.93$ and $g = 2.26$ signals reflect a different geometry of the bis-hitidinyl coordinated heme iron. The second is that these two *g* values indicate that the ligation of the heme iron is His/Met. Two lines of evidence support this interpretation. First, $g = 2.92$ and $g = 2.26$ are seen for the oxidized form of "semi-apo" cytochrome cd_1; the latter retains the *c*-type heme but has lost the d_1 heme. This "semi-apo" form of the enzyme is known to have His/Met coordination of the *c* heme for two reasons: first because of a characteristic absorption band at 695 nm (*31*), and second because of an upfield shifted of the methyl resonance, characteristic of an axial methyl ligand, in the nuclear magnetic resonance spectrum (*34*). In addition, similar *g* values are observed for the *c*-type heme center in the oxidized *P. aeruginosa* enzyme, which is known to have His/Met ligation (*35*).

Although the matter cannot be regarded as completely settled, it seems likely that during the oxidase reaction a His/Met ferric *c* heme occurs. This might be expected to be more readily reduced by ascorbate than the His/His form, thus explaining the more rapid reduction of the enzyme by ascorbate after an initial oxidation by oxygen. Support for this view comes from the finding that the semi-apo enzyme is much more rapidly reduced by ascorbate than the "as prepared" oxidized holoenzyme.

Until 77 ms after mixing reduced enzyme with oxygen, no EPR signal attributable to the d_1 heme could be detected. However, unexpectedly, a signal characteristic of an organic radical species was identified. The *g*-values suggest that this radical resides on an amino acid side chain, tentatively assigned as tyrosine. The decay of this radical species correlated with the appearance of an EPR signal that could be assigned to a hydroxide ligand on low-spin ferric d_1 heme.

The combined results from rapid visible absorption and EPR spectroscopies on cytochrome cd_1 from *P. pantotrophus* suggest the following sequence of events (*31*). First, the fully reduced enzyme with five-coordinate d_1 heme iron binds oxygen, an event that is followed by very rapid electron transfer from the d_1 heme to the bound oxygen. This is followed almost immediately by the arrival of electrons from the *c* heme and a tyrosine side chain. This results in the cleavage of the oxygen molecule, release of water, and generation of an EPR silent Fe(IV)–oxo species; three electrons are required for this process. In parallel, the *c* heme center, which has retained His/Met coordination, can be reduced by ascorbate. Electrons derived in this way by ascorbate are used to reduce the amino acid side chain and to generate a ferric hydroxyl species at the d_1 heme active site. This mechanism clearly needs to be tested by further work. It is of general interest because the Fe(IV)–oxo species with attendant amino acid radical species appears to accumulate on this enzyme, but not in others, such as yeast cytochrome *c* peroxidase or mitochondrial cytochrome oxidase, where it has been proposed. The reason for the difference is probably related to the small k_{cat} value for the oxidase reaction of cytochrome cd_1, which may, therefore, provide an important model system for characterizing the role of amino acid radicals in other oxygen-dependent enzymic reactions.

Unfortunately there are not as yet any reports of rapid reaction studies on the *P. pantotrophus* enzyme when it is catalyzing nitrite reduction. The oxidase reaction and the properties of the "semi-apo" enzyme suggests that the His/His ligation of the *c*-type cytochrome center in the *P. pantotrophus* enzyme would only be observed when Tyr 25 is anchored to the d_1 heme iron. Given the close proximity of Tyr 25 to His 17 in the sequence, such a connection between the binding of these amino acid side chains to the two heme irons is reasonable. However, it should be recalled that in the crystal structures of the enzyme with nitrite or nitric oxide bound, His 17 is still a ligand to the heme *c* center. Hence, although work on the oxidase reaction may suggest that the His/His coordinated state of the *c* heme is not catalytically relevant, this may not be true of the physiological reduction of nitrite to nitric oxide.

F. Relationship of Cytochromes cd_1 to the Copper Nitrite Reductases

The copper-containing nitrite reductase is a trimer of identical subunits. In each subunit there is a type 1 copper, which acts analogously to the *c*-type heme in cytochrome cd_1, and thus is the point of entry of electrons into the enzyme. The three catalyzic sites have type 2 copper and are located on interfaces between two subunits. The type 2 copper ligands are provided by amino acid side chains; there is no organic cofactor. Current mechanisms for the copper-type nitrite reductase envisage that nitrite binds in bidentate fashion, via its two oxygen atoms, to copper(II) with concomitant displacement of a hydroxide ion. Subsequently it is proposed that an electron is received from the type 1 copper center. The reduction of the type 2 copper, together with protonation and loss of hydroxide, generates nitric oxide bound via its oxygen to copper (*22, 27*). There is therefore a debate as to whether there is rearrangement of the bound nitric oxide so as to give a copper–nitrogen bond before product release (*36–38*). In any event, it is clear that the mechanism of nitrite reduction catalyzed by the copper enzyme is distinct from that of cytochrome cd_1.

G. Outstanding Issues

It is remarkable that the oxidized states of the cytochromes cd_1 from *P. pantotrophus* and *P. aeruginosa* have different structures. It is not clear at present whether one of these structures is superior for catalyzing nitrite reduction. Certainly in the *P. pantotrophus* enzyme the ligand switching at both ligand centers upon changing oxidation state is novel. The X-ray structures of enzyme with bound substrate and product would suggest that these switches occur during catalysis, but this is not definitely demonstrated. If these switches do not relate to catalysis, then their role is not clear. Arguably, the oxidized structure of the enzyme may represent a resting or protected state and that it is the reduced conformer, which is similar in both enzymes, and is important in catalysis (*39*). If this is the case, then why is a similar protected state seemingly not needed for the *P. aeruginosa* enzyme?

It is interesting that copper or iron can often be used to catalyze the same reaction in biology, and nitrite reduction is one of many examples. In the case of the cytochromes cd_1, the chemical rational for the adoption of the d_1 heme ring has not been rigorously determined, but it is striking that no organic cofactor is needed for the copper enzyme. It is also unclear why in both cytochrome cd_1 and copper nitrite reductase there is a requirement for a metal center to act as a receiver from an electron donor protein. Since a one-electron reaction is catalyzed, a single metal center might have been expected to be sufficient. In the case of cytochrome cd_1, regulation of the electron transfer from the c to the d_1 heme may be necessary in order to avoid the formation of a dead-end Fe(II)–d_1 heme nitric oxide complex.

Finally, this may not yet be the end of the story concerning structural variation in cytochromes cd_1. The sequence of cytochrome cd_1 from *P. stutzeri* shows no counterpart of either Tyr 10 or Tyr 25, and thus the c heme domain may be quite distinct from those observed to date.

III. Diheme Cytochrome *c* Peroxidases

A. Properties of Bacterial Diheme Cytochrome *c* Peroxidases

It is nearly 50 years since a c-type cytochrome was shown to catalyze peroxidase activity in crude extracts of *Pseudomonas fluorescens* (*40*). The enzyme responsible was first purified some 20 years later by Ellfolk and Soninen from the closely related *P. aeruginosa* and shown to be a diheme cytochrome c peroxidase (CCP) (*41*). These bacterial diheme CCPs are quite distinct from the superfamily of plant and yeast peroxidases (*42*) and are widely distributed among the Gram-negative bacteria (*41, 43–46*). Diheme CCPs are located in the periplasm (Fig. 2), where they catalyze the two-electron reduction of H_2O_2 to H_2O by soluble one-electron donors such as cytochromes c and cupredoxins.

The physiological function of bacterial cytochrome c peroxidase is not completely clear, but usually the enzyme seems to be expressed at the highest levels following growth under microaerobic conditions. In *P. denitrificans* expression of CCP is controlled by the FNR system of global regulation (*47*), suggesting that CCP is an enzyme of microaerobic metabolism linking an essential detoxification pathway to respiration. This may reflect the availability of hydrogen peroxide in soils and aquatic environments under some conditions (*48*). Richardson and Ferguson (*49*) have suggested that hydrogen peroxide respiration (via CCP) occurs in anaerobically grown cells of *P. pantrotrophus* and *Rhodobacter capsulatus*. This was deduced from an inhibitory effect of hydrogen peroxide, but not of oxygen, on nitrate respiration via the periplasamic nitrate reductase (*49*). This differential effect of hydrogen peroxide and oxygen excludes a role for catalase in converting added hydrogen peroxide to oxygen. It would, therefore, appear that in a number of important human pathogens, including *P. aeruginosa, Helicobacter pylori,* and *Vibrio cholera,* CCP may be an essential element of the organisms's response to toxic oxygen species, and as such a potential target for antibacterial agents. Note, however, that Koutny *et al.* (*50*), using biosensor methodology, concluded that exogenous hydrogen peroxide was not acted upon by cytochrome *c* peroxidase, despite being present in high levels, in anaerobically grown cells of *P. denitrificans*. Instead it was converted by catalase to oxygen, which in turn acted as a terminal electron acceptor.

In contrast to the extensively studied superfamily of plant peroxidases, diheme CCPs contain two heme *c* moieties covalently attached to a single polypeptide chain. Thus far, the two best characterized CCPs are the enzymes isolated from *P. aeruginosa* and from *P. denitrificans*. A number of early studies of both enzymes have demonstrated the inequivalence of the two hemes, in particular their midpoint redox potentials. These have been determined to be about +320 and −320 mV (*51, 52*). The similarity of the latter value to the reduction potentials reported for heme *b* in the active sites of plant peroxidases (*53, 54*), led to the assumption that this low-potential heme is the catalytic center. The high-potential heme would then serve to receive electrons from external donors and mediate subsequent electron transfer to the active site. Thus, there are similarities to the cytochromes cd_1 described earlier in terms of the strategy used to accept electrons from their redox partners. Also, both types of enzyme need to become part reduced before they can function as catalysts—reduction initiating a change in the coordination number of the active site heme to allow substrate binding. It must be stressed that in the CCPs it is reduction of electron accepting heme that precipitates change at the active site heme, whereas in cytochrome cd_1 the active site itself must be reduced (*21*).

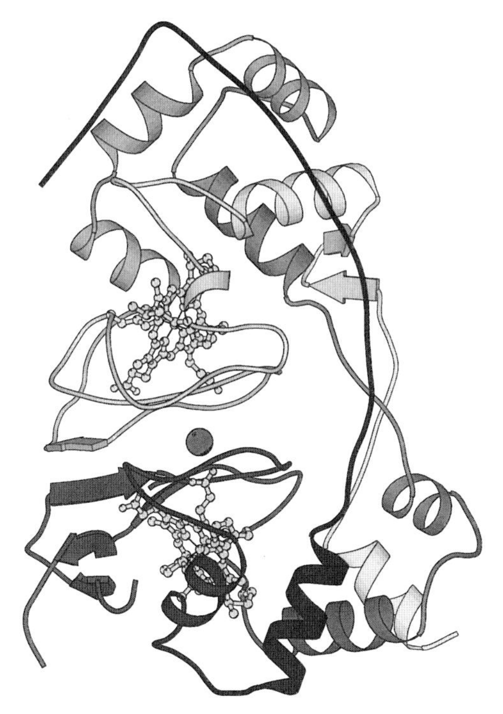

FIG. 7. The structure of the oxidized state of cytochrome *c* peroxidase from *P. aeruginosa* (only a monomer is shown here). The small sphere between the low-potential heme (top) and the high-potential heme (bottom) domains shows the position of the Ca^{2+} ion.

B. STRUCTURE OF CYTOCHROME *c* PEROXIDASE FROM *PSEUDOMONAS AERUGINOSA*

Protein molecules of *P. aeruginosa* CCP form dimers both in solution and in the crystal, and the crystallographic asymmetric unit contains a monomer. The three-dimensional structure of *P. aeruginosa* CCP at 2.4 Å resolution (*55*) shows that the 34-kDa monomer is organized into two structurally distinct domains around the two covalent *c* heme centers (Fig. 7). The structures of both domains resemble the structure of class 1 *c*-type cytochromes (*56*). However, both domains are somewhat larger than is usual for cytochromes because of additional α-helices, β-strands, and loops (Fig. 8). The crystallized enzyme is in the fully oxidized state and shows, in agreement with numerous spectroscopic measurements (*57–59*), that the low-potential heme iron at the N-terminal domain has two axial ligands (Fig. 8a). Both ligands are histidines (His 55 and His 71), as in the *c* domain of *P. pantotrophus* cytochrome cd_1 nitrite

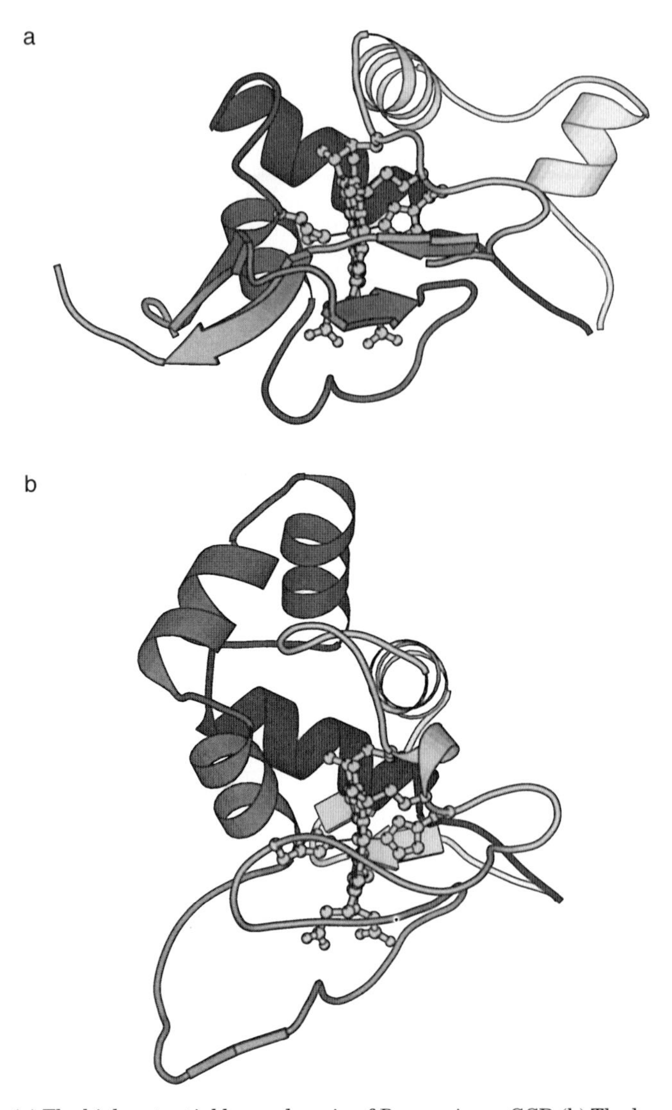

FIG. 8. (a) The high-potential heme domain of *P. aeruginosa* CCP. (b) The low-potential heme domain of *P. aeruginosa* CCP.

reductase (*55*). The immediate environment of this heme is predominantly hydrophobic. This structure is of the catalyzically inert "resting" form of the enzyme in which both hemes are Fe(III) six-coordinate and unable to bind substrate. Numerous studies of both *P. aeruginosa* CCP and *P. denitrificans* CCP have demonstrated that for catalysis to take place, the enzyme must be activated by accepting an electron at the high-potential heme (*52, 57, 60–63*). This event must be accompanied by the dissociation of one of the histidine ligands from the low-potential heme. His 55 is unlikely to leave because it is buried and positioned next to Cys 54, which covalently links the heme to the protein in a rigid structural unit. His 71, on the other hand, sits on the tip of a flexible loop (Figs. 7, 8a). This arrangement suggests that His 71 is the more likely ligand to leave the heme prior to catalysis. The activation of the enzyme might be coupled to domain movement controlled by changes in the spin and oxidation state of the two hemes, as was found in cytochrome cd_1 (*12*). However, these subsequent structural changes at the heme iron are not well understood. The His/Met (residues 201 and 275) ligated C-terminal domain contains the high-potential heme (Fig. 8b), which is site of electron entry from donors (*60, 61, 64, 65*). Residues 223–228 are disordered in the structure and these residues are often the target for proteolytic cleavage by *P. aeruginosa* elastase (*66, 67*). The two heme groups are perpendicular to each other and separated by an iron–iron distance of 20.9 Å. One of the propionates of one heme points toward the a propionate of the second. The shortest distance between the two hemes is via these propionates (9.6 Å).

The two domains in each *P. aeruginosa* CCP monomer are connected by three strands of chain and an antiparallel β-sheet (Fig. 7). The domain interface is hydrophobic and provides a Ca^{2+} ion site with unusual ligation (Fig. 9). A detailed biochemical study of *P. denitrificans* CCP has suggested an essential role for Ca^{2+} ions in catalysis. This enzyme appears to bind two Ca^{2+} ions, the first with high affinity in

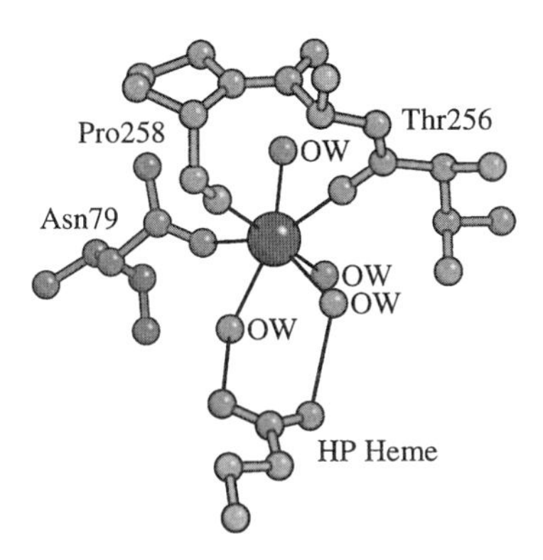

FIG. 9. The calcium binding site in *P. aeruginosa* CCP. One of the propionates of the high-potential (HP) heme is hydrogen bonded to two bound water (OW) molecules.

both the resting and mixed valence (MV) states and the second with lower affinity (*52, 63, 68*). The second site appears to promote dimerization of the active, MV form of the enzyme and stabilizes the high-spin (five-coordinate) form of the low-potential peroxidatic (where hydrogen peroxide is reduced) heme (see Section III,D). However, reexamination of the dimeric interface region did not reveal any putative metal binding site in the *P. aeruginosa* CCP enzyme. The high-affinity site is almost certainly that seen in the crystal structure of the *P. aeruginosa* CCP (Fig. 9) (*55*). Although all these ligands are oxygen atoms and there are seven of them in a pentagonal bipyramidal arrangement, reminiscent of calcium sites in other proteins, what is rather unusual, but not completely unprecedented, is that there is no negatively charged residue coordinated to the calcium (*55*). The ligands are the amide oxygen of Asn 79, the main chain carbonyls of Thr 256 and Pro 258 and four molecules of water (Fig. 9). The nearest carboxylate to the calcium ion is one of the propionates of the high-potential heme center whose oxygen atoms form hydrogen bonds to two bound water ligands. By contrast, the distal Ca^{2+} binding site of horseradish peroxidase *c* (HRP) uses three carboxylate ligands (*69*). In spite of these differences the distal Ca^{2+} binding site of class III plant peroxidases (HRPC and peanut peroxidase) may provide a clue as to the role of the calcium ion in *P. aeruginosa* CCP. The distal Ca^{2+} in HRPC is intimately liked to a histidine residue in the heme distal pocket that is essential for catalysis. Moreover, this calcium is required to maintain the structure of the heme distal pocket. In the structure of resting *P. aeruginosa* CCP the calcium binding site is adjacent to a conserved histidine residue that when chemically modified in *P. denitrificans* CCP leads to a loss of catalytic activity (*70*). Hence, it may be that the bound calcium in *P. aeruginosa* CCP fulfils a similar role in MV-CCP. Another possible role of the calcium might be to maintain the structural integrity of the enzyme, as the metal is found at the interface between the two domains.

C. Solution Spectroscopy of CCP from *Pseudomonas aeruginosa*

Before the availability of a high-resolution structure of *P. aeruginosa* CCP, the properties and environments of the two hemes had been probed using a range of solution spectroscopies. These include electron paramagnetic resonance (EPR) (*51, 57, 58, 61*), resonance Raman (*59*), circular dichroism (CD) (*71, 72*), MCD (*58, 61, 73, 74*). Until the demonstration by Ellfolk and colleagues that it is the mixed-valence form of the enzyme that reacts with hydrogen peroxide to form an active complex (*60*), these spectroscopic studies focused almost exclusively on the oxidized and fully reduced forms of the enzyme.

Subsequently, the emphasis shifted to using spectroscopy to probe the structural changes that occur in the immediate vicinity of the hemes upon half-reduction. An understanding of the structure of the MV form of CCP is a key element in any description of the mechanism of this class of enzyme, since it is the state that can bind substrate. In order to do this, one of the hemes must become both five-coordinate and accessible to hydrogen peroxide. In the absence of an X-ray structure of the MV state of the enzyme, one must rely on spectroscopic evidence to understand the nature of the structural changes. Of these spectroscopic studies, perhaps the most complete and arguably the most informative are the variable-temperature MCD studies and EPR of Greenwood and colleagues.

MCD spectroscopy in range 300 to 2000 nm at both ambient and liquid helium (4.2 K) temperatures can yield information about the spin, oxidation, and coordination states of each heme in a multiheme protein such as CCP (*75*). This technique, in combination with low-temperature X-band EPR spectroscopy, was used to great effect in characterizing the properties of the fully oxidized and MV forms of the *P. aeruginosa* CCP in solution. At 4.2 K, both hemes in the oxidized enzyme are low-spin ferric, with diagnostic features in the near infrared–MCD (NIR-MCD) spectrum consistent with one heme with His/Met axial coordination and the other with bis-histidine axial coordination; this is entirely consistent with the crystal structure. In contrast, at room temperature only the low-potential (bis-histidine coordinated) heme in the C-terminal domain remains completely low-spin, whereas the high-potential (His/Met coordinated) heme exists as mixture of high- and low-spin forms (*58*).

The MV state of two forms of *P. aeruginosa* CCP, termed “active” and “inactive,” which describes their behavior in the standard assay, has also been studied (*61*). The X-band EPR spectrum at 10 K of the MV state of both “active” and “inactive” forms contain a rhombic spectrum arising from a single low-spin ferric heme, such as the low-potential heme in the active site. In the MV state the high-potential (His/Met coordinated) heme is reduced and therefore EPR inactive. In the “inactive” enzyme the rhombic trio ($g_z = 3.00, g_y = 2.26, g_x = \sim 1.4$) arising from the ferric state of the peroxidatic heme is associated with a ligand–metal charge transfer band in the NIR-MCD spectrum recorded at 4.2 K. The maximum of this charge-transfer (CT) band lies at 1450 nm, confirming that the heme is a bis-histidine ferric species (*75, 76*). The rhombic spectrum ($g_z = 2.85, g_y = 2.35, g_x = \sim 1.54$) in the X-band EPR spectrum associated with the single low-spin (at 10 K) ferric heme in the MV state of the "active" enzyme is both well defined and intense (*57, 61*). Interestingly, this EPR spectrum is associated with a most unusual NIR-MCD spectrum that is both rather weak in intensity and very broad (*61*). The position of this NIR charge transfer band is consistent with bis-histidine ligation (*75, 76*), rather than the His/His$^-$ axial ligation proposed by Aasa *et al.* (*57*), which would give rise to a NIR-MCD charge transfer band of higher energy (*75*).

At room temperature the NIR-MCD spectrum of the "inactive" MV form is similar to that recorded at liquid helium temperatures, suggesting the same spin and ligation states. In contrast, the NIR-MCD spectrum of the active MV species is quite different. The broad band has gone and is replaced by two weaker features: a trough at ~630 nm and a derivative-shaped feature ($\lambda_{cross} \simeq 1150$ nm). The position of this pair of features is indicative of the nature of the distal ligand, in this case water, for high-spin hemes whose proximal ligand is a histidine (*77, 78*). Thus, it would appear that the loss of the distal histidine ligand and formation of the high-spin (substrate binding) state in the "active" form of *P. aeruginosa* CCP is temperature dependent. Similar temperature dependence has been shown for the formation of the high-spin heme in the Ca^{2+}-loaded CCP from *P. denitrificans* (*52*).

Taken together, the X-ray structure of *P. aeruginosa* CCP and the magneto-optical studies of the oxidized and MV forms of the "active" enzyme can be interpreted in terms of the following activation mechanism. Reduction of the high-potential heme causes the peroxidatic heme to shed His 71 (*55, 61*). This would not only provide an open face at the heme to bind H_2O_2, but would perhaps position His 71 in the distal pocket, allowing it to promote heterolytic peroxide cleavage during compound I formation. However, alignment of the primary sequences of 10 CCPs and three examples of the closely related MauG proteins shows His 71 is not conserved (Watmough and Fadhl, unpublished data). This raises the intriguing possibility that some CCPs, for example, the enzyme from *H. pylori,* do not adopt a "resting state" and maintain an open peroxidatic site. This is the case for the CCP purified from *Nitrosomonas europea,* where a combination of EPR spectroscopy and ligand binding experiments clearly show that that the oxidized form of the enzyme contains a high-spin heme that can bind substrate (*46*). However, rather surprisingly the primary amino acid sequence of this enzyme reveals that a histidine residue equivalent to His 71 of *P. aeruginosa* CCP is present.

D. THE ROLE(S) OF THE BOUND CA^{2+} ION(S) IN THE ACTIVATION OF THE CCP FROM *PARACOCCUS DENITRIFICANS*

The position of the Ca^{2+} binding site at the domain interface of each monomer of *P. aeruginosa* CCP suggests that it will be sensitive to any structural changes associated with formation of the MV-CCP. The residues, Asn 79, Thr 256, and Pro 258, that contribute to this Ca^{2+} binding site are highly conserved among all the bacterial diheme CCPs, including the *P. denitrificans* enzyme. This suggests that the work of Pettigrew and colleagues on the role of Ca^{2+} ions in the activation of *P. denitrificans* CCP (*17, 52, 63, 68*) might be extrapolated to other members of the family.

The CCP purified from the periplasm of *P. denitrificans* (like *P. aeruginosa* CCP) is a dimer in solution, each monomer having a molecular weight of 38 kDa. The isoelectric point ($pI = 4.8$) of *P. denitrificans* CCP reflects the basic nature of its preferred electron donors cytochrome c_{552} and horse-heart cytochrome *c* (*63, 79*). Since the enzyme is isolated from a periplasmic fraction (*79*), rather than a whole cell homogenate or an acetone powder (as is the case for *P. aeruginosa* CCP), it is exposed both to slightly alkaline buffers and high concentrations of EDTA, which is a strong chelator of divalent cations. This apparently depletes at least some of the bound Ca^{2+} ions from the enzyme during the course of purification (*52*). As a result it was realized that to obtain maximum catalytic activity it was necessary to add exogenous Ca^{2+} ions to the MV form of *P. denitrificans* CCP (*63*). The addition of Ca^{2+} ions not only led to changes in the UV/Vis absorption spectrum, but showed that these changes were associated with a shift in the reduction potential of the high-potential heme (*52*). In particular, on half reduction of the *Paracoccus* enzyme there are slow time-dependent changes in the UV/Vis absorption spectrum spectra that are associated with formation of high-spin ferric heme (*52*). The extent of these changes, which are maximal in the presence of added Ca^{2+} ions, is diagnosed by a decrease in absorption at 410 nm with a concomitant gain at 380 nm and the formation of a 640 nm ligand-to-metal charge transfer band. The intensity of the latter feature, which corresponds to the "630 nm" band in the room temperature MCD spectrum of *P. aeruginosa* CCP (see Section III,C), is temperature dependent, almost disappearing on cooling of the enzyme to 4°C (*52*). Similar spectral changes have recently been reported for the CCP isolated from *Pseudomonas nautica* (*43*).

It has also been reported that Ca^{2+} ions appear to promote dimerization of *P. denitrificans* CCP. This is argued on the basis of monomeric (inactive) CCP and dimeric (active) CCP being in rapid equilibrium, such that at low concentrations the inactive monomer is favored. Concentration of previously diluted enzyme restores catalyzic activity. To account for all the Ca^{2+} induced spectroscopic changes and the monomer–dimer equilibrium, Pettigrew and colleagues have proposed a model for the *Paracoccus* enzyme in which Ca^{2+} ions bind at two different sites. The first site is present in *each* monomer and binds Ca^{2+} with a K_d of 1.2 μM in both the resting and mixed valence states. The second, low affinity site, which is proposed to promote dimerization in the *P. denitrificans* CCP is not apparent in the structure of the *P. aeruginosa* enzyme (see section III B).

The presence or absence of Ca^{2+} ions in one or both sites also appears to effect the reduction potential of the high-potential heme. In equilibrium redox titrations monitored spectroscopically, done in the presence of Ca^{2+} ions, this is shifted positive by about 50 mV ($E^{o\prime} = +226$ mV) compared with titrations done in the presence of a chelator ($E^{o\prime} = +176$ mV) (*52*). This former value is close to the reduction potentials reported for the high-potential heme in the CCP from *P. aeruginosa* (*51*), but about 200 mV lower than reported for the high-potential hemes in the enzymes from *N. europea* (*46*) and *Methylococcus capsulatus* Bath (*80*). In contrast, the reduction potential of the peroxidatic heme is unaffected by the presence or absence of Ca^{2+} ions (*16, 52*).

The fully reduced state of CCP is probably of no physiological significance. However, the visible region spectrum of the fully reduced enzyme is very sensitive to the presence or absence of bound Ca^{2+} ions. In the absence of Ca^{2+} ions the α-band maximum of the reduced enzyme, recorded during the course of the redox titrations described earlier, is at 551 nm with a shoulder at 557 nm. In contrast, in fully reduced Ca^{2+}-loaded enzyme the maximum is at 557 nm with a shoulder 551 nm (*52*). These spectra should be compared with those of the fully reduced *Pseudomonas* CCP in its "inactive" and "active" forms, respectively, which show similar differences (*81*). Further similarities exist between the X-band EPR spectra of the "active" CCP from *P. aeruginosa* in the MV state and that of the Ca^{2+}-loaded CCP from *P. denitrificans* and, moreover, the X-band EPR spectrum of MV *P. denitrificans* CCP that had previously been treated with EDTA (Table I).

The work of Foote and colleagues (*61, 81*) was done before the realization of the importance of Ca^{2+} ions for this class of enzymes. Instead, they suggested that the "inactive" enzyme is identical to the two-fragment complex that arises from elastase cleavage of *P. aeruginosa* CCP (*66, 67*). Unfortunately, no direct evidence was shown to support this view other than a statement to the effect that chromatography

TABLE I

SPECTROSCOPIC CHARACTERISTICS OF DIFFERENT FORMS OF THE CCPs FROM *P. AERUGINOSA* AND *P. DENITRIFICANS* IN THE MIXED-VALENCE (HALF-REDUCED) STATE

	Pseudomonas aeruginosa		*Paracoccus denitrificans*	
CCP origin	"Active"	"Inactive"	Ca^{2+} bound	Ca^{2+} depleted
Visible absorption (maxima)				
Soret (both hemes)	420 nm (~395 nm)	421 nm (410 nm)	419 nm (~395 nm)	419 nm (411 nm)
Visible (both hemes)	557 nm; 525 nm	557 nm; 525 nm	557 nm; 525 nm	557 nm; 525 nm
Near IR (LP Fe(III) heme)	~630 nm	Absent	~640 nm	Absent
X-band EPR (LP Fe (III) heme)	$g_z = 2.85$, $g_y = 2.26$, $g_x = \sim 1.54$	$g_z = 3.00$, $g_y = 2.26$, $g_x = \sim 1.4$	$g_z = 2.89$, $g_y = 2.32$, $g_x = 1.51$	$g_z = 3.00$
NIR-MCD 4.2 K	$\lambda_{max} = \sim 1400$ nm	$\lambda_{max} = \sim 1450$ nm	n.d.	n.d.
298 K	$\lambda_{min} = \sim 630$ nm; $\lambda_{cross} = \sim 1150$ nm	$\lambda_{max} = \sim 1450$ nm	n.d.	n.d.
NMR	n.d.	n.d.	−3.1 ppm (HP); 64–50 ppm (LP)	−3.7 ppm (HP); 30–22 ppm (LP)

of the "inactive" enzyme on a strong cation-exchange column resolves three colored fractions, rather than the single fraction observed with "active" protein (*61, 81*). This is surprising, since if the 23-kDa diheme containing fragment was resolved from the native enzyme and a colorless 10-kDa fragment, only two colored fractions might be expected. Thus, it is not clear whether the fragmentation observed by Foote *et al.* occurred by elastase cleavage prior to chromatography or is a consequence of the chromatographic conditions.

E. The Nature of the Peroxidatic Heme Distal Pocket in the Mixed-Valence Form of Diheme CCPs

In the extensively studied monoheme yeast CCP (YCCP) and HRPC, the distal cavity of the peroxidatic heme contains conserved arginine and histidine residues, positioned to maximize the rate of heterolytic oxygen cleavage in compound I formation. In both YCCP and HRPC, mutation of the distal histidine lowers the bimolecular rate constant for peroxide binding by five orders of magnitude (*82, 83*). Assuming the peroxidatic heme in *P. aeruginosa* CCP uses a similar strategy to promote O–O bond cleavage, it is interesting to contemplate which residues may perform similar roles in the distal pocket after activation. Simplistically, it might be assumed that after His 71 is shed from the distal face of the peroxidatic heme, this residue remains close enough to serve as a base during catalysis. However, this residue is not absolutely conserved, nor does it necessarily serve as an axial ligand in the oxidized enzyme—for example, in the CCP from *N. europea* (*46*).

In contrast, the histidine at position 261 (*P. aeruginosa* CCP numbering) is absolutely conserved. Moreover, specific modification of the equivalent residue (His 275) in *P. denitrificans* CCP with diethylpyrocarbonate leads to loss of catalytic activity and an inability to form a high-spin heme in the MV enzyme (*70*). It has been argued, based on sequence conservation and these chemical modification experiments, that it is this histidine rather than His 71 that serves as a base into the peroxidatic heme. In this light it is worth noting that His 261 lies in a highly conserved region of the protein that includes Arg 251 and Trp 266, which could be functional analogues of Arg 48, Trp 51, and His 52 at the distal side of the YCCP heme. However, this triad of residues lies a long way from the presumed peroxidatic heme and it is difficult to see how the shedding of His 71 would introduce them to heme environment.

To overcome these objections, Prazeres *et al.* (*17*) have proposed an alternative model in which the half-reduction of CCP leads to a switch of axial ligands at both hemes reminiscent of the changes known to occur in *P. pantotrophus* cd_1. In this model, when the high-potential heme (center I) is reduced it induces a ligand switch at the low-potential heme (center II) such that His 71 is replaced by the totally conserved methionine at position 115. This, it is presumed, causes a change in the midpoint redox potential (cf. the *c* heme domain of *P. pantotrophus* cytochrome cd_1, Section II,A). The electron moves to center II, and simultaneously Met 275 is shed from center I. This leaves a five-coordinate heme able to bind substrate, with the proposed catalyzic triad of His 261, Arg 251, and Trp 266 (*P. aeruginosa* CCP numbering), presumably positioned above the distal face of the heme to participate in catalysis. Such a scheme cannot be argued against on the basis of any spectroscopic or electrochemical data. In fact, such a scheme makes it easier to interpret the apparent inequivalence of the oxidative and reductive waves associated with the "high-potential" heme in cyclic voltammetry (CV) experiments (*16*). However, these residues also lie a long way from center II, and it is also difficult to see how they would be introduced to the heme environment.

In considering this alternative model, it is important to remember three things. First of all, there is, as yet, no evidence that diheme CCPs will use exactly the same chemistry to promote O–O bond cleavage as their eukaryotic counterparts. Secondly, there is not yet a structure of a half-reduced CCP, nor are there any solution studies that can give any indication of the extent of the changes that take place in the protein structure upon half reduction. Finally, His 261 is adjacent to Thr 256 and Pro 258, and modification of this imidazole group with a bulky substituent may sterically impair Ca^{2+} binding. In this context it is interesting to note that the EPR spectrum of the MV form of the chemically modified *P. denitrificans* CCP is unaffected by Ca^{2+} ions (*70*) and resembles that of the MV form of the unmodified enzyme that has previously been treated with EDTA. Consequently, in the continued absence of a structure of a MV CCP, a proper evaluation of the role of His 261 will require substitution of this residue by site-directed mutagenesis.

F. Mechanistic Studies on *P. aeruginosa* CCP

The detailed mechanism of *P. aeruginosa* CCP has been studied by a combination of stopped-flow spectroscopy (*64, 65, 84, 85*) and paramagnetic spectroscopies (*51, 74*). These data have been combined by Foote and colleagues (*62*) to yield a quantitative scheme that describes the activation process and reaction cycle. A version of this scheme, which involves four spectroscopically distinct intermediates, is shown in Fig. 10. In this scheme the "resting" oxidized enzyme (structure in Section III,B) reacts with 1 equiv of an electron donor (Cu(I) azurin) to yield the "active" mixed-valence (half-reduced) state. The "active" MV form reacts productively with substrate, hydrogen peroxide, to yield compound I. Compound I reacts sequentially with two further equivalents of Cu(I) azurin to complete the reduction of peroxide (compound II) before returning the enzyme to the MV state. A further state, compound 0, that has not been shown experimentally but would precede compound I formation is proposed in order to facilitate comparison with other peroxidases.

The formation of the MV state has been studied using azurin as an electron donor (*64, 65*). Azurin has the advantage over cytochrome c_{551} in that even at high concentrations there is little spectral overlap with the Soret or visible region transitions of CCP. The reaction of azurin with CCP under pseudo-first-order conditions is described by two independent exponential processes, with the faster process predominating

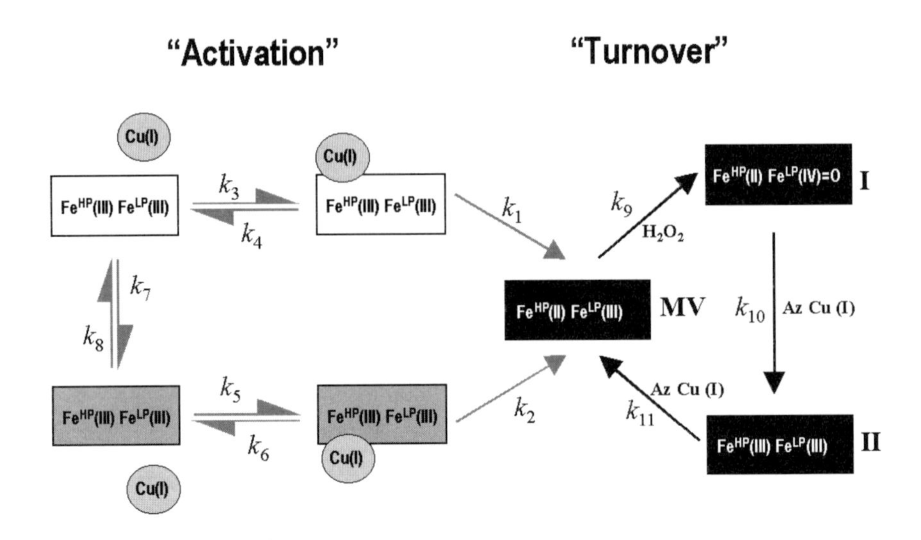

FIG. 10. A model of the mechanism of activation and turnover in *P. aeruginosa* CCP after Foote *et al.* (*62*). In order to successfully model steady-state turnover, Foote *et al.* used the following experimentally derived rate constants: $k_1 = 7\ s^{-1}, k_2 = 0.2\ s^{-1}, k_9 = 5.6 \times 10^7\ M^{-1}s^{-1}, k_{10} = 2.0 \times 10^6\ M^{-1}s^{-1}, k_{11} = 1.5 \times 10^6\ M^{-1}s^{-1}$ (*62*). In addition the following rate constants were determined by kinetic modeling; $k_3 = 5 \times 10^4\ M^{-1}s^{-1}, k_4 = 3.4\ s^{-1}, k_5 = 1 \times 10^6\ M^{-1}s^{-1}, k_6 = 4.2\ s^{-1}, k_7 = 0.125\ s^{-1}, k_8 = 0.1\ s^{-1}$ (*62*). Note that the origin of the equilibrium (k_7/k_8) between conformers of CCP that gives rise to the fast phase of activation (white infilled rectangles) and that giving rise to the slow phase of activation (gray infilled rectangles) is unknown. If, however, it arises from Ca^{2+} binding to the enzyme, a more complex model would be required, as presumably an equilibrium would exist between catalytically active (Ca^{2+} bound) and inactive (Ca^{2+} free) states of the MV state. Such models have been put forward for the *P. denitrificans* CCP by Prazeres *et al.* (*17*) and Lopes *et al.* (*16*).

at high azurin concentrations (*62, 64, 65*). These observations have been interpretated in terms of the CCP existing in two conformers, both of which can interact with Cu(I) azurin to form a bimolecular complex prior to electron transfer to the high-potential heme. The formation of the bimolecular complex by both CCP conformers is strongly pH dependent (*64*). The equilibrium between the two reactive forms of CCP has been proposed to depend upon the spin state of the high-potential heme (*62*), which in turn may depend upon the occupancy of the Ca^{2+} binding site, although this remains to be tested experimentally.

Once activated, MV-CCP reacts with 1 equiv of H_2O_2 in a bimolecular reaction, presumably to form compound 0. In YCCP and HRP this species is referred to as compound ES or compound I, respectively, and contains oxyferryl heme and either a porphyrin π-cation radical (HRP) or an amino acid radical (YCCP). However, the presence of an extra reducing equivalent on the second heme in CCP suggests that such an oxidizing radical species close to the active site heme will be very short-lived and readily form compound I (Fig. 10), which is formally Fe(III) Fe(IV)=O. The bimolecular rate constant for compound I formation is reported to be very close to the diffusion limit (*84*).

P. aeruginosa CCP compound I has been further characterized by rapid scan stopped-flow spectrophotometry (*85*), low-temperature EPR spectroscopy (*51, 74*), and variable-temperature MCD (*74*). The single low-spin Fe(III) heme observed in the low-temperature X-band EPR spectrum of compound I ($g_z = 3.15$) (*51, 74*) differs from any of the low-spin Fe(II) hemes seen in the EPR spectra of the "resting" and MV states. Since compound I can only be observed using freeze–quench methodologies, the MCD spectrum has only been recorded at liquid helium temperatures and shows the Fe(III) to have His/Met axial ligands (*74*). The second heme, an Fe(IV)=O species, is EPR silent, but is readily observed in the visible region of the low-temperature MCD spectrum (*74*). The form of this spectrum suggests that it is related to compound II of HRP — for example, neither the porphryrin macrocycle nor a nearby amino acid are oxidized as is the case for HRP compound I and YCCP compound ES, respectively. That said, a radical species observed in the EPR spectrum 4 ms after mixing MV CCP with a slight molar excess of hydrogen peroxide (*51*), but that is much reduced after a few seconds (*51, 74*), has yet to be assigned.

Compound I will accept a further electron from azurin and decays to a species known as compound II with a bimolecular rate constant of $6 \times 10^6\ M^{-1}s^{-1}$ (*84*). It is anticipated that compound II is a form of CCP containing two ferric hemes, but possibly not identical to the structurally defined fully oxidized enzyme "as isolated." This is because during turnover at room temperature there is no obvious reason for the histidine ligand displaced from the peroxidatic heme iron to return. Consequently, it might be assumed that compound II is structurally related to the MV conformer rather than the resting enzyme.

Simplistically, it might be assumed that X-band EPR spectrum of compound II would contain signals arising from two isolated Fe(III) hemes, one low-spin species with His/Met ligation (cf. compound I) and a high-spin Fe(III) species. In fact, the EPR spectrum of compound II exhibits a single derivative-shaped feature at $g = 3.45$, which intensifies on cooling the sample from 10 K to 5 K that was ascribed to an integer-spin signal arising from the magnetic coupling of two ($S = \frac{1}{2}$) ferric hemes (*74*). This observation is of considerable current interest, as similar signals from pairs of interacting hemes have been reported in a number of multi-cytochrome *c* containing enzymes. These include hydroxylamine oxidase (HAO) and its redox partner cytochrome c_{554} from *N. europea* (*86, 87*), and the pentaheme bacterial cytochrome c_{552} nitrite reductases from a number of organisms (*88–91*). The structures of HAO (*92*) and cytochrome *c* nitrite reductase (quite distinct from cytochrome cd_1) (*93*) reveal that the spatial organization of the heme groups is conserved, but it is not clear which pair(s) of hemes are coupled to give rise to these signals, or what the mechanism of coupling may be. However, recent work on the origin of related integer-spin systems from the dinuclear center (high-spin Fe(III) heme::Cu(II)) of *Escherichia coli* cytochrome bo_3 (*94*) suggests that these integer spin EPR signals arise from the weak coupling of a pair of hemes within these multiheme cytochromes, although which pair(s) of hemes are involved remains to be established.

G. Unresolved Issues

The principal and most pressing issue remains the nature of the structural changes that take place upon half-reduction. Ellfolk *et al.* (*67*) reported substantial changes in the far-UV CD spectrum of *P. aeruginosa* CCP associated with the formation of the MV state, but unfortunately did not elaborate. Given the extent of the changes proposed in the model of Prazeres *et al.* (*17*), the issue may only be resolved by solving the structure of a CCP in the MV state. This could be done by obtaining diffracting crystals of previously half-reduced enzyme or by reducing a single crystal *in situ* with ascorbate, in a manner analogous to the experiments done on cytochromes cd_1 to observe the structural changes that take place on reduction (*12, 20*). Such experiments should also indicate the nature of any residues in the distal pocket of the substrate binding heme that are involved in peroxidatic cleavage. However, the participation of such residues in catalysis will have to be tested by substituting them individually, experiments that are now tractable with the development of systems for the heterologous expression of *c*-type cytochromes at high levels (*95*).

The next issue to be addressed experimentally is the number and role(s) of the Ca^{2+} binding site(s) in the *P. aeruginosa* CCP in solution. The use of $^{45}Ca^{2+}$ in equilibrium binding assays should determine the stoichiometry of calcium binding. However, further experiments that relate the number of bound Ca^{2+} ion(s) in *P. aeruginosa* CCP to the spectroscopically defined "active" and "inactive" conformers described by Foote *et al.* (*61, 81*) are also needed. Such experiments will also provide insight into how applicable the results on the Ca^{2+}-dependent activation of *P. denitrificans* CCP are to other members of this family of diheme peroxidases, including the MauG proteins. These proteins are proposed to participate in the biosynthesis of tryptophan tryptoquinone, which is the cofactor in bacterial methyamine dehydrogenase (*96*).

There is also uncertainty as to the identity of the physiological electron donors to CCP in both *P. aeruginosa* and *P. denitrificans*. *In vitro P. aeruginosa* CCP will accept electrons from cytochrome c_{551} and from azurin. *P. denitrificans* CCP will accept electrons from basic cytochromes *c in vitro,* but its activity toward pseudoazurin has not yet been reported. Pettigrew *et al.* (*97*) have used molecular docking and ^{1}H NMR spectroscopy to study the putative electron transfer complex between *P. denitrificans* CCP and cytochromes *c*. For the natural redox partner, cytochrome c_{550}, their results are consistent with a complex in which the heme of the cytochrome *c* lies above the high-potential heme of cytochrome *c* peroxidase. In contrast, the nonphysiological but kinetically competent horse cytochrome *c* is predicted to bind between the two heme domains of the peroxidase. Reexamination of the surface of *P. aeruginosa* CCP structure (*55*) did not reveal any negative hydrophobic surface patch that would be similar to that found in cytochrome cd_1 (12).

Acknowledgments

The authors' work was supported by the BBSRC (grants B05860 and 83/B10987) and the European Union (BIO4-CT96-0281). V.F. is a Royal Society University Research Fellow. N.J.W. is a Wellcome Trust University Award Lecturer. We thank János Hajdu, James W. B. Moir, and Pamela A. Williams for their important contributions to the structural work on *P. pantotropha* cytochrome cd_1. Colin Greenwood initiated the work on the *P. aeruginosa* CCP at UEA, where Nicholas Foote and Christopher Ridout made important contributions. N.J.W. particularly thanks Tom Brittain and Andrew Thomson for useful discussions concerning the similarities and differences between the *P. aeruginosa* and *P. denitrificans* CCPs.

Note in Proof

Details of the unusual histeretic oxidation/reduction behaviour of *P. pantotrophus* cd_1 have recently been published: Koppenhöfer, A., Turner, K. L., Allen, J. W. A., Chapman, S. K., Ferguson, S. J., *Biochemistry* **2000,** *39,* 4243.

References

1. Richardson, D. J. *Microbiology* **2000,** *146,* 551.
2. Barker, P. D.; Ferguson, S. J. *Structure* **1999,** *7,* R281.

3. Rainey, F. A.; Kelly, D. P.; Stackebrandt, E.; Burghardt, J.; Hiraishi, A.; Katayama, Y.; Wood, A. P. *Int. J. Syst. Bact.* **1999,** *49,* 645.
4. Fülöp, V.; Moir, J. W. B.; Ferguson, S. J.; Hajdu, J. *Cell* **1995,** *81,* 369.
5. Page, C. C.; Moser, C. C.; Chen, X.; Dutton, P. L. *Nature* **1999,** *402,* 47.
6. Baker, S. C.; Saunders, N. F.; Willis, A. C.; Ferguson, S. J.; Hajdu, J.; Fülöp, V. *J. Mol. Biol.* **1997,** *269,* 440.
7. Berks, B. C.; Ferguson, S. J.; Moir, J. W. B.; Richardson, D. J. *Biochim. Biophys. Acta* **1995,** *1232,* 97.
8. Averill, B. A. *Chem. Rev.* **1996,** *96,* 2951.
9. Van Spanning, R. J.; Wansell, C.; Harms, N.; Oltmann, L. F.; Stouthamer, A. H. *J. Bacteriol.* **1990,** *172,* 986.
10. Moir, J. W.; Ferguson, S. J. *FEMS Microbiol. Lett.* **1993,** *113,* 321.
11. Williams, P. A.; Fülöp, V.; Leung, Y. C.; Chan, C.; Moir, J. W. B.; Howlett, G.; Ferguson, S. J.; Radford, S. E.; Hajdu, J. *Nat. Struct. Biol.* **1995,** *2,* 975.
12. Williams, P. A.; Fülöp, V.; Garman, E. F.; Saunders, N. F.; Ferguson, S. J.; Hajdu, J. *Nature* **1997,** *389,* 406.
13. Stenkamp, R. E.; Jensen, L. H. *Acta Cryst. A* **1984,** *40,* 251.
14. Scheidt, W. R.; Ellison, M. K. *Acc. Chem. Res.* **1999,** *32,* 350.
15. Frazee, R. W.; Orville, A. M.; Dolbeare, K. B.; Yu, H.; Ohlendorf, D. H.; Lipscomb, J. D. *Biochemistry* **1998,** *37,* 2131.
16. Lopes, H.; Pettigrew, G. W.; Moura, I.; Moura, J. J. G. *J. Biol. Inorg. Chem.* **1998,** *3,* 632.
17. Prazeres, S.; Moura, J.; Moura, I.; Gilmour, R.; Goodhew, C.; Pettigrew, G.; Ravi, N.; Huynh, B. *J. Biol. Chem.* **1995,** *270,* 24264.
18. Nurizzo, D.; Silvestrini, M. C.; Mathieu, M.; Cutruzzola, F.; Bourgeois, D.; Fülöp, V.; Hajdu, J.; Brunori, M.; Tegoni, M.; Cambillau, C. *Structure* **1997,** *5,* 1157.
19. Cutruzzola, F.; Arese, M.; Grasso, S.; Bellelli, A.; Brunori, M. *FEBS Lett.* **1997,** *412,* 365.
20. Nurizzo, D.; Cutruzzol, F.; Arese, M.; Bourgeois, D.; Brunori, M.; Cambillau, C.; Tegoni, M. *Biochemistry* **1998,** *37,* 13987.
21. Nurizzo, D.; Cutruzzola, F.; Arese, M.; Bourgeois, D.; Brunori, M.; Cambillau, C.; Tegoni, M. *J. Biol. Chem.* **1999,** *274,* 14997.
22. Cutruzzola, F. *Biochim. Biophys. Acta.* **1999,** *1411,* 231.
23. Wilson, E. K.; Bellelli, A.; Liberti, S.; Arese, M.; Grasso, S.; Cutruzzola, F.; Brunori, M.; Brzezinski, P. *Biochemistry* **1999,** *38,* 7556.
24. Vijgenboom, E.; Busch, J. E.; Canters, G. W. *Microbiology* **1997,** *143,* 2853.
25. Cheesman, M. R.; Walker, F. A. *J. Am. Chem. Soc.* **1996,** *118,* 7373.
26. Cheesman, M. R.; Ferguson, S. J.; Moir, J. W. B.; Richardson, D. J.; Zumft, W. G.; Thomson, A. J. *Biochemistry* **1997,** *36,* 16267.
27. Watmough, N. J.; Butland, G.; Cheesman, M. R.; Moir, J. W. B.; Richardson, D. J.; Spiro, S. *Biochim. Biophys. Acta.* **1999,** *1411,* 456.
28. Kobayashi, K.; Koppenhöfer, A.; Ferguson, S. J.; Tagawa, S. *Biochemistry* **1997,** *36,* 13611.
29. Wang, Y. N.; Averill, B. A. *J. Am. Chem. Soc.* **1996,** *118,* 3972.
30. Makinen, M. W.; Schichman, S. A.; Hill, S. C.; Gray, H. B. *Science* **1983,** *222,* 929.
31. Koppenhöfer, A.; Little. R.H.; Lowe, D. J.; Ferguson, S. J.; Watmough, N. J. *Biochemistry* **2000,** *39*, 4028.
32. Silvestrini, M. C.; Tordi, M. G.; Musci, G.; Brunori, M. *J. Biol. Chem.* **1990,** *265,* 11783.
33. Greenwood, C.; Barber, D.; Parr, S. R.; Antonini, E.; Brunori, M.; Colosimo, A. *Biochem. J.* **1978,** *173,* 11.
34. Koppenhöfer, A. D.Phil. thesis, University of Oxford, **1999**.

35. Walsh, T. A.; Johnson, M. K.; Greenwood, C.; Barber, D.; Springall, J. P.; Thomson, A. J. *Biochem. J.* **1979,** *177,* 29.
36. Dodd, F. E.; Van Beeumen, J.; Eady, R. R.; Hasnain, S. S. *J. Mol. Biol.* **1998,** *282,* 369.
37. Murphy, M. E.; Turley, S.; Adman, E. T. *J. Biol. Chem.* **1997,** *272,* 28455.
38. Strange, R. W.; Murphy, L. M.; Dodd, F. E.; Abraham, Z. H.; Eady, R. R.; Smith, B. E.; Hasnain, S. S. *J. Mol. Biol.* **1999,** *287,* 1001.
39. Richardson, D. J.; Watmough, N. J. *Curr. Opin. Chem. Biol.* **1999,** *3,* 207.
40. Lenhoff, H. M.; Kaplan, N. O. *Nature* **1953,** *172,* 730.
41. Ellfolk, N.; Soininen, R. *Acta Chem. Scand.* **1970,** *24,* 2126.
42. Welinder, K. G. *Biochim. Biophys. Acta* **1991,** *1080,* 215.
43. Alves, T.; Besson, S.; Duarte, L. C.; Pettigrew, G. W.; Girio, F. M.; Devreese, B.; Vandenberghe, I.; Van Beeumen, J.; Fauque, G.; Moura, I. *Biochim. Biophys. Acta* **1999,** *1434,* 248.
44. Pettigrew, G. *Biochim. Biophys. Acta* **1991,** *1058,* 25.
45. Hanlon, S. P.; Holt, R. A.; McEwan, A. G. *FEMS Microbiol. Lett.* **1992,** *97,* 283.
46. Arciero, D. M.; Hooper, A. B. *J. Biol. Chem.* **1994,** *269,* 11878.
47. Van Spanning, R. J.; De Boer, A. P.; Reijnders, W. N.; Westerhoff, H. V.; Stouthamer, A. H.; Van Der Oost, J. *Mol. Microbiol.* **1997,** *23,* 893.
48. Kontchou, C. Y.; Blondeau, R. *FEMS Microbiol. Lett.* **1990,** *68,* 323.
49. Richardson, D. J.; Ferguson, S. J. *FEMS Microbiol. Lett.* **1995,** *132,* 125.
50. Koutny, M.; Kriz, L.; Kucera, I.; Pluhacek, I. *Biochim. Biophys. Acta.* **1999,** *1410,* 71.
51. Ellfolk, N.; Rönnberg, M.; Aasa, R.; Andreasson, L.; Vanngard, T. *Biochim. Biophys. Acta* **1983,** *743,* 23.
52. Gilmour, R.; Goodhew, C.; Pettigrew, G.; Prazeres, S.; Moura, I.; Moura, J. *Biochem. J.* **1993,** *294,* 745.
53. Harbury, H. A. *J. Biol. Chem.* **1957,** *225,* 1009.
54. Conroy, C. W.; Tyma, P.; Daum, P. H.; Erman, J. E. *Biochim. Biophys. Acta* **1978,** *537,* 62.
55. Fülöp, V.; Ridout, C.; Greenwood, C.; Hajdu, J. *Structure* **1995,** *3,* 1225.
56. Moore, G. R.; Pettigrew, G. W. "Cytochromes *c*: Evolutionary, Structural and Physicochemical Aspects"; Springer-Verlag: Berlin, 1990.
57. Aasa, R.; Ellfolk, N.; Ronnberg, M.; Vanngard, T. *Biochim. Biophys. Acta* **1981,** *670,* 170.
58. Foote, N.; Peterson, J.; Gadsby, P.; Greenwood, C.; Thomson, A. *Biochem. J.* **1984,** *223,* 369.
59. Rönnberg, M.; Osterlund, K.; Ellfolk, N. *Biochim. Biophys. Acta* **1980,** *626,* 23.
60. Araiso, T.; Rönnberg, M.; Dunford, H.; Ellfolk, N. *FEBS Lett.* **1980,** *118,* 99.
61. Foote, N.; Peterson, J.; Gadsby, P.; Greenwood, C.; Thomson, A. *Biochem. J.* **1985,** *230,* 227.
62. Foote, N.; Turner, R.; Brittain, T.; Greenwood, C. *Biochem. J.* **1992,** *283,* 839.
63. Gilmour, R.; Goodhew, C.; Pettigrew, G.; Prazeres, S.; Moura, J.; Moura, I. *Biochem. J.* **1994,** *300,* 907.
64. Brittain, T.; Greenwood, C. *J. Inorg. Biochem.* **1992,** *48,* 71.
65. Rönnberg, M.; Araiso, T.; Ellfolk, N.; Dunford, H. *J. Biol. Chem.* **1981,** *256,* 2471.
66. Soininen, R.; Ellfolk, N. *Acta Chem Scand [B]* **1975,** *29,* 134.
67. Ellfolk, N.; Ronnberg, M.; Osterlund, K. *Biochim Biophys Acta* **1991,** *1080,* 68.
68. Gilmour, R.; Prazeres, S.; McGinnity, D.; Goodhew, C.; Moura, J.; Moura, I.; Pettigrew, G. *Eur. J. Biochem.* **1995,** *234,* 878.
69. Gajhede, M.; Schuller, D.; Henriksen, A.; Smith, A.; Poulos, T. *Nat. Struct. Biol.* **1997,** *4,* 1032.

70. McGinnity, D.; Devreese, B.; Prazeres, S.; Van Beeumen, J.; Moura, I.; Moura, J.; Pettigrew, G. *J. Biol. Chem.* **1996,** *271,* 11126.
71. Rönnberg, M.; Ellfolk, N. *Biochim. Biophys. Acta.* **1979,** *581,* 325.
72. Rönnberg, M.; Ellfolk, N.; Soininen, R. *Biochim. Biophys. Acta.* **1979,** *578,* 392.
73. Foote, N.; Gadsby, P.; Field, R.; Greenwood, C.; Thomson, A. *FEBS Lett* **1987,** *214,* 347.
74. Greenwood, C.; Foote, N.; Gadsby, P. M. A.; Thomson, A. J. *Chem. Scripta* **1988,** *28A,* 79.
75. Cheesman, M. R.; Greenwood, C.; Thomson, A. J. *Adv. Inorg. Chem.* **1991,** *36,* 201.
76. Gadsby, P. M. A.; Thomson, A. J. *J. Am. Chem. Soc.* **1990,** *112,* 5003.
77. Watmough, N. J.; Cheesman, M. R.; Butler, C. S.; Little, R. H.; Greenwood, C.; Thomson, A. J. *J. Bioenerg. Biomembr.* **1998,** *30,* 55.
78. Seward, H. Ph.D. thesis, University of East Anglia, 1999.
79. Goodhew, C.; Wilson, I.; Hunter, D.; Pettigrew, G. *Biochem. J.* **1990,** *271,* 707.
80. Zahn, J. A.; Arciero, D. M.; Hooper, A. B.; Coats, J. R.; DiSpirito, A. A. *Arch. Microbiol.* **1997,** *168,* 362.
81. Foote, N. Ph.D. thesis, University of East Anglia, 1985.
82. Erman, J. E.; Vitello, L. B.; Millar, M. A.; Kraut, J. *J. Am. Chem. Soc.* **1992,** *114,* 6592.
83. Rodriguez-Lopez, J. N.; Smith, A. T.; Thorneley, R. N. F. *J. Biol. Inorg. Chem.* **1996,** *1,* 136.
84. Rönnberg, M.; Araiso, T.; Ellfolk, N.; Dunford, H. *Arch. Biochem. Biophys.* **1981,** *207,* 197.
85. Rönnberg, M.; Lambeir, A.; Ellfolk, N.; Dunford, H. *Arch. Biochem. Biophys.* **1985,** *236,* 714.
86. Hendrich, M. P.; Logan, M.; Andersson, K. K.; Arciero, D. M.; Lipscomb, J. D.; Hooper, A. B. *J. Am. Chem. Soc.* **1994,** *116,* 11961.
87. Andersson, K. K.; Lipscomb, J. D.; Valentine, M.; Munck, E.; Hooper, A. B. *J. Biol. Chem.* **1986,** *261,* 1126.
88. Costa, C.; Moura, J. J.; Moura, I.; Liu, M. Y.; Peck, H. D., Jr.; LeGall, J.; Wang, Y. N.; Huynh, B. H. *J. Biol. Chem.* **1990,** *265,* 14382.
89. Pereira, I. A.; Pacheco, I.; Liu, M. Y.; Legall, J.; Xavier, A. V.; Teixeira, M. *Eur. J. Biochem.* **1997,** *248,* 323.
90. Schumacher, W.; Hole, U.; Kroneck, P. M. *Biochem. Biophys. Res. Commun.* **1994,** *205,* 911.
91. Blackmore, R. S.; Brittain, T.; Gadsby, P. M.; Greenwood, C.; Thomson, A. J. *FEBS Lett.* **1987,** *219,* 244.
92. Igarashi, N.; Moriyama, H.; Fujiwara, T.; Fukumori, Y.; Tanaka, N. *Nat. Struct. Biol.* **1997,** *4,* 276.
93. Einsle, O.; Messerschmidt, A.; Stach, P.; Bourenkov, G. P.; Bartunik, H. D.; Huber, R.; Kroneck, P. M. *Nature* **1999,** *400,* 476.
94. Oganesyan, V. S.; Cheesman, M. R.; Butler, C. S.; Watmough, N. J.; Greenwood, C.; Thomson, A. J. *J. Am. Chem. Soc.* **1998,** *120,* 4232.
95. Reincke, B.; Thony-Meyer, L.; Dannehl, C.; Odenwald, A.; Aidim, M.; Witt, H.; Ruterjans, H.; Ludwig, B. *Biochim. Biophys. Acta.* **1999,** *1411,* 114.
96. McIntire, W. S.; Wemmer, D. E.; Chistoserdov, A.; Lidstrom, M. E. *Science* **1991,** *252,* 817.
97. Pettigrew, G. W.; Prazeres, S.; Costa, C.; Palma, N.; Krippahl, L.; Moura, I.; Moura, J. J. *J. Biol. Chem.* **1999,** *274,* 11383.
98. Kraulis, P. J. *J. Appl. Crystallogr.* **1992,** *24,* 946.
99. Esnouf, R. M. *J. Mol. Graph. Model.* **1997,** *15,* 132.

BINDING AND TRANSPORT OF IRON-PORPHYRINS BY HEMOPEXIN

WILLIAM T. MORGAN and ANN SMITH

Division of Molecular Biology and Biochemistry, School of Biological Sciences,
University of Missouri — Kansas City, Kansas City,
Missouri 64110

I. Introduction

Iron plays many vital roles in sustaining life, and a large portion of the human body's iron is in the form of heme, ferri-protoporphyrin

0898-8838/01 $35.00

IX.[1] The tetrapyrrole moiety provides iron with four equatorial nitrogen ligands and an electron-rich ring system that can participate in electron transfer reactions. The axial iron ligands and the overall environment provided by the protein enable fine-tuning of the properties of the heme and account for the wide range of chemical properties and biological activities of heme proteins. The heme proteins have essential roles that involve not only diverse redox reactions (*1*), including protection against oxidative stress (catalase and GSH peroxidase), fatty acid metabolism (cytochrome b_5 and cytochrome b_5 reductase), and electron transport in mitochondria (cytochrome *c*), but also oxygen storage and transport (hemoglobin and myoglobin), signaling processes (nitric oxide synthase and guanyl cyclase), and steroid and xenobiotic metabolism (cytochrome P450 systems).

Consequently, a variety of heme–protein linkages, heme binding sites, and axial coordinations to the heme iron have evolved. In a few heme proteins such as the mitochondrial electron transporter cytochrome *c*, heme is bound covalently via thioether bonds, but in the majority of cases heme is noncovalently bound through a variety of interactions, including hydrophobic contacts between the tetrapyrrole ring and nonpolar amino acid residues in the heme binding site, salt bridges between heme propionates and Lys or Arg residues, and axial coordination between the central heme iron and one or more His, Met, Cys, Lys, or Met residues. Two strong-field ligands, such as bis-His coordination in cytochrome b_5 (*2*), produce a low-spin complex, and one or more weak-field ligands induces a high-spin complex, such as His–hydroxide in metmyoglobin.

Understanding the roles of the diverse heme–protein interactions in producing specific structures and functions has been increased by the ongoing quest for three-dimensional structures in addition to the classic examples of myoglobin, hemoglobin, cytochrome *c*, and others. Among the novel prokaryotic heme proteins whose structures have been analyzed are cytochrome P450*cam* (*3*), the secreted heme binder Has A in *Serratia marsecens* (*4*), bacterioferritin (*5*), and the iron-responsive transcription factor Fur (*6*) of *Escherichia coli*. Eukaryotic examples include cytochrome P450 isozymes (*7*), nitric oxide synthase (*8*),

[1] *Abbreviations*: Heme, ferri-protoporphyrin IX; N-domain, N-terminal domain of hemopexin; C-domain, C-terminal domain of hemopexin; mesoheme, iron-2,4-ethyl-deuteroporphyrin IX; SnPP; Sn-protoporphyrin IX; CoPP, Co-protoporphyrin IX; FTIR, Fourier transform infrared; SPDS, solvent perturbation difference spectra; CD, circular dichroism; MCD, magnetic CD; DSC; differential scanning calorimetry; EPR, electron paramagnetic resonance; NMR, nuclear magnetic resonance; MHBP membrane heme binding protein; HO-1, heme oxygenase-1; MT-1, metallothionein-1.

guanylate cyclase (*9*), and the heme binding transcription factor Hap1 from yeast (*10*).

However, taking advantage of the chemical versatility of iron and heme also requires considerable effort in obtaining, transporting, and storing iron as well as in preventing oxidative damage from both iron and heme. This presents a different challenge from those faced in accomplishing other biological roles, namely the need to bind heme reversibly and to target its fate in the body. Hemopexin, which acts as a reversible binder for heme transport, has a novel heme binding site (*11*), perhaps as a consequence of these unique requirements. For example, the heme in many of the proteins noted above lies between one or more helices, buried in the interior with the propionic acid groups projecting outward, none of which obtains in hemopexin.

Hemopexin participates in specific receptor-mediated transport of heme (iron-protoporphyrin IX) and binds heme noncovalently but avidly ($K_d \sim$ pM; *12*) in a 1:1 low-spin (*13*), nonreactive complex and effects the specific receptor-mediated transport of heme. The activity of hemopexin in binding and transporting heme plays a central role in the complex task of iron and heme homeostasis, as it contributes to minimizing loss of biological iron stores, preventing toxic heme-catalyzed oxidations, sequestering iron from pathogens, and generating coordinated cellular responses, including activation of signaling cascades (the N-terminal c-Jun kinase) (*14*) and coordinated expression of the heme oxygenase-1 (HO-1) and metallothionein-1 (MT-1) genes (*15, 16*).

Here, the structure and function of hemopexin, the mechanisms of hemopexin-mediated heme transport, and the biological consequences of this specific transport system are reviewed and questions for future research proposed. This is an opportune time for review since recent advances not only provide new viewpoints and directions but also enable new insights to be derived from earlier work.

II. Biological Properties of Hemopexin

Hemopexin was first identified as a heme binding β-globin in electrophoretograms of plasma of patients with hemolysis (*17–19*). The protein is synthesized and secreted by the liver (*20–22*), and during secretion the signal peptide is removed and the protein is glycosylated (*23*). Tissue forms of hemopexin are expected due to the presence of mRNA in brain (*24*), peripheral neurons (*25*), and neural retina (*26*), pointing to a function of hemopexin in barrier tissues.

Although normally present in normal human plasma in abundance (~0.6 mg/ml, 10 μM) (*27–30*), concentrations of hemopexin are sensitive to a variety of pathological conditions. Decreased levels have been noted in chronic and severe hemolytic states (*31*) and in heme infusion of acute intermittent porphyria patients (*32*). On the other hand, hemopexin levels increase in the acute-phase response (*33–36*), and hemopexin has been designated as a type II acute-phase reactant. Plasma hemopexin also increases in certain conditions of muscle breakdown and neuromuscular disease (*37*).

The human hemopexin gene has been cloned (*38*) and is located near the β-globin gene cluster on human chromosome 11 (*39*). The promoters of the human, rat, and mouse hemopexin genes have been cloned, and the human gene contains a liver specific element (*40*) and an interleukin-6 responsive element (*41*), consistent with the positive acute phase response of hemopexin (*33, 34*).

The complete mouse hemopexin gene has also been cloned (A. Smith *et al.*, unpublished data), and computer analyses of both the human and mouse genes show evidence for gene duplication and intron-mediated evolution, leading to the two-homologous-domain structure discussed later. Although the hemopexin gene appears to have evolved by duplication, no direct duplication of domain function occurred, unlike transferrin, which can form (2 Fe) : (1 protein) complexes. Whether during the evolution of hemopexin a heme binding site was lost or gained, or whether, considering the crystal structure described later, two half-sites became linked remains an interesting question.

III. Biological Activities of Hemopexin

A. Heme Transfer to Hemopexin and Direct Antioxidant Effects

In vivo heme is released into the plasma by erythrocyte lysis in the form of hemoglobin and by tissue trauma in the form of myoglobin, and both heme proteins are quickly oxidized to their ferric heme forms (methemoglobin and metmyoglobin) at the oxygen tension found in tissue capillary beds.

Haptoglobin binds hemoglobin in a 1 : 1 molar complex (two αβ dimers per haptoglobin) (*42*). This complex is quickly removed from the circulation via a suicidal receptor-mediated endocytosis (*43, 44*), which consequently depletes haptoglobin. Even a short bout of exercise can deplete haptoglobin (*45*), whose total plasma amount is equivalent to the hemoglobin in about 4.5 cm^3 of red cells.

The kidney is known to be particularly vulnerable to the toxic effects of heme and hemoglobin in disseminated intravascular hemolysis, and recent studies on transgenic mice lacking hemopexin reinforce this (*46*). The null mice were healthy, but induced hemolysis caused kidney damage, even in the presence of normal albumin levels, but no appreciable damage to heart, liver, brain, skeletal muscle, or spleen. The healthy state of the null mice and lack of general organ damage is presumably due to the presence of haptoglobin, which is sufficient for normal levels of erythrocyte lysis. However, the role of hemopexin in protection of cells against hemolytic events and circulating heme (*47*) is corroborated by the deleterious effects of enhanced hemolysis. Interestingly, haptoglobin null mice (*48*) did not exhibit decreased hemoglobin clearance, but did suffer increased tissue damage, again particularly in kidney. Clearance of hemoglobin heme by rapid transfer to hemopexin in the absence of haptoglobin could account for the unimpaired clearance. The viability and relative good health of both hemopexin null and haptoglobin null mice *under normal conditions* can be attributed to two factors: Both systems act primarily in times of stress; and, as in other biologically important systems, there is redundancy, overlap, and backup in heme transport provided by hemopexin, haptoglobin, and albumin.

The physiological and pathophysiological situation involves several factors. A simplified view of the various binding interactions of heme involved in hemopexin-mediated heme transport is shown in Fig. 1, and more information is given in Section III, D. Haptoglobin is limited to being effective only under mild stress due to its suicidal mechanism of transport. Albumin provides a means to keep heme from adsorbing to cells and infiltrating membranes, but albumin, too, has limitations. It is abundant, but this makes blockade by apo-protein a problem,

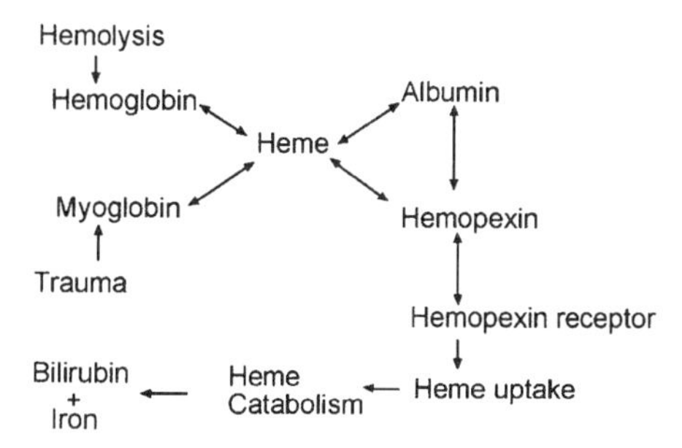

FIG. 1. Overview of intravascular heme catabolism. Hemoglobin, myoglobin, and other heme proteins are released into the circulation upon cellular destruction, and the heme moiety is oxidized by O_2 to the ferric form (e.g., methemoglobin and metmyoglobin). Haptoglobin can bind a substantial amount of hemoglobin, but is readily depleted. Ferric heme dissociates from globin and can be bound by albumin or more avidly by hemopexin. Hemopexin removes heme from the circulation by a receptor-mediated transport mechanism, and once inside the cell heme is transported to heme oxygenase for catabolism.

and in any event albumin lacks a specific cellular receptor system for heme, which makes targeting impossible. Finally, albumin binds a variety of hydrophobic ligands and thus lacks specificity, but this of course makes albumin ideally suited to be a circulating reservoir for several significant compounds. The abundance of albumin likely explains why the 100-fold difference in heme affinity between bovine ($K_d \sim \mu M$) and human albumin (~10 nM) (*49*) is not evolutionarily selective.

Heme dissociates from methemoglobin or metmyoglobin in the circulation and can be bound by hemopexin or albumin, a heme binding plasma protein of lower avidity than hemopexin (*49*). It is important that the heme be controlled, since this amphipathic, oxidatively active compound can nonspecifically associate with membrane lipids or lipoproteins and cause oxidative damage of vital biomolecules, including DNA (*50, 51*).

If bound first by albumin, heme circulates until it is transferred to hemopexin (*52*). *In vitro* in the absence of hemopexin, nonspecific cellular uptake of heme by diffusion is facile (*53*), but as expected, the presence of hemopexin greatly slows uptake (*54*), since receptor-mediated uptake is necessarily slower and of lower capacity than diffusion-limited uptake. There is currently no evidence that either receptors for albumin or membrane transporters for heme, like those in prokaryotes, are present in the plasma membrane of mammalian cells, although such transport proteins may be present in the membranes of organelles.

The dissociation rate of heme from methemoglobin and consequent formation of the heme–hemopexin complex is facile at 37°C, and the presence of small amounts of H_2O_2 (even below levels obtained from the "respiratory burst" of neutrophils) dramatically increases heme dissociation from oxyhemoglobin (*55*). The binding of heme by hemopexin prevents the oxidation of lipoprotein (*50, 55, 56*) and lipid and membrane damage (*57–59*).

Heme bound to liposomes (sonicated phosphatidylcholine) also rapidly transfers to hemopexin as well as to albumin and apo-myoglobin (*60*). The transfer is independent of protein concentration, and heme dissociation is the rate-determining step. There is a slow and a fast transfer phase, attributable to different heme orientations in the lipid bilayer, and the transfer rate decreases with increasing positive charge and cholesterol content of the liposome. In all cases, there was a nearly complete transfer. In a biological context, once heme dissociates from a cell membrane, in accord with the results just discussed, it would soon be bound by hemopexin for transport and catabolism. Without the plasma heme binding proteins, heme-catalyzed oxidative damage to cellular components, including DNA, would be extensive and have severe consequences. For example, micromolar levels of heme–hemopexin produce cell arrest in cultured cells, but upon removal of hemopexin the cells continue to grow and divide (*14*). Hemopexin is also present in barrier tissues such as the neural retina and is considered to provide antioxidant protection there also (*26, 61, 62*).

B. Iron Conservation and Nutritional Effects

After intravenous injection, heme from hemopexin is taken up principally by hepatocytes via a specific, membrane receptor-mediated process in which the protein is recycled and the heme degraded (*63, 64*). This action is analogous to that of the transferrin system (*65, 66*). A second membrane protein, membrane heme binding protein (MHBP), which appears to be related to cytochrome b_5 (A. Smith, unpublished), participates in the heme uptake process (*67*). The heme is catabolized by heme oxygenase and the iron released is stored on ferritin (*68*). Without this action of hemopexin, body iron stores would be rapidly depleted. In the absence of iron–transferrin, heme–hemopexin can serve as the sole source of iron for cultured cells expressing hemopexin receptors, including lymphocytes (*69*) and hepatocytes. At 5 μM heme–hemopexin, mouse Hepa cells no longer even secrete transferrin into the culture medium (A. Smith, unpublished).

The mechanism of iron and heme uptake by the intestine is becoming better understood (*70–72*), but clearly heme-iron is more efficiently absorbed from the gastrointestinal tract than inorganic iron (*73–75*), and there is a receptor for heme in the duodenal brush border membrane (*76*). Duodenal mucosal cells efficiently catabolize heme, and iron–transferrin can be detected in the plasma of blood vessels draining the intestinal segment shortly after ^{55}Fe–heme–histidine is administered (*75*).

C. Antipathogen Effects

Invading microbes require iron for growth, and transferrin, haptoglobin, and hemopexin all have bacteriostatic effects by sequestering iron and heme. Unfortunately, pathogens have evolved a wide array of methods to obtain iron, including siderophores, hemolysins, and receptors for heme (*77*), hemoglobin (*78*), transferrin (*79–81*), haptoglobin (*82*), and hemopexin (*83–87*). The expression by pathogens of mechanisms to obtain heme iron from heme–hemopexin may point to means to target drugs to these microbes using hemopexin as a delivery vehicle.

D. Hemopexin-Mediated Gene Regulation

The expression of several genes is induced or repressed by hemopexin-mediated heme transport. Most of these are simple responses of the cell to the increased heme (or iron derived from heme) in the cell. For example, HO-1 is induced (*15, 88*), ferritin levels rise (*14, 61, 89*), the transferrin receptor is down-regulated (*15*), and hemopexin mRNA itself is induced (A. Smith, unpublished). However, MT-1 is also induced, apparently to prepare the cell for oxidative stress; thus, in addition to sequestering heme in a low-spin, non-oxidatively active form, hemopexin also indirectly exerts antioxidant effects by inducing MT-1 (*16, 61, 90*).

As described later, hemopexin interacts with a variety of heme analogs, two of which, Sn-protoporphyrin IX (SnPP) and Co-protoporphyrin IX (CoPP), helped clarify the mechanism of MT-1 gene regulation by hemopexin. SnPP–hemopexin interacts with the hemopexin receptor and the SnPP (an inhibitor of HO-1, (*91*)) enters the cell and induces HO-1 (*92*). In contrast, CoPP–hemopexin interacts with the receptor, but CoPP is not internalized. Interestingly, the mere occupancy of the receptor by hemopexin is sufficient to activate signaling pathways and consequent induction of MT-1 (*90, 92, 93*), whereas heme uptake is required for activation of HO-1 gene transcription (*92, 93*).

The hemopexin heme transport system thus acts as an early warning system for cells by activating signaling pathways (including the N-terminal c-Jun kinase, kinases to release RelA/NFκB family members for nuclear translocation) and transcription of the HO-1 and MT-1 genes. The details of this aspect of hemopexin function with redox-active copper as an initial event in the coordinated induction of HO-1 and MT-1 by heme–hemopexin have recently been reviewed (*89*) and are not presented in detail here.

IV. Physical–Chemical Properties of Hemopexin

A. General Structural Features

Hemopexin is a single polypeptide chain (439 residues in humans) and circulates in the blood as a monomer. There are three or four (depending on the species) N-linked complex oligosaccharides, and the human congener has an O-linked monosaccharide on its N-terminal Thr (*94*), but detailed descriptions of the oligosaccharides are lacking and may account for the marked electrophoretic heterogeneity (*95*). The molecular weight of the mature, glycosylated protein is ~58,000 (*96*) and its p*I* is ~5.8 (*97, 98*), making it a plasma β1-globin. There are six

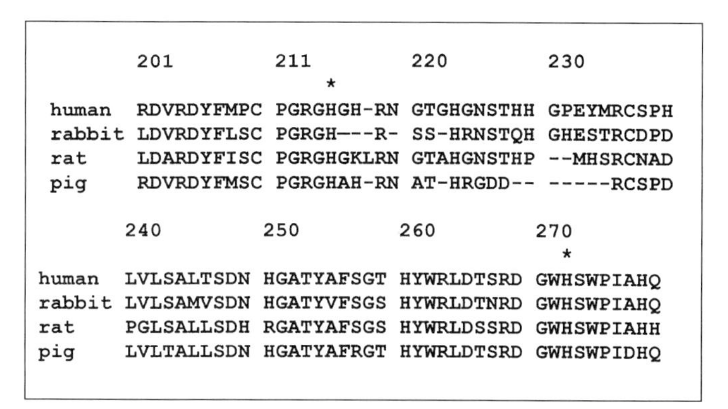

```
        201        211        220        230
                     *
human   RDVRDYFMPC PGRGHGH-RN GTGHGNSTHH GPEYMRCSPH
rabbit  LDVRDYFLSC PGRGH---R- SS-HRNSTQH GHESTRCDPD
rat     LDARDYFISC PGRGHGKLRN GTAHGNSTHP --MHSRCNAD
pig     RDVRDYFMSC PGRGHAH-RN AT-HRGDD-- -----RCSPD

        240        250        260        270
                                           *
human   LVLSALTSDN HGATYAFSGT HYWRLDTSRD GWHSWPIAHQ
rabbit  LVLSAMVSDN HGATYVFSGS HYWRLDTNRD GWHSWPIAHQ
rat     PGLSALLSDH RGATYAFSGS HYWRLDSSRD GWHSWPIAHH
pig     LVLTALLSDN HGATYAFRGT HYWRLDTSRD GWHSWPIDHQ
```

FIG. 2. Sequences of the heme binding regions of human, rabbit, rat, and porcine hemopexins. Human mature hemopexin residue numbers are shown. His215 and His272, by analogy with the crystal structure of the rabbit congener, coordinate with the heme iron and are marked by asterisks. The PIR database accession numbers for the human, rabbit, rat, and porcine proteins are OQHU, OQRB, OQRT, and A55486, respectively. (The porcine congener has been identified as a hyaluronidase (*103*), but this has not been substantiated and was apparently an artifact. Highly purified human (*95*) and rabbit hemopexin preparations display no such activity (P. Weigle, personal communication), and hemopexin does not bind hyaluronic acid in solution (C. Saez and W. T. Morgan, unpublished observations) but may in a more complex agarose milieu (*95*).

conserved disulfide bonds (three in each domain) and no free sulfhydryl groups.

The amino acid sequences (Fig. 2) of human (*99, 100*), rabbit (*101*), rat (*102*), pig (*103*), and mouse (*101*) hemopexins are known. All exhibit strong sequence homology, e.g., the human congener has 79% identity with rabbit, 75% with rat, and murine, and 74% with pig hemopexin. As noted earlier, a domain structure with two homologous structures derived by gene duplication is predicted from the sequence (*38, 99*), and the hinge region between the two domains is strikingly less conserved.

Hemopexin is notably rich in Trp residues (15–17 residues, nearly all of which are conserved), and FTIR spectroscopy indicated a high β-sheet secondary structure (*104*). The two-domain structure was demonstrated by cleavage of rabbit apo-hemopexin into two approximately equal domains, a heme binding N-terminal (N-domain) and nonbinding C-terminal domain (C-domain) (*105*). Heme–hemopexin is refractory to cleavage, and the association affinity of the isolated domains (interdomain interaction) is enhanced more than 50-fold by heme binding (*106*). The "pexin" domain structure also occurs in vitronectin (*107, 108*), interstitial collagenase (*107*), and certain metalloproteinases (*109*), and the structural information now available on hemopexin may prove helpful in elucidating the functions of pexin domains in other proteins.

B. Heme–Hemopexin Characteristics

The heme–hemopexin complex has a stoichiometry of 1 : 1 (*110, 111*) and a remarkably high stability (K_d near pM) (*12*), accounting for the ready transfer of heme to hemopexin from human albumin (*52*). The redox potential of the heme–hemopexin complex is +60 mV (*112*), ensuring that in oxygenated plasma the heme iron is in the ferric state. In addition, a low redox potential is characteristic of heme groups exposed to solvent, as substantiated by solvent perturbation difference spectra (SPDS) (*113*). Heme bound to hemopexin exhibits absorbance spectra indicative of a low-spin complex (Fig. 3), and reduction causes a redshift and intensification in signal consistent with retention of both strong field axial ligands. The heme–hemopexin complex was shown to be low-spin by EPR spectroscopy (*13, 114*), and chemical modification studies (*115, 116*), and near-IR magnetic circular dichroism (MCD) spectra also indicate that intact heme–hemopexin contains bis-histidyl heme iron coordination and significantly that heme–N-domain has similar coordination (*114*).

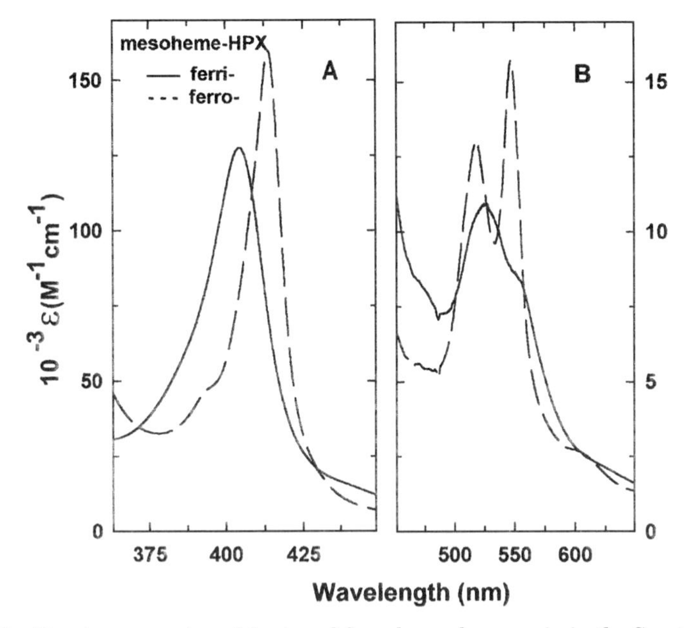

FIG. 3. Absorbance spectra of ferri- and ferro-heme–hemopexin in the Soret and visible regions. The spectra of rabbit ferri- and ferro-mesoheme–hemopexin (solid line and dashed line, respectively) are shown and are typical of low-spin bis-histidyl heme proteins. Other species display similar spectra, and a variety of other 2,4-substituted heme analogs are also bound by hemopexin.

Hemopexin is capable of binding a wide variety of natural and synthetic porphyrins and metallo-porphyrins, including protoporphyrin IX, coproporphyrin I, SnPP, CoPP, and tetra (4-sulfonatophenyl) porphinatoferrate (FeTPPS) (*92, 111, 117–121*). This is consistent with the relatively open heme binding pocket indicated by SPDS. Also, there is a requirement for negatively charged peripheral substituents, an early suggestion of a residual positive charge in the heme binding site (*118, 120*), and kinetic data show a dependence on ionic strength consistent with a residual 2+ charge in the heme binding site (*122*).

Binding of heme is rapid and can be described as a single, second-order process with a $k = 1.8 \times 10^6\ M^{-1}\ s^{-1}$ in 40% Me_2SO–water, pH 7.4, $\mu = 0.2$ M (NaCl), and $k = 3.9 \times 10^7\ M^{-1}\ s^{-1}$ in water, pH 7.4, $\mu = 0.2$ M (NaCl), with 25 mM caffeine (*122*). (The tendency of heme to aggregate in aqueous solution at physiologic pH (*123*) requires the use of mixed solvent systems to obtain reliable kinetic measurements, but raises the possibility that solvent effects on the protein perturb heme binding.) Although dimeric iron-porphyrin is bound initially, the complex is unstable, and the kinetic data show a rapid conversion to a stable 1 : 1 complex (*120*).

Formation of the bis-histidyl–heme complex also produces characteristic alterations in the protein's conformation, particularly in the environment of aromatic amino acids, notably tryptophan. Exposure of Trp residues to solvent decreases (*113*), Trp fluorescence is quenched (*111*), and an unusual band of positive ellipticity at ~230 nm attributable to Trp is nearly doubled in intensity (Fig. 4) (*104, 111, 124*).

Binding of heme by isolated N-domain causes a change in sedimentation coefficient consistent with a more compact conformation and leads to the more avid association with the C-domain (*125*). Sedimentation equilibrium analysis showed that the K_d decreases from 55 μM to 0.8 μM (Fig. 5) (*106*). In addition, the calorimetric ΔH (+11 kcal/mol) and ΔS (+65 kcal/mol K) for the heme–N-domain–C-domain interaction and the ΔH (−3.6 kcal/mol) and ΔS (+8.1 kcal/mol K) derived from van't Hoff analysis of ultracentrifuge data for the interaction in the absence of heme indicate that hydrophobic interactions predominate in the presence of heme and a mix (e.g., hydrophobic and van der Waals forces) drives the interaction in the absence of heme. However, FTIR spectra (Fig. 6) indicate that little change in the secondary structure of domains or intact hemopexin occurs upon heme binding (*104*).

Many metal-free porphyrins or porphyrins lacking a metal and therefore unable to ligate with His residues are nonetheless also bound by hemopexin, and changes in absorbance and fluorescence of both hemopexin and the ligand result. However, the affinity of binding

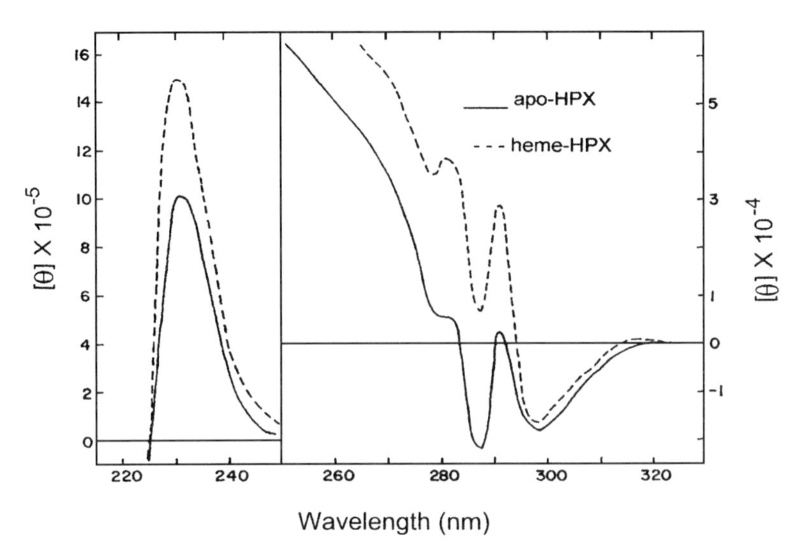

FIG. 4. CD spectra in the near and far UV of apo- and heme–hemopexin. The CD spectra of rabbit apo- and heme–hemopexin (solid line and dashed line, respectively) at pH 7.4 in 0.05 M sodium phosphate buffer are shown. The increases in ellipticity in the near UV are attributable to changes in tertiary conformation leading to altered environments of aromatic residues, particularly tryptophan. The unusual positive ellipticity in the far UV is attributable to tryptophan–tryptophan interactions that are perturbed by heme binding (*124, 130*). This positive signal precludes analysis of the secondary structure of hemopexin using current CD-based algorithms.

is $\sim 10^6$ fold less avid, and no detectable changes in hemopexin conformation are induced (*126, 127*). As expected, chemical modification of histidine residues of hemopexin abolishes normal heme coordination, and heme binding is consequently weakened and conformational changes are absent (*115*). Thus, the bis-histidyl–heme coordination complex of hemopexin is of fundamental functional and structural importance.

Defining in detail the conformational changes in hemopexin caused by heme is important for understanding its mechanism of action. For example, hemopexin circulates under normal conditions as the apoprotein. Thus, when some heme appears in the circulation for hemopexin to bind, a change in conformation is necessary for the hemopexin receptor to recognize the heme complex and "fish it from the sea" of apo-hemopexin. Furthermore, conformational changes in hemopexin induced by its interaction with receptor could help effect the release of heme from its tight complex, although heme release, as discussed later, probably involves reduction of the heme iron and exposure to the acidic pH of the endosome.

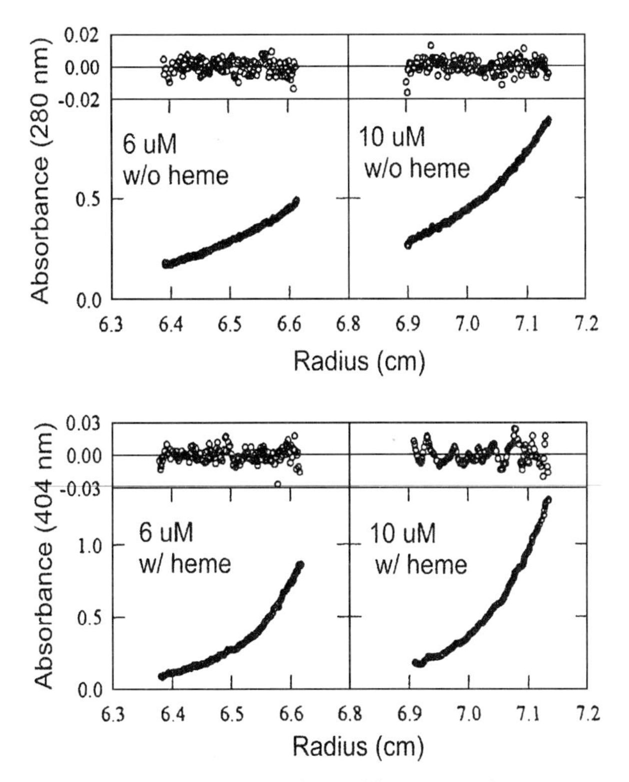

FIG. 5. Sedimentation equilibrium analysis of hemopexin domain interactions. Mixtures of N- and C-domain were centrifuged to equilibrium at 25°C in the absence (upper panels) and presence (lower panels) of heme. Nonlinear fitting procedures were performed to obtain apparent K_d values, and residuals of the fits are shown in the top portion of each panel (*106*).

C. CRYSTAL STRUCTURE OF HEMOPEXIN

The structure of the C-domain of hemopexin was determined first (*128*). The structure is a four-bladed β-propeller (Fig. 7), the smallest β-propeller known, and serves as the paradigm for the several proteins known to have a pexin domain, including vitronectin (*108*), and several metalloproteinases (*107*). The repeats evident in the sequence of hemopexin (*99–101*), for instance DAAV/F motifs and WD repeat, form a large part of the β-strands of the four blades, which are connected by short loops and α-helices.

This structure "looks" stable because of the compact, symmetric structure with multiple reinforcing interactions, and direct examination of hemopexin (*vide infra*) has borne this out. Smith *et al.* (*129*) have

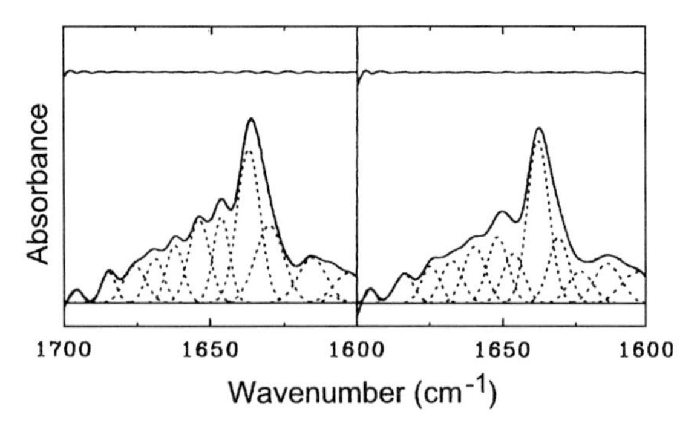

FIG. 6. Deconvolved amide I' region FTIR spectra of apo- and heme–hemopexin. The amide I' FTIR spectra of apo- and heme–hemopexin in D_2O were recorded and curve-fitted to resolve the individual bands. The differences between the original and fitted curves are shown in the upper traces in the panels. The estimated helix (15%), beta (54%), turn (19%), and coil (12%) content of the apo-protein are not significantly changed upon heme binding (*104*). This analysis was required because of the positive 231-nm ellipticity band in hemopexin and is consistent with the derived crystal structure results.

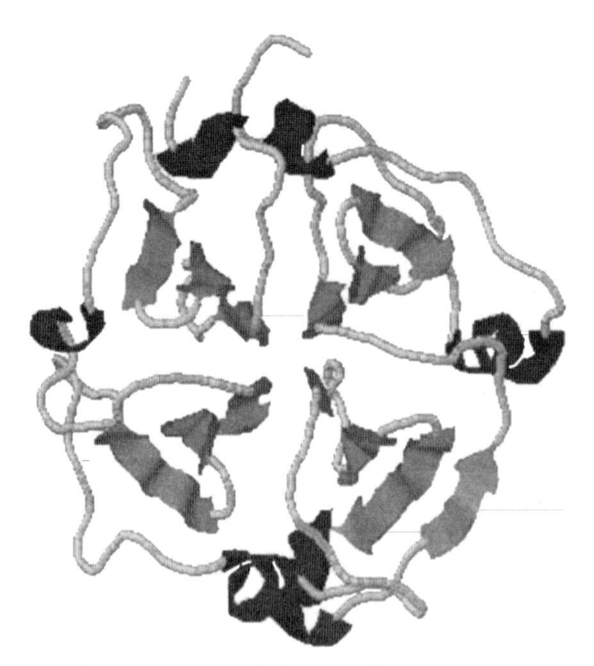

FIG. 7. The crystal structure of the C-domain of hemopexin (PDB accession number 1HXN) (*128*) showed a four-bladed β-propeller structure, which because of sequence similarity was also expected in the N-domain. The high degree of beta structure and limited α-helix content agrees with the earlier FTIR analysis.

depicted the WD repeat propeller structures as stable platforms with three potentially interacting surfaces: the top, bottom, and circumference. Binding ions within the central tunnel of the pexin domain also stabilize the protein, as do disulfide bonds, particularly that anchoring the N-terminal blade to the C-terminal blade. Notably, the other pexin family members also are characterized by their ability to form reversible complexes involving sequential or simultaneous interactions with ligands and proteins, as hemopexin does with heme and its receptor. Interestingly, despite the extensive data available on prokaryotic genome sequences, no recognizable WD-repeat proteins have been noted.

The isolated C-domain recombines with heme–N-domain and alters the conformation of its partner domain (*106, 125, 130*), but has no intrinsic heme or porphyrin binding activity (N. Shipulina *et al.*, unpublished).

Determination of the structure of the entire rabbit heme–hemopexin complex (*11*) (Fig. 8) clearly revealed the expected complementary structures of the N- and C-domains. More importantly, several fundamental

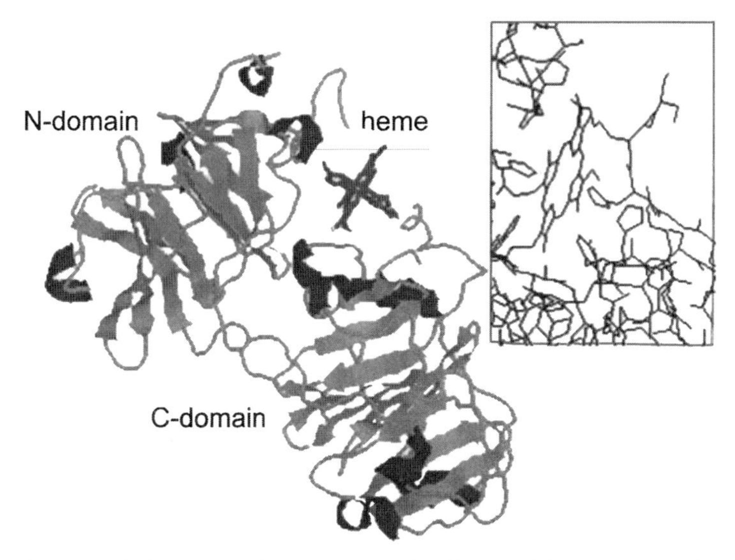

FIG. 8. Crystal structure of heme–hemopexin. The crystal structure of the rabbit mesoheme–hemopexin complex (PDB accession number 1QHU) (*11*) showed heme to be bound in a relatively exposed site between the N- and C-domains with one axial His ligand being contributed by the hinge or linking region between the domains and the other by the C-domain. Also noteworthy is the disposition of the heme with its propionate residues pointing inward and neutralized by positive charges in the binding site.

and unusual features of the heme binding site were discovered: the heme is bound between the two domains in intact hemopexin and is coordinated to His 213 near the junction of the N-domain and the hinge region and to His 265 about 25 residues from the beginning of the C-domain; the heme propionate side chains, unlike those of most known heme proteins, point inward and are neutralized by arginyl side chains; the heme binding site has an overall positive character and several histidine and tryptophan residues surround the bound heme; and the heme can bind in either of two conformers differing by a 180° rotation about the α and γ meso carbon axis, as found in cytochrome b_5 (*131*).

This structure is in good agreement with previous chemical (*115, 124*) and absorbance, EPR, NMR, and natural and magnetic circular dichroism (CD) spectroscopic results (*13, 104, 110, 114, 116, 126, 132*) regarding bis-histidyl coordination and conformational linkage with tryptophan residues that are prevalent in the vicinity of the heme as well as a positive charge in the binding site (*122*). A notable exception is that several studies showed that the isolated N-domain itself forms a bis-histidyl heme binding site (*101, 105, 106, 114, 125*). A complex is also formed when the N-terminal 22 residues are removed from the N-domain (*133*). However, comparison of the EPR (Fig. 9) and NMR spectra (Fig. 10) of heme–N-domain and heme–hemopexin (*114*) clearly indicate that the heme environments differ, consistent with two modes of heme binding.

The existence of two independent but exclusive (since the stoichiometry of binding is always 1 : 1) heme binding sites in hemopexin merits careful consideration. Histidine residues 56 and 127 of the N-domain were proposed to be heme coordinating residues based on chemical modification (His 127) and conservation of sequence (His 56) (*101*). However, site-directed mutagenesis showed that mutation to Thr at His 127, but not at His 56, reduced the affinity of mutant hemopexin for heme by an order of magnitude relative to wild-type (*134, 135*). Molecular modeling indicates that heme could bind in a tunnel-like depression in the rabbit hemopexin N-domain structure and coordinate with His 127 and His 80. The possibility of two binding modes raises interesting possibilities regarding structure–function mechanisms of hemopexin. Some of these are described next.

D. Dynamics of Heme–Protein–Hemopexin Interactions

A dynamic system of heme binding exists in the bloodstream, and to function in heme transport hemopexin must be able to (a) obtain heme from methemoglobin (released by hemolysis) and metmyoglobin (introduced by trauma); (b) compete with serum albumin for heme in the

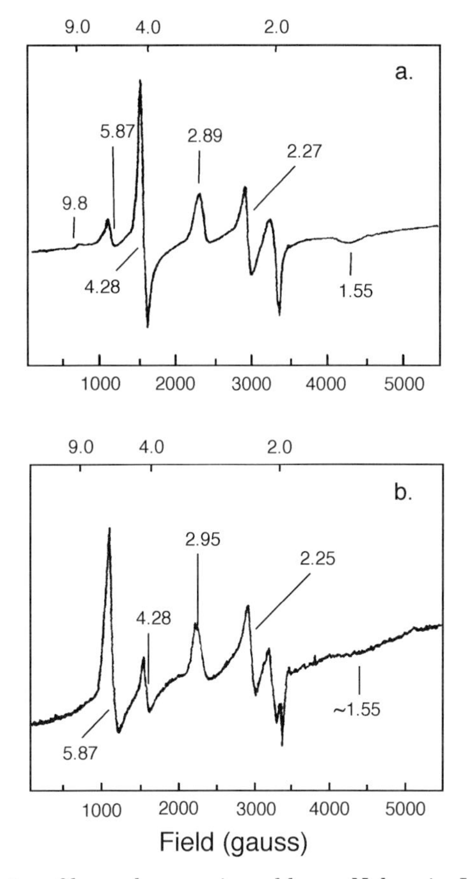

FIG. 9. EPR spectra of heme–hemopexin and heme–N-domain. X-band EPR spectra at 4 K of ferri-mesoheme–hemopexin (a) and ferri-mesoheme–N-domain (b) are shown. The concentration of both heme complexes was 0.15 mM in 50 : 50 (v/v) 10 mM sodium phosphate/150 mM NaCl (pH 7.2) : glycerol. The *g*-value scale is noted at the top and the *g*-values observed are noted in each spectrum. Although both complexes are low-spin (some adventitious high-spin iron is present), the differences in *g*-values indicate nonidentical heme environments in the two complexes (*114*).

circulation; and (c) interact with the hemopexin receptor. A simplified diagram of the various interactions is shown in Fig. 1. The overall process is governed by the relative concentrations and affinities for heme of the various protein components and the kinetics of binding and heme uptake.

The affinity of hemopexin for heme (estimated to be ~pM (*12*)) is among the highest reported for a noncovalent heme complex, and direct examinations of heme binding support the effectiveness of hemopexin

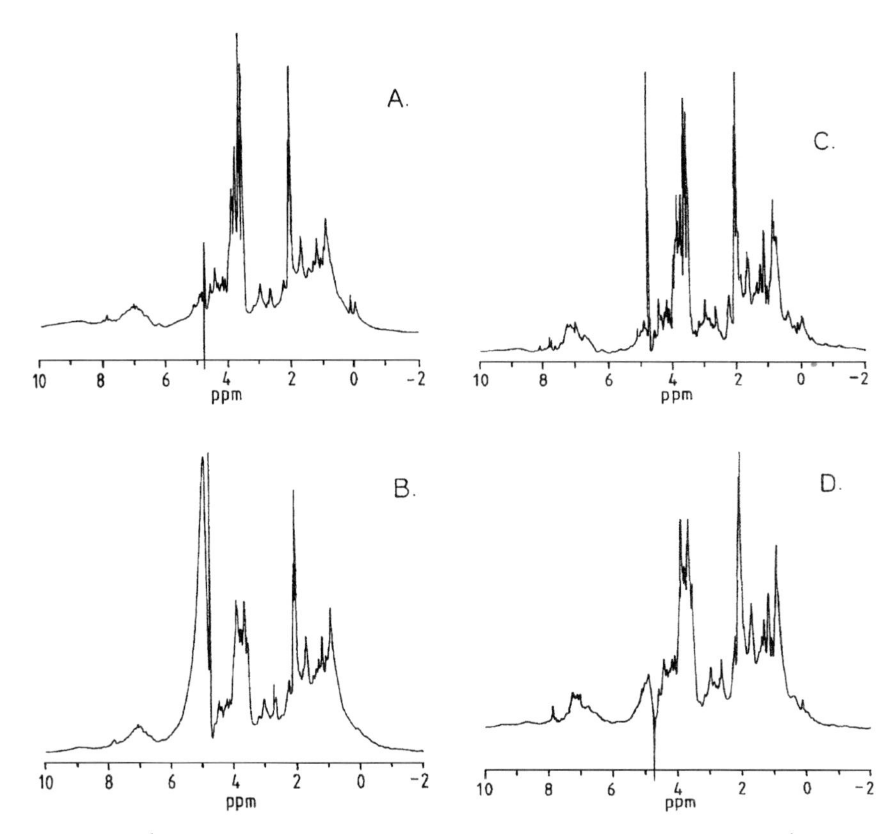

FIG. 10. ^{1}H NMR spectra of rabbit hemopexin and domain. The 400 MHz ^{1}H NMR spectra at 298 K of rabbit apo- and ferri-mesoheme–hemopexin are shown in panels A and B, and the spectra of apo- and mesoheme–N-domain in panels C and D, respectively. The spectra demonstrate that the heme environment in the intact protein is distinct from that in the N-domain. The heme resonances are broadened in the N-domain, consistent with a greater accessibility to solvent (*114*).

in scavenging heme in the circulation. Heme transfers readily from ferri-hemoglobin to hemopexin (*12*), and this transfer is much more rapid at 37°C than at 20°C and is accelerated further by the presence of micromolar amounts of H_2O_2 (*55, 56*).

Adding heme to human serum or to a mixture of 70 human albumin: 1 hemopexin showed that heme is rapidly bound by hemopexin, whereas heme transfer from a heme–albumin complex (treated to remove loosely bound heme) is much slower (*52*) and requires 24 hours to complete. This is in accord with the existence of one strong ($K_d \sim 10$ nM) and several weak heme binding sites on human albumin (*49*), from which heme readily dissociates. The presence of imidazole catalyzes the transfer via a ligand mediated pathway (*122*); interestingly, reduction of albumin-bound heme causes a much more rapid transfer of heme to hemopexin ($K_d \sim$ pM), even from the tight albumin site (*52, 136*).

These *in vitro* findings on the effectiveness of hemopexin in managing intravascular heme are bolstered by experiments with animals (*47, 63, 68*), including hemopexin null mice (*46*), cells (*137*), and isolated membranes (*64, 67*) as well as the function of the specific hemopexin membrane receptor (discussed later). Nonetheless, several reports have concluded that hemopexin does not play a vital role in heme transport and have even questioned the idea of the existence of specific hemopexin receptors. In our view, several key points are often overlooked. First, heme is amphipathic and lipophilic and readily associates nonspecifically with biological membranes, and the exposure of animals, cells, or membranes to a bolus of free heme is not physiologically relevant. The fact that hemopexin slows the uptake of heme by cells expressing hemopexin receptors (*54*), rather than mitigating against the function of hemopexin, actually substantiates it, since a specific receptor-mediated process requiring endocytosis is inevitably a slower process than one limited only by diffusion. Second, exposing cells to heme–albumin to assess uptake also is not a good model of most physiological processes, since the binding dynamics in the circulation (Fig. 1) together with the affinities and concentrations of the various heme binding components precludes formation of appreciable amounts of stable heme–albumin complexes. The exceptions are extreme hemolysis or trauma or heme-therapy regimes, which eventually deplete hemopexin, since the recycling mechanism must inevitably be less than perfectly efficient, and for this situation heme and heme–albumin are suitable experimental systems. Additional considerations on hemopexin-mediated cell uptake of heme are presented later.

E. Reduction of Heme–Hemopexin and Binding of Exogenous Ligands

Exposure of ferri-heme–hemopexin to imidazole or KCN can displace one or both of the heme coordinating His residues, but millimolar concentrations are required (*138*). Other potential ligands such as azide or fluoride are inactive. This coordination stability of the ferri-heme–hemopexin bis-histidyl complex, despite the exposed heme site, is borne out by thermal unfolding studies (Section IV,F). Reduction of the heme–hemopexin complex, however, has dramatic effects on its stability.

Heme–hemopexin is readily reduced by $Na_2S_2O_4$, and the rate k_{obs} for reduction (2.5×10^{-3} $[S_2O_4^{2-}]^{1/2}$) is much higher than that of heme–albumin reduction (*136*). This is also consistent with a relatively open heme site that has some positive charge character. The ferro-heme–hemopexin complex remains intact upon reduction of the heme iron (Fig. 3); nonetheless, CO or ˙NO readily displaces one histidyl axial ligand and binds to ferro-heme–hemopexin (Fig. 11). This ligation produces an unusual bisignate CD spectrum (Fig. 11) (*139*), which is also seen in the ˙NO-ferro-mesoheme–hemopexin complex (*140*). The affinity and kinetics of CO binding are pH dependent. Binding at pH 7 is biphasic, and the extent of the fast CO binding reaction decreases as pH increases (*141*), presumably because of histidyl protonation effects.

The biphasic reaction with CO points to the existence of multiple heme–hemopexin conformers, and this is borne out by spectral analyses. The absorbance spectra of rabbit ferri-, ferro-, and CO-ferro-mesoheme–hemopexin are entirely analogous to those of other bis-histidyl heme proteins such as cytochrome b_5 (*142*), but the CD spectra exhibit unusual features (Fig. 11). Of particular interest are the weak signal of the ferro complex and the bisignate signal of the CO-ferro complex (also seen in the ˙NO-ferro-mesoheme–hemopexin complex (*140*) and in human ferri-protoheme–hemopexin (*139*)).

The ferro-complex CD spectrum shows that reduction of the heme iron alters the heme environment. Redox-induced protein conformation changes could alter the symmetry in the heme pocket or produce two binding modes for the reduced complex whose asymmetries nearly cancel each other. Redox-linked conformational changes are especially interesting in view of recent findings of oxido-reductase activity associated with the heme–hemopexin–receptor interaction (*89*).

Upon reduction, not only are the molecular orbitals altered, but the heme iron is forced from the plane of the tetrapyrrole ring and different bonding energies are produced with the heme axial ligands. Tension from movement of the heme-coordinating histidyl ligands would induce conformational changes in the attached segments of the hemopexin structure, analogous to the Perutz (*143*) mechanism for O_2 binding by hemoglobin; subsequent CO binding would stabilize one or the other of the conformers. Furthermore, hemopexin, unlike hemoglobin with one proximal and one distal histidine, has two equally probable sites for the incoming CO to occupy (Fig. 12). (The absorbance spectra clearly show that the heme remains bound to hemopexin upon heme reduction and CO-coordination.)

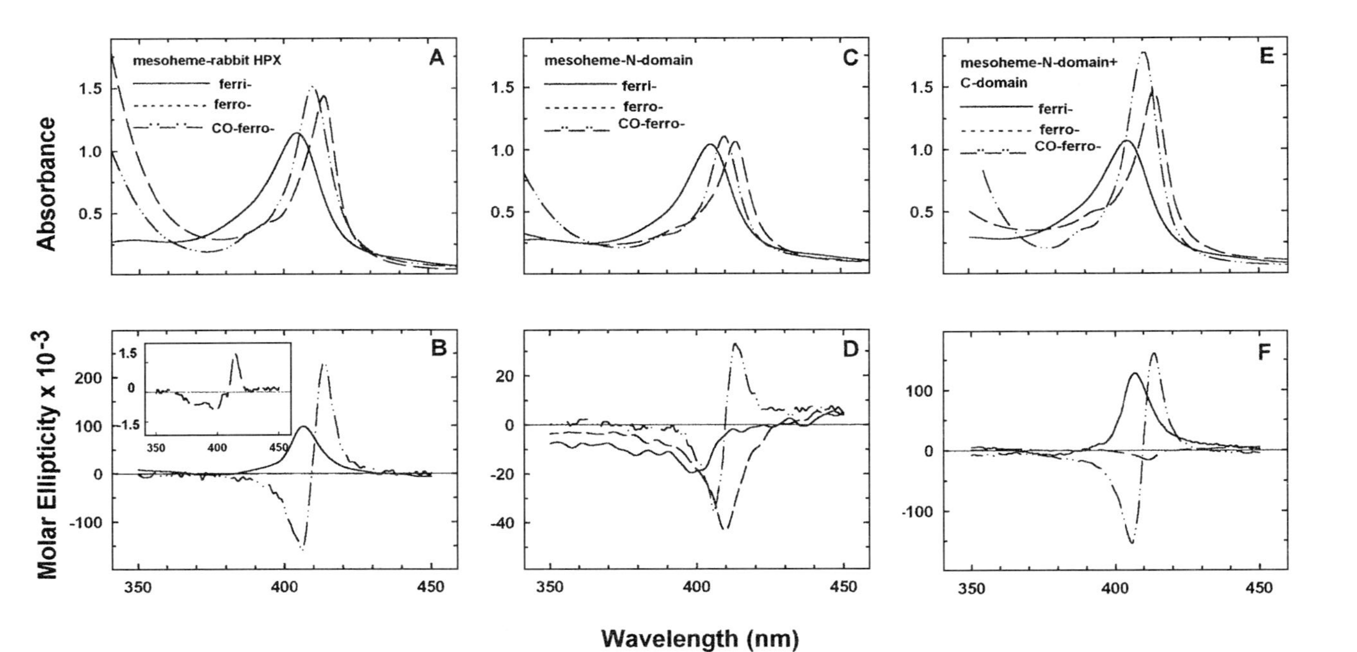

FIG. 11. Absorbance and CD spectra of ferri-, ferro-, and CO-ferro-heme complexes of hemopexin and its domains. Panels A, C, and E show the Soret region absorbance spectra, and panels B, D, and F the corresponding CD spectra. Rabbit hemopexin (panels A and B), N-domain (panels C and D), and N-domain–C-domain (panels E and F) complexes with mesoheme in the ferri- (solid lines), ferro- (dashed lines), and CO-ferro- (dash-double dot lines) states in phosphate-buffered saline, pH 7.4, are presented. The differences between the spectra of hemopexin and the N-domain point to multiple heme binding modes in the protein (*139*).

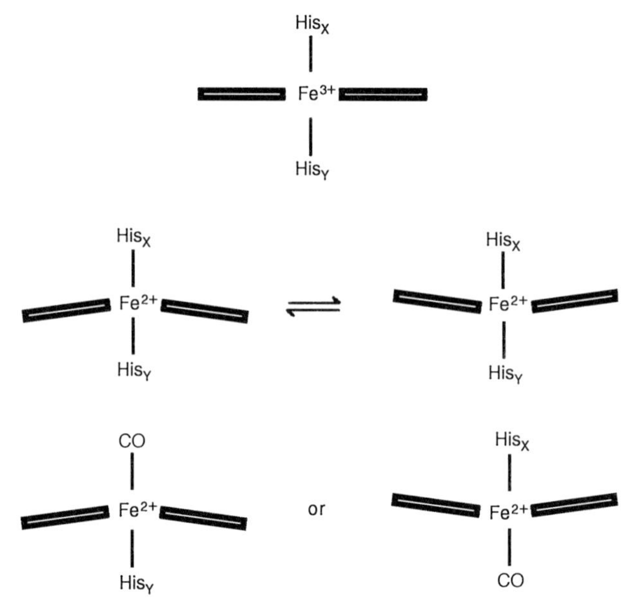

FIG. 12. Schematic views of bis-histidyl ferri-, ferro-, and CO-ferro-heme–hemopexin. Unlike myoglobin with one open distal site, heme bound to hemopexin is coordinated to two strong field ligands, either of which *a priori* may be displaced by CO. This may well produce coupled changes in protein conformation like the Perutz mechanism for O_2-binding by hemoglobin (*143*). The environment of heme bound to hemopexin and to the N-domain may be influenced by changes in the interactions of porphyrin-ring orbitals with those of aromatic residues in the heme binding site upon reduction and subsequent CO binding.

However, this explanation is not sufficient to account for the biphasic CD spectrum of human ferri-protoheme–hemopexin (with 2,4-vinyl substituents), as well as the much weaker human CO-ferro-heme–hemopexin bisignate signal compared to the rabbit congener (*139*), and hence other factors must be involved. Several potential effectors exist: (a) exciton coupling; (b) the conformers produced by a 180° rotation about the α- and γ-meso-carbon axis and consequent nonisometric interactions of the asymmetric 2,4- and 9,10-substituents; (c) the aromatic tryptophan residues near the heme binding site (s); and (d) two independent binding modes or sites.

A bisignate spectrum is characteristic of exciton coupling between identical chromophores on one molecule (*144*), an impossibility with the 1 : 1 heme–hemopexin complex. Bisignate CD spectra have been observed with bilirubin–albumin; however, bilirubin, unlike the planar molecule heme, can adopt an extended, helical stacked conformation allowing intramolecular dipole interactions (*145*). The effects of meso-carbon conformers and the 2,4-tetrapyrrole ring substituents may also play a role in affecting the observed CD spectra. Two conformers (180° rotation about the α and γ meso carbon axis of heme) were found in the crystal structure of rabbit ferri-mesoheme–hemopexin (*11*). Asymmetric interactions between the electron-rich vinyl groups of heme and aromatic residues near the binding site could produce the unusual spectra.

An intriguing possibility is that heme can be bound in two distinct modes by hemopexin, presumably at one of two sites: the site shown by the crystal structure and one involving the N-domain site. As expected from the obviously different heme environments, the EPR, NMR, and CD spectra of heme–N-domain complexes differ extensively from those of heme hemopexin (Figs. 9–11), even though the absorbance spectra are similar. Notably, addition of C-domain to heme–N-domain restores much of the CD characteristics of the holo-protein, particularly the bisignate CO-reduced-heme–hemopexin spectrum, suggesting a shift from one heme binding mode to another influenced by interdomain interactions. *In toto,* the data support the idea of multiple, functionally significant heme states in hemopexin. Further studies of site-directed mutants and determination of additional crystal structures are under way to resolve exactly how hemopexin binds heme.

F. Effects of pH and Temperature on Heme–Hemopexin

The thermodynamic stability of hemopexin has been examined using DSC (Fig. 13), which showed apo-hemopexin to be a stable protein with a single T_m of 54°C (ΔH 185 kcal/mol), which increases to 66.5°C (ΔH 290 kcal/mol) upon binding heme (*130*). The N-domain of hemopexin is less stable (T_m 52°C, ΔH 95 kcal/mol) but is even more strikingly stabilized by ferri-heme (T_m 78°C, ΔH 370 kcal/mol). The presence of C-domain (T_m 49.5°C, ΔH 140 kcal/mol) slightly destabilizes heme–N-domain (T_m 75°C, ΔH 320 kcal/mol) (*130*), showing another effect of interdomain interactions that may act in heme release.

Temperature and pH effects on hemopexin, its domains, and the respective heme complexes have also been examined using absorbance and CD spectroscopy, which reflect stability of the heme iron–bis-histidyl coordination of hemopexin and of the conformation of protein, rather than overall thermodynamic unfolding of the protein. Using these spectral methods to follow temperature effects on hemopexin stability yielded results generally comparable to the DSC findings, but also revealed interesting new features (Fig. 14) (N. Shipulina *et al.*, unpublished). Melting experiments showed that apo-hemopexin loses tertiary

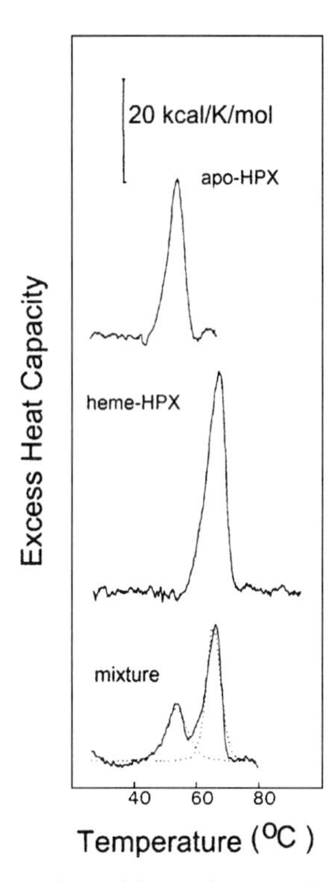

FIG. 13. DSC of apo-hemopexin and heme–hemopexin. DSC recordings of apo- and heme–hemopexin in 0.05 M sodium phosphate at pH 7.4 and of a mixture of the two are presented. The stabilization of hemopexin upon heme binding (*106*) is evident. Other results established an even greater stabilization of N-domain by heme.

structure, as monitored by second derivative absorbance spectra, with a T_m of 52°C in 0.05 M sodium phosphate buffer (comparable to DSC results) and is stabilized dramatically by adding 150 mM NaCl to the phosphate buffer (T_m 63°C). This presumably reflects the binding of ions in the central pocket of the pexin domain as noted earlier.

Importantly, heme coordination in ferri-heme–hemopexin, as monitored by Soret absorbance, is strikingly more stable than in the reduced form, both with and without NaCl (T_m 70°C and 51°C; T_m 55.5°C and T_m 48°C; respectively). This is consistent with a reduction step aiding heme release by weakening the complex prior to, or during, the actual release of heme effected by the hemopexin receptor. (The nondestructive mode of transport by hemopexin with recycling of intact protein demands mechanisms to reversibly lower the affinity of the complex.) With or without NaCl, lowering the pH from 7.4 to 6.5 has little effect

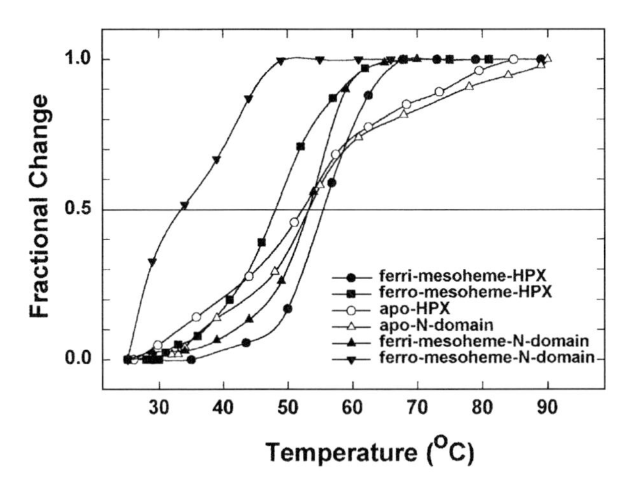

FIG. 14. Effects of temperature on the absorbance of hemopexin and the N-domain of hemopexin. The unfolding of hemopexin and N-domain in 25 mM sodium phosphate, pH 7.4, was examined using absorbance spectroscopy (N. Shipulina *et al.*, unpublished). The second derivative UV absorbance spectra of the protein moieties were used to follow protein unfolding and the Soret and visible region spectra to monitor the integrity of the heme complexes, as done with cytochrome b_{562} (*166*). The ferri-heme complex is more stable than the apo-protein moiety, but the T_m is slightly lower than that assessed by DSC, indicating that changes in conformation occur before thermodynamic unfolding. Reduction causes a large decrease in heme-complex stability, which is proposed to be a major factor in heme release from hemopexin by its cell membrane receptor, and addition of 150 mM sodium chloride enhanced the stability of all forms of hemopexin.

on the apo-protein's stability but increases the T_m of the three heme complexes by 2–8°C (N. Shipulina *et al.*, unpublished). This suggests either that any effect of lower pH in endosomes on heme release is less important than the effect of heme reduction, or that a more extreme acidification (e.g., to pH 5.5 or 5) is needed.

The effects of pH on ferri-heme binding by hemopexin monitored by heme absorbance are described by a broad plateau extending from pH 6 to 9 (Fig. 15) (*111*), with apparent pK_a values of pH 5 and 10.5. This is consistent with the thermal stability at pH 6.5 noted earlier. A similar profile is exhibited by the ellipticity at 231 nm due to Trp residue asymmetry, but the pK_a values are slightly shifted (Fig. 15) indicating some uncoupling of heme-coordination and conformational effects. The heme binding exhibits notably sharper titrations, suggesting that the disruption of the heme–histidine ligation, with the pK_a in the acidic region likely reflecting His residue protonation, is more sensitive to pH than the tertiary conformation of the Trp residues and is subject to more cooperative forces.

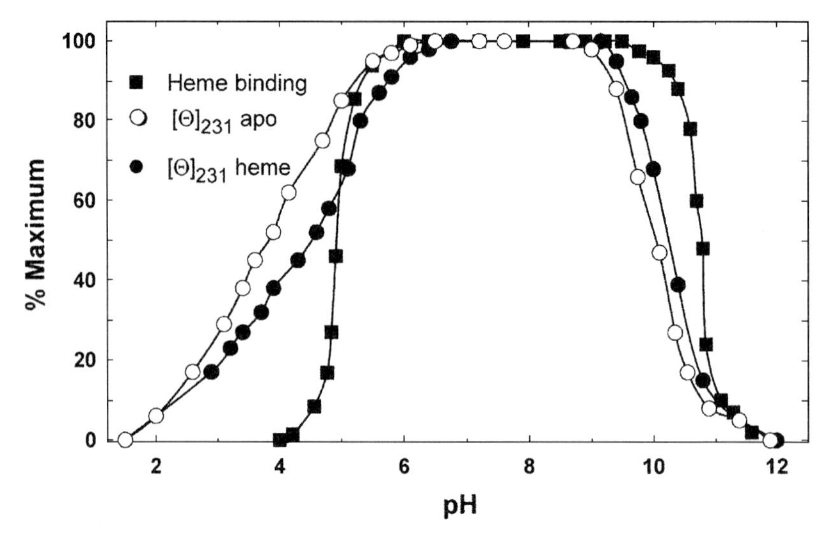

FIG. 15. Effects of pH on apo- and heme–hemopexin. The Soret region absorbance (filled squares) of rabbit heme–hemopexin was monitored in two separate titrations, from pH 7.4 to 11.8 in one and from pH 7.4 to 3.8 in the other. Similarly, the ellipticity at 231 nm of apo-hemopexin (open circles) and of heme–hemopexin (filled circles) was assessed from pH 7.4 to 11.8 and from pH 7.4 to 1.7 (*111*). The heme complex and the tertiary structure are unaffected by pH in the region from pH 6 to 9, and other values are normalized to these.

V. Hemopexin Receptor Properties and Activities

A. HEME UPTAKE *IN VIVO* AND *IN VITRO*

From the earliest observations on the plasma clearance of radiolabeled apo-hemopexin and heme–hemopexin complexes (*146, 147*), a surface hemopexin receptor has been inherent in the presumed function of hemopexin in heme transport, and action of a specific receptor is implicit in the results of many studies of transport in animals and cells. *In vitro* binding experiments generating saturable, high-affinity binding also confirm a specific molecular recognition between hemopexin and a cell membrane entity. Yet little molecular information has been gleaned, and some have questioned the concept and existence of hemopexin receptor-mediated heme transport, while accepting that hemopexin avidly binds heme, and have proposed albumin as the relevant transporter. However, no alternative models for the fate of the heme–hemopexin complex and no data supporting an active transport role for heme–albumin beyond simple dissociation of the hydrophobic ligand and diffusion to cell membrane sites have been reported. A review of the findings, however, indicates to us that the lack of information and concerns regarding the hemopexin receptor are due to technical difficulties and that a concerted, persevering effort is likely to capture this elusive molecule and enable its characteristics to be elucidated. Some of the pertinent observations follow.

In patients clearance of intravenous heme is rapid until hemopexin levels are depleted (*148*), and lack of interaction with hemopexin may explain the higher clinical efficacy of heme–arginate compared with hemin itself (*149, 150*). In intact animals, i.v. heme causes rapid association of hemopexin but not albumin with the liver (*47, 63, 68*), and heme uptake from heme–albumin complexes into isolated rat hepatocytes occurs via diffusion of heme released from its loose complex with BSA (*137*). Moreover, unlike uptake from heme–hemopexin, free heme uptake by cells occurred even at 4°C, as expected for nonspecific membrane association and in total disagreement with a membrane-receptor-mediated or active transport uptake process.

Some have considered uptake of heme from heme–albumin a receptor-mediated process (*54, 151*), but dissociation of heme from its weak complex with albumin is ignored. Thus, apparent saturation at higher heme–BSA levels is not necessarily due to saturation of receptor but is more likely due to lower concentrations of free heme. The lower capacity of hemopexin-mediated heme transport compared with diffusion of free heme is expected since there is a finite number of receptors per cell; a transport protein's job is to target a ligand, not to maximize the amount transported. Further, relevant comparisons must be made at short times on the order of minutes, not hours.

Early accounts in humans of the effects on hemopexin of elevated heme levels in disease (*152–155*) or experimental (*156*) conditions showed that hemopexin obviously bound heme and, more importantly, that a connection existed between that binding and heme clearance from the circulation. Subsequent work in animals made clear that the removal was targeted, for example, to the liver hepatocytes (*47, 63, 68, 146*), and hence must involve a specific recognition mechanism. This was confirmed in isolated or cultured cell experimental systems, and high-affinity, specific hemopexin receptors as well as a variety of biochemical responses of cells to heme–hemopexin have been noted in a variety of cell types, including human promyelocytic (HL-60), K562, retinal pigment epithelial cells, U937, $HepG_2$, placental syncytic BeWo (*26, 69, 137, 157–159*), and, recently, rat pheochromocytoma cells that differentiate into neurons (J. Eskew and A. Smith, unpublished observations). Not unexpectedly the number of receptors varies with cell type, but the reported apparent K_d values are all in the nanomolar range. The saturability of the hemopexin–receptor interaction has been shown *in vivo,* and three different species of hemopexin gave the same saturation level in rats, indicative of a single molecular species on the liver that binds the protein (*47, 63*).

B. Hemopexin Receptor Properties and Activities

Several interesting facts have emerged. First, hemopexin is not degraded as an obligate step in transport (*63, 137, 160*), like transferrin and unlike haptoglobin. Second, iron derived from heme delivered via the hemopexin system is stored on ferritin for reutilization and hence remains biologically available (*68*). Third, hemopexin can serve as the sole source of iron for cultured cells expressing the hemopexin receptor, readily replacing transferrin in defined serum-free medium. The specific routing of heme by the hemopexin receptor is evident from the drastically different fate of heme iron bound to asialo–hemopexin and hence taken into liver cells via the gal-receptor (*161*). Not only is asialo–hemopexin degraded, but heme iron is also not efficiently routed to ferritin. In contrast to asialo–hemopexin, heme–hemopexin undergoes receptor-mediated endocytosis and recycling by an uptake and intracellular route entirely congruent with that taken by iron–transferrin (*160*).

Nevertheless, little direct information has been obtained concerning the molecular properties of the hemopexin receptor. The receptor isolated from human placenta (*162*) had an apparent molecular mass on SDS-PAGE of 80–100 kDa. The murine hepatic hemopexin receptor was identified by crosslinking as a (*163*) disulfide-linked two-subunit protein with an apparent total mass of 85 kDa (Fig. 16). Using a heterobifunctional crosslinking agent, hemopexin was crosslinked to the smaller (20-kDa) subunit, indicating it has a free sulfhydryl group. However, the membrane disposition and roles of each unit remain to be defined. Recent work indicates that the human hemopexin receptor is a homodimer (R. Vanacore, W. T. Morgan, and A. Smith, unpublished observations). Moreover, we have determined by N-terminal sequence analysis that an 80 kDa protein identified as a putative human hemopexin receptor (*162*) is a soluble form of the transferrin receptor.

Another factor operating in the cellular uptake of heme by the hemopexin system is a membrane heme binding protein (MHBP) that accumulates heme as the system cycles (*67*). This protein ($M_r \sim 30$ kDa) binds heme and may serve as an intracellular vehicle to transport heme

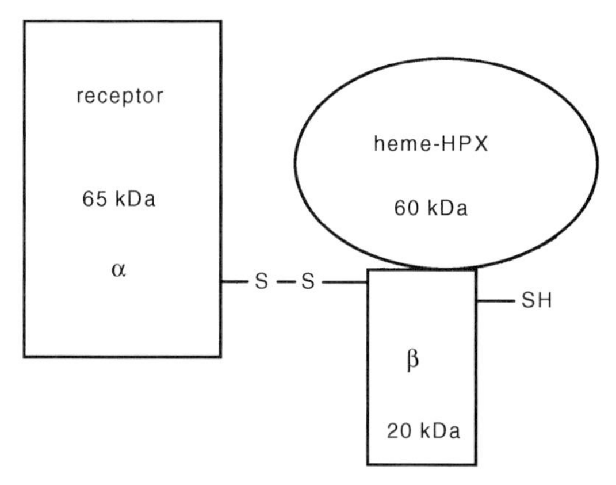

FIG. 16. Schematic depiction of the murine hemopexin receptor. The results of crosslinking studies (*163*) suggested that the murine receptor was composed of two subunits (α and β). The larger (65 kDa) is designated as α and is disulfide-linked to the smaller β subunit, which contains a free sulfhydryl. Hemopexin (which has no free sulfhydryl groups) can be crosslinked to the α-subunit but may also bind to the β-subunit.

to heme oxygenase. The protein has some properties in common with, but is distinct from, cytochrome b_5 (A. Smith, unpublished). For example, the absorbance spectra of MHBP and cytochrome b_5 are similar, as are the apparent molecular weights under reducing conditions on SDS-PAGE. Additional information on the molecular properties of this interesting molecule are needed to clarify its mechanism of action in cellular heme transport.

C. Working Model of the Hemopexin-Mediated Heme Transport Process

A working model of the mechanism of hemopexin-mediated heme uptake (Fig. 17) has been formulated that not only incorporates the known facts concerning the structure of hemopexin, the interaction of hemopexin with heme and its receptor, the factors that influence these interactions, and the cellular responses to hemopexin, but also postulates additional features likely to be needed for the system to function efficiently. As such this scheme represents a testable hypothesis and basis for future study of the hemopexin system.

The initial event is the binding of ferri-heme to hemopexin in the plasma, which induces a conformational change in the protein and enables high-affinity molecular recognition by cells expressing the

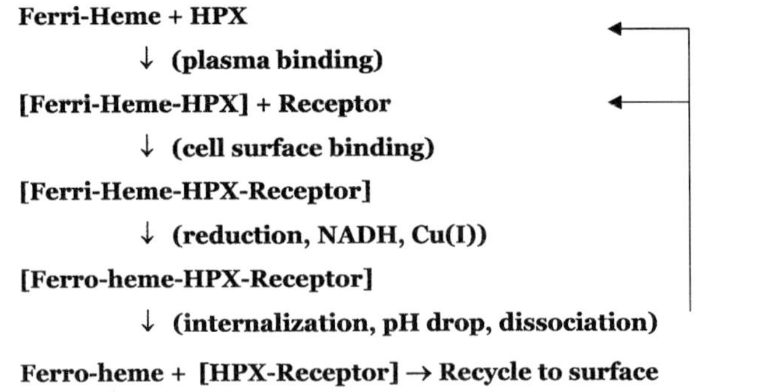

FIG. 17. Hemopexin-mediated heme transport. The following is a proposed scenario for the transport of iron-porphyrins by hemopexin. Evidence is cited throughout the text that supports the major features of this working model. However, questions remain, and the molecular details of several features will require further efforts to elucidate. The role of the hemopexin receptor in activating signaling pathways is presented elsewhere (*89*) and thus is not considered here.

hemopexin receptor. This is mandatory if the heme protein is to be efficiently retrieved from the circulation, since more apo- than heme–hemopexin is present under all but the most extreme conditions of hemolysis.

Upon interacting with the receptor, further conformational changes in heme–hemopexin are likely to occur together with concomitant or subsequent reduction of heme iron. This reduction, which appears to involve a Cu(II)–Cu(I) redox couple, is likely to be a key to understanding how the tight heme–hemopexin complex is broken without degrading the protein. Possibly, the multiple heme binding modes (sandwiched between the domains or on the N-domain) may play a role in moving the protein from a transport-stable site to a site poised for release. Functionally important conformational changes linked to reduction of heme have been noted in other systems, such as cytochrome *c* (*164*) and cytochrome *c* oxidase (*165*), but migration of heme from site to site, if validated, would be unique.

The heme–hemopexin–receptor complex is internalized via receptor-mediated endocytosis (*160*), presumably with the heme in the reduced state, and experiences a lower pH in the endosome. In the ferro-form, the acidic pH in the endosome may further weaken the association between heme and hemopexin, and heme is likely to be released at this stage without degradation of the protein. However, two prerequisites must be met. First, heme released into an acidic environment requires some means to maintain the heme in solution, and second a means to target heme to heme oxygenase inside the cell is needed. Both of these functions may be played by the MHBP, perhaps with some as yet to be identified intracellular partners.

After transfer of heme to MHBP, either directly from hemopexin or from the hemopexin receptor, hemopexin and the receptor both recycle to the surface to undergo further rounds of transport. The heme inside the cell requires further intracellular trafficking to deliver heme to regulatory sites and to HO-1 for catabolism to biliverdin and iron, making intracellular transport an interesting focus of future research. The biliverdin is reduced and excreted as bilirubin, and the iron released, which can also have regulatory effects, is reutilized or stored on ferritin.

VI. Conclusion

This examination of the available information on the structure and function of hemopexin, the mechanisms of hemopexin-mediated heme transport, and the biological consequences of this specific transport system is intended to convey a sense of the considerable information available on this system, to show how past and present results combine to provide a coherent picture of the system, and to point to the many exciting new directions for continued research.

The combination of the broader perspective on the biological function of hemopexin as more than a simple heme scavenger, together with the advances in delimiting the structure of hemopexin, suggest new opportunities for advances in understanding the mechanisms of hemopexin action in preventing nephrotoxicity, oxidative damage, and loss of biologically available iron. Further, the emerging structure–function relationships of hemopexin also may contribute to understanding how the functionally and mechanistically similar transferrin system acts as well as the role of the pexin domains of other proteins.

It is tempting to speculate that the next comprehensive examination of hemopexin-mediated heme transport will have exciting new sections devoted to the molecular properties of MHBP and the hemopexin receptor, to the mechanism of heme release from hemopexin, to new intracellular heme transport partners, and to the links provided by the hemopexin system among heme, iron, and copper at the cellular level.

Acknowledgments

The research presented here has been supported in part by grants to W.T.M. from the National Science Foundation, the National Institutes of Health, the University of Missouri Research Board, and the Marion-Merrill-Dow Scientific Education Partnership and in part by grants to A.S. from the National Institutes of Health, the Marion-Merrill-Dow Scientific Education Partnership, and the University of Missouri Research Board.

The authors thank the many friends, scientific colleagues, and students who have provided so much personal and professional enlightenment, support, and pleasure over the years. Many are coauthors in the publications referred to here. A special dedication is made to Dr. John P. Riehm (1935–1999).

References

1. Chapman, S. K.; Daff, S.; Munro, A. W. *Structure and Bonding* **1997,** *88,* 39–70.
2. Banci, L.; Pierattelli, R.; Turner, D. L. *Eur. J. Biochem.* **1995,** *232,* 522–527.
3. Poulos, T. L.; Howard, A. J. *Biochemistry* **1987,** *26,* 8165–8174.
4. Arnoux, P.; Haser, R.; Izadi, N.; Lecroisey, A.; Delepierre, M.; Wandersman, C.; Czjzek, M. *Nature Struct. Biol.* **1999,** *6,* 516–520.
5. Frolow, F.; Kalb, A. J.; Yariv, J. *Nature Struct. Biol.* **1994,** *1,* 453–460.
6. Saito, T.; Williams, R. J. P. *Eur. J. Biochem.* **1991,** *197,* 43–47.
7. Lewis, D. F.; Bird, M. G.; Dickins, M.; Lake, B. G.; Eddershaw, P. J.; Tarbit, M. H.; Goldfarb, P. S. *Xenobiotica* **2000,** *30,* 1–25.
8. Poulos, T. L.; Raman, C. S.; Li, H. *Structure* **1998,** *6,* 255–258.
9. Koesling, D. *Methods* **1999,** *19,* 485–493.
10. King, D. A.; Zhang, L.; Guarente, L.; Marmorstein, R. *Nature Struct. Biol.* **1999,** *6,* 64–71.
11. Paoli, M.; Anderson, B. F.; Baker, H. M.; Morgan, W. T.; Smith, A.; Baker, E. N. *Nature Struct. Biol.* **1999,** *6,* 926–931.
12. Hrkal, Z.; Vodrazka, Z.; Kalousek, I. *Eur. J. Biochem.* **1974,** *43,* 73–78.
13. Bearden, A. J.; Morgan, W. T.; Muller-Eberhard, U. *Biochem. Biophys. Res. Commun.* **1974,** *61,* 265–272.
14. Eskew, J. D.; Vanacore, R. M.; Sung, L.; Morales, P. J.; Smith, A. *J. Biol. Chem.* **1999,** *274,* 638–648.
15. Alam, J.; Smith, A. *J. Biol. Chem.* **1989,** *264,* 17637–17640.
16. Alam, J.; Smith, A. *J. Biol. Chem.* **1992,** *267,* 16379–16384.
17. Liang, C.-C. *Br. J. Haemat.* **1957,** *66,* 552–558.
18. Neale, F. C.; Aber, G. M.; Northam, B. E. *J. Clin. Path.* **1958,** *11,* 206–219.
19. Nyman, M. *Scand. J. Clin. Lab. Invest.* **1960,** *12,* 121–130.
20. Jeejeebhoy, K. N.; Bruce-Robertson, A.; Ho, J.; Kida, S.; Muller-Eberhard, U. *Can. J. Biochem.* **1976,** *54,* 74–78.
21. Katz, N. R.; Goldfarb, V.; Liem, H. H.; Muller-Eberhard, U. *Eur. J. Biochem.* **1985,** *146,* 155–159.
22. Satoh, H.; Nagae, Y.; Immenschuh, S.; Satoh, T.; Muller-Eberhard, U. *J. Biol. Chem.* **1994,** *269,* 6851–6858.
23. Bernard, N.; Lombart, C.; Strecker, G.; Montreuil, J.; Van Halbeek, H.; Vliegenthart, J. F. *Biochimie* **1983,** *65,* 185–192.

24. Tolosano, E.; Cutufia, M. A.; Hirsch, E.; Silengo, L.; Altruda, F. *Biochem. Biophys. Res. Commun.* **1996,** *218,* 694–703.
25. Morris, C. M.; Candy, J. M.; Edwardson, J. A.; Bloxham, C. A.; Smith, A. *Neurosci. Lett.* **1993,** *149,* 141–144.
26. Hunt, R. C.; Handy, I.; Smith, A. *J. Cell Physiol.* **1996,** *168,* 81–86.
27. Diotallevi, P.; Balducci, E.; Canapa, A.; Danni, M.; Giamagli, C. A.; Lucesoli, S.; Ravaglia, P.; Milani-Comparetti, M. *Boll. Soc. Ital. Biol. Sper.* **1988,** *64,* 531–538.
28. Luft, S.; Glinska-Urban, D.; Brzezinska, B.; Rosnowska, M. *Reumatologia* **1973,** *11,* 279–286.
29. Meiers, H. G.; Lissner, R.; Mawlawi, H.; Bruster, H. *Klin. Wochenschr.* **1974,** *52,* 453–454.
30. Foidart, M.; Liem, H. H.; Adornato, B. T.; Engel, W. K.; Muller-Eberhard, U. *J. Lab. Clin. Med.* **1983,** *102,* 838–846.
31. Eyster, M. E.; Edgington, T. S.; Liem, H. H.; Muller-Eberhard, U. *J. Lab. Clin. Med.* **1972,** *80,* 112–116.
32. Muller-Eberhard, U.; Liem, H. H.; Mathews-Roth, M.; Epstein, J. H. *Proc. Soc. Exp. Biol. Med.* **1974,** *146,* 694–697.
33. Kushner, I.; Edgington, T. S.; Trimble, C.; Liem, H. H.; Muller-Eberhard, U. *J. Lab. Clin. Med.* **1972,** *80,* 18–25.
34. Grieninger, G.; Liang, T. J.; Beuving, G.; Goldfarb, V.; Metcalfe, S. A.; Muller-Eberhard, U. *J. Biol. Chem.* **1986,** *261,* 15719–15724.
35. Baumann, H.; Jahreis, G. P.; Gaines, K. C. *J. Cell Biol.* **1983,** *97,* 866–876.
36. Immenschuh, S.; Nagae, Y.; Satoh, H.; Baumann, H.; Muller-Eberhard, U. *J. Biol. Chem.* **1994,** *269,* 12654–12661.
37. Adornato, B. T.; Engel, W. K.; Foidart-Desalle, M. *Arch. Neurol.* **1978,** *35,* 577–580.
38. Altruda, F.; Poli, V.; Restagno, G.; Silengo, L. *J. Molec. Evol.* **1988,** *27,* 102–108.
39. Law, M. L.; Cai, G. Y.; Hartz, J. A.; Jones, C.; Kao, F. T. *Genomics* **1988,** *3,* 48–52.
40. Poli, V.; Silengo, L.; Altruda, F.; Cortese, R. *Nucleic Acids Res.* **1989,** *17,* 9351–9365.
41. Nagae, Y.; Muller-Eberhard, U. *Biochem. Biophys, Res. Commun.* **1992,** *185,* 420–429.
42. Bunn, H. F. *Semin. Hematol.* **1972,** *9,* 3–17.
43. Tsunoo, H.; Higa, Y.; Kino, K.; Nakajima, H.; Hamaguchi, H. *Proc. Soc. Japan Acad.* **1977,** *53(B),* 18–25.
44. Takami, M. *J. Biol. Chem.* **1993,** *268,* 20335–20342.
45. Cordova Martinez, A. C.; Escanero, J. F. *Physiol. Behav.* **1992,** *51,* 719–722.
46. Tolosano, E.; Hirsch, E.; Patrucco, E.; Camaschella, C.; Navone, R.; Silengo, L.; Altruda, F. *Blood* **1999,** *94,* 3906–3914.
47. Smith, A.; Morgan, W. T. *Biochem. J.* **1979,** *182,* 47–54.
48. Lim, S. K.; Kim, H.; bin Ali, A.; Lim, Y. K.; Wang, Y.; Chong, S. M.; Costantini, F.; Baumann, H. *Blood* **1998,** *92,* 1870–1877.
49. Beaven, G. H.; Chen, S. H.; d'Albis, A.; Gratzer, W. B. *Eur. J. Biochem.* **1974,** *41,* 539–546.
50. Vincent, S. H.; Grady, R. W.; Shaklai, N.; Snider, J. M.; Muller-Eberhard, U. *Arch. Biochem. Biophys.* **1988,** *265,* 539–550.
51. Miller, Y. I.; Altamentova, S. M.; Shaklai, N. *Biochemistry* **1997,** *36,* 12189–12198.
52. Morgan, W. T.; Liem, H. H.; Sutor, R. P.; Muller-Eberhard, U. *Biochim. Biophys. Acta* **1976,** *444,* 435–445.
53. Noyer, C. M.; Immenschuh, S.; Liem, H. H.; Muller-Eberhard, U.; Wolkoff, A. W. *Hepatology* **1998,** *28,* 150–155.

54. Taketani, S.; Immenschuh, S.; Go, S.; Sinclair, P. R.; Stockert, R. J.; Liem, H. H.; Muller Eberhard, U. *Hepatology* **1998,** *27,* 808–814.
55. Miller, Y. I.; Smith, A.; Morgan, W. T.; Shaklai, N. *Biochemistry* **1996,** *35,* 13112–13117.
56. Miller, Y. I.; Shaklai, N. *Biochim. Biophys. Acta* **1999,** *1454,* 153–164.
57. Shaklai, N.; Solar, I. *Prog. Clin. Biol. Res.* **1989,** *319,* 575–585.
58. Solar, I.; Dulitzky, J.; Shaklai, N. *Arch. Biochem. Biophys.* **1990,** *283,* 81–89.
59. Solar, I.; Muller-Eberhard, U.; Shviro, Y.; Shaklai, N. *Biochim. Biophys. Acta* **1991,** *1062,* 51–58.
60. Cannon, J. B.; Kuo, F. S.; Pasternack, R. F.; Wong, N. M.; Muller-Eberhard, U. *Biochemistry* **1984,** *23,* 3715–3721.
61. Hunt, R. C.; Hunt, D. M.; Gaur, N.; Smith, A. *J. Cell Physiol.* **1996,** *168,* 71–80.
62. Chen, W.; Lu, H.; Dutt, K.; Smith, A.; Hunt, D. M.; Hunt, R. C. *Exp. Eye Res.* **1998,** *67,* 83–93.
63. Smith, A.; Morgan, W. T. *Biochem. Biophys. Res. Commun.* **1978,** *84,* 151–157.
64. Smith, A.; Morgan, W. T. *J. Biol. Chem.* **1984,** *259,* 12049–12053.
65. De Jong, G.; Van Dijk, J. P.; Van Eijk, H. G. *Clin. Chim. Acta* **1990,** *190,* 1–46.
66. Thorstensen, K.; Romslo, I. *Biochem. J.* **1990,** *271,* 1–9.
67. Smith, A.; Morgan, W. T. *J. Biol. Chem.* **1985,** *260,* 8325–8329.
68. Davies, D. M.; Smith, A.; Muller-Eberhard, U.; Morgan, W. T. *Biochem. Biophys. Res. Commun.* **1979,** *91,* 1504–1511.
69. Smith, A.; Eskew, J. D.; Borza, C. M.; Pendrak, M.; Hunt, R. C. *Exp. Cell Res.* **1997,** *232,* 246–254.
70. Hallberg, L. *Biol. Trace Elem. Res.* **1992,** *35,* 25–45.
71. Lash, A.; Saleem, A. *Ann. Clin. Lab. Sci.* **1995,** *25,* 20–30.
72. Conrad, M. E. *Prog. Clin. Biol. Res.* **1993,** *380,* 203–219.
73. Hallberg, L.; Solvell, L. *Acta Med. Scand.* **1967,** *181,* 335–354.
74. Bjorn-Rasmussen, E.; Hallberg, L.; Isaksson, B.; Arvidsson, B. *J. Clin. Invest.* **1974,** *53,* 247–255.
75. Smith, A. "Transport of Tetrapyrroles: Mechanisms and Biological and Regulatory Consequences"; Dailey, H. A., Jr., Ed.; McGraw-Hill Publishing Co.: New York, 1990, pp. 435–490.
76. Grasbeck, R.; Majuri, R.; Kouvonen, I.; Tenhunen, R. *Biochim. Biophys. Acta* **1982,** *700,* 137–142.
77. Otto, B. R.; Verweij-van Vught, A. M. J. J.; MacLaren, D. M. *CRC Crit. Rev. Microbiol.* **1992,** *18,* 217–233.
78. Fouz, B.; Mazoy, R.; Vazquez, F.; Lemos, M. L.; Amaro, C. *FEMS Microbiol. Lett.* **1997,** *156,* 187–191.
79. Schryvers, A. B.; Gray-Owen, S. *J. Infect. Dis.* **1992,** *165 Suppl 1,* S103–104.
80. Schryvers, A. B.; Gonzalez, G. C. *Can. J. Microbiol.* **1990,** *36,* 145–147.
81. Cornelissen, C. N.; Sparling, P. F. *Mol. Microbiol.* **1994,** *14,* 843–850.
82. Eaton, J. W.; Brandt, P.; Mahoney, J. R.; Lee, J. T., Jr. *Science* **1982,** *215,* 691–693.
83. Cope, L. D.; Thomas, S. E.; Hrkal, Z.; Hansen, E. J. *Infect. Immun.* **1998,** *66,* 4511–4516.
84. Hanson, M. S.; Pelzel, S. E.; Latimer, J.; Muller-Eberhard, U.; Hansen, E. J. *Proc. Natl. Acad. Sci. USA* **1992,** *89,* 1973–1977.
85. Wong, J. C.; Holland, J.; Parsons, T.; Smith, A.; Williams, P. *Infect. Immun.* **1994,** *62,* 48–59.
86. Wong, J. C.; Patel, R.; Kendall, D.; Whitby, P. W.; Smith, A.; Holland, J.; Williams, P. *Infect. Immun.* **1995,** *63,* 2327–2333.

87. Cope, L. D.; Yogev, R.; Muller-Eberhard, U.; Hansen, E. J. *J. Bacteriol.* **1995,** *177,* 2644–2653.
88. Alam, J.; Shibahara, S.; Smith, A. *J. Biol. Chem.* **1989,** *264,* 6371–6375.
89. Smith, A. *Anti-Oxidants and Redox Signaling,* **2000,** in press.
90. Ren, Y.; Smith, A. *J. Biol. Chem.* **1995,** *270,* 23988–23995.
91. Breslow, E.; Chandra, R.; Kappas, A. *J. Biol. Chem.* **1986,** *261,* 3135–3141.
92. Morgan, W. T.; Alam, J.; Deaciuc, V.; Muster, P.; Tatum, F. M.; Smith, A. *J. Biol. Chem.* **1988,** *263,* 8226–8231.
93. Smith, A.; Alam, J.; Escriba, P. V.; Morgan, W. T. *J. Biol. Chem.* **1993,** *268,* 7365–7371.
94. Takahashi, N.; Takahashi, Y.; Putnam, F. W. *Proc. Natl. Acad. Sci., USA* **1984,** *81,* 2021–2025.
95. Hrkal, Z.; Kuzelova, K.; Muller-Eberhard, U.; Stern, R. *FEBS Lett.* **1996,** *383,* 72–74.
96. Seery, V. L.; Hathaway, G.; Muller-Eberhard, U. *Arch. Biochem. Biophys.* **1972,** *150,* 269–272.
97. Plancke, Y.; Dautrexvaux, M.; Biserte, G. *FEBS Lett.* **1977,** *78,* 291–294.
98. Morgan, W. T. *Ann. NY Acad. Sci.* **1975,** *244,* 624–650.
99. Altruda, F.; Poli, V.; Restagno, G.; Argos, P.; Cortese, R.; Silengo, L. *Nucleic Acids Res.* **1985,** *13,* 3841–3859.
100. Takahashi, N.; Takahashi, Y.; Putnam, F. W. *Proc. Natl. Acad. Sci. USA* **1985,** *82,* 73–77.
101. Morgan, W. T.; Muster, P.; Tatum, F.; Kao, S. M.; Alam, J.; Smith, A. *J. Biol. Chem.* **1993,** *268,* 6256–6262.
102. Nikkilä, H.; Gitlin, J. D.; Muller-Eberhard, U. *Biochemistry* **1991,** *30,* 823–829.
103. Zhu, L.; Hope, T. J.; Hall, J.; Davies, A.; Stern, M.; Muller-Eberhard, U.; Stern, R.; Parslow, T. G. *J. Biol. Chem.* **1994,** *269,* 32092–32097.
104. Wu, M. L.; Morgan, W. T. *Proteins* **1994,** *20,* 185–190.
105. Morgan, W. T.; Smith, A. *J. Biol. Chem.* **1984,** *259,* 12001–12006.
106. Wu, M. L.; Morgan, W. T. *Protein Sci.* **1995,** *4,* 29–34.
107. Hunt, L. T.; Barker, W. C.; Chen, H. R. *Protein Seq. Data Anal.* **1987,** *1,* 21–26.
108. Stanley, K. K. *FEBS Lett.* **1986,** *199,* 249–253.
109. Murphy, G.; Knauper, V. *Matrix Biol.* **1997,** *15,* 511–518.
110. Hrkal, Z.; Muller-Eberhard, U. *Biochemistry* **1971,** *10,* 1746–1750.
111. Morgan, W. T.; Muller-Eberhard, U. *J. Biol. Chem.* **1972,** *247,* 7181–7187.
112. Hrkal, Z.; Suttnar, J.; Vodrazka, Z. *Studia Biophysica* **1977,** *63,* 55–58.
113. Morgan, W. T.; Sutor, R. P.; Muller-Eberhard, U. *Biochim. Biophys. Acta* **1976,** *434,* 311–323.
114. Cox, M. C.; Le Brun, N.; Thomson, A. J.; Smith, A.; Morgan, W. T.; Moore, G. R. *Biochim. Biophys. Acta* **1995,** *1253,* 215–223.
115. Morgan, W. T.; Muller-Eberhard, U. *Arch. Biochem. Biophys.* **1976,** *176,* 431–441.
116. Morgan, W. T.; Vickery, L. E. *J. Biol. Chem.* **1978,** *253,* 2940–2945.
117. Morgan, W. T.; Smith, A.; Koskelo, P. *Biochim. Biophys. Acta* **1980,** *624,* 271–285.
118. Conway, T. P.; Muller-Eberhard, U. *Arch. Biochem. Biophys.* **1976,** *172,* 558–564.
119. Seery, V. L.; Muller-Eberhard, U. *J. Biol. Chem.* **1973,** *248,* 3796–3800.
120. Gibbs, E.; Skowronek, W. R.; Morgan, W. T.; Muller-Eberhard, U.; Pasternack, R. F. *J. Amer. Chem. Soc.* **1980,** *102,* 3939–3944.
121. Smith, A.; Tatum, F. M.; Muster, P.; Burch, M. K.; Morgan, W. T. *J. Biol. Chem.* **1988,** *263,* 5224–5229.
122. Pasternack, R. F.; Gibbs, E. J.; Hoeflin, E.; Kosar, W. P.; Kubera, G.; Skowronek, C. A.; Wong, N. M.; Muller-Eberhard, U. *Biochemistry* **1983,** *22,* 1753–1758.

123. Brown, S. B.; Lantzke, I. R. *Biochem. J.* **1969,** *115,* 279–285.
124. Morgan, W. T.; Muller-Eberhard, U. *Enzyme* **1974,** *17,* 108–115.
125. Morgan, W. T.; Muster, P.; Tatum, F. M.; McConnell, J.; Conway, T. P.; Hensley, P.; Smith, A. *J. Biol. Chem.* **1988,** *263,* 8220–8225.
126. Morgan, W. T. *Ann. Clin. Res.* **1976,** *8,* 223–232.
127. Morgan, W. T.; Sutor, R. P.; Muller-Eberhard, U.; Koskelo, P. *Biochim. Biophys. Acta* **1975,** *400,* 415–422.
128. Faber, H. R.; Groom, C. R.; Baker, H. M.; Morgan, W. T.; Smith, A.; Baker, E. N. *Structure* **1995,** *3,* 551–559.
129. Smith, T. F.; Gaitatzes, C.; Saxena, K.; Neer, E. J. *Trends Biochem. Sci.* **1999,** *24,* 181–185.
130. Wu, M. L.; Morgan, W. T. *Biochemistry* **1993,** *32,* 7216–7222.
131. Mortuza, G. B.; Whitford, D. *FEBS Lett.* **1997,** *412,* 610–614.
132. Vickery, L.; Nozawa, T.; Sauer, K. *J. Am. Chem. Soc.* **1976,** *98,* 351–357.
133. Muster, P.; Tatum, F.; Smith, A.; Morgan, W. T. *J. Prot. Chem.* **1991,** *10,* 123–128.
134. Satoh, T.; Satoh, H.; Iwahara, S.; Hrkal, Z.; Peyton, D. H.; Muller-Eberhard, U. *Proc. Natl. Acad. Sci. USA* **1994,** *91,* 8423–8427.
135. Wu, M. L. "Interaction of heme with hemopexin: Structural characterization and functional implications"; Ph.D. Dissertation, University of Missouri—Kansas City, 1994, pp. 1–156.
136. Pasternack, R. F.; Gibbs, E. J.; Mauk, A. G.; Reid, L. S.; Wong, N. M.; Kurokawa, K.; Hashim, M.; Muller-Eberhard, U. *Biochemistry* **1985,** *24,* 5443–5448.
137. Smith, A.; Morgan, W. T. *J. Biol. Chem.* **1981,** *256,* 10902–10909.
138. Hrkal, Z.; Kalousek, I.; Vodrazka, Z. *Studia Biophysica* **1981,** *82,* 69–73.
139. Shipulina, N.; Smith, A.; Morgan, W. T. *J. Prot. Chem.* **2000,** in press.
140. Shipulina, N.; Hunt, R. C.; Shaklai, N.; Smith, A. *J. Prot. Chem.* **1998,** *17,* 255–260.
141. Shaklai, N.; Sharma, V. S.; Muller-Eberhard, U.; Morgan, W. T. *J. Biol. Chem.* **1981,** *256,* 1544–1548.
142. Ozols, J.; Strittmatter, P. *J. Biol. Chem.* **1964,** *239,* 1013–1023.
143. Perutz, M. F.; Wilkinson, A. J.; Paoli, M.; Dodson, G. G. *Annu. Rev. Biophys. Biomol. Struct.* **1998,** *27,* 1–34.
144. Fasman, G. D. "Circular Dichroism and the Conformational Analysis of Biomolecules"; Plenum Press, New York, 1996.
145. Lightner, D. A.; Reisinger, M.; Landen, G. L. *J. Biol. Chem.* **1986,** *261,* 6034–6038.
146. Muller-Eberhard, U.; Bosman, C.; Liem, H. H. *J. Lab. Clin. Med.* **1970,** *76,* 426–431.
147. Lane, R. S.; Rangeley, D. M.; Liem, H.; Wormsley, S.; Muller-Eberhard, U. *J. Lab. Clin. Med.* **1972,** *79,* 935–941.
148. Petryka, Z. J.; Dhar, G. J.; Bossenmaier, I. In "Hematin Clearance in Porphyria"; Doss, M., Ed. Karger: Basel, 1976, pp. 259–265.
149. Tokola, O.; Tenhunen, R.; Volin, L.; Mustajoki, P. *Br. J. Clin. Pharmacol.* **1986,** *22,* 331–335.
150. Volin, L.; Rasi, V.; Vahtera, E.; Tenhunen, R. *Blood* **1988,** *71,* 625–628.
151. Sinclair, P. R.; Bement, W. J.; Healey, J. F.; Gorman, N.; Sinclair, J. F.; Bonkovsky, H. L.; Liem, H. H.; Muller-Eberhard, U. *Hepatology* **1994,** *20,* 741–746.
152. Muller-Eberhard, U.; Javid, J.; Liem, H. H.; Hanstein, A.; Hanna, M. *Blood* **1968,** *32,* 811–815.
153. Lundh, B.; Oski, F. A.; Gardner, F. H. *Acta Paediatr. Scand.* **1970,** *59,* 121–126.
154. Braun, H. J.; Aly, F. W. *Klin. Wochenschr.* **1971,** *49,* 451–458.
155. Braun, H. J. *Klin. Wochenschr.* **1971,** *49,* 445–451.
156. Sears, D. A. *Proc. Soc. Exp. Biol. Med.* **1969,** *131,* 371–373.

157. Taketani, S.; Kohno, H.; Tokunaga, R. *J. Biol. Chem.* **1987,** *262,* 4639–4643.
158. Taketani, S.; Kohno, H.; Tokunaga, R. *Biochem. Int.* **1986,** *13,* 307–312.
159. Okazaki, H.; Taketani, S.; Kohno, H.; Tokunaga, R.; Kobayashi, Y. *Cell Struct. Funct.* **1989,** *14,* 129–140.
160. Smith, A.; Hunt, R. C. *Eur. J. Cell Biol.* **1990,** *53,* 234–245.
161. Smith, A. *Biochem. J.* **1985,** *231,* 663–669.
162. Taketani, S.; Kohno, H.; Naitoh, Y.; Tokunaga, R. *J. Biol. Chem.* **1987,** *262,* 8668–8671.
163. Smith, A.; Farooqui, S. M.; Morgan, W. T. *Biochem. J.* **1991,** *276,* 417–425.
164. Qi, P. X.; Beckman, R. A.; Wand, A. J. *Biochemistry* **1996,** *35,* 12275–12286.
165. Yoshikawa, S.; Shinzawa-Itoh, K.; Nakashima, R.; Yaono, R.; Yamashita, E.; Inoue, N.; Yao, M.; Fei, M. J.; Libeu, C. P.; Mizushima, T.; Yamaguchi, H.; Tomizaki, T.; Tsukihara, T. *Science* **1998,** *280,* 1723–1729.
166. Fisher, M. T. *Biochemistry* **1991,** *30,* 10012–10018.

STRUCTURES OF GAS-GENERATING HEME ENZYMES: NITRIC OXIDE SYNTHASE AND HEME OXYGENASE

THOMAS L. POULOS, HUIYING LI, C. S. RAMAN, and DAVID J. SCHULLER

Departments of Molecular Biology and Biochemistry and of Physiology and Biophysics and the Program in Macromolecular Structure, University of California, Irvine, Irvine, California 92697-3900

0898-8838/01 $35.00

I. Introduction

Two simple and noxious gases, nitric oxide (NO) and carbon monoxide (CO), are receiving increasing attention in biology. A Medline search shows 5036 publications on nitric oxide in 1998, whereas carbon monoxide is somewhat less popular with 440 listings. The stature of NO is evidenced by *Science* magazine choosing NO as molecule of the year in 1992; the discovery of NO as the causative agent in smooth muscle relaxation was awarded the Nobel Prize in Physiology and Medicine in 1998. The seemingly far less interesting CO molecule, once relegated the status of a mere pollutant, is now being viewed by some as an important signaling molecule operating in a manner similar to NO. It is remarkable that Nature has selected such apparently simple molecules for such important biological functions.

Although the role of CO in biological signaling processes remains somewhat controversial, the roles played by NO are well established. NO was first recognized as EDRF (endothelial-derived relaxing factor) in the mid-1980s (*1, 2*). Since then, NO has been implicated in a variety of physiological responses. In addition to the role of NO in vascular regulation, NO also is involved in neural communication and in the immune host-defense system (*3, 4*). Controlling NO levels also is an important pharmaceutical goal, since diminished levels of NO contribute to chronic hypertensive diseases, while NO overproduction contributes to pathological conditions related to primary neurodegenerative, inflammatory, and vascular disorders (*5*). In septic shock, inhibitors of NO production can restore vascular tone and blood pressure (*6*). Blocking NO production also limits ischemia-elicited infarct size in animal models (*7*).

Evidence is accumulating that CO also may operate as a signaling molecule similar to the way NO functions (*8*). The enzyme responsible for producing CO is heme oxygenase (HO), which catalyzes the first step in the degradation of free heme. The enzyme responsible for NO formation is nitric oxide synthase (NOS). The enzymology and structure of both HO and NOS are areas of active investigation. The focus of this article is on the current state of structure–function relationships in both HO and NOS, with an emphasis on crystallographic studies. Before reviewing specific recent advances in NOS and HO, short preliminary overviews on the biological role of NO and CO, and heme enzymes in general is warranted.

II. Biological Targets of NO and CO Action

The primary target for NO is guanylate cyclase (GC). GC converts guanosine 5′-triphosphate (GTP) to cyclic guanosine 3′,5′-monophosphate (cGMP) (*9*), the signaling molecule leading to the observed physiological effects of NO. There are two forms of GC, one membrane-bound (pGC, p = particulate) and one soluble (sGC). sGC consists of a 70–80 kDa $\alpha 1$ subunit and a $\approx$70 kDa $\beta 1$ subunit (*10, 11*) that also provides the His ligand for the signal heme group (*12–14*).

Activation of GC by NO (*15*) is directly related to well-known porphyrin–NO chemistry. NO binds tightly to heme, which is known to significantly weaken the axial ligand *trans* to NO (*16*). This unique feature of NO complexes is related to the fact that NO is a paramagnetic molecule with one unpaired spin. Upon formation of the NO–Fe complex, the NO unpaired electron resides in an orbital with substantial metal dz^2 character that serves to weaken the His–Fe axial bond (*16*). This effect is not observed in other diatomic ligands with paired spins, such as CO. Not too surprisingly, NO also breaks the His–Fe bond in GC (*17*) and is thought to be the basis for GC activation (*18*), although this view has been challenged (*19*). Precisely how cleavage of the His–Fe bond could be connected to activation is unknown.

Evidence is accumulating that CO can mimic some of the physiological properties of NO (*20, 21*). This view is somewhat tempered by the fact that CO is able to stimulate GC activity only about 4-fold compared to 100- to 200-fold for NO. However, it has been shown that (3-(5′-(hydroxymethyl-2′-furyl)-1-benzylindazole (YC-1) relaxes smooth muscle (*22*) and inhibits platelet aggregation (*23, 24*) by elevating cGMP levels, presumably by direct interaction with GC. Furthermore, YC-1 potentiates CO activation of GC to levels observed for NO activation (*19, 25*). This raises the possibility that some natural analog of YC-1 may act in concert with CO for potent GC activation (*19*). Many of these observations on CO performing NO-like functions are intriguing but unsettled. Overall, the precise connection between CO and the various proposed physiological roles remains an area of investigation still in its infancy. Further understanding on the mechanism of CO action will require more detailed information on GC structure–function relationships.

NO generated in the immune system does not operate through GC. Instead, NO appears to act directly as a cytotoxic molecule or as some other reactive product such as $ONOO^-$ (peroxynitrite). During the immune oxidative burst in macrophage cells, reactive oxygen intermediates, including NO and the O_2^- (superoxide anion), are produced. NO reacts with O_2^- at a near diffusion controlled rate to generate the oxidant, $ONOO^-$ (*26*), suggesting that most of the NO generated is captured as $ONOO^-$, which can serve as a potent cytotoxic molecule (*27*) capable of destroying invading organisms. The ability of $ONOO^-$ to nitrate proteins, especially tyrosine residues, is often taken as evidence for the presence of $ONOO^-$ and may provide a mechanism for $ONOO^-$ toxicity. Despite the attractive features of $ONOO^-$ operating as an important cytotoxic molecule, the relatively complex chemistry of NO in the presence of activated oxygen intermediates makes the precise role of any one intermediate uncertain. The question of how NO exerts its cytotoxic effects and whether or not $ONOO^-$ acts directly as an indiscriminate cytotoxic oxidant remains open (*28*).

III. Overview of Oxygen Activating Heme Enzymes

Both NOS and HO are designed to regioselectively hydroxylate their respective substrates. In this regard, one would expect some similarity to other related enzyme systems that catalyze similar reactions. Of the various heme enzymes that have been under investigation, the most thoroughly studied from the perspective of both structure and function are the peroxidases and P450s. Of these two, the catalytic mechanism of peroxidase is more thoroughly understood primarily because the various intermediates formed in the catalytic cycle are stable enough to be analyzed by a variety of biophysical probes. The overall catalytic mechanism of a typical heme peroxidase is as follows:

1. $Fe^{3+}R + H_2O_2 \longrightarrow \underset{\text{Compound I}}{Fe^{4+}{=}O\ R\cdot}$

2. $\underset{\text{Compound I}}{Fe^{4+}{=}O\ R\cdot} + S \longrightarrow \underset{\text{Compound II}}{Fe^{4+}{=}O\ R} + S\cdot$

3. $Fe^{4+}{=}O\ R + S \longrightarrow Fe^{3+}R + S\cdot$

In the first of the preceding steps; peroxide removes two electrons from the protein. One derives from the iron and one from an organic group, R in the scheme, giving compound I. Compound I thus has two fewer electrons than the native enzyme, so both oxidizing equivalents of peroxide now are stored in the enzyme active site. In most peroxidases the organic group that is oxidized is the porphyrin leading to a porphyrin π cation radical (*29*). In cytochrome *c* peroxidase (CCP), R = tryptophan

(*30, 31*). The two oxidizing equivalents in compound I are next utilized to oxidize two substrate molecules. In the second step in the scheme, the first reducing substrate (S) delivers one electron to compound I, which reduces the porphyrin π cation radical, thereby generating compound II. A second substrate molecule reduces compound II back to the resting state.

Compound I or its electronic equivalent is thought to be the active species in a number of heme enzymes, but only in peroxidases has it been possible to provide a detailed picture of both its formation and its structure. The reaction with peroxide is a second-order process that typically occurs with a rate constant between 10^7 and 10^8 $M^{-1}s^{-1}$ (*32*). Compound I has been studied by a variety of methods, including electron paramagnetic resonance (*33*), Mössbauer (*33*), and X-ray absorption spectroscopies (*34*). In addition, crystal structures have been determined (*35–37*). From these studies it is clear that the iron is in the Fe^{4+} oxidation, which forms a short ($\approx$1.6 Å) bond with a single peroxide-derived oxygen atom. The peroxide O–O bond is heterolytically cleaved (*38*) via an acid–base catalytic mechanism (*39, 40*). Mutagenesis studies have shown that the distal His is very likely the key acid base catalytic group (*41, 42*). This leaves behind a single Fe-linked O atom with only six valence electrons, a potent oxidizing agent (*43*). Two oxidizing equivalents are transferred to the Fe-linked O atom, one from the iron and one from an organic group, which is the heme macrocycle in most peroxidase. Thus, the oxidizing power of the single O atom now is stored on the iron and heme, where useful redox chemistry can be controlled.

Most well-studied peroxidases are designed to oxidize small aromatic molecules, with the exception of cytochrome *c* peroxidase. It generally is thought that such aromatic molecules bind near the heme edge where an electron can transfer directly to the heme edge (*44*), which is supported by both crystal structures (*45, 46*) and NMR studies (*47*). However, recent work suggests that some physiologically important substrates may utilize other sites on the enzyme surface (*48, 49*).

Cytochromes P450 are designed to hydroxylate aliphatic and aromatic compounds by introducing one O_2^- derived oxygen atom into the substrate molecule. Like heme peroxidases, P450s have a single heme group, but unlike peroxidases, the proximal heme ligand is not His but Cys (*50*). Unlike peroxidases, P450s do not form stable intermediates, so the homologue to compound I has not been directly observed. It generally is thought that P450s must use an $(Fe–O)^{3+}$ intermediate similar to peroxidases, but this has only been inferred and never directly proven. One simple and reasonable view holds that a major difference between P450s and peroxidases is access to the $(Fe–O)^{3+}$ center. Crystal structures support this view, since P450-like substrates are sterically prevented from direct access to the iron-linked activated oxygen in peroxidases. In sharp contrast to peroxidases, the P450 active site architecture is such that substrates diffuse through an access channel where they are held by specific interactions directly over the heme near the $(Fe–O)^{3+}$ center for stereo- and regioselective hydroxylation. This forcing of direct contact between the reactive $(Fe–O)^{3+}$ center and substrate is one reason P450s are able to attack unactivated C–H bonds.

Of course, simple access to the heme center is not the entire story. There also must be a fundamental difference in reactivity between $(Fe–O)^{3+}$ in peroxidases and P450s. This is underscored by the fact that compound I and the $(Fe–O)^{3+}$ center are quite stable in many peroxidases, whereas if the $(Fe–O)^{3+}$ center exists at all in P450s, its lifetime is fleetingly short. The structure of compound I further illustrates the stability of the $(Fe–O)^{3+}$ center. The peroxidase distal pocket, where peroxide binds, has a conserved distal His, the acid–base catalytic group, and an Arg. The crystal structure of cytochrome *c* peroxidase compound I (*35, 36*) shows that the Arg moves close to the iron-linked ferryl oxygen where it can form an H-bond. Most interestingly, the location of the distal Arg is positioned close to the $(Fe–O)^{3+}$ O atom, similar to what is expected for the positioning of substrates in P450 complexes and NOS, yet the Arg in peroxidases is not oxidized. It is clear that in peroxidases, $(Fe–O)^{3+}$ is quite unreactive, and part of peroxidase function is to stabilize this intermediate. This is just the opposite of P450s, where $(Fe–O)^{3+}$ is designed to be very reactive, as it must be in order to hydroxylate an unreactive stable carbon center. Despite the attractive features of considering peroxidases and P450s as forming quite similar reactive intermediates, this clearly is not the case. One would anticipate that NOS and HO might more closely resemble P450, since the job of these enzymes is not to store oxidizing equivalents but to rapidly transfer oxidizing equivalents to the substrate.

Another quite striking difference between peroxidase and P450 is how the acid–base catalytic machinery operates to cleave the O–O bond. Although P450s bind O_2, it generally is thought that the species that undergoes bond cleavage is at the peroxide oxidation state. As noted earlier, it has been believed that peroxidases utilize a distal His as the acid–base catalytic group (*39, 40*), which is supported by mutagenesis studies (*41, 42*). The situation with P450s is much less clear, since there is no similarly positioned residue in the P450 active site that is an obvious candidate for the acid–base catalytic group. Instead, it has been necessary to postulate rather complex proton delivery mechanisms involving ordered solvent molecules (*51–55*). It also has been argued (*55*) that no single acid–base mechanism may operate in all P450s, since the polarity, H-bonding possibilities, and dehydration of the active site will vary widely depending on the substrate.

Peroxidases and P450s thus represent two extreme ends of the spectrum with respect to reactivity of the $(Fe–O)^{3+}$ intermediate and requirements for acid–base catalysis. There is, however, a middle ground represented by the P450–peroxidase hybrid, chloroperoxidase (CPO). The function of CPO is to utilize peroxide oxidizing equivalents to chlorinate organic molecules in a mechanism closely resembling traditional peroxidases (*56*). However, CPO exhibits spectral properties characteristic of P450 (*57*), suggesting an axial Cys ligand, and distal ligand binding properties similar to those of peroxidases (*58*), indicating a polar distal pocket as opposed to a P450-like hydrophobic pocket. The X-ray structure (*59*) confirmed this picture, showing that the axial ligand is, indeed, Cys29 as predicted (*60*), and that the distal pocket is polar. Although His (*61*) and Arg (*62*) were thought to be part of the distal pocket as in peroxidases, instead the structure shows that a Glu residue serves the same acid–base catalytic function as the distal His in traditional peroxidases (*63, 64*), while there is no distal Arg or His capable of direct interaction with Fe-linked ligands. This wealth of information on peroxidase and P450 provides the necessary framework to guide, but not unduly influence, similar studies on NOS and HO, which undoubtedly will increase with renewed vigor now that X-ray structures are available.

IV. Background on NOS

NOS catalyzes the O_2 and NADPH dependent oxidation of L-Arg to NO and L-citrulline (Fig. 1). The first major advance in NOS structure–function relationships was made by cloning of nNOS, which revealed the close similarity to the cytochrome P450 monooxygenase system (*65*). Although the N-terminal half of the nNOS sequence has no similarity to cytochromes P450, the C-terminal half containing the FMN, FAD, and NADPH binding sites exhibits 36% identity and 58% homology with P450 reductase (*65*).

Figure 1 provides a schematic representation of the three major NOS isoforms. The variation in size is due primarily to extensions and modifications at the N-terminal region where the greatest heterogeneity is observed. This variation is related to targeting of NOS to specific cellular

$$\text{L-arginine} \xrightarrow{\text{NADPH} + O_2} \text{N-hydroxy-L-arginine} \xrightarrow{0.5\ \text{NADPH} + O_2} \text{L-citrulline} + NO\cdot$$

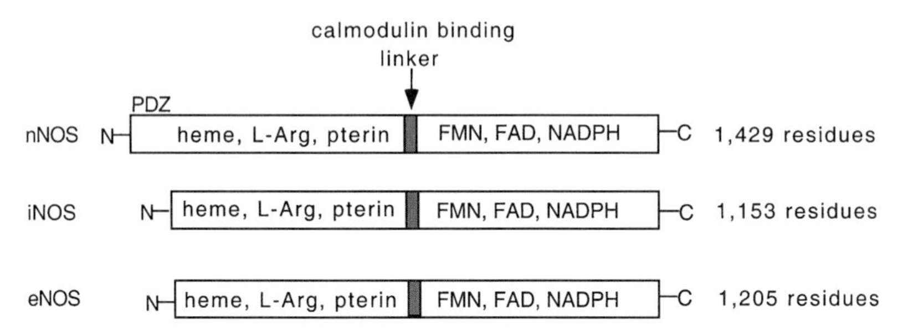

FIG. 1. The overall reaction catalyzed by NOS and a schematic representation of the three main NOS isoforms. L-Arg is first converted to *N*-hydroxy-L-Arg in a P450-like monooxygenation reaction requiring two electrons and one O_2 molecule. In the second step, *N*-hydroxy-L-Arg to NO and L-citrulline, only one electron and one O_2 molecule are required. nNOS, neuronal NOS (neural system); iNOS, inducible NOS (immune system); eNOS, endothelial NOS (cardiovascular system).

locations. For example, eNOS is membrane bound (*66*) owing to myristoylation near the N-terminus, which targets eNOS to plasmalemmal caveolae (*67*). Interactions between eNOS and caveolin are critical for the regulation of eNOS activity in vascular endothelial cells (*68, 69*). In a second example, nNOS contains an N-terminal PDZ domain that interacts with a similar motif of the postsynaptic density-95 protein in neurons and syntrophin in muscle (*70*). The crystal structures of the nNOS PDZ domain alone (residues *1–130*) and in a complex with the syntrophin PDZ domain (residues *77–171*) have been determined (*71*), which provides the first structural details of how NOS is targeted to specific cell types.

Although the heme-binding N-terminal half of NOS bears no sequence similarity to P450s, the spectral properties of the Fe^{2+}–CO complex with the characteristic 450 nm Soret maximum (*72–74*) clearly places NOS in the category of P450-like thiolate enzymes. The architecture of NOS also is strikingly similar to cytochrome P450BM-3 (*75*) in that the catalytic heme domain is covalently attached to the P450 reductase. The flow of electrons is NADPH→FAD→ FMN→heme. Constitutively expressed NOS isoforms, eNOS and nNOS, are regulated by Ca^{2+}/calmodulin (*76*). Calmodulin binds to the linker between the heme and reductase domains, resulting in a conformational switch that enables the flow of electrons from FMN to heme to proceed (*76*). iNOS, however, is controlled at the level of transcription and calmodulin is a permanently bound subunit (*77*).

V. NOS Structure

A. Overall NOS Structure

Crane *et al.* first established the three-dimensional fold of NOS by solving the structure of a monomeric form of the mouse iNOS heme domain (*78*). This version of iNOS was missing the first 114 residues, which are known to be critical for dimer formation and activity (*79*). The monomer structure was soon followed by the dimeric heme domain structures of mouse iNOS (*80*), bovine eNOS (*81*), and the human isoforms of iNOS (*82, 83*) and eNOS (*82*). A comparison of eNOS and iNOS reveals that the structures are essentially the same with an overall root-mean-square deviation in backbone atoms of ≈1.1 Å (*83*). The sequence identity between human iNOS and bovine eNOS is ≈60% for 420 residues compared in the crystal structures (*83*).

A schematic diagram of the bovine eNOS heme domain dimer is shown in Fig. 2. NOS is a mix of sheets and helices with a unique arrangement not previously seen in other enzymes. A very notable difference between NOS and other well-known heme enzymes such as peroxidases and P450s is the essential role that beta sheets play in forming the active site. In the globins, peroxidases, and P450s, a distal helix runs over the surface of the heme where ligands such as O_2 and substrates bind. With NOS the roof of the active site above the L-Arg substrate (Fig. 3) is formed exclusively of sheets and extended sections of polypeptide.

A comparative analysis of the dimer interface between NOS isoforms is important since there are questions on significant variation in dimer stability between isoforms (*84, 85*). The dimer interface is extensive with approximately 2700 $Å^2$ of surface area buried per monomer. The interface contacts involve a mix of nonpolar and polar interactions, including hydrogen bonding. Approximately 60% of the interface in both iNOS and eNOS is hydrophobic, although the higher resolution eNOS

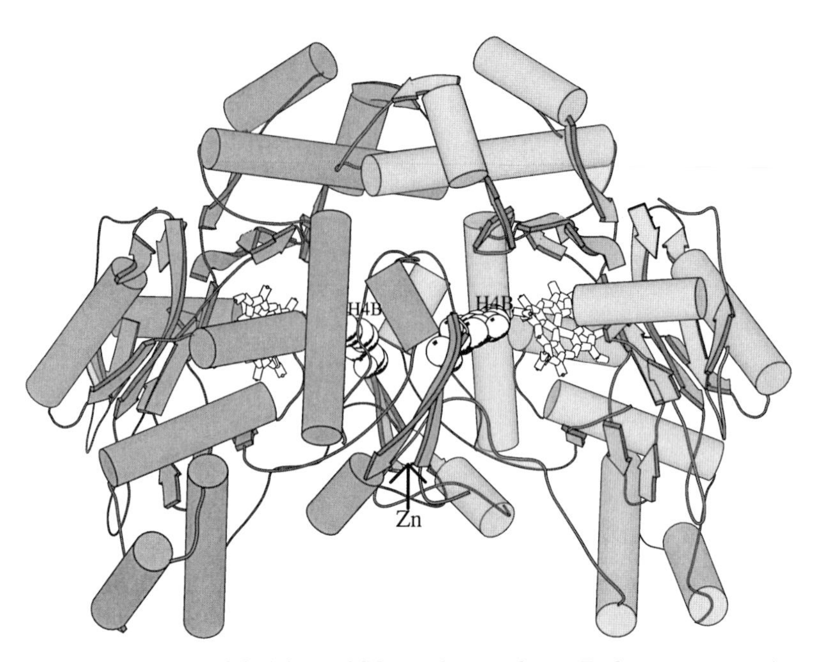

FIG. 2. Molecular model of the eNOS heme domain dimer. Each monomer consists of residues 69–482. The pterin is depicted as a space-filled model and the hemes are the white stick models. The single Zn ion situated at the lower part of the dimer interface is depicted as the white ball. The overall fold is unique to NOS.

structure reveals several bridging water molecules at the interface (*81, 83*). The overall similarity between the interface regions in iNOS and eNOS is striking. Another important consideration is the novel Zn ion situated at the dimer interface (see next section). The Zn undoubtedly plays an important role in dimer stability, and the crystal structures suggest that Zn may be labile in iNOS.

B. Zinc Ion

An unusual and unexpected feature of NOS is the presence of a Zn ion tetrahedrally coordinated to pairs of symmetry-related Cys residues along the dimer interface (*81, 82*) (Fig. 4). The original structure of the mouse iNOS dimer did not have the Zn but instead two symmetry-related Cys residues formed a disulfide bond (*80*). The structure of the human iNOS heme domain also had a disulfide, but the disulfide could readily be broken and Zn reconstituted to give a ZnS_4 center indistinguishable from that found in eNOS (*83*). An independent structure

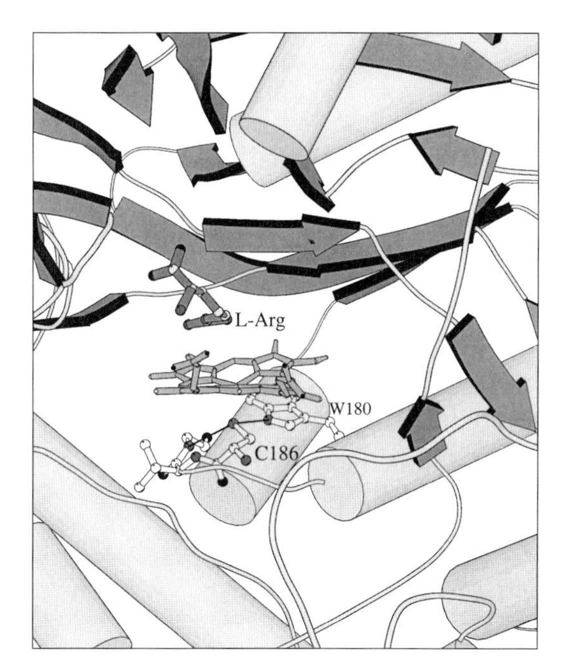

FIG. 3. Close-up view of the L-Arg binding site and surrounding protein structure. Note the sheet structure forming the roof of the active site, which is distinct from peroxidases and P450s, where helical structures form the distal cavity. Note, too, the H-bonds between the Cys ligand, Trp 180, and a peptide NH group. Donation of two H-bonds to the Cys ligand is a common feature found in other iron–thiolate proteins.

solution of both the human eNOS and iNOS heme domains (*82*) as well as the nNOS heme domain (C. S. Raman and H. Li, unpublished) also revealed the same ZnS_4 center. More recently, Crane *et al.* (*86*) obtained a new structure of the mouse iNOS heme domain dimer with Zn bound. Finally, an analysis of *Escherichia coli* expressed holo-nNOS, holo-eNOS, and the nNOS heme domain generated by proteolysis of holo-nNOS gave one Zn per dimer (*87*), leaving little doubt that the "natural" form of NOS contains Zn.

It was originally hypothesized that one role of the Zn center is to help form and stabilize the pterin binding site (*81*). As shown in Fig. 4, the conserved Ser104, which is only three residues from one Zn ligand, directly H-bonds to the pterin. A disruption of the Zn site should, therefore, perturb interactions at the pterin site. A direct comparison between the Zn-bound and disulfide forms supports this hypothesis. In order to form the S—S bond, the entire section of polypeptide involving Ser104 must slide down and away from the pterin site, which

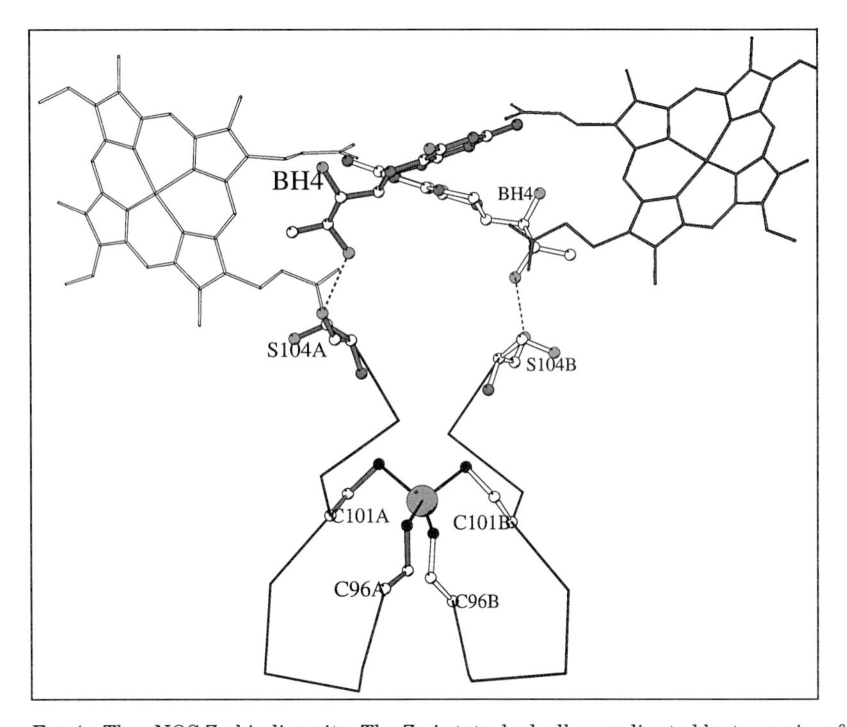

FIG. 4. The eNOS Zn binding site. The Zn is tetrahedrally coordinated by two pairs of symmetry-related Cys residues. Because the Zn is located exactly along the dimer axis of symmetry, the Zn is equidistant from each heme and pterin. Three residues apart from one Cys ligand in each subunit is Ser104, which directly H-bonds with the pterin. The Zn helps to stabilize these interactions in the pterin site; the pterin, in turn, helps to shape the L-Arg binding site, since both L-Arg and the pterin H-bond with the same heme propionate. Hence, there is long-range structural communication among the Zn, pterin, and substrate.

lengthens and weakens the Ser104–pterin H-bond (*83*). Moreover, the disulfide bond form of iNOS loses eight intersubunit hydrogen bonds present in the Zn-bound form. Hence, the Zn not only is important for pterin–protein interactions, but also is important for dimer interactions. In addition, various mutagenesis studies of Cys ligands reveal a perturbation in pterin binding, protein stability, and catalytic activity (*79, 87–91*), further supporting the interplay among the ZnS_4 site, pterin, and L-Arg binding.

Relevant to the role that Zn plays in NOS, updated versions of the mouse iNOS structure illustrate two forms of mouse iNOS termed "swapped" and "unswapped" (*86*). In the unswapped conformation, Zn is bound just as in eNOS and nNOS. However, in the swapped conformation, the Zn is absent and two symmetry-related Cys residues form an S–S bridge across the dimer interface as in the original mouse iNOS structure (*80*). The more recent Zn-free mouse iNOS structure shows domain swapping resulting from the conversion of intrasubunit interactions in the Zn-bound structure into intersubunit interactions in the Zn-free structure near the Zn site. This analysis is somewhat complicated by the relatively low resolution of the Zn-free swapped data set (2.7 Å) compared to the Zn-bound unswapped data set (2.35 Å) and by the observed mixture of swapped and unswapped conformers in some crystals. In addition, the human iNOS Zn-free structure with the same intersubunit S–S bond found in the Zn-free mouse iNOS structure is partially disordered with no clearly defined electron density in the swap region (*83*). Therefore, mouse iNOS is the only NOS out of the four known structures that exhibits the swapped conformation. Part of the complication may have its origin in the various expression systems employed. Miller *et al.* (*87*) found that the nNOS heme domain generated from *E. coli* expressed holo-nNOS has one Zn per dimer while the recombinantly expressed heme domain has about 0.2 per dimer. Nonetheless, various mutants that lead to disruption of mouse iNOS dimers can be satisfactorily explained by alterations of the swapped dimer interface, which is consistent with the Zn-free, swapped conformer being significantly populated and functionally important (*92*). At present, it appears that the Zn plays primarily a structural role in helping to stabilize the pterin binding site, which, in turn, stabilizes the L-Arg binding site. Whether or not NOS cycles between the Zn-bound and Zn-free disulfide domain swapped conformer remains an intriguing possibility.

C. Proximal Pocket

As expected from spectroscopic studies (*72–74*) the proximal ligand coordinated to the heme iron is a Cys thiolate. Although the structure of NOS bears little similarity to other heme thiolate enzymes, the local interactions between the Cys ligand and protein are very similar to those in other heme thiolate enzymes. In P450s and chloroperoxidase (*59*), the thiolate ligand accepts H-bonds from nearby peptide NH groups. The same is true in NOS, with the exception that one H-bond is provided by the indole ring NH of a Trp residue (Fig. 3). Thiolate iron ligands accepting peptide NH H-bonds is not limited to heme enzymes but also is found in ferredoxin-like iron–sulfur clusters (*93, 94*). In many of these, the peptide NH groups donating the H-bonds are situated at the N-terminal end of a helix where peptide NH groups have no helical H-bonding partner. Very often an amino acid side chain serves as an “N-capping” H-bond acceptor for these NH groups (*95, 96*), which, in the case of the heme thiolate enzymes, is the thiolate ligand. These H-bonds plus the partial positive charge at the N-terminal end of the helical dipole aid in stabilizing the thiolate form of the ligand. This stabilization also can serve the purpose of redox potential regulation. Without attenuating the negative charge on the Cys ligand, the redox potential of the iron would likely be too low to be physiologically useful (*97*). Model studies also have shown that H-bonding to the thiolate ligands results in a large positive increase in metal redox potential (*97, 98*). Therefore, the protein aids in increasing the redox potential by providing a local positive electrostatic environment. This is just the opposite to peroxidases, where interactions between the His heme ligand and a nearby Asp residue impart a greater anionic character to the His ligand, which decreases the redox potential (*40, 99, 100*). Both extremes, decreasing the redox potential in peroxidases and increasing the redox potential in thiolate-iron proteins, underscore the importance of electrostatic interactions in controlling redox properties in metalloproteins (*101*).

NOS affords a rare opportunity to directly test the role of the Cys–HN H-bonds using mutagenesis, since one of the H-bonds is provided by the indole ring of a Trp residue (Fig. 3). During anaerobic titration with NADPH, the Fe^{2+}/Fe^{3+} equilibrium in the Trp-to-Phe nNOS mutant shifted toward Fe^{3+}, which would be expected if removal of the Trp–Cys H-bond lowers the redox potential of the heme iron (*102*). Therefore, the effect of removing one H-bond donor to the Cys is consistent with the expected role of such H-bonds in modulating the heme iron redox potential. A lower redox potential, however, also should decrease the rate of heme iron reduction, which, in turn, should slow turnover, but just the opposite is observed, since the Trp-to-Phe mutant is more active than wild-type enzyme (*102*). This apparent inconsistency can be explained by the role that the product, NO, plays in inhibiting NOS. During steady-state turnover, nNOS accumulates substantial amounts of the iron–NO complex, which blocks enzyme activity (*102*). Increasing the electronegativity of the Cys ligand by removal of the Cys–Trp H-bond would decrease the strength of the Fe–NO bond and hence, increase activity by relieving NO inhibition (*102*).

D. Substrate and Pterin Binding Sites

The active site pocket is defined by a deep channel formed in part by dimer interactions (Fig. 5). Access to the L-Arg binding pocket is open, and there appears to be little need for conformational adjustments to

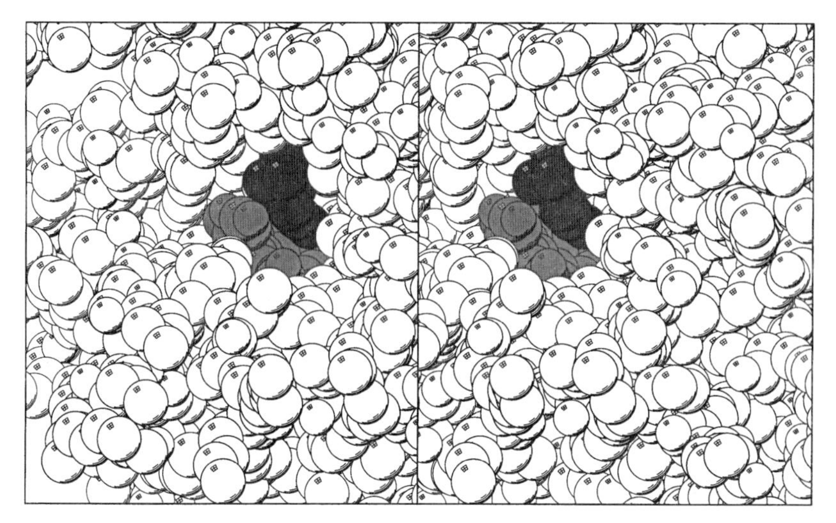

FIG. 5. Stereo view showing the substrate access channel in eNOS. The heme is lightly shaded and the substrate, L-Arg, is darkly shaded. The channel is deep yet solvent accessible for ready entry of substrate and exit of product. Part of the access channel is shaped by the second molecule (shaded) in the dimer. There appears to be no requirement for major structural changes upon substrate binding or release.

enable substrate to enter and product to leave. This architecture is strikingly different than in P450s. A long-standing problem with P450s is how the substrate enters the active site, since the available P450–substrate complex crystal structures show an essentially buried substrate with no obvious route into the heme pocket (*103*). Certain regions of P450cam were identified as providing a possible access route based primarily on enhanced thermal factors in the substrate-free enzyme (*104*) and inhibitor complexes (*105*). The true magnitude of required structural rearrangements was not fully appreciated until the structure of P450BM-3 complexed with a fatty acid substrate was compared to the substrate-free structure (*106*). Here fairly large movements of helices and loops are involved in closing down the protein around the substrate. Such motions are not necessary with NOS. This possibly might be due to the rather extensive polar interactions required for binding L-Arg to NOS, as opposed to P450 substrates, which are typically quite nonpolar. The energetic cost of sequestering the active site polar groups involved in L-Arg binding to NOS possibly precludes a P450-like buried active site.

Interactions holding the substrate, L-Arg, in place are primarily polar and conserved (Fig. 6). Glu363 forms a pair of H-bonds with the

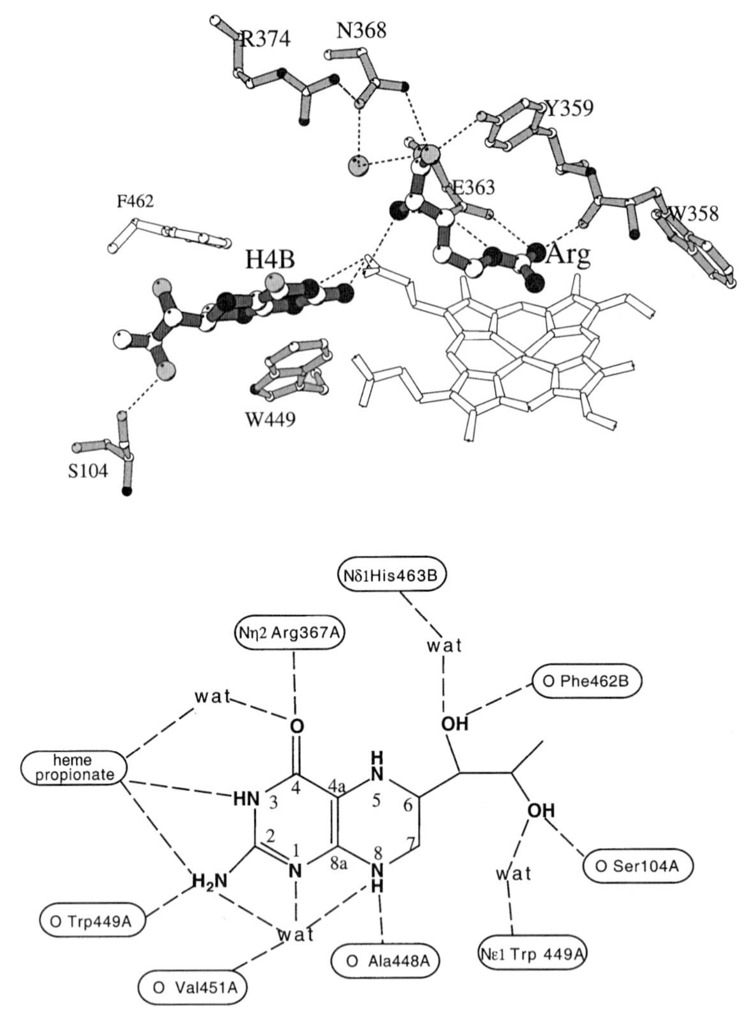

FIG. 6. The active site architecture of eNOS in the A subunit. L-Arg is held in place by several H-bonds with conserved groups. The H_4B is sandwiched between aromatic groups contributed from each subunit: Trp449 in subunit A and Phe462 in subunit B. Note that the amino group of L-Arg and the pterin donate an H-bond to the same heme propionate, which helps to understand the observed interdependence of pterin and substrate binding. The schematic diagram illustrates the extensive contacts between pterin and protein groups in each subunit.

guanidinium group of L-Arg, which aids in positioning a guanidinium N atom ≈4 Å from the iron. The amino/carboxyl end of L-Arg is similarly held in place by H-bonding groups, including an H-bond between the heme propionate and the L-Arg amino group.

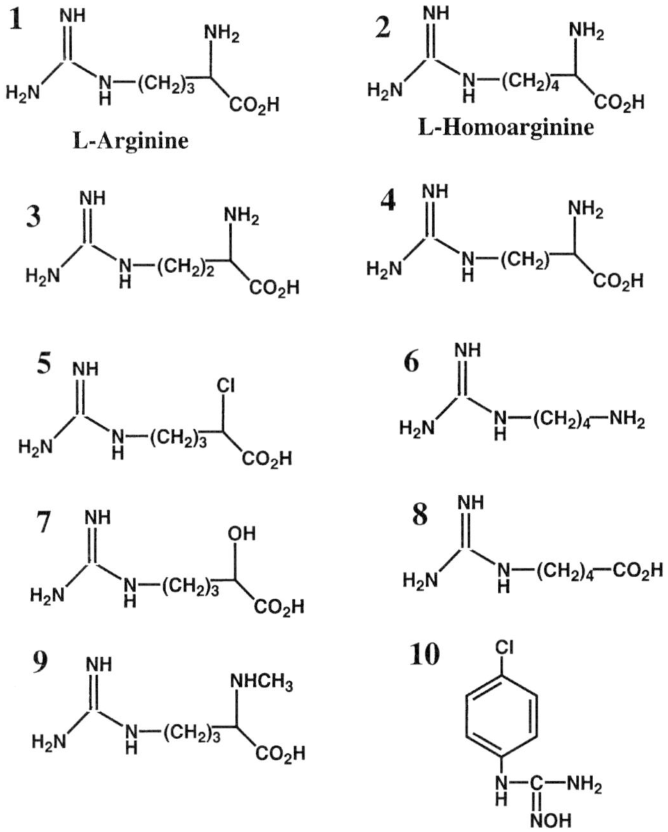

1 L-Arginine

2 L-Homoarginine

FIG. 7. Structures of various substrates and inhibitors for NOS. Only **1**, **2**, and **10** are substrates for NOS. An important conclusion from studies on such substrate analogs is the requirement for the α-amino group for catalysis.

H_4B is located at the dimer interface sandwiched between Trp 449 in one subunit and Phe 462 in the other (Fig. 6). In addition to these aromatic stacking and other nonbonded contacts, there is an array of H-bonding interactions involving the peptide backbone and side chains from both subunits and solvent molecules (Fig. 6). Note, too, that the pterin H-bonds directly with the same heme propionate that interacts with the L-Arg substrate. The observed mutual dependence of pterin and L-Arg binding to NOS (*107*) undoubtedly is mediated through the heme propionate. Relevant to this discussion are studies carried out on a range of Arg analogs that have been tested as substrates (*108*). Of these, only L-Arg and L-Homoarg (Fig. 7) are good substrates, although there is very low activity for **6** in Fig. 7. The rest of the Arg analogs are not substrates. The main conclusion from this work is the critical importance of the L-Arg free amino group (*108*), which further underscores the importance of the L-Arg–propionate H-bond. Although the precise function remains unknown, the NOS structures limit the possibilities. While the amino group no doubt contributes to the overall substrate binding affinity to NOS, it is unlikely that the amino group of L-Arg is needed to properly orient the guanidinium group for catalysis. However, the L-Arg amino–propionate H-bond is very likely important for correctly positioning the propionate for interactions with H_4B. The direct H-bonded connection between the pterin and L-Arg via the heme propionate might also provide an electron transfer link to the heme iron.

Another property of H_4B is its interplay with the heme–thiolate bond. Like P450s, NOS can convert to an inactive state that gives an absorption band at 420 nm in the reduced-CO form. This inactive form of P450 or NOS is called P420. It appears the P420 may be a form of the enzymes where the S–Fe bond is broken or the Cys is protonated (*109*). Huang *et al.* (*110*) have found that NOS P420 can be slowly reactivated by the addition of H_4B. Quite clearly there is no direct structural link between H_4B and the Cys ligand. From the current NOS structures, the main structural effect of H_4B binding would be to stabilize the heme propionate to which H_4B H-bonds in addition to displacing solvent from the H_4B hydrophobic pocket. Whether or not the absence of H_4B results in sufficient instability of the heme pocket to alter the S–Fe bond is not clear from the structures now in hand.

E. Pterin Function

An intriguing puzzle in NOS catalysis is the precise role of H_4B. The traditional function of H_4B is in aromatic amino acid metabolism where H_4B directly participates in the hydroxylation reaction via a nonheme iron. However, the NOS pterin site has no similarity to the pterin site in the hydroxylases, nor does NOS have a nonheme iron to assist pterin in substrate hydroxylation as in the amino acid hydroxylases (*111*). NOS more closely resembles pterin-containing enzymes that have a redox function (*81*). In particular, N3 and the C3 amino group form H-bonds with either Glu or Asp residues in a series of pterin enzymes (*112–116*) similar to NOS, except that NOS utilizes the heme propionate (Fig. 6).

Support for a redox role for H_4B stems from studies carried out in several laboratories. Using low-temperature rapid reaction methods, Bec *et al.* (*117*) obtained evidence for reductive activation of the NOS–oxy complex that correlated with the oxidation of L-Arg to L-OH-Arg.

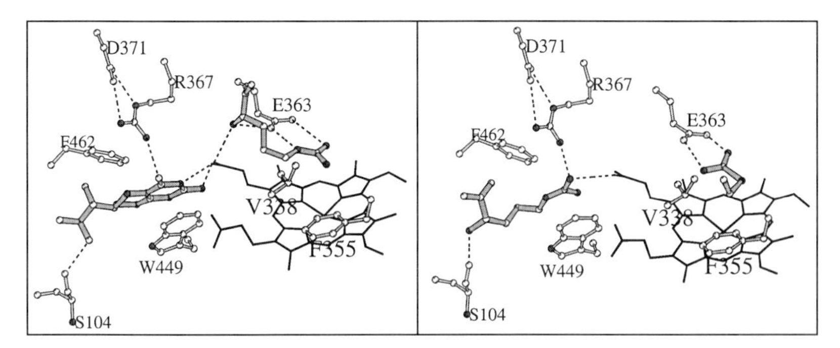

FIG. 8. Comparison of the pterin-bound and -free structures. In the absence of pterin with the SEITU inhibitor in the active site (right), L-Arg binds in the pterin site. L-Arg is an excellent mimic for H_4B owing to the similar H-bonding pattern with the heme propionate and Ser104, and the ability of the L-Arg guanidinium group to aromatically stack with Trp449.

Since this requires a second electron transfer to the heme iron, it was concluded that H_4B must be the source of this second electron, thus forming a pterin radical. Witteveen *et al.* (*118*) analyzed the state of H_4B during steady-state turnover conditions and found that H_4B is oxidized while bound to NOS, a finding consistent with the pterin forming a radical and serving as a source of reducing equivalents to the heme iron. The most direct evidence to date for a pterin radical derives from freeze–quench EPR experiments (*119*). Mixing O_2 with dithionite reduced iNOS heme domain followed by freeze–quenching gives an intermediate with a strong ^{15}N-pterin sensitive EPR signal, clearly indicating a pterin radical. Moreover, the amount of pterin radical formed per heme, 0.8 equiv, was the same as the amount of L-OH-Arg formed, indicating that pterin oxidation is coupled to L-Arg oxidation and hence is part of the productive catalytic cycle. A pterin radical also is supported by a comparison between the pterin-free and -bound eNOS heme domain structures (*81*). In the absence of pterin, the site either fills with solvent molecules or the substrate, L-Arg (Fig. 8). The presence of L-Arg in the pterin site is surprising until one appreciates the similarity between L-Arg and H_4B. The guanidinium group of L-Arg replaces N3 and the C2 amino group in H-bonding with the propionate, while the L-Arg amino group replaces the pterin side chain OH interactions with Ser104 (Fig. 8). The planar guanidinium group also is able to form π stacking interactions with Trp 449 as does the pterin. That L-Arg is able to replace H_4B further suggests that the NOS pterin site may be designed to stabilize a cationic pterin radical (*81*). Presumably such stabilization is achieved electrostatically, as has been proposed for stabilization of the cationic Trp radical in cytochrome *c* peroxidase (*120*). By analogy with peroxidase compound I that utilizes $(FeO)^{3+}$ and an organic radical (heme or side chain), NOS may utilize an $(FeO)^{3+}$/pterin radical intermediate as the active species toward hydroxylation of L-Arg. It should be cautioned, however, that the traditional $(FeO)^{3+}$ intermediate known to be involved in peroxidase and thought to be involved in P450s has not been demonstrated in NOS. In addition, there is only one known example of an enzyme that forms a pterin radical, and in this case, the radical is centered on molybdopterin (*121*), although non-protein-bound pterin radicals are well known, especially under acidic conditions (*122*). The extensive aromatic stacking interactions in the pterin site, similar to the FMN site in flavoproteins, could further stabilize a pterin radical.

Since *E. coli* does not biosynthesize tetrahydrobiopterin, it is possible to prepare pterin-free NOS in *E. coli* recombinant expression systems. This is important for probing pterin-independent NOS functions as well as studying the role of various pterin analogs with altered redox and stereochemical properties. Pterin-free NOS is unable to catalyze the conversion of L-Arg to N^G-hydroxy-L-Arg (*123*). In contrast, pterin-free NOS can convert N^G-hydroxy-L-Arg to products using either H_2O_2 or NADPH/O_2 to support the reaction. However, the rates in the pterin-free NOS are five to eightfold slower than in holo-NOS, and the NADPH-supported reaction in pterin-free NOS does not generate NO but instead NO^- (*123*). Therefore, pterin is essential for the first step in the production of N^G-hydroxy-L-Arg and important in controlling the conversion of N^G-hydroxy-L-Arg to the correct product, NO. A variety of pterin analogs also have provided important insights into the role of pterin (*124*). In these studies it was shown that reduced pterin is required for NO production but not heme reduction. The strict requirement for reduced pterin involves some step after the initial one-electron reduction of the heme iron and most likely after O_2 binding. It is known that reduced pterin destabilizes the oxy-complex (*125*), indicating, possibly, that reduced pterin is involved in oxygen activation to the intermediate responsible for the initial hydroxylation of L-Arg. These observations are consistent with H_4B donating one electron to Fe^{2+}–O_2, resulting in formation of a pterin radical and $(Fe–O)^{3+}$.

Some of these ideas incorporating a pterin radical are outlined in Fig. 9. Here the pterin serves as an electron donor to the oxy intermediate to give the peroxy di-anion. The substrate, L-Arg, itself serves as the proton donor to the peroxo intermediate as proposed by Crane *et al.* (*80*), which is required for heterolytic cleavage of the peroxide O–O bond. The proton delivery machinery outlined in Fig. 9 could also help to explain why peroxide itself cannot support the conversion of L-Arg to N^G-hydroxy-L-Arg (*126*). The reaction may require a potent base like

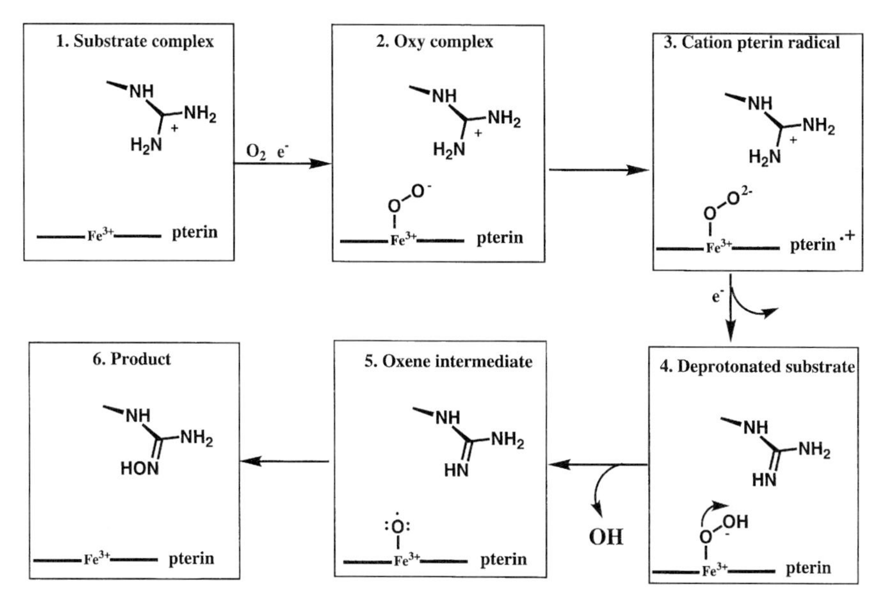

FIG. 9. Possible electron transfer mechanism for NOS utilizing a pterin radical. The oxy-complex in **2** is shown as the ferric (Fe^{3+})–superoxide complex. The role of the pterin then is to donate an electron to the iron, thus giving the peroxy dianion in **3**. The dianion is a potent base that abstracts a proton from the substrate, giving **4**. The system is now set up for a peroxidase-like heterolytic cleavage of the O–O bond to give the active hydroxylating intermediate in **5** and, finally, the first product in **6**.

the peroxide dianion in order to prepare the substrate for hydroxylation by abstracting an L-Arg proton. Peroxide would enter the heme pocket and most likely bind as the monoanion incapable of deprotonating L-Arg leading to a short-lived, dead-end complex. The precise nature of the hydroxylating intermediate (**5** in Fig. 9) is unknown but is shown as the oxene oxygen linked to Fe^{3+} iron. The more traditional picture based on peroxidases is the oxyferryl/radical intermediate, $Fe^{4+}{=}O/R\cdot$, where $R\cdot$ is either a heme or amino acid radical. In the case of NOS, the pterin could form the radical and be one of two sites where oxidizing equivalents are stored.

F. NOS INHIBITORS

An important practical goal in NOS research is the development of isoform-specific inhibitors (*127*). Diminished levels of NO contribute to chronic hypertensive diseases. In contrast, NO overproduction contributes to pathological conditions related to primary neurodegenerative, inflammatory, and vascular disorders (*5*). In septic shock NOS

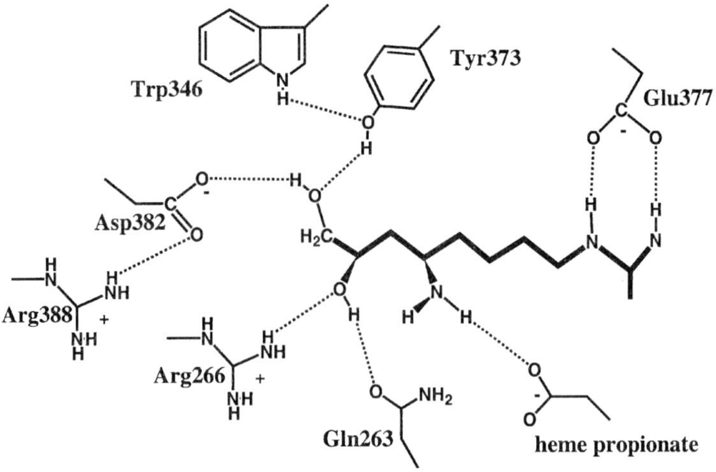

FIG. 10. Hypothetical model of inhibitor binding to the active site of iNOS. This inhibitor binds ≈700-fold more tightly to iNOS than eNOS. One obvious and significant difference between the two isoforms is that the residue corresponding to Asp382 in iNOS is Asn in eNOS. Asn would not be able to donate an H-bond to the inhibitor OH group as shown, which could contribute to the difference in binding affinity.

inhibitors can restore vascular tone and blood pressure *(6)*. Blocking NO production also limits ischemia-elicited infarct size in animal models *(7)*, and the production of NO by nNOS can cause further tissue damage following a stroke. As a result, it would be useful to block nNOS without altering the vasoregulatory properties of eNOS.

The attractive aspect of NOS from the perspective of structure-based inhibitor design is that only three isoforms are involved. This is in sharp contrast to P450s, where a myriad of isoforms are involved in drug metabolism. Obtaining high-resolution structures of the various pharmaceutically interesting P450s is a daunting task, whereas this problem is nearly solved with the NOS heme domains. Despite this apparent advantage with NOS, it is quite clear from a comparison of eNOS and iNOS that the active sites are very similar with no obvious structural variations in the immediate vicinity of the active site that could be exploited for drug design, with one possible exception. Based on a comparative modeling study between iNOS and eNOS *(83)*, a structural rationale has been provided for why one particular NOS inhibitor, *N*-(5(*S*)-amino-6,7-dihydroxyheptyl)ethanimidamide, is a ≈700-fold better inhibitor of iNOS than eNOS *(128)*. As shown in Fig. 10, the inhibitor was modeled into the iNOS active site to mimic substrate binding. The ethanimidamide group mimics the guanidinium group of

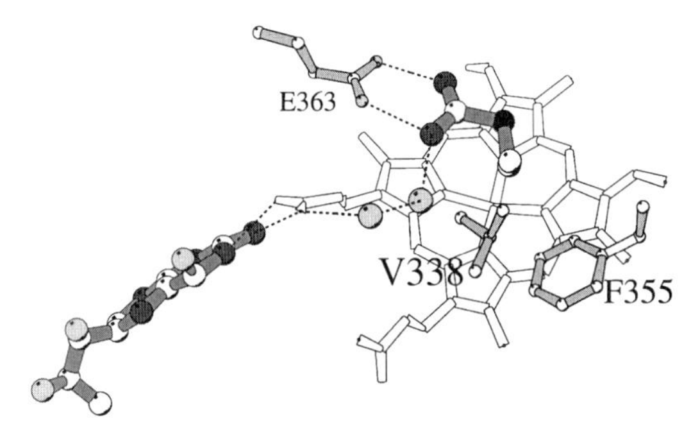

FIG. 11. The structure of eNOS complexed with SEITU. The ureido function H-bonds with Glu363 while the ethyl group interacts with both Val338 and Phe355.

L-Arg to H-bond with Glu 377 while the 5-amino group interacts with the heme propionate. The inhibitor glycol moiety extends into the polar pocket occupied by the carboxylate of L-Arg. A key feature of this hypothetical complex is the H-bond formed between the inhibitor and Asp 382. The corresponding residue in eNOS is Asn, whose amide group cannot serve as an H-bond acceptor as does Asp382 in iNOS. Therefore, the difference in binding affinity could be due to this missing H-bond interaction in eNOS.

The published structural data on inhibitor complexes is rather scant and to date, only the iNOS structure complexed with a product analog, thiocitrulline (*80*), and eNOS (*81*) and iNOS (*82*) complexed with *S*-ethylisothiourea or SEITU have been published. Despite this lack of structural information, which will no doubt change in the near future, a good deal is known about the binding of various inhibitors to NOS, and especially the thiourea class of NOS inhibitors (*129*). SEITU binds to eNOS and iNOS with inhibition constants, K_i, values of 19 and 39 nM, respectively (*129*). The alkyl group with *1–3* carbon atoms showed the best affinity, whereas total removal of the ethyl group to give thiourea is not an inhibitor. Hence, the alkyl chain is critical for binding.

The crystal structure of the eNOS–SEITU complex is shown in Fig. 11. As expected, the ureido group H-bonds with Glu363 similar to the way L-Arg interacts with Glu363. Two water molecules form H-bonding bridges between the inhibitor and heme propionates. The inhibitor sulfur is $\approx$4 Å from the heme iron but appears not to coordinate with iron, since there is no continuous electron density between the inhibitor and heme iron, which is consistent with the high-spin shift triggered by SEITU binding. Aside from alterations in solvent structure at the active site, there is essentially no difference between the SEITU and L-Arg complexes. SEITU forms far fewer direct H-bonds with the protein than does L-Arg, yet SEITU binds to eNOS with a $K_i = 39$ nM while L-Arg binds in the 1 μM range (*129*). The strict requirement for an alkyl group attached to the sulfur provides at least one clue why SEITU is such a good inhibitor. The crystal structure (Fig. 11) shows that the ethyl group interacts with both Phe355 and Val338. Nonpolar interactions in addition to entropically favored solvent release could be factors that contribute to tight binding. It also is possible that SEITU forms a stronger set of H-bonds with Glu363. Unfortunately, the inhibitor binding data are mostly based on the ability to block the NOS reaction, with no information on the thermodynamic state functions. The prospects for a detailed understanding of inhibitor binding, however, look very promising. It is quite clear that NOS crystals bind a variety of compounds at the active site, so there should soon be a wealth of structural information. This, together with detailed kinetic and thermodynamic analyses, should provide a clear picture on what factors control inhibitor binding.

VI. NOS Catalytic Cycle

A. Heme Iron Reduction

Similar to P450s, reduction of the NOS heme occurs via electron transfer from the FMN of the reductase domain. The electron transfer process is especially important in NOS, since this is the step regulated by Ca^{2+}/calmodulin. Binding of calmodulin to the linker connecting the heme and flavin domains results in an ill-defined switch, most likely a conformational change, which allows electrons to flow from FMN to heme (*76*). It has been postulated that the nature of the Ca^{2+}/calmodulin switch relates to the relative orientation of the FMN and FAD reductase domains (*130, 131*). The crystal structure of P450 reductase, a close homolog and model for the NOS reductase, shows that the FMN and FAD domains are connected by a flexible linker and that the edges of the FMN and FAD cofactors are in contact (*132*). The crystal structure of P450BM-3 complexed with its FMN domain shows that the edge of the FMN is oriented toward the P450 heme (*133*), which (Fig. 12) is compatible with docking of the entire P450 reductase only if significant movement of the FMN domain occurs from the orientation observed in the mammalian P450 reductase structure. Changes in the fluorescence properties of the NOS flavins induced by Ca^{2+}/calmodulin (*130*) are consistent with such a rearrangement of the FAD and FMN domains. Moreover, work with the isolated NOS reductase domain indicates that

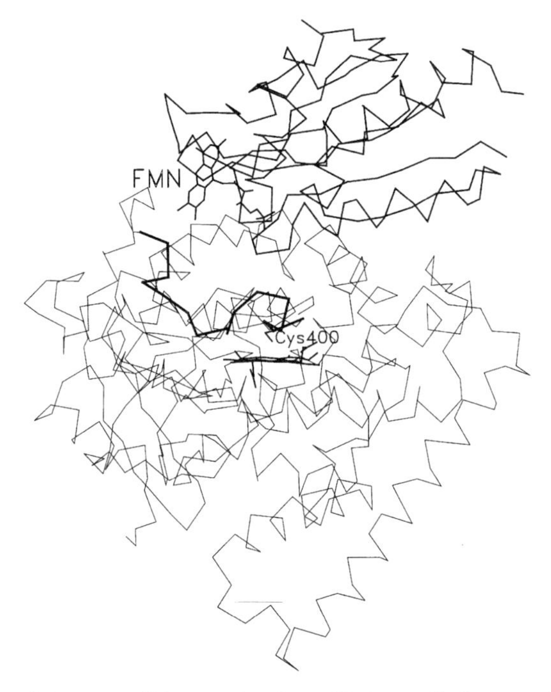

FIG. 12. Crystal structure of the complex formed between the heme and FMN domains of cytochrome P450BM-3 (*133*). The FMN domain docks on the proximal surface of the heme domain. The thicker trace in the heme domain highlights residues 387 to the heme ligand, Cys400. This section of polypeptide contacts the FMN and might provide an electron transfer conduit to the heme ligand. The two interacting surfaces are electrostatically complementary, with similar complementarity expected for HO and NOS.

of the Ca^{2+}/calmodulin effects are centered on the reductase domain (*134*). In this picture Ca^{2+}/calmodulin binding results in the reorientation of the FMN and FAD domains such that the reductase domain can properly dock to the heme domain.

Utilization of a domain linker to control electron flow is not unique to NOS. Like NOS, P450BM-3 has the heme and reductase domains fused to give a heme–FMN–FAD architecture (*75*). In addition, the linker between the heme and FMN domains is critical for electron transfer. Engineering studies on the P450BM-3 linker reveals that the length of the linker but not the sequence is critical in controlling the FMN-to-heme electron transfer reaction (*135, 136*). Similar experiments with flavocytochrome b2 (*137*) illustrate the importance of the linker in interdomain electron transfer, presumably by assisting in proper orientation of redox partners. The same appears to be true for NOS, with the important

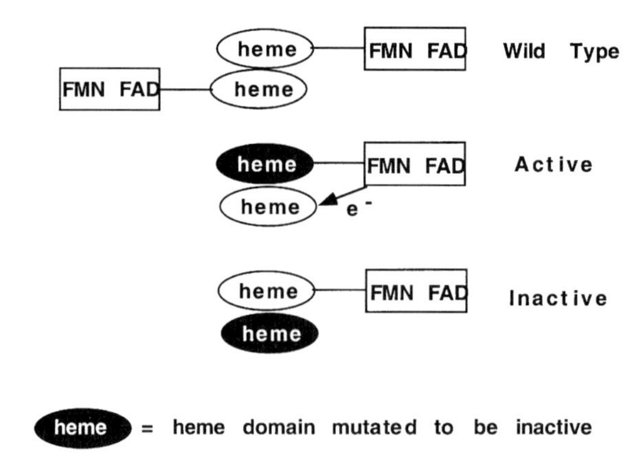

FIG. 13. Schematic representation of the overall NOS architecture and summary of work presented in (*139*). Heterodimers were generated to test if electron transfer from the FMN domain proceeds via an inter- or intrasubunit process. When holo-NOS containing an inactive heme domain was dimerized with an active heme domain, activity was observed. However, when active holo-NOS was dimerized with the inactive heme domain, no activity was observed. These results indicate that the flavin domain of monomer A transfers electrons to the heme domain of monomer B.

exception that calmodulin binding regulates heme–reductase domain interactions. Chimeric NOS constructs where the reductase domain of one NOS isoform is fused to the heme domain of another isoform further illustrate the importance of electron transfer in controlling NO biosynthesis (*138*). The order of NOS activity for the various isoforms is iNOS $\gg$ nNOS $>$ eNOS. Fusing the reductase domain of nNOS to the heme domain of eNOS or iNOS confers nNOS-like activity to both chimeras, which underscores the importance of electron flow in regulating the rate of NO production.

Although NOS is a dimer, it is reasonable to assume that electron transfer is confined to intrapolypeptide electron transfer within one subunit. However, some clever engineering studies indicate that this may not the case, and that the reductase of one subunit is the electron donor to the neighboring subunit (*139*) (Fig. 13). This should provide some limitation on the possible models that can be proposed based on current structural information, including the estimated distance between FMN and heme as $>$15 Å (*140*). At least two quite different sites have been proposed for docking of the reductase domain (*80, 81*), each with its own attractive features, although neither has as yet been directly tested by suitably designed experiments. Another puzzle with respect to electron transfer is the requirement for three electrons to take L-Arg to NO and L-citrulline. Since NADPH is a two-electron carrier, one electron must be "off-loaded" to continue the cycle. However, NOS is a dimer, so it may be more appropriate to consider a single cycle requiring six NADPH electrons per two NO molecules formed. In this case, it might be beneficial and less wasteful of reducing equivalents for the enzyme to be able to disproportionate reducing equivalents between reductase domains. If so, then the reductase domains would need to be in close proximity. The docking site proposed by Raman *et al.* (*81*) requires that both reductase domains dock along the dyad axis of symmetry near the Zn site, which would enable the reductase domains to directly interact. In this model, however, the electron transfer path to each heme is the same based on the crystal structures, and because of this symmetry, there appears to be no obvious way for a reductase domain to select one heme over the other.

B. Conversion of L-Arg to N^G-Hydroxy-L-Arg

Once the heme iron is reduced, O_2 binds followed by a second electron transfer to the heme and monooxygenation of L-Arg. This is similar to P450s with one important exception. In P450s, two-electron transfers and oxygenation effectively reduce O_2 to the level of peroxide. The peroxide O–O bond is cleaved, leaving behind an iron-linked oxygen that is introduced into the substrate as OH. The electron transfer steps in P450s can be bypassed by direct reaction of ferric P450 with peroxide. In contrast, peroxide is not effective in supporting the hydroxylation of L-Arg in NOS (*126*). As noted in Section V,D, one possible explanation is that NOS may require an unprotonated peroxy anion whose pK_a is sufficiently high to abstract a proton from the substrate, L-Arg. H_2O_2 might not be a strong enough base even when coordinated to the heme to remove the L-Arg proton. It also is possible that despite the apparent similarities between NOS and P450s, NOS does not utilize a P450-like mechanism, but rather a dioxy intermediate as the direct oxidant of L-Arg. In this case cleavage of the O–O bond is not a prerequisite to substrate hydroxylation. A quite different non-P450 mechanism also has been proposed. Based on the observed stimulation of nNOS activity upon addition of nonheme iron (*141*), an argument has been made for NOS operating more like an amino acid hydroxylase where the pterin and nonheme iron are in close proximity such that both participate in oxygen activation and hydroxylation. However, the recent structure determination of tyrosine hydroxylase (*111*) reveals little similarity to NOS in pterin binding. In addition, NOS contains no nonheme metal site or ligands in a position analogous to those found in the aromatic amino acid hydroxylase, and the only nonheme metal located in the known NOS structures is the Zn.

In addition to the similar positions of the substrate relative to the heme iron in both NOS and P450, another reason for drawing close analogies between NOS and P450 hydroxylation chemistry is the close similarity in environment of the Cys ligand, which suggests similar roles in formation of the active hydroxylating intermediate. Theoretical work (*142, 143*) indicates that the delocalization of spin-density in the S–Fe–O intermediate in thiolate heme systems favors oxygen transfer, whereas utilization of His ligands favors electron transfer. One would therefore expect both NOS and P450 to promote similar active intermediates, which means $(Fe{-}O)^{3+}$ or its electronic equivalent.

Despite the attractive features of equating NOS and P450, the major unique feature in NOS remains the role of H_4B. That reduced pterin is required for oxidation of L-Arg to N^G-hydroxy-L-Arg (*123*) separates NOS from P450. Whether the connection between pterin and substrate hydroxylation is in electron transfer, oxygen activation, or some other process is still not settled; the weight of the evidence now favors an electron transfer function for the pterin (Section V,E) (*81, 117, 119, 144*). H_4B is known to be required for some step after heme reduction and O_2 binding (*124*), and it is also known that H_4B destabilizes the oxy-complex (*125*). Presta *et al.* have proposed that pterin binding might increase the redox potential of the heme, which enables electrons to flow from FMN to heme (*124*). Considering the H-bonded structural link between H_4B and L-Arg (Fig. 6), H_4B could well help to stabilize L-Arg binding, which, in turn, could alter the heme redox properties and sterically compromise O_2 binding, which leads to the observed destabilization of the oxy complex (*125*). There must, however, be more than a structural connection, since any suitable pterin analog could satisfy the structural link but only reduced pterin supports function. This is partly the basis for proposing an electron transfer role for H_4B (*81, 117, 119*). If H_4B donates an electron to the iron center, it would do so via the heme propionate, since this would provide the most direct continuous covalent and H-bonded link to the iron (Fig. 6). The use of heme propionates to carry reducing equivalents is not without precedent, since manganese peroxidase appears to operate in this manner (*145*).

C. N^G-Hydroxy-L-Arg to NO

Although the mechanistic details of this step are not well understood, there are, unlike the oxygenation of L-Arg, sound enzymatic and

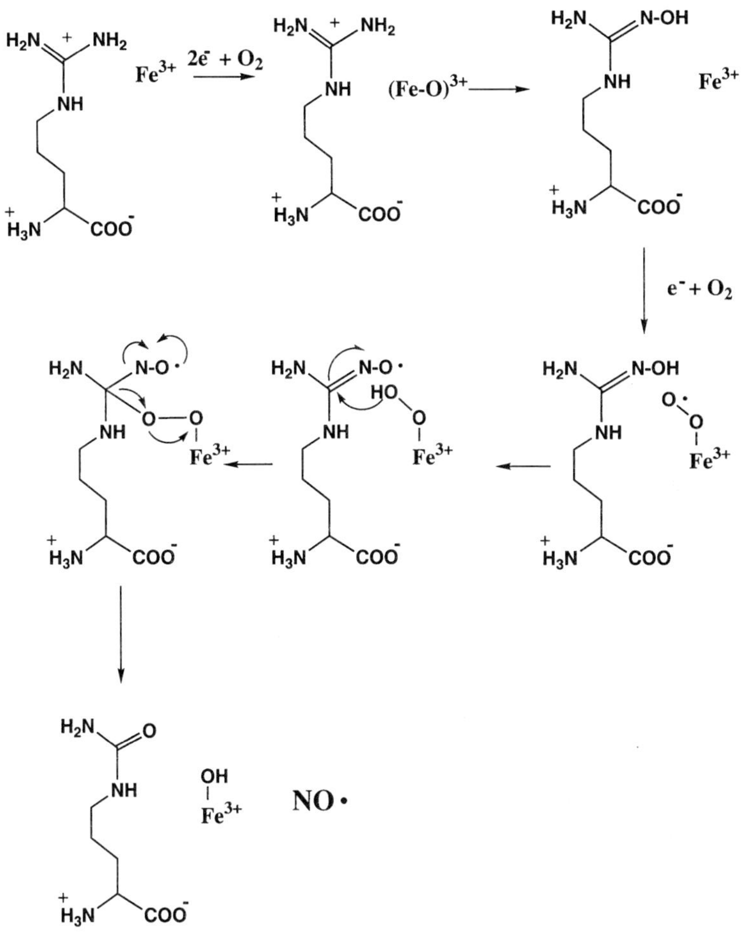

FIG. 14. Overall mechanism for the conversion of L-Arg to NO and L-citrulline. The first part of the reaction in forming N^G-hydroxy-L-Arg is fashioned after P450, which is useful for discussion but still unproven. The conversion of N^G-hydroxy-L-Arg to product is taken after (*147*).

chemical models to support the proposed mechanism (Fig. 14). Unlike oxidation of L-Arg, oxidation of N^G-hydroxy-L-Arg by NOS is supported by H_2O_2 but not hydroperoxides or agents that form the $(Fe–O)^{3+}$ intermediate (*126*). This argues for a dioxy intermediate as the hydroxylating agent and not the P450-like oxyferryl intermediate. In addition, single turnover experiments show that only one NADPH-derived electron is required to convert N–OH–L-Arg to NO (*146*), which is inconsistent with cleavage of the O–O bond. This has led to the proposed mechanism

(*147*) shown in Fig. 14, which is similar to that proposed for aromatase (*148*) and supported by chemical model systems (*149*). Quite recently compound **12** (Fig. 7) has been shown to serve as a substrate (*150*). This intriguing observation opens the way for probing the reactivity of the N–OH group by, possibly, modifying reactivity using suitable aromatic substituents. As in the hydroxylation of L-Arg, a major factor in the production of NO from *N*-hydroxy-L-Arg is the requirement for H_4B (*146*). Here, too, the precise role of H_4B is not known. It should be clear that since the conversion of L-Arg to products and H_4B are so intimately linked, the major challenge in understanding the conversion of L-Arg to NO is in deciphering the precise functional role of H_4B.

VII. Background on HO

Heme oxygenase catalyzes the first and rate limiting step in the degradation of heme to give biliverdin (Fig. 15). Humans and other mammals have two HO isoforms termed HO-1 and HO-2. Human HO-1 consists of 288 residues (*151*), whereas rat HO-2 has 316 residues (*152*). HO contains a C-terminal membrane anchoring sequence (*153*) that is

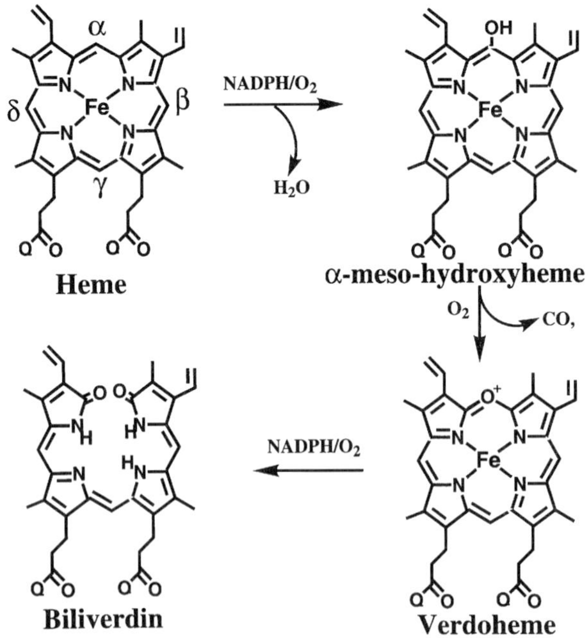

Heme

NADPH/O_2

H_2O

α-meso-hydroxyheme

O_2

CO,

Verdoheme

NADPH/O_2

Biliverdin

FIG. 15. Overall reaction catalyzed by heme oxygenase.

not necessary for enzyme activity (*154*). The two human isoforms share about 45% sequence identity and 56% for the conserved core. More recently a bacterial HO has been characterized (*155*) and exhibits about 36% identity with human HO-1. The high sequence homology, especially in the catalytic core, indicates quite similar structures and mechanism for all known HO isoforms.

HO-1 is controlled at the level of transcription by oxidative stress and several inducers such as porphyrins, metals, and progesterone *(8)*. The absence of HO-1 is associated with severe growth retardation, anemia, and enhanced endothelial cell injury (*156, 157*). In addition, HO-1 has been implicated in protection against transplant rejection (*158, 159*). HO-2 is constitutively expressed, and its presence in brain and the noted similar effects of CO and NO has led to the proposal that CO-generated by HO-2 is involved in signaling processes (*8, 160–162*).

VIII. HO Structure

A. Overall Heme Oxygenase Structure

The crystal structure of human HO-1 complexed with heme has been determined to 2.07 Å (*163*). In order to obtain crystals for high-resolution studies (*164*), it was necessary to use an *E. coli* expressed version of HO-1 consisting of residues 1–233 (missing residues 234–288, which includes the C-terminal membrane anchor). This shorter version of HO-1 is soluble and retains about 50% wild-type activity (*165*). As shown in Fig. 16, HO-1 is formed exclusively by helices and connecting segments of "random" coil. Although the rich helical content is similar to that of

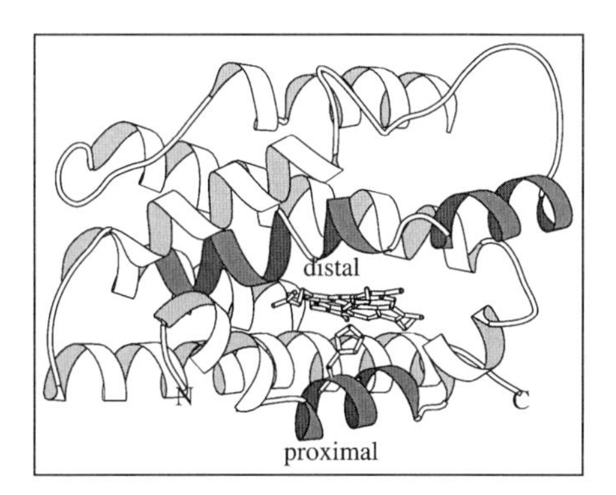

Fig. 16. Crystal structure of human HO-1 (*163*). The heme and His25 ligand are shown. The proximal and distal helices are highlighted. Note the break in the normal helical pattern in the distal helix directly over the heme.

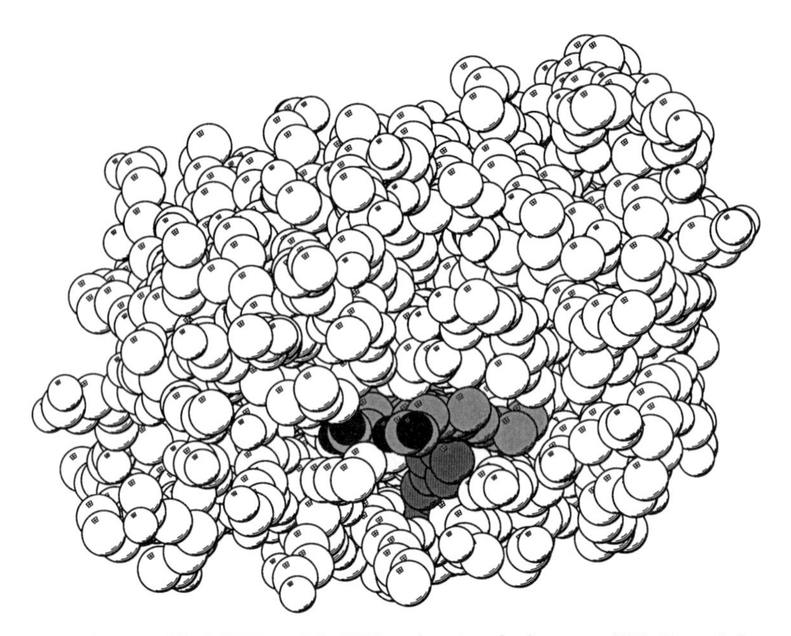

FIG. 17. A space-filled CPK model of HO-1 showing the heme and His ligand. One propionate is quite exposed, as is one heme edge. This surface of the protein also is the most electropositive, which probably serves as the docking site for the electronegative P450 reductase. This would enable close approach of the FMN to heme for efficient electron transfer.

the globins and peroxidases, the overall topology bears no similarity to other helical proteins (*163*). The relatively exposed heme (Fig. 17) is situated between two helices, termed the proximal and distal helices.

B. PROXIMAL HEME REGION

The proximal helix provides the axial heme ligand, His25, which was correctly predicted from mutagenesis studies (*166, 167*). His25 is approximately 2.8 Å from Glu29 for a possible hydrogen bond. This is similar to peroxidases, where the proximal His ligand hydrogen bonds with a conserved Asp residue. However, in peroxidases the Asp–His interaction imparts considerable anionic character to the His ligand (*40, 168*), which is thought to help regulate the heme iron redox potential (*99, 100*). In HO, however, resonance Raman studies revealed that the His ligand is neutral (*169*). This is understandable owing to the longer H-bonding distance between the His and Glu in HO-1 compared to peroxidases. In addition, the His–Asp pair in peroxidases is buried in the proximal pocket, whereas Glu29 in HO-1 is exposed on the molecular surface where access to solvent should weaken the His–Glu interaction.

The role of the proximal His ligand has been probed by replacing His25 with Ala (*170*). The resulting HO-1 binds heme but is inactive (*169*). This indicates that the proximal heme ligand is not critical for heme recognition or binding, but is necessary for activity. This, too, is consistent with the structure. The primary interactions holding the heme in place are propionate ion pairs/hydrogen bonds with Lys and Arg residues and interactions between the distal surface of the heme and the distal helix (discussed in next section). The methyl and vinyl substituents appear not to be critical, since NMR studies have shown that the heme is orientationally disordered about the α–γ axis (*171*). This, too, is consistent with the crystal structure, since flipping of the heme about the α–γ axis would not lead to any obvious steric problems.

Even though the His25Ala mutant is inactive, activity can be restored upon addition of imidazole, which occupies the proximal ligand cavity to coordinate with the heme iron (*170*). Similar "cavity complementation" experiments have been carried out with cytochrome *c* peroxidase (*172, 173*). In addition, replacement of the proximal His in cytochrome *c* peroxidase with Gln does not eliminate the ability of the enzyme to react with peroxide to form compound I (*99*). The parallels between HO and peroxidase are interesting, since the loss in activity by removal of the proximal His ligand can be restored by cavity complementation. Nonetheless, one very clear difference between peroxidase and HO is the requirement for peroxidases to cleave the peroxide O–O bond and generate the $(Fe–O)^{3+}$ intermediate. In contrast, HO does not form the $(Fe–O)^{3+}$ intermediate, but instead the key catalytic steps require binding of O_2, reduction to peroxide, and a mechanism to ensure that the terminal O atom of the peroxide reacts with the correct *meso* carbon (*174–176*). Since activity can be restored to the proximal cavity mutant of HO-1 (*170*), it might be possible to reconstitute with other nonimidazole but isosteric ligands such as amines or alcohols, or even replace the proximal His ligand with Gln and Glu. The activity profiles of such HO variants would be of considerable interest.

C. Distal Pocket

In the HO-1 crystal structure, the heme is hexacoordinate with a water molecule/hydroxide ion $\approx$2.0 Å from the iron. This is consistent with spectral studies showing that the HO-1 heme converts from high to low spin with a pK_a 7.6–8.0 (*169, 177*), presumably due to titration of a water molecule resulting in hydroxide binding to heme iron. The

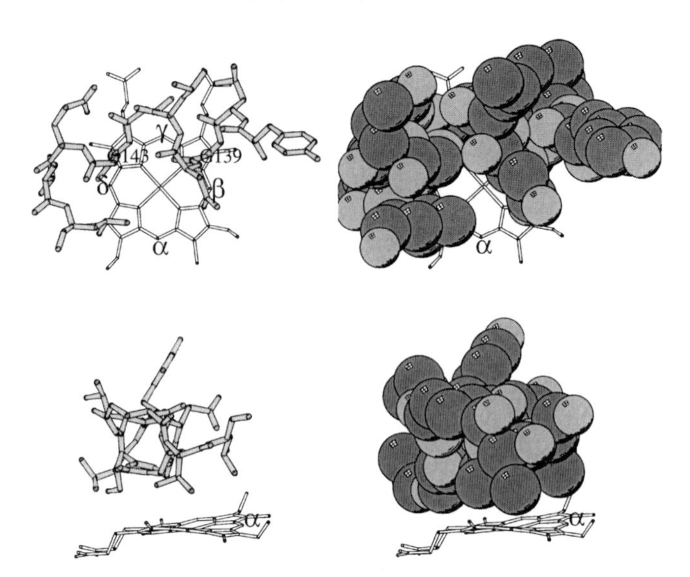

FIG. 18. Two views of the distal helix and its relation to the heme. Each view (top and bottom) is represented as a stick model and a CPK model. Note that the distal helix approaches quite close to the heme such that backbone atoms make direct contact with the heme. The top view shows that the α-*meso* position is the most exposed while the remaining *meso* carbons are masked by the distal helix, leaving the α-*meso* position open for reaction. The locations of G139 and G143, which provide flexibility, are indicated.

crystals are obtained at pH 7.5, which is near the pK_a of the spin transition, so what is seen in the crystal structure may be a mix of low- and high-spin heme.

As in peroxidases, globins, and P450s, HO-1 has a helix over the distal surface of the heme (Figs. 16, 18). In other heme proteins side chains from the distal helix provide the primary contacts with the heme as well as side chains that interact with heme ligands. In sharp contrast, the distal helix in HO-1 lies much closer to the heme such that backbone atoms form the primary heme contacts. In addition, there is no neighboring residue that could serve the same function as the distal His in the globins and peroxidases for interaction with iron-linked ligands.

A notable feature of the HO-1 distal helix is its flexibility. The HO-1 crystal form used for solving the structure has two molecules in the asymmetric unit, which provides two independent views of the HO-1 structure. As shown in Fig. 19, a plot of the rms (root-mean-square) deviation in backbone atoms between molecules A and B in the asymmetric unit reveals a large deviation in the distal helix as well as in the loop immediately following the distal helix. In addition, the distal

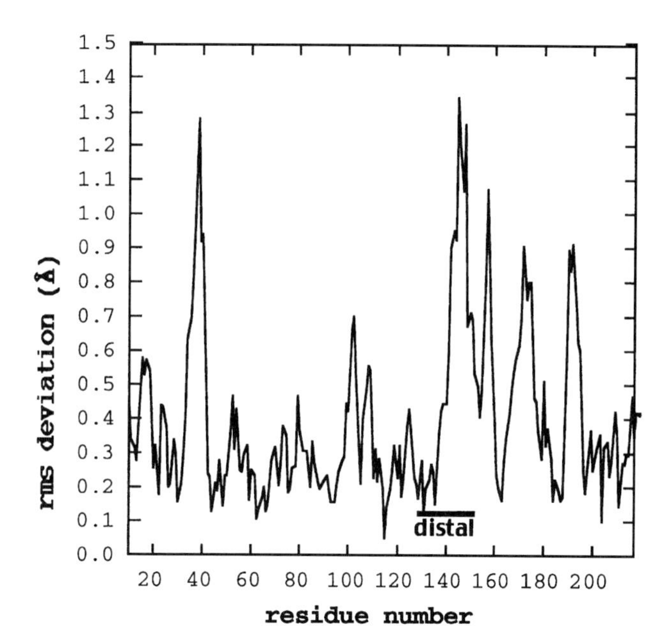

FIG. 19. A plot of the root-mean-square (rms) deviation of backbone atoms between HO-1 molecules A and B in the asymmetric unit. There is a large deviation in the distal helix just above the heme. In molecule B the distal helix is farther from the heme, thereby providing a more open pocket. Such flexibility may be required for substrate binding and product release. The required flexibility is very likely due to the conserved Gly residues in the distal helix.

helix is kinked, with a local disruption in the normal α-helical hydrogen bonding pattern. This flexibility is due to the conserved Gly residues in the G_{139}-D-L-S-G_{143}-G sequence. Both Gly139 and Gly143 make direct backbone–heme contacts. However, in molecule B Gly143 is 5.9 Å from the heme, whereas in molecule A, the same distance is only 3.7 Å. Hence, in molecule B the heme pocket is more open, owing to movement of the Gly-rich region of the distal helix. This results in a less tightly bound heme, as evidenced by the relatively poor electron density for the heme in molecule B compared to that in molecule A. This motion of the distal helix may have mechanistic relevance for substrate binding and product release.

Conspicuously absent in HO-1 is any residue analogous to the globin and peroxidase distal His that can directly interact with iron-linked ligands. Nevertheless, the HO-1 distal pocket is polar (Fig. 20) with the potential for diatomic ligands to interact with the kinked region of the distal helix, as well as water molecules that could bridge between

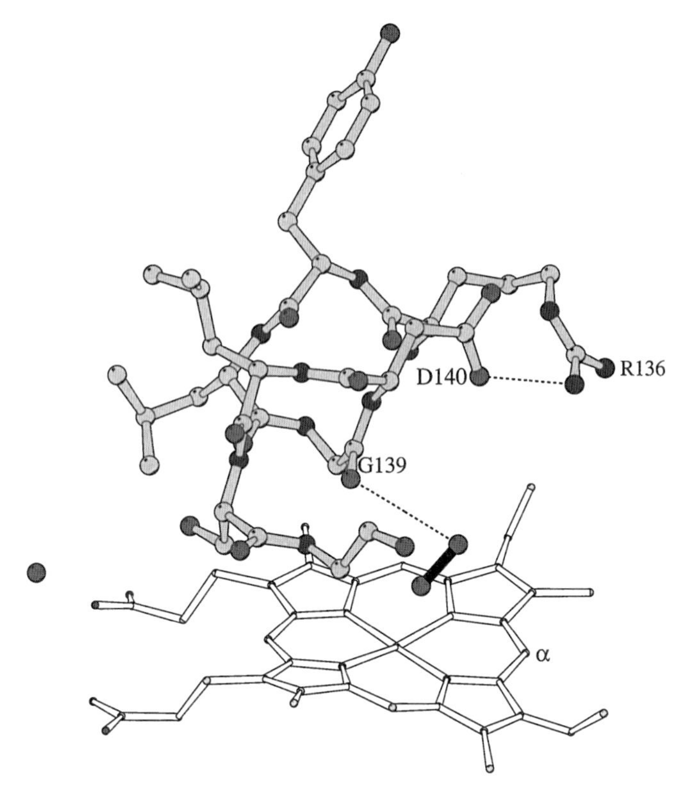

FIG. 20. A hypothetical model of one possible binding mode for a diatomic ligand to HO-1. Both Asp140 and Arg136 should interact with a ligand via an ordered water molecule. Whether or not these residues are critical in an acid–base catalytic process or are simply used to sterically orient an iron-linked peroxo intermediate remains unknown.

the ligand and polar protein groups (Poulos, unpublished; Fig. 20). For example, Arg136 and Asp140 form an ion pair/H-bond. Asp140 also forms an H-bond with an ordered solvent molecule. Modeling peroxide in the active site indicates that with some repositioning of solvent, the peroxide could be linked to Asp140 via a water molecule. In addition, peroxide or other diatomic ligands could H-bond with the peptide carbonyl oxygen of Gly139. Such interactions could "steer" the peroxide toward the α *meso* heme carbon.

D. Ligand Discrimination

An important feature of heme proteins is the ability to discriminate between ligands, especially O_2 and CO. In some model heme systems

TABLE I

PARAMETERS FOR CO AND O_2 BINDING TO HEME, MYOGLOBIN, AND HEME OXYGENASE

Sample	k_{on} CO ($M^{-1}s^{-1}$)	k_{off} CO s^{-1}	K_{eq} CO	k_{on} O_2 ($M^{-1}s^{-1}$)	k_{off} O_2 s^{-1}	K_{eq} O_2	K_{eq} CO/K_{eq} O_2	Ref.
Protoheme (benzene)	1.1×10^7	0.025	4.4×10^8	6.2×10^7	4000	1.6×10^4	2.8×10^4	*202*
Myoglobin (human)	7.6×10^5	0.22	3.5×10^7	1.9×10^7	22	8.6×10^5	41	*182*
HO-1(fast)	1.3×10^6	0.009	1.4×10^8	6.9×10^6	0.25	2.8×10^7	5	*203*
HO-1(slow)	0.31×10^6	Same	3.4×10^7				1.3	

the binding of CO over O_2 is favored by several orders of magnitude (*178–182*), whereas in myoglobin CO binds only ≈41 times more tightly (Table I). This discrimination can be ascribed to changes in relative on/off rates for the various ligands (Table I). HO is even more discriminating than myoglobin, primarily because of the slow off-rate of O_2. It was at one time thought that the main discriminating factor was steric, since CO prefers to form a linear Fe–C–O bond, whereas distal crowding forces the CO to bend or tilt. This view now has fallen into disfavor (*182*), although recent work has revived the steric argument with a new and detailed twist. A 1.15-Å X-ray structure of the CO–myoglobin complex shows a linear Fe–C–O bond with small but significant changes in active site geometry (*183*). These authors conclude that the required changes in structure are the main source of lowered affinity for CO. However, these authors fail to take into account the quantitative differences between myoglobin and model heme systems. Myoglobin binds CO about as well as protoheme (Table I). Indeed, protoheme in aqueous solution binds CO only about 16-fold more tightly than myoglobin and about the same as R-state hemoglobin (*179*). The main difference is in slower on-rates for myoglobin, which could simply be due to the requirement that CO must move through the protein first before binding to the heme iron rather than some energetically costly structural change. Hence, it is not the altered affinity for CO that requires an explanation, but rather the discrimination between CO and O_2.

Extensive myoglobin engineering work supports a major role for the distal His in stabilizing the oxy complex relative to the CO complex (*182*). O_2 is a more polar molecule than CO and hence will interact more favorably with neighboring polar groups such as the distal His. This explanation is supported by theoretical work showing that even if CO does tilt or bend, the energetic cost is minimal (*184*), while the oxy complex does form a stronger H-bond with the distal His than does CO (*185*).

It is not surprising that HO-1 developed an even greater level of discrimination between O_2 and CO, since CO is a product of the HO reaction. Therefore, unlike other hemoproteins, HO is constantly exposed to CO formed at the active site. Without a mechanism for ligand discrimination, the HO reaction would be significantly impaired by the competition between CO and O_2 for the heme iron. Unlike the globins, however, HO has no distal side chain that could directly H-bond with an iron-linked ligand. As illustrated in Fig. 20, however, a diatomic ligand could interact with the solvent in the distal polar pocket above the α-*meso* heme edge. Another possibility is an interior "escape" route for CO. A large internal chamber near the α-*meso* heme edge lined with aromatic and aliphatic groups could provide a pocket for CO and an eventual escape route to the external milieu (*163*).

IX. HO Catalytic Mechanism

A. Reduction of the Heme Iron

A prerequisite to forming the oxy complex is reduction of the HO heme iron to the ferrous (Fe^{2+}) state. This is achieved by electron transfer from P450 reductase. P450 reductase functions by utilizing reducing equivalents derived from NADPH and two flavins, FAD and FMN, to deliver electrons to the heme. If the pattern of HO reduction is similar to that of P450s, then the flow of electrons is NADPH→FAD→FMN→heme. Presumably HO has a docking site for interaction with the reductase. Such complexes are generally thought to form owing to complementary electrostatic interactions between molecular surfaces. Some hints as to what such a complex might look like derive from the first crystal structure of a P450 complexed with a redox partner: P450BM-3 complexed with the FMN domain of its reductase (*133*). Even though P450BM-3 is a bacterial enzyme, it resembles microsomal P450s in that it utilizes a diflavin P450 reductase. The important difference is that the reductase is covalently tethered to the C-terminal end of the P450 (*75*). The crystal structure of the electron transfer complex is of a recombinant version of P450BM-3 missing the FAD domain. As shown in Fig. 12, the FMN domain docks on the proximal surface of the heme, which is the closest approach of the heme to the molecular surface. In this case, the electropositive surface of the P450 interacts with the electronegative surface of the FMN domain. The crystal structure of microsomal P450 reductase (*132*) reveals that the FMN is even more electronegative than the FMN domain in P450BM-3, so one can expect the reductase to dock to an electropositive surface of HO. The HO-1 structure shows that electrostatic potential is quite asymmetric with a large positive electrostatic potential near the exposed heme edge (Fig. 17). Docking of the reductase on this surface would provide a relatively short route of electron transfer.

B. Oxy Complex

The visible absorption spectrum of oxy-HO-1 closely matches that of myoglobin indicating a very similar oxy complex. However, a detailed analysis of the resonance Raman data reveals subtle but important differences (*186*). In HO-1 the low-frequency Fe–O_2 vibrational mode is more widely coupled to other porphyrin modes and also exhibits a slightly larger isotope shift than in oxymyoglobin. These observations are consistent with a more highly bent Fe–O–O angle estimated to be $\approx 110^\circ$ (*186*). Bending of the oxygen causes the Fe–O_2 vibrational mode to oscillate perpendicular to the porphyrin ring, which accounts for the greater coupling observed with porphyrin modes. These effects are not due to the proximal His, since the His–Fe stretching frequencies in HO-1 and myoglobin are very similar (*186*). Therefore, it was concluded that bending is due to local steric factors in the distal pocket (*186*), which is consistent with the crystal structure. This has important mechanistic consequences since a greater bending of Fe–O–O will enable the terminal O atom to directly contact the α-*meso* heme edge.

C. Heme to α-*meso*-Hydroxyheme

Once the oxy complex is formed, a second electron transfer to the HO heme effectively reduces the oxy complex to the peroxide level. From this point many heme enzymes catalyze the heterolytic fission of the peroxide O–O bond, leaving behind the well known oxyferryl center, $(Fe–O)^{3+}$, characteristic of peroxidase compound I and similar to the active hydroxylating intermediate thought to operate in P450s. However, in HO the active oxidizing intermediate is peroxide. Peracids that form the $(Fe–O)^{3+}$ intermediate do not support the HO reaction, whereas H_2O_2 addition to Fe^{3+} HO does support substrate hydroxylation (*187, 188*). EPR and ENDOR spectroscopy have been used to analyze the cryogenically reduced oxy–HO complex (*189*). In these studies reduction of the oxy complex gave rise to the peroxy complex, whereas annealing of the peroxy-complex to 238 K for 1 min generates α-*meso*-hydroxyheme. These results are consistent with earlier views that the active intermediate is peroxide (*174, 175*). This also makes stereochemical sense. The terminal O atom in a bent Fe–O–O complex would be able to directly contact the correct *meso* carbons, whereas it is difficult to envision how an O atom directly linked to the iron could attack any *meso* carbon.

Reducing equivalents plus oxygen also can lead to nonenzymatic *meso* heme oxidation (*190*), so there is nothing unusual about the chemistry of the HO reaction. The main role of the enzyme, in addition to enhancing rates, is in controlling regioselectivity. The same factors that help in ligand discrimination also help in controlling regioselectivity. Tilting or bending of Fe–O–O toward the α-*meso* edge will favor polar interactions with O_2 in the open region of the distal cavity. In addition, the distal helix sterically restricts access to all but the α-*meso* carbon. However, the two molecules in the asymmetric unit show considerable heterogeneity in the distal helix and hence, heme access. Molecule B is in the "open" conformation with sufficient room for interactions between peroxide and the δ-*meso* carbon. The "closed" conformation in molecule A has access to only the α-*meso* carbon. If the closed conformation represents the catalytically active conformer, then steric control is central to ensuring that the α-*meso* carbon is hydroxylated, although the possibility remains open that other factors besides steric contribute.

Here recent mutagenesis work with cytochrome *b*5 (*191*) and myoglobin (*192*) is informative but also complicated. It has been possible to engineer both proteins to carry out HO-type heme hydroxylation with a significant level of regioselectivity. However, the basis for selectivity is not easily understood. In the case of cyt.*b*5 where the axial heme ligand, His63, was converted to Met (*191*), the preferred hydroxylation of the α-*meso* carbon is inconsistent with the close approach (3.4 Å) of Phe358 to the α-*meso* carbon in the wild-type structure. This indicates that a structural rearrangement or enhanced dynamics in the mutant opens up the α-*meso* position for hydroxylation. With myoglobin a series of distal mutants gives various levels of regioselectivity at different positions (*192*) with no clear pattern emerging. These authors attribute regioselectivity in the myoglobin mutants to polar, steric, and other electronic factors. A further complication arises when one examines the metabolism of modified hemes by HO-1. *meso*-Methylmesohemes are hydroxylated by HO-1 exclusively at the *meso* carbon to which the methyl group is attached (*193*). These results support electronic control, since the methyl group is an electron-donating moiety and will favor electrophilic attack of the *meso*-methyl carbon on the iron-linked peroxide. There could, of course, be structural differences with the modified hemes that open up sites to hydroxylation that would otherwise be blocked. Since no consistent picture emerges from these various studies, it appears that there is more than one way to control regioselectivity, with a major unknown factor being the dynamics and conformational changes in the engineered and synthetic heme systems. Fortunately, the situation with HO-1 is becoming clearer. Although the precise role of electronic control still remains uncertain, it is clear that the distal helix exercises a significant level of steric control and that polar interactions between Fe–O–O and the surrounding environment may assist in steering the peroxide toward the α-*meso* carbon.

D. α-*meso*-Hydroxyheme to Verdoheme

Addition of O_2 to the α-*meso*-hydroxyheme HO-1 generates an EPR signal characteristic of an organic radical (*194*). In the presence of CO, the EPR signal increases (*195*). This indicates that the Fe^{3+} form of α-*meso*-hydroxyheme is in equilibrium between resonance structures, as indicated in Fig. 21, and that CO traps the Fe^{2+}/heme radical form, which accounts for the increase in EPR signal in the presence of CO. This much is generally agreed upon, but a point of controversy exists over the requirement of reducing equivalents to produce verdoheme. Liu *et al.* (*195*) generated α-*meso*-hydroxyheme by the addition of peroxide to the HO-1 heme complex under anaerobic conditions. The introduction of O_2 gives verdoheme, indicating that reducing equivalents are not required to produce verdoheme. Matera *et al.* (*194*) generated α-*meso*-hydroxyheme by reconstituting synthetically generated α-*meso*-hydroxyheme with apo-HO-1. These authors found that the addition of electrons to Fe^{3+} α-*meso*-hydroxyheme HO-1 or the addition of O_2 to Fe^{2+} α-*meso*-hydroxyheme gives verdoheme. Moreover, the yield of verdoheme was stoichiometric with respect to reducing equivalents with 1.06 verdoheme formed per 1.0 electron added as dithionite to Fe^{3+} α-*meso*-hydroxyheme HO-1. These results argue for the requirement of one reducing equivalent to generate verdoheme. It is difficult to reconcile these quite different conclusions. In one case α-*meso*-hydroxyheme was generated *in situ* from the heme HO-1 complex by addition of peroxide (*195*), whereas in the other HO-1 was reconstituted with Fe^{3+} α-*meso*-hydroxyheme HO-1. The anaerobic titration of Fe^{3+} α-*meso*-hydroxyheme HO-1 with dithionite to give verdoheme was carried out under CO in order to stop the reaction at the verdoheme stage by

1. Heme to α-*meso*-hydroxyheme

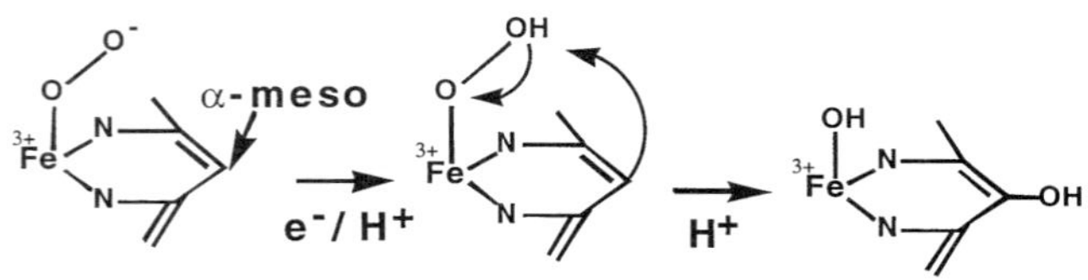

α-meso

e^-/H^+

H^+

2. α-*meso*-hydroxyheme to verdoheme

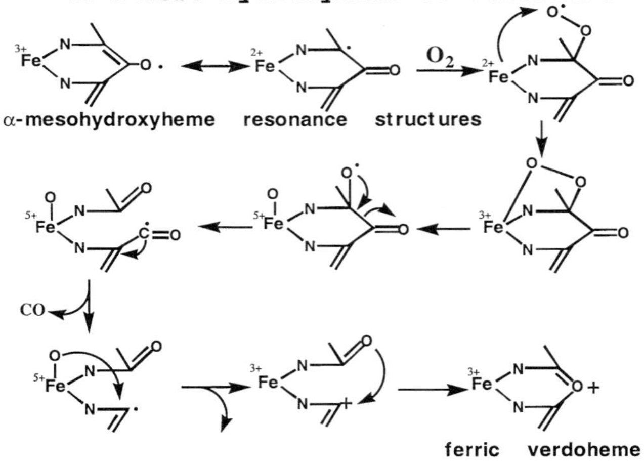

α-mesohydroxyheme resonance structures

O_2

CO

ferric verdoheme

3. Verdoheme to biliverdin

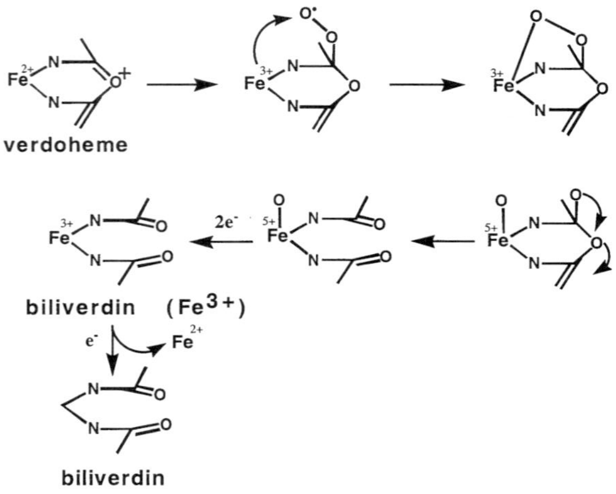

verdoheme

$2e^-$

biliverdin (Fe^{3+})

e^-

Fe^{2+}

biliverdin

complexing the iron with CO (*194*). It has been argued that the presence of CO in these experiments may actually have trapped the system at the Fe^{2+}/radical stage in Fig. 21, thereby requiring an additional electron to "restart" the reaction (*195*). More recently, Sakamoto *et al.* (*196*) found that titration of reduced HO-1 reconstituted with α-hydroxyheme with O_2 gives verdoheme without the need for a second electron. A similar experiment has been carried out by Migita *et al.* (*197*), but in this case it was concluded that the addition of oxygen leads to porphyrin oxidation and not catalytically relevant oxygenation of α-hydroxyheme to give verdoheme. One obvious difference in the two experiments is that Sakamoto *et al.* (*196*) titrated with O_2 to generate a series of spectra showing a decrease in the main Soret band and an increase in bands at 640 and 685 nm, consistent with partial formation of the CO–ferrous verdoheme complex with the CO having been generated by oxidation of α-hydroxyheme. In contrast, Migita *et al.* (*197*) exposed the anaerobic reduced α-hydroxyheme to atmospheric O_2 and analyzed the resulting products. Hence, there remains some discrepancy on the requirement of an additional electron in the conversion of α-hydroxyheme to verdoheme. Nevertheless, the most detailed proposal is from Ortiz de Montellano and colleagues (*176*) as outlined in Fig. 21, which assumes that no reducing equivalents are required for verdoheme formation.

E. Verdoheme to Biliverdin

This step in the reaction cycle requires one O_2 molecule and two electrons to give Fe^{3+} biliverdin with a third electron needed to reduce Fe^{3+} to Fe^{2+} followed by release of the iron from biliverdin (*176, 188, 198, 199*). A hydrolytic mechanism for the incorporation of oxygen in the conversion of verdoheme to biliverdin is ruled out because ^{18}O labeling experiments show incorporation of ^{18}O into biliverdin (*200*). Details on the mechanism remain sketchy, and one possible pathway is illustrated in Fig. 21, taken after Ref. (*176*).

Fig. 21. Detailed mechanism for HO-1 catalysis. In **1,** oxygenation and electron transfer forms the ferric (Fe^{3+})–peroxy complex. Steric factors and H-bonding help to bend the peroxide toward the α-*meso*-heme position for regio-selective hydroxylation. One proposed mode of forming verdoheme is shown in part **2.** A key part of step **2** is the resonance structures between Fe^{3+} and Fe^{2+}/radical, which enable the porphyrin ring to be oxygenated. Although the mechanism shown does not require any reducing equivalents (*176*), there remain experimental inconsistencies on the requirement of an additional electron in step **2.** However, reduction of the verdoheme iron is necessary to prepare the substrate for step **3,** verdoheme to biliverdin.

X. Outlook

Various isoforms of both HO and NOS can be expressed in recombinant systems. As a result, the immediate future will undoubtedly witness a wealth of mutagenesis experiments guided by the crystal structures. It also may be possible to trap in crystalline form the various intermediates of the HO reaction cycle, which will greatly facilitate a deeper understanding of the catalytic mechanism. Conformational dynamics appear to be quite important in HO, and hence, a variety of spectral probes such as NMR and fluorescence should prove especially useful in studying the role of protein dynamics in function. Overall there should be considerable optimism for understanding HO at the level of detail achieved for peroxidases and other well-studied enzyme systems.

NOS, however, is much more complicated. Of central importance is the role of H_4B. Significant steps have been made in this area, with the weight of the current data favoring a redox role for the pterin in oxygen activation. NOS also faces limitations that have plagued the P450 field. Unlike peroxidases, neither NOS nor P450 forms stable intermediates similar to compound I in peroxidases, making identification of critical intermediates quite difficult. One possible approach is to find suitable mutants or substrates that alter the rate-limiting step in order to allow important intermediates to build up for sufficient periods of time to be analyzed using spectral methods. In P450s it has been possible, after some extensive work and searching, to find a mutant that has an altered rate-limiting step that enables a potentially new intermediate to build up (*201*). A similar approach might work with NOS. If the substrate itself, L-Arg, serves as the proton donor in order to achieve O–O bond cleavage, then it might be possible to design substrates with altered pK_as to test this hypothesis.

With respect to understanding the fundamental energetic properties that control substrate and inhibitor recognition and binding to NOS, the future looks bright. Since it appears that the structure does not undergo large adjustments in response to inhibitor binding, it should be possible to couple many NOS–inhibitor crystal structures with thermodynamic binding data to develop a detailed picture of NOS–inhibitor interactions. NOS, too, provides a near ideal system for understanding inhibitor selectivity, since all three NOS isoform heme domain structures will soon be available, and despite the close similarity in active site architecture, it should be possible to understand the basis for selectivity.

What we now know about NOS structure is only half the picture. The complex control of NOS activity involving Ca^{2+}/calmodulin will require the holo-NOS structure. This also addresses a problem in common with many heme enzymes: the nature of intermolecular redox complexes. Both NOS and HO utilize very similar P450 reductase-like proteins to deliver electrons, and both HO and NOS have positive electrostatic patches proposed to serve as the docking site. We know very little about the detailed structure of such complexes. It does appear, however, that the electron transfer field and structural biology in general are moving toward determining the structures of larger intermolecular complexes. The structure of complexes such as the holo-NOS or HO/P450 complexed with P450 reductase is an area where X-ray crystallography is likely to have a very significant impact in, we may hope, the not-too-distant future.

References

1. Ignarro, L. J.; Byrns, R. E.; Wood, K. S. In "Vasodilatation: Vascular Smooth Muscle, Peptides, Autonomic Nerves, and Endothelium"; Anhoutte, P. M., Ed.; Raven Press: New York, 1988; p. 427.
2. Furchgott, R. F. In "Vasodilatation: Vascular Smooth Muscle, Peptides, Autonomic Nerves, and Endothelium"; Anhoutte, P. M., Ed.; Raven Press: New York, 1988; p. 401.
3. Bredt, D. S.; Snyder, S. H. *Annu. Rev. Biochem.* **1994,** *63,* 175.
4. Griffith, O. W.; Stuehr, D. J. *Annu. Rev. Physiol.* **1995,** *57,* 707.
5. Gross, S. G.; Wolin, M. S. *Annu. Rev. Physiol.* **1995,** *57,* 737.
6. Kilbourn, R. G.; Griffith, O. W. *J. Natl. Cancer Inst.* **1992,** *84,* 1671.
7. Patel, V. C.; Yellon, D. M.; Singh, K. J.; Neild, G. H.; Woolfson, R. G. *Biochem. Biophys. Res. Commun.* **1993,** *194,* 234.
8. Maines, M. D. *Annu. Rev. Toxicol.* **1997,** *37,* 517.
9. Hobbs, A. J. *Trends Pharmacol. Sci.* **1997,** *18,* 484.
10. Kamisaki, Y.; Saheki, S.; Nakane, M.; Palmieri, J. A.; Kuno, T.; Chang, B. Y.; Waldman, S. A.; Murad, F. *J. Biol. Chem.* **1986,** *261,* 7236.
11. Humbert, P.; Niroomand, F.; Fischer, G.; Mayer, B.; Koesling, D.; Hinsch, K., D.; Gausepohl, H.; Frank, R.; Schultz, G.; Bohme, E. *Eur. J. Biochem.* **1990,** *190,* 273.
12. Gerzer, R.; Hofmann, F.; Schultz, G. *Eur. J. Biochem.* **1981,** *116,* 479.
13. Brandish, P. E.; Buechler, W.; Marletta, M. A. *Biochemistry* **1998,** *37,* 16898.
14. Zhao, Y.; Schelvis, J. P.; Babcock, G. T.; Marletta, M. A. *Biochemistry* **1998,** *37,* 4502.
15. Ignarro, L. J.; Degnan, J. N.; Baricos, W. H.; Kadowitz, P. J.; Wolin, M. S. *Biochim. Biophys. Acta* **1982,** *718,* 49.
16. Schedit, W. R.; Ellison, M. K. *Acc. Chem. Res.* **1999,** *32,* 350.
17. Zhao, Y.; Hoganson, C.; Babcock, G. T.; Marletta, M. A. *Biochemistry* **1998,** *37,* 12458.
18. Dierks, E. A.; Hu, S. Z.; Vogel, K. M.; Yu, A. E.; Spiro, T. G.; Burstyn, J. N. *J. Am. Chem. Soc.* **1997,** *119,* 7316.
19. Stone, J. R.; Marletta, M. A. *Chem. Biol.* **1998,** *5,* 255.
20. Venema, R. C.; Ju, H.; Zou, R.; Ryan, J. W.; Venema, V. J. *J. Biol. Chem.* **1997,** *272,* 1276.
21. Ingi, T.; Ronett, G. V. *J. Neurosci.* **1995,** *15,* 8214.

22. Mulsch, A.; Bauersachs, J.; Schafer, A.; Stasch, J. P.; Kast, R.; Busse, R. *Br. J. Pharmacol.* **1997,** *120,* 681.
23. Wu, C. C.; Ko, F. N.; Kuo, S. C.; Lee, F. Y.; Teng, C. M. *Br. J. Pharmacol.* **1995,** *116,* 1973.
24. Ko, F. N.; Wu, C. C.; Kuo, S. C.; Lee, F. Y.; Teng, C. M. *Blood* **1994,** *84,* 4226.
25. Friebe, A.; Schultz, G.; Koesling, D. *EMBO J.* **1996,** *15,* 6863.
26. Huie, R. E.; Padmaja, S. *Free Radical Res. Commun.* **1993,** *18,* 195.
27. Beckman, J. S.; Beckman, T. W.; Chen, J.; Marshall, P. A.; Freeman, B. A. *Proc. Natl. Acad. Sci. USA* **1990,** *87,* 1620.
28. Fukuto, J. M.; Ignarro, L. J. *Acc. Chem. Res.* **1997,** *30,* 149.
29. Dolphin, D.; Forman, A.; Borg, D. C.; Fajer, J.; Felton, R. H. *Proc. Natl. Acad. Sci. USA* **1971,** *68,* 614.
30. Coulson, A. F.; Yonetani, T. *Biochem. Biophys. Res. Commun.* **1972,** *49,* 391.
31. Sivaraja, M.; Goodin, D.; Smith, M.; Hoffman, B. *Science* **1989,** *345,* 738.
32. Dunford, H. B. In "Horseradish Peroxidase: Structure and Kinetic Properties"; Everse, J., Everse, K. E., Grisham, M. B., Eds.; CRC Press: Boca Raton, FL, 1991; Vol. II; p. 1.
33. Rutter, R.; Valentine, M.; Hendrich, M. P.; Hager, L. P.; Debrunner, P. G. *Biochemistry* **1983,** *22,* 4769.
34. Chance, B.; Powers, L.; Ching, Y.; Poulos, T.; Schonbaum, G. R.; Yamazaki, I.; Paul, K. G. *Arch. Biochem. Biophys.* **1984,** *235,* 596.
35. Edwards, S. L.; Nguyen, H. X.; Hamlin, R. C.; Kraut, J. *Biochemistry* **1987,** *26,* 1503.
36. Fulop, V.; Phizackerley, R. P.; Soltis, S. M.; Clifton, I. J.; Wakatuski, S.; Erman, J.; Hajdu, J.; Edwards, S. L. *Structure* **1994,** *2,* 201.
37. Gouet, P.; Jouve, H. M.; Williams, P. A.; Andersson, I.; Andreoletti, P.; Nussaume, L.; Hajdu, J. *Nature Struct. Biol.* **1996,** *3,* 951.
38. Schonbaum, G. R.; Lo, S. *J. Biol. Chem.* **1972,** *247,* 3353.
39. Poulos, T. L.; Kraut, J. *J. Biol. Chem.* **1980,** *255,* 8199.
40. Poulos, T. L.; Finzel, B. C. In "Peptide and Protein Reviews"; Mearn, M. T., Ed.; Marcel Dekker: New York, 1984; Vol. 4; p. 115.
41. Erman, J. E.; Vitello, L. B.; Miller, M. A.; Shaw, A.; Brown, K. A.; Kraut, J. *Biochemistry* **1993,** *32,* 9798.
42. Rodriguez-Lopez, J. N.; Smith, A. T.; Thorneley, R. N. *J. Biol. Inorg. Chem.* **1996,** *1,* 136.
43. Hamilton, G. In "Chemical Models and Mechanisms for Oxygenases"; Hayaishi, O., ed.; Academic Press: New York, 1974; p. 405.
44. Ortiz de Montellano, P. R. *Annu. Rev. Toxicol.* **1992,** *32,* 89.
45. Itakura, H.; Oda, Y.; Fukuyama, K. *FEBS Lett.* **1997,** *412,* 107.
46. Henriksen, A.; Schuller, D. J.; Meno, K.; Welinder, K. G.; Smith, A. T.; Gajhede, M. *Biochemistry* **1998,** *37,* 8054.
47. Veitch, N. C.; Gao, Y.; Smith, A. T.; White, C. G. *Biochemistry* **1997,** *36,* 14751.
48. Doyle, W. A.; Blodig, W.; Veitch, N. C.; Piontek, K.; Smith, A. T. *Biochemistry* **1998,** *37,* 15097.
49. Mandelman, D.; Jamal, J.; Poulos, T. L. *Biochemistry*, *37,* 17610.
50. Poulos, T. L.; Finzel, B. C.; Howard, A. J. *J. Mol. Biol.* **1987,** *195,* 687.
51. Raag, R.; Martinis, S. A.; Sligar, S. G.; Poulos, T. L. *Biochemistry* **1991,** *30,* 11420.
52. Gerber, N. C.; Sligar, S. G. *J. Biol. Chem.* **1994,** *269,* 4260.

53. Park, S. Y.; Shimizu, H.; Adachi, S.; Nakagawa, A.; Tanaka, I.; Nakahara, K.; Shoun, H.; Obayashi, E.; Nakamura, H.; Iizuka, T.; Shiro, Y. *Nature Struct. Biol.* **1997,** *4,* 827.
54. Vidakovic, M.; Sligar, S. G.; Li, H.; Poulos, T. L. *Biochemistry* **1998,** *37,* 9211.
55. Cupp-Vickery, J. R.; Han, O.; Hutchinson, C. R.; Poulos, T. L. *Nature struct. Biol.* **1996,** *3,* 632.
56. Morris, D. R.; Hager, L. P. *J. Biol. Chem.* **1966,** *241,* 1763.
57. Hollenberg, P. F.; Hager, L. P. *J. Biol. Chem.* **1973,** *248,* 2630.
58. Sono, M.; Dawson, J. H.; Hall, K.; Hager, L. P. *Biochemistry* **1986,** *25,* 347.
59. Sundaramoorthy, M.; Terner, J.; Poulos, T. L. *Structure* **1995,** *3,* 1367.
60. Blanke, S. R.; Hager, L. P. *J. Biol. Chem.* **1988,** *263,* 18739.
61. Blanke, S. R.; Hager, L. P. *J. Biol. Chem.* **1990,** *265,* 12454.
62. Dugad, L. B.; Wang, X.; Wang, C.-C.; Lukat, G. S.; Goff, H. M. *Biochemistry* **1992,** *31,* 1651.
63. Sundaramoorthy, M.; Terner, J.; Poulos, T. L. *Chem. Biol.* **1998,** *5,* 461.
64. Wagenknecht, H. A.; Woggon, W. D. *Chem. Biol.* **1997,** *4,* 367.
65. Bredt, D. S.; Hwang, P. M.; Glatt, C. E.; Lowenstein, C.; Reed, R. R.; Snyder, S. H. *Nature* **1991,** *351,* 714.
66. Pollock, J. S.; Forstermann, U.; Mitchell, J. A.; Warner, T. D.; Schmidt, H. H.; Nakane, M.; Murad, F. *Proc. Natl. Acad. Sci. USA* **1991,** *88,* 10480.
67. Shaul, P. W.; Smart, E. J.; Robinson, L. J.; German, Z.; Yuhanna, I. S.; Ying, Y.; Anderson, R. G.; Michel, T. *J. Biol. Chem.* **1996,** *271,* 6518.
68. García-Cardeña, G.; Martásek, P.; Masters, B. S. S.; Skidd, P. M.; Couet, J.; Li, S.; Lisanti, M. P.; W.C., S. *J. Biol. Chem.* **1997,** *272,* 25437.
69. Ju, H.; Zou, R.; Venema, V. J.; Venema, R. C. *J. Biol. Chem.* **1997,** *272,*.
70. Brenman, J. E.; Chao, D. S.; Gee, S. H.; McGee, A. W.; Craven, S. E.; Santillano, D. R.; Wu, Z.; Huang, F.; Xia, H.; Peters, M. F.; Froehner, S. C.; Bredt, D. S. *Cell* **1996,** *84,* 757.
71. Hillier, B. J.; Christopherson, K. S.; Prehoda, K. E.; Bredt, D. S.; Lim, W. A. *Science* **1999,** *284,* 812.
72. McMillan, K.; Bredt, D. S.; Hirsch, D. J.; Snyder, S. H.; Clark, J. E.; Masters, B. S. *Proc. Natl. Acad. Sci. USA* **1992,** *89,* 11141.
73. Stuehr, D. J.; Ikeda-Saito, M. *J. Biol. Chem.* **1992,** *267,* 20547.
74. White, K. A.; Marletta, M. A. *Biochemistry* **1992,** *31,* 6627.
75. Ruettinger, R. T.; Wen, L. P.; Fulco, A. J. *J. Biol. Chem.* **1989,** *264,* 10987.
76. Abu-Soud, H. M.; Stuehr, D. J. *Proc. Natl. Acad. Sci. USA* **1993,** *90,* 10769.
77. Cho, H. J.; Xie, Q. W.; Calaycay, J.; Mumford, R. A.; Swiderek, K. M.; Lee, T. D.; Nathan, C. *J. Exp. Med.* **1992,** *176,* 599.
78. Crane, B. R.; Arvai, A. S.; Gachhui, R.; Wu, C.; Ghosh, D. K.; Getzoff, E. D.; Stuehr, D. J.; Tainer, J. A. *Science* **1997,** *278,* 425.
79. Ghosh, D. K.; Wu, C.; Pitters, E.; Moloney, M.; Werner, E. R.; Mayer, B.; Stuehr, D. J. *Biochemistry* **1997,** *36,* 10609.
80. Crane, B. R.; Arvai, A. S.; Ghosh, D. K.; Wu, C.; Getzoff, E. D.; Stuehr, D. J.; Tainer, J. A. *Science* **1998,** *279,* 2121.
81. Raman, C. S.; Li, H.; Martasek, P.; Kral, V.; Masters, B. S. S.; Poulos, T. L. *Cell* **1998,** *95,* 939.
82. Fischmann, T. O.; Hruza, A.; Niu, X. D.; Fossetta, J. D.; Lunn, C. A.; Dolphin, E.; Prongay, A. J.; Reichert, P.; Lundell, D. J.; Narula, S. K.; Weber, P. C. *Nature Struct. Biol.* **1999,** *6,* 233.

83. Li, H.; Raman, C. S.; Glaser, C. B.; Blasko, E. B.; Young, T. A.; Parkinson, J. F.; Whitlow, M.; Poulos, T. L. *J. Biol. Chem.* **1999**.

84. Ghosh, D. K.; Abu-Soud, H. M.; Stuehr, D. J. *Biochemistry* **1996,** *35,* 1444.

85. Mayer, B.; Wu, C.; Gorren, A. C.; Pfeiffer, S.; Schmidt, K.; Clark, P.; Stuehr, D. J.; Werner, E. R. *Biochemistry* **1997,** *36,* 8422.

86. Crane, B. C.; Rosenfeld, R. J.; Arvai, A. S.; Ghosh, D. K.; Ghosh, S.; Tainer, J. A.; Stuehr, D. J.; Getzoff, E. D. *EMBO J.* **1999,** *18,* 6271.

87. Miller, R. T.; Martasek, P.; Raman, C. S.; Masters, B. S. *J. Biol. Chem.* **1999,** *274,* 14537.

88. Chen, P. F.; Tsai, A. L.; Wu, K. K. *Biochem. Biophys. Res. Commun.* **1995,** *215,* 1119.

89. Miller, R. T.; Martásek, P.; Roman, L. J.; Nishimura, J. S.; Masters, B. S. *Biochemistry* **1997,** *36,* 15277.

90. Rodríguez-Crespo, I.; Moënne-Loccoz, P.; Loehr, T. M.; Ortiz de Montellano, P. R. *Biochemistry* **1997,** *36,* 8530.

91. Martasek, P.; Miller, R. T.; Liu, Q.; Roman, L. J.; Salerno, J. C.; Migita, C. T.; Raman, C. S.; Gross, S. S.; Ikeda-Saito, M.; Masters, B. S. S. *J. Biol. Chem.* **1998,** *273,* 34799.

92. Ghosh, D. K.; Crane, B. R.; Ghosh, S.; Wolan, D.; Gachhui, R.; Crooks, C.; Presta, A.; Tainer, J. A.; Getzoff, E. D.; Stuehr, D. J. *EMBO J.* **1999,** *18,* 6260.

93. Adman, E.; Watenpaugh, K. D.; Jensen, L. H. *Proc. Nat. Acad. Sci. USA* **1975,** *72,* 4854.

94. Georgiadis, M. M.; Komiya, H.; Chakrabarti, P.; Woo, D.; Kornuc, J. J.; Rees, D. C. *Science* **1992,** *257,* 1653.

95. Richardson, J. S.; Richardson, D. C. *Science* **1988,** *240,* 1648.

96. Presta, L. G.; Rose, G. D. *Science* **1988,** *240,* 1632.

97. Ueyama, N.; Terakawa, T.; Nakata, M.; Nakamura, A. *J. Am. Chem. Soc.* **1983,** *105,* 7098.

98. Ueyama, N.; Okamura, T.-A.; Nakamura, A. *J. Am. Chem. Soc.* **1992,** *114,* 8129.

99. Choudhury, K.; Sundaramoorthy, M.; Hickman, A.; Yonetani, T.; Woehl, E.; Dunn, M. F.; Poulos, T. L. *J. Biol. Chem.* **1994,** *269,* 20239.

100. Goodin, D. B.; McRee, D. E. *Biochemistry* **1993,** *32,* 3313.

101. Langen, R.; Jensen, G. M.; Jacob, U.; Stephens, P. J.; Warshel, A. *J. Biol. Chem.* **1992,** *267,* 25625.

102. Adak, S.; Crooks, C.; Wang, Q.; Crane, B. R.; Tainer, J. A.; Getzoff, E. D.; Stuehr, D. J. *J. Biol. Chem.* **1999,** *274,* 26907.

103. Poulos, T. L.; Cupp-Vickery, J.; Li, H. In "Structural Studies on Prokaryotic Cytochromes P450"; Ortiz de Montellano, P., Ed.; Plenum: New York and London, 1995; p. 125.

104. Poulos, T. L.; Finzel, B. C.; Howard, A. J. *Biochemistry* **1986,** *25,* 5314.

105. Raag, R.; Li, H.; Jones, B. C.; Poulos, T. L. *Biochemistry* **1993,** *32,* 4571.

106. Li, H.; Poulos, T. L. *Nature Struct. Biol.* **1997,** *4,* 140.

107. Klatt, P.; Schmid, M.; Leopold, E.; Schmidt, K.; Werner, E. R.; Mayer, B. *J. Biol. Chem.* **1994,** *269,* 13861.

108. Grant, S. K.; Green, B. G.; Wilusz, J. S.; Durette, P. L.; Shah, S. K.; Kozarcih, J. W. *Biochemistry* **1998,** *37,* 4174.

109. Hanson, L. K.; Eaton, W. A.; Sligar, S. G.; Gunsalus, I. C.; Gouterman, M.; Connel, C. R. *J. Am. Chem. Soc.* **1976,** *98,* 2672.

110. Huang, L.; Abu-Soud, H. M.; Hille, R.; Stuehr, D. J. *Biochemistry* **1999,** *38,* 1912.

111. Goodwill, K. E.; Sabatier, C.; Stevens, R. C. *Biochemistry* **1998,** *37,* 13437.

112. Auerbach, G.; Herrmann, A.; Gutlich, M.; Fischer, M.; Jacob, U.; Bacher, A.; Huber, R. *EMBO J.* **1997,** *16,* 7219.
113. Birdsall, D. L.; Finer-Moore, J.; Stroud, R. M. *J. Mol. Biol.* **1996,** *255,* 522.
114. Chan, M. K.; Mukund, S.; Kletzin, A.; Adams, M. W.; Rees, D. C. *Science* **1995,** *267,* 1463.
115. Hennig, M.; D'Arcy, A.; Hampele, I. C.; Page, M. G.; Oefner, C.; Dale, G. E. *Nature Struct. Biol.* **1998,** *5,* 357.
116. McTigue, M. A.; Davies, J. F. D.; Kaufman, B. T.; Kraut, J. *Biochemistry* **1992,** *31,* 7264.
117. Bec, N.; Gorren, A. C.; Voelker, C.; Mayer, B.; Lange, R. *J. Biol. Chem.* **1998,** *273,* 13502.
118. Witteveen, C. F. B.; Giovanelli, J.; Kaufman, S. *J. Biol. Chem.* **1999,** *274,* 29755.
119. Hurshman, A. R.; Krebs, C.; Edmondson, D. E.; Huynh, B. H.; Marletta, M. A. *Biochemistry* **1999**.
120. Bonagura, C. A.; Sundaramoorthy, M.; Pappa, H. S.; Patterson, W. R.; Poulos, T. L. *Biochemistry* **1996,** *35,* 6107.
121. Luykx, D. M.; Duine, J. A.; de Vries, S. *Biochemistry* **1998,** *37,* 11366.
122. Kappock, T. J.; Caradonna, J. P. *Chem. Rev.* **1996,** *96,* 2659.
123. Rusche, K. M.; Spiering, M. M.; Marletta, M. A. *Biochemistry* **1998,** *37,* 15503.
124. Presta, A.; Siddhanta, U.; Wu, C.; Sennequier, N.; Huang, L.; Abu-Soud, H. M.; Erzurum, S.; Stuehr, D. J. *Biochemistry* **1998,** *37,* 298.
125. Abu-Soud, H.; Gachhui, R.; Raushel, F.; Stuehr, D. *J. Biol. Chem.* **1997,** *272,* 17349.
126. Pufahl, R. A.; Wishnok, J. S.; Marletta, M. A. *Biochemistry* **1995,** *34,* 1930.
127. Hobbs, D. J.; Higgs, A.; Moncada, S. *Annu. Rev. Pharmacol. Toxicol.* **1999,** *39,* 191.
128. Hallinan, E. A.; Tsymbalov, S.; Finnegan, P. M.; Moore, W. M.; Jerome, G. M.; Currie, M. G.; Pitzele, B. S. *J. Medicinal Chem.* **1998,** *41,* 775.
129. Garvey, E. P.; Oplinger, J. A.; Tanoury, G. J.; Sherman, P. A.; Fowler, M.; Marshall, S.; Harmon, M. F.; Paith, J. E.; Furfine, E. S. *J. Biol. Chem.* **1994,** *269,* 26669.
130. Brunner, K.; Tortschanoff, A.; Hemmens, B.; Andrew, P. J.; Mayer, B.; Kungl, A. J. *J. Biol. Chem.* **1998,** *37,* 17545.
131. Adak, S.; Ghosh, S.; Abu-Soud, H. M.; Stuehr, D. J. *J. Biol. Chem.* **1999,** *274,* 22313.
132. Wang, M.; Roberts, D. L.; Paschke, R.; Shea, T. M.; Masters, B. S.; Kim, J. J. *Proc. Nat. Acad. Sci. USA* **1997,** *94,* 8411.
133. Sevrioukova, I. F.; Li, H.; Zhang, H.; Peterson, J. A.; Poulos, T. L. *Proc. Natl. Acad. Sci. USA* **1999,** *96,* 1863.
134. Gachhui, R.; Presta, A.; Bentley, D. F.; Abu-Soud, H. M.; McArthur, R.; Brudvig, G.; Ghosh, D. K.; Stuehr, D. J. *J. Biol. Chem.* **1996,** *271,* 20594.
135. Govindaraj, S.; Poulos, T. L. *Biochemistry* **1995,** *34,* 11221.
136. Govindaraj, S.; Poulos, T. L. *Prot. Sci.* **1996,** *5,* 1389.
137. Sharp, R. E.; White, P.; Chapman, S. K.; Reid, G. A. *Biochemistry* **1994,** *33,* 5115.
138. Nishida, C. R.; Ortiz de Montellano, P. R. *J. Biol. Chem.* **1998,** *273,* 5566.
139. Siddhantha, U.; Presta, A.; Baochen, F.; Wolan, D.; Rousseau, D.; Steuhr, D. *J. Biol. Chem.* **1998,** *273,* 18950.
140. Perry, J.; Moon, N.; Zhao, Y.; Dunham, W.; Marletta, M. *Chem. Biol.* **1998,** *5,* 355.
141. Perry, J.; Marletta, M. *Proc. Natl. Acad. Sci. USA* **1998,** *95,* 1101.
142. Du, P.; Loew, G. H. *Intl. J. Quantum Chem.* **1992,** *44,* 251.
143. Rietjens, I. M. C. M.; Osman, A. M.; Veeger, C.; Zakharieva, O.; Antony, J.; Grodzicki, M.; Trautwein, A. X. *J. Biol. Inorg. Chem.* **1996,** *1,* 372.
144. Witteveen, C. F.; Giovanelli, J.; Kaufman, S. *J. Biol. Chem.* **1996,** *271,* 4143.

145. Sundaramoorthy, M.; Kishi, K.; Gold, M. H.; Poulos, T. L. *J. Biol. Chem.* **1994,** *269,* 32759.
146. Abu-Soud, H.; Presta, A.; Mayer, B.; Stuehr, D. *Biochemistry* **1998,** *36,* 10811.
147. Marletta, M. A. *J. Biol. Chem.* **1993,** *268,* 12231.
148. Vaz, A. D. N.; Roberts, E. S.; Coon, M. J. *J. Am. Chem. Soc.* **1991,** *113,* 5886.
149. Cole, P. A.; Robinson, C. H. *J. Am. Chem. Soc.* **1991,** *113,* 8130.
150. Renodon-Corniere, A.; Boucher, J.-L.; Dijols, S.; Stuehr, D. J.; Mansuy, D. *Biochemistry* **1999,** *38,* 4663.
151. Yoshida, T.; Biro, P.; Cohen, T.; Mueller, R. M.; Shibahara, S. *Eur. J. Biochem.* **1988,** *171,* 457.
152. Rotenberg, M. O.; Maines, M. D. *J. Biol. Chem.* **1990,** *265,* 7501.
153. Yoshida, T.; Sato, M. *Biochem. Biophys. Res. Comm.* **1989,** *163,* 1086.
154. Yoshida, T.; Ishikawa, K.; Sato, M. *Eur. J. Biochem.* **1991,** *199,* 729.
155. Schmitt, M. P. *J. Bacteriol.* **1997,** *179,* 838.
156. Yachie, A.; Niida, Y.; Wada, T.; Igarashi, N.; Kaneda, H.; Toma, T.; Ohta, K.; Kasahara, Y.; Koizumi, S. *J. Clin. Invest.* **1998,** *103,* 129.
157. Poss, K. D.; Tonegawa, S. *Proc. Natl. Acad. Sci. USA* **1997,** *94,* 10919.
158. Hancock, W. W.; Buelow, R.; Sayegh, M. H.; Turka, L. A. *Nature Med.* **1998,** *4,* 1392.
159. Soares, M. P.; Lin, Y.; Anrather, J.; Csizmadia, E.; Takigami, K.; Sato, K.; Grey, S. T.; Colvin, R. B.; Choi, A. M.; Poss, K. D.; Bach, F. H. *Nature Med.* **1998,** *4,* 1073.
160. Zakhary, R.; Gaine, S. P.; Dinerman, J. L.; Ruat, M.; Flavahan, N. A.; Snyder, S. H. *Proc. Natl. Acad. Sci. USA* **1996,** *93,* 795.
161. Zakhary, R.; Poss, K. D.; Jaffrey, S. R.; Ferris, C. D.; Tonegawa, S.; Snyder, S. H. *Proc. Natl. Acad. Sci. USA* **1997,** *94,* 14848.
162. Zhuo, M.; Small, S. A.; Kandel, E. R.; Hawkins, R. D. *Science* **1993,** *260,* 1946.
163. Schuller, D. J.; Wilks, A.; Ortiz de Montellano, P. R.; Poulos, T. L. *Nature Struct. Biol.* **1999,** *6,* 860.
164. Schuller, D. J.; Wilks, A.; Ortiz de Montellano, P. R.; Poulos, T. L. *Protein Sci.* **1998,** *7,* 1836.
165. Wilks, A.; Medzihradszky, K. F.; Ortiz de Montellano, P. R. *Biochemistry* **1998,** *37,* 2889.
166. Ishikawa, K.; Sato, M.; Ito, M.; Yoshida, T. *Biochem. Biophys. Res. Commun.* **1992,** *182,* 981.
167. Sun, J.; Loehr, T. M.; Wilks, A.; Ortiz de Montellano, P. R. *Biochemistry* **1994,** *33,* 13734.
168. Valentine, J. S.; Sheridan, R. P.; Allen, L. C.; Kahn, P. C. *Proc. Nat. Acad. Sci. USA* **1979,** *76,* 1009.
169. Sun, J.; Wilks, A.; Ortiz de Montellano, P. R.; Loehr, T. M. *Biochemistry* **1993,** *32,* 14151.
170. Wilks, A.; Sun, J.; Loehr, T. M.; Ortiz de Montellano, P. R. *J. Am. Chem. Soc.* **1995,** *117,* 2925.
171. Hernandez, G.; Wilks, A.; Paolesse, R.; Smith, K. M.; Ortiz de Montellano, P. R.; La Mar, G. N. *Biochemistry* **1994,** *33,* 6631.
172. McRee, D. E.; Jensen, G. M.; Fitzgerald, M. M.; Siegel, H. A.; Goodin, D. B. *Proc. Natl. Acad. Sci. USA* **1994,** *91,* 12847.
173. Goodin, D. B. *J. Biol. Inorg. Chem.* **1996,** *1,* 360.
174. Brown, S. B.; Chabot, A. A.; Enderby, E. A.; North, A. C. T. *Nature* **1981,** *289,* 93.
175. Noguchi, M.; Yoshida, T.; Kikuchi, G. *J. Biochem.* **1983,** *93,* 1027.
176. Ortiz de Montellano, P. R. *Acc. Chem. Res.* **1998,** *31,* 543.

177. Takahashi, S.; Wang, J.; Rousseau, D. L.; Ishikawa, K.; Yoshida, T.; Host, J. R.; Ikeda-Saito, M. *J. Biol. Chem.* **1994,** *269,* 1010.
178. Collman, J. P.; Brauman, J. I.; Iverson, B. L.; Sessler, J. L.; Morris, R. M.; Gibson, Q. H. *J. Am. Chem. Soc.* **1983,** *105,* 3052.
179. Traylor, T. G.; Koga, N.; Dearduff, L. A.; Sweptson, P. N.; Ibers, J. A. *J. Am. Chem. Soc.* **1984,** *106,* 5132.
180. Traylor, T. G.; Koga, N.; Dearduff, L. A. *J. Am. Chem. Soc.* **1985,** *107.*
181. Lavalette, D.; Tetreau, C.; Mispelter, J.; Momenteau, M.; Lhoste, J.-M. *Eur. J. Biochem.* **1984,** *145,* 555.
182. Springer, B. A.; Sligar, S. G.; Olson, J. S.; Phillips, G. N. *Chem. Rev.* **1994,** *94,* 669.
183. Kachalova, G. S.; Popov, A. N.; Bartunk, H. D. *Science* **1999,** *284,* 473.
184. Ghosh, A.; Bocian, D. F. *J. Phys. Chem.* **1996,** *100,* 6363.
185. Sigfridsson, E.; Ryde, U. *J. Biol. Inorg. Chem.* **1999,** *4,* 99.
186. Takahashi, S.; Ishikawa, K.; Takeuchi, N.; Ikeda-Saito, M.; Yoshida, T.; Rousseau, D. J. *J. Am. Chem. Soc.* **1995,** *117,* 6002.
187. Ishikawa, K.; Takeuchi, N.; Takahashi, S.; Matera, K. M.; Sato, M.; Shibahara, S.; Rousseau, D. L.; Ikeda-Saito, M.; Yoshida, T. *J. Biol. Chem.* **1995,** *270,* 6345.
188. Wilks, A.; Ortiz de Montellano, P. R. *J. Biol. Chem.* **1993,** *268,* 22357.
189. Davydov, R. M.; Yoshida, T.; Ikeda-Saito, M.; Hoffman, B. M. *J. Am. Chem. Soc.* **1999,** *121,* 10656.
190. Sano, S.; Sano, T.; Morishima, I.; Shiro, Y.; Maeda, Y. *Proc. Natl. Acad. Sci. USA* **1986,** *83,* 531.
191. Rodríquez, J. C.; Rivera, M. *Biochemistry* **1998,** *37,* 13082.
192. Murakami, T.; Morishima, I.; Matsui, T.; Ozaki, S.; Hara, I.; Yang, H.-J.; Watanabe, Y. *J. Am. Chem. Soc.* **1999,** *121,* 2007.
193. Torpey, J.; Ortiz de Montellano, P. R. *J. Biol. Chem.* **1996,** *271,* 26067.
194. Matera, K. M.; Takahashi, S.; Fujii, H.; Zhou, H.; Ishikawa, K.; Yoshimura, T.; Rousseau, D. L.; Yoshida, T.; Ikeda-Saito, M. *J. Biol. Chem.* **1996,** *271,* 6618.
195. Liu, T.; Moënne-Locoz, P.; Loehr, T. M.; Ortiz de Montellano, P. R. *J. Biol. Chem.* **1997,** *272,* 6909.
196. Sakamoto, H.; Omata, Y.; Palmer, G.; Noguchi, M. *J. Biol. Chem.* **1999,** *274,* 18196.
197. Migita, C. T.; Fujii, H.; Matera, K. M.; Takahashi, S.; Zhou, H.; Yoshida, T. *Biochim. Biophys. Acta* **1999,** *1432,* 203.
198. Saito, S.; Itano, H. *Proc. Natl. Acad. Sci. USA* **1982,** *79,* 1393.
199. Yoshida, T.; Noguchi, M. *J. Biochem.* **1984,** *96,* 563.
200. Docherty, J. C.; Schacter, B. A.; Firneisz, G. D.; Brown, S. B. *J. Biol. Chem.* **1984,** *259,* 13066.
201. Benson, D. E.; Suslick, K. S.; Sligar, S. G. *Biochemistry* **1997,** *36,* 5104.
202. Traylor, T. G.; Berzinis, A. P. *Proc. Natl. Acad. Sci. USA* **1980,** *77,* 3171.
203. Migita, C. T.; Matera, K. M.; Ikeda-Saito, M.; Olson, J. S.; Fujii, H.; Yoshimura, T.; Zhou, H.; Yoshida, T. *J. Biol. Chem.* **1998,** *273,* 945.

THE NITRIC OXIDE-RELEASING HEME PROTEINS FROM THE SALIVA OF THE BLOOD-SUCKING INSECT *Rhodnius prolixus*

F. ANN WALKER and WILLIAM R. MONTFORT

Departments of Chemistry and Biochemistry,
University of Arizona, Tucson, Arizona 85721

0898-8838/01 $35.00

I. Introduction

In early 1998, we prepared a chapter entitled "Novel Nitric Oxide-Liberating Heme Proteins from the Saliva of Bloodsucking Insects," which covered the (then) known properties of the nitrophorins ("nitro" = nitric oxide, "phorin" = carrier) from *Rhodnius prolixus* (a member of the kissing bug family) and *Cimex lectularius* (the bedbug), as well as the historical background related to the spectroscopy, electrochemistry, kinetics, and structural features of NO binding to heme proteins (*1*). That chapter was published a little over a year ago, and it is remarkable to see the progress that has been made on this project since that time. Because of the publication of that recent chapter, we now focus our attention on the accomplishments since its publication, and only summarize the necessary background.

There have been two periods of intensive research on nitric oxide in recent years. The first was initiated by chemists in the 1960s as a tool for studying metal–ligand interactions and their molecular orbital structure. Nitric oxide, which has an unpaired electron, was of interest since it could be monitored by electron paramagnetic resonance (EPR) spectroscopy. Its chemistry was of further interest because of the wide range of metals to which it can ligate, due to the "back-bonding" possible through the antibonding orbital occupied by the unpaired electron, in addition to the usual ligation through a nonbonding lone pair of electrons. Those interested in molecular orbital theory also turned to nitric oxide, because the ordering of the *d*-orbital energies could be predicted based on the shape of the NO–metal bond, which ranges from linear (180°) to severely bent (115°) depending on the total number of metal *d* plus NO π electrons (the $\{MNO\}^n$ formalism (*2*)). Considerable understanding of NO chemistry was developed during this period.

Biologists discovered NO in the early 1980s when it was identified as the endothelium-derived relaxing factor (EDRF) that is responsible for smooth-muscle relaxation and vasodilation, among other functions (reviewed in (*3, 4*)). Since then, NO has been confirmed as a signaling molecule used in virtually all vertebrate cells (possibly all eukaryotic cells), and which has numerous physiological roles, including involvement in neurotransmission and memory formation in the brain, the killing of invading cells by macrophages, and the previously mentioned vasodilation (*3, 4*). In blood vessels, NO is synthesized by endothelial nitric oxide synthase (eNOS) and diffuses to smooth muscle cells, where it binds to and stimulates soluble guanylyl cyclase (sGC), leading to vasodilation. Some of the actions of NO are summarized in Fig. 1. Like

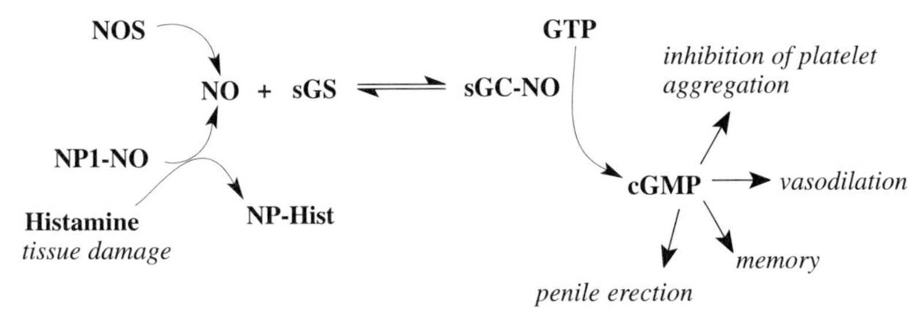

FIG. 1. Nitric oxide (NO) synthesis by nitric oxide synthase (NOS) (upper left), NO reaction with soluble guanylate cyclase (sGC) (middle), and formation of cyclic GMP, which causes tissue-specific signaling (right). The roles of the salivary nitrophorins from *Rhodnius prolixus* in storing and releasing NO and binding histamine are included (lower left).

the nitrophorins, both NOS and sGC use a heme center in their NO interactions.

In 1998 the Nobel Prize in Medicine was awarded jointly to Robert F. Furchgott, Louis J. Ignarro, and Ferid Murad for their discoveries concerning "nitric oxide as a signaling molecule in the cardiovascular system" (*5*). Furchgott first named the signal molecule that made vascular smooth muscle cells relax EDRF. Ignarro and Furchgott, together and independently, showed that EDRF was identical to nitric oxide, NO. Murad separately analyzed how nitroglycerin and related vasodilating compounds act, and discovered in 1977 that they release nitric oxide, which relaxes smooth muscle cells (*3–5*). More recently, nitric oxide has been shown to have importance in many areas of human health, including treatment of atherosclerosis by relaxing blood vessels and thereby lowering blood pressure, treatment of the lungs of patients in intensive care and of premature babies to dilate the lung tissue and make it possible for more oxygen to be absorbed, induction of programmed cell death (apoptosis) as desired in the treatment of cancer, penile erection, causing precipitous drop in blood pressure during septic shock (a negative role), and leading to inflammatory autoimmune diseases such as asthma and colitis (also a negative role) (*3–5*). As we show in this chapter, some blood-sucking insects have learned the beneficial properties of the gas NO, and they utilize it to ensure that they obtain a sufficient meal (Fig. 1).

A. Biochemistry of the Saliva of Blood-Sucking Insects

Blood-sucking insects have in their saliva a number of agents that are designed to help them obtain a sufficient blood meal. These substances allow them either to minimize the time necessary for obtaining their meal or to prevent the victim from detecting their presence while they feed in a leisurely manner. Among these salivary components are anticoagulants (*6*), antiplatelet aggregation compounds with such diverse activities as apyrases (*7*) and peptides that prevent collagen (*8*) or fibrinogen (*9*) binding, prostaglandins PGE_2 and PGF_2 (potent skin vasodilators) (*10–14*), various vasoactive peptides (*15–19*) and tachykinins (*20*), and catechol oxidases/peroxidases that destroy vasoactive amines (*21*), as well as the nitrovasodilators (NO-releasing substances) (*22–24*) and histamine-binding proteins (*25, 26*) of interest to this chapter.

Histamine, another signaling molecule in addition to NO, is released by mast cells of the victim in response to tissue trauma, which occurs while the insect is probing in search of a blood vessel (*25*). Release of histamine gives rise to inflammation, tissue swelling, and itching, and signals the cells of the victim's immune system to assemble at the wound site, all of which can lead to the victim discovering the insect while it feeds. Although histamine initially causes local vasodilation, the inflammatory response leads to a "walling off" of the affected area and restriction of blood flow. Potential outcomes for the insect due to these factors range from the simple loss of a meal to complete disaster.

In general, a given blood-sucking insect has a minimum of three different substances from the classes mentioned above that aid in its obtaining a sufficient blood meal. Typically, when the insect feeds, it spits its saliva into the tissues of the victim, thereby allowing these salivary components to act on the tissues to carry out their functions in preventing coagulation of blood and aggregation of platelets, dilating skin pores, dilating blood vessels to allow more blood to come to the region of the bite, and/or delaying swelling and the onset of the immune response.

The insect of interest to this report is *Rhodnius prolixus,* a member of the "kissing bug" family of the order *Hemiptera. Rhodnius* is prevalent in South America, but is also found in North America as far north as southern Texas and Arizona. It feeds on the blood of rodents and larger mammals, including humans. The insect is a relatively slow feeder and pokes around in the tissue until a suitable internal wound is created, or a blood vessel is tapped (*27*). While probing, the insect secretes saliva that contains numerous proteins to aid in this process. Eight such proteins have so far been identified, and all eight have antihemostatic activities (*7*). They include the four nitrophorins (NO-carrying heme proteins), NP1–4 (*24*), of interest to this chapter; three proteins called salivary antiplatelet lipocalins, SAPL1-3, or RPAI1-3 (*Rhodnius prolixus* anti-inflammatory proteins) that inhibit clotting at

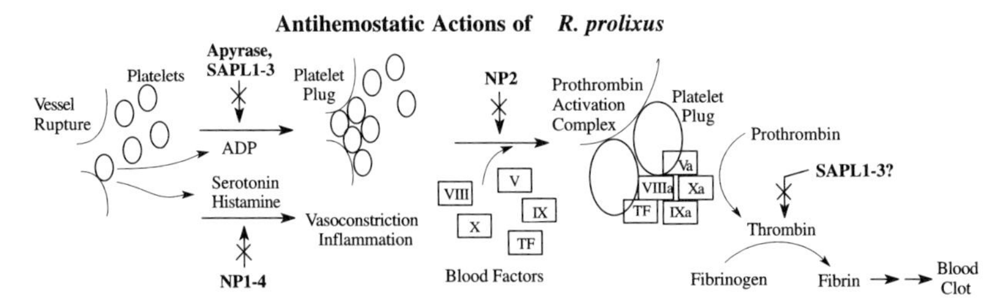

FIG. 2. Simplified illustration of the steps in blood coagulation inhibited by the salivary proteins of *Rhodnius prolixus*. The inhibited steps are indicated with crossed arrows. SAPL1-3 are also called RPAI1-3.

both the platelet aggregation and blood coagulation steps to ensure the blood vessel and insect mouthparts do not become clogged (*28*); and an apyrase that hydrolyzes ADP and thus provides another means of interfering with platelet aggregation. One of the nitrophorins, NP2, also has anti-clotting activity (*29, 30*). All four nitrophorins also have antihistaminic activity (*26, 31*) by binding histamine and thus deactivating its ability to cause inflammation during feeding and begin the victim's immune response, thereby preventing detection of the insect. All of these activities are summarized in Fig. 2. As we will show, all of the proteins except the apyrase exhibit a common structural motif, a beta-barrel structure called a lipocalin fold (*32*), although only the nitrophorins have heme as a cofactor.

Nitric oxide is a reactive molecule, especially in the presence of molecular oxygen or superoxide ion, and thus, if an insect such as *Rhodnius* (or the bedbug, *Cimex lectularius,* which has an unrelated nitrophorin in its saliva (*33, 34*)) wants to utilize it in order to obtain a better meal, it must synthesize NO in its salivary glands and store it in a stable form for perhaps up to a month's time until it finds a victim. At this point, the NO must be ready to be released when the saliva is spit into the tissues of the victim. *R. prolixus* has a nitric oxide synthase (NOS) in the cells of its saliva glands (*35–38*), so clearly, NO is produced at the site where it is to be stored. And what better way of storing NO than binding it to a heme protein?! Heme–NO complexes have been investigated for many years, and the NO complexes of ferrous hemes are known to be extremely stable, with dissociation constants in the picomolar range (*39*) or lower. Herein lies a major problem for the insect: A Fe(II)–NO complex is *too stable* to allow dissociation of NO upon injection into the tissues of the victim! Solution: Stabilize the heme–NO center as a Fe(III)–NO complex, which should have dissociation constants in the millimolar to micromolar range (*39*), i.e., making them NO carriers (phorins) that bind NO when the concentration of the NO–heme protein is high, as in the saliva glands, and release it when the concentration of the NO–heme protein is low, as in the tissues. As we shall show, this is exactly what the *Rhodnius* insect does! And at the same time, the heme center is designed to bind histamine, thus delaying swelling and the onset of the immune response, both of which would alert the victim to the insect's activities. Both of these functions are included in Fig. 2, in addition to the anticlotting activity of NP2.

B. *Rhodnius prolixus* and Chagas' Disease

R. prolixus often carries the parasite *Trypanosoma cruzi*, the vector of Chagas' disease. This protozoan is passed through the feces, dropped while the insect is feeding. The parasite is thought to enter the bloodstream when the victim scratches the feeding site after the insect has departed. Once in the blood, the parasite most commonly infects the heart muscle, leading to reduced cardiac output and frequently death (*40*). Chagas' disease is endemic in Central and South America, and 16–18 million Latin Americans are chronically infected with *T. cruzi* at the present time; some 43,000 die annually. The disease is now present in the United States, with some 300,000 migrants from Latin America being infected (*40*). And although there are antibody markers used in South America to screen blood donations, such screens are not currently used in the United States.

Investigations of the DNA of mummies from the Atacama desert on the west coast of South America indicate that Chagas' disease was a frequent killer of New World peoples as far back as at least 5000 years ago (*41*), and current hypotheses being tested by archaeoparasitologists include the possibility that the origins of the disease in humans relate to the domestication of animals, including dogs and guinea pigs, or that it predates these events (*42*). Even at present, there is no drug that is effective against Chagas' disease. From the manner in which Chagas' disease is transmitted, it is clear that the eight proteins present in the saliva of *R. prolixus* contribute to this transmission by allowing the insect to linger long enough to leave behind its feces.

C. Early Work on the Nitrophorins from *Rhodnius prolixus*

In this chapter, we summarize our recent work that has characterized the nitrophorins from *R. prolixus* by spectroscopic, electrochemical, kinetic, and structural methods. The early work is summarized briefly only for continuity and to put the current work in perspective.

1. Discovery of and Early Information Obtained about the Proteins

As part of a long-standing interest in blood-sucking insects, the disease vectors they carry, and the mechanisms that they use in order to achieve their goal of obtaining a sufficient blood meal, Professor José M. C. Ribiero, then of the Department of Entomology at the University of Arizona, discovered the bright cherry red colored saliva glands of *Rhodnius prolixus* and showed that a homogenate of the saliva glands had an optical spectrum typical of a heme protein, with an intense absorption (Soret) band at 422 nm (*24*). He further showed that serial dilution of the salivary gland homogenate at pH 5 did not change the optical spectrum, whereas serial dilution at pH 7.35 caused a shift of the Soret band to 403 nm (*24*). Blowing argon over the solution caused the same shift of the Soret band at pH 7.35, and the Soret band shifted back to 422 nm when the solution was exposed to gaseous nitric oxide, but not to carbon monoxide or molecular oxygen. As is typical of other vasodilators that release NO, the salivary gland homogenate produced reversible vasodilation of endotheliumless rabbit aortic rings preconstricted with norepinephrine to an extent that correlated with the optical spectral changes (*24*).

In the spring of 1992, just after delivery of the new Bruker EPR spectrometer, the EPR spectrum of the salivary gland homogenate in the absence and presence of NO was obtained. It was found that in the presence of NO there was no EPR signal, whereas if the homogenate were pretreated by blowing argon over the solution, an EPR signal typical of a high-spin Fe(III) heme center was obtained (*24*), as shown in Fig. 3D. In order to obtain the EPR signal so typical of the spectra frequently published of heme–NO centers (*43*), it was necessary to prereduce the salivary gland homogenate with dithionite, and then blow NO over the sample (*24*). The nature of the EPR signal of this reduced Fe–NO center strongly suggested that an imidazole nitrogen of a histidine was the protein-provided ligand. This key experiment told us that the nitrophorins were Fe(III) heme centers, to which NO could bind with dissociation constants in the millimolar to micromolar range (*39*), as suggested above. Hence, these proteins appeared to be unique and worthy of detailed study. We thus expanded the small colony of *Rhodnius* insects that Ribeiro was raising in the Center for Insect Sciences insectary in the Biochemistry Department at the University of Arizona, in the hope of obtaining enough protein to be able to do more

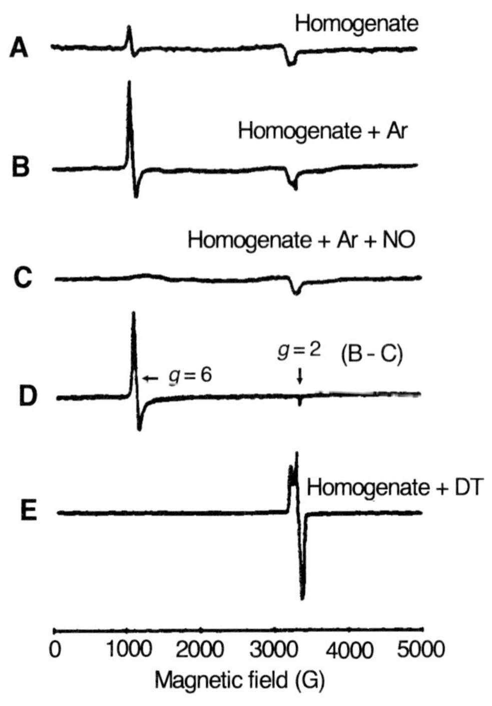

FIG. 3. EPR spectra of 100 pairs of *Rhodnius* salivary glands homogenized in 125 μl of phosphate-buffered saline at pH 7.2 (A) before argon equilibration; (B) after equilibration in an argon atmosphere for 4 h; (C) after equilibration of (B) with NO for 2 min. (D) difference spectrum, that is; B – C. (E) homogenate as in (B) treated with dithionite (DT) to reduce Fe(III) to Fe(II), followed by equilibration with NO for 2 min. (The small signal at $g \approx 2$ in A–C is due to copper oxide in the liquid helium which had been condensed at the University of Arizona in a copper-plumbed helium liquification apparatus!) All spectra are plotted on the same scale except (E), which is reduced in amplitude by a factor of 3. Reproduced with permission from Ref. (*24*).

detailed spectroscopic characterizations, as well as to crystallize the protein for structure determination.

2. *Beheading Kissing Bugs*

One nice thing about working with proteins from insects is that there are no Insect Rights groups that are concerned if insects, especially blood-sucking ones, are beheaded. Hence, except for the problem of *feeding* the *Rhodnius* insects, which required allowing them to suck blood from the shaved hindquarters of anesthetized rabbits once a month, there were no barriers to obtaining their saliva glands. The cardboard boxes, each containing a group of the insects, were weighed both before and after insect feeding, in order to ensure that the rabbits were not stressed by losing too much blood. At fifth instar (between the fifth and sixth molting), their heads were simply pulled off and the cherry-red saliva gland pairs exposed. (A graduate student reports that upon beheading, the headless bodies and legs continued walking. Clearly this was not a higher life form!) The glands were removed with stainless steel tweezers and collected in a petri dish. Batches of a thousand gland pairs were then homogenized and passed through an HPLC equipped with a size exclusion column, followed by a strong cation exchange column. This second column separated the red proteinaceous fraction into four components, NP1–4. From a thousand gland pairs, 3 mg of nitrophorins were obtained; half of this was NP1. A total of 5000 insects were sacrificed for one dilute NMR sample of NP1, and it was not considered productive to try to obtain enough protein for NMR samples of the other three nitrophorins. As it turns out, NP1 is not the best behaved protein for NMR studies (see Section II,E), so our first efforts were discouraging. Likewise, although small needles were obtained, crystals of the size necessary for X-ray crystallography could not be obtained from the purified fractions of native NP1. Thus, the necessary time and effort had to be expended to successfully express the genes for NP1–NP4.

3. *Cloning and Sequence Homology*

The background work on cloning the gene for NP1 was completed by the summer of 1994 by Champagne, Nussenzveig, and Ribiero, who determined the amino-terminal protein sequences of the four nitrophorins, developed DNA probes for NP1, and cloned its gene from the salivary gland cDNA library (*44*). Later, the genes for the other three nitrophorins were also cloned (*45*); each gene contains a signal peptide at the N-terminus, as is typical of secreted proteins. For NP1 this signal peptide is 23 residues in length, 14 of which are hydrophobic (*44*). For all four proteins, the signal peptides are not present in the mature protein. The amino acid sequences to which these four genes give rise are shown in Fig. 4.

Initially, it was suspected that the nitrophorins were insect hemoglobins. Indeed, they showed 45–48% homology with monomeric hemoglobins from insects, annelids, mollusks, nematodes, and even human β chains and leghemoglobin (*44*). However, in due time it became clear that these proteins were not globins at all, but rather, beta-barrel proteins called lipocalins (see Section III). As for the four nitrophorins, the sequences of NP1 and NP4 are 90% identical, whereas those of NP2 and NP3 are 79% identical; NP1 and NP2, however, are only 38% identical.

Protein Sequences of *Rhodnius prolixus* Nitrophorins

```
      1            __α1__      _βA__ |--A-B loop--| ___βB___
NP1   KCTKNALAQT GFNKDKYFNG DVWYVTDYLD LEPDDVPKRY CAALAAGTAS
NP4   A*****I*** ********** ********** ********** **********
NP2   D*ST*ISPKQ *LD*A***S* .K****HF** KD*.Q*TDQ* *SSFTPRESD
NP3   *********K ********** .T*****Y** ***.****P* ******K**G

     51 __βC____          _____βD______    ___βE_____
NP1   GKLKEALYHY DPKTQDTFYD VSELQEESPG .KYTANFKKV EKNGNVKVDV
NP4   ********** ********** *****V**** .********* D*******A*
NP2   *TV******* NANKKTS**N IG*GKL**S* LQ***KY*T* **KKA*LKEA
NP3   *********F *SK******* ******G*** V******N** ***RKEIEP*

    101   ___βF_____     _____βG___         __βH__      ____
NP1   TSGNYYTFTV MYADDSSALI HTCLHKGNKD LGDLYAVLNR NKDTNAGDKV
NP4   A********* ********** ********** ********** ***AAA****
NP2   DEK*S**L** LE*******V *I**RE*S** *****T**TH Q***EPSA**
NP3   *P*D****** ********** *******P** ********S* **TG****T*

    151_ α2__                      ___α3___
NP1   KGAVTAASLK FSDFISTKDN KCEYDNVSLK SLLTK
NP4   *S**S**T*E **K*****E* N*A***D*** *****
NP2   ****TQ*G*Q L*Q*VG**DL G*Q**.*QFT **~~~
NP3   *N**A****K *ND**D**T* S*T**.**** *M~~~
```

FIG. 4. Protein sequences of *Rhodnius* NP1–4 obtained from the gene sequences (*45*) and confirmed for NP1, NP2, and NP4 by X-ray crystallography. Identical residues are marked*. Alpha and beta structural features are also marked, as is the A-B loop that becomes ordered upon binding of NO (see Section III).

These levels of identity suggest a scenario in which the original single gene was duplicated, each of the two genes differentiated for some time until one of the genes was again duplicated and each of those genes continued differentiating, producing NP2 and NP3. The other originally duplicated gene continued differentiating for a longer period of time than the first, but was eventually also duplicated and the two new genes continued differentiating for a shorter period of time until the present, producing NP1 and NP4.

4. Expression, Isolation, Renaturation, Reconstitution, and Purification of the Nitrophorins

The initial procedures involved in expression of the *Rhodnius* nitrophorins in *Escherichia coli* were worked out for NP1 by Drs. Donald E. Champagne and John F. Andersen at the University of Arizona during the spring and summer of 1996, and the purification protocols were developed thereafter (*46*). The cDNA for NP1 was modified by removal of the oligonucleotide coding for the putative signal sequence for secretion to the salivary gland lumen, and for insertion into the expression vector pET17b by PCR mutagenesis. After verification of the sequence, the plasmid was moved into *E. coli* (BL21DE3) to give the expression strain. The bacteria were grown overnight, expression was initiated by adding IPTG, and incubation was continued for 15–18 hours. The cells were centifuged, washed, and lysed using a French press, and the lysate was incubated with DNase I and RNase A, followed by centrifugation to obtain the insoluble inclusion bodies containing NP1. They were washed and then solublized in a pH 7.4 denaturing buffer containing the reducing agent dithiothreitol (DTT). The solublized, denatured protein was refolded by slow addition to a large volume of pH 7.4 buffer containing DTT but not the denaturant. After the protein solution was concentrated by ultafiltration through a 10 kDa membrane, it was dialyzed to remove the remaining DTT, and then hemin (in DMSO (*46*) or 0.01 M NaOH (*47*)) was added slowly. Incorporation of heme was monitored using the ratio of the Soret absorbance at 404 nm to the absorbance at 280 nm. A ratio of approximately 3.0, despite repeated addition of hemin followed by centrifugation to remove precipitated, unincorporated hemin, indicated full titration of the refolded protein to its native state. The pH of the refolded heme protein was adjusted to 5.0, the solution was centrifuged at $100,000 \times g$, and the supernatant was further purified chromatographically on a SP-Sepharose column. Optical spectra in the absence and presence of NO were compared to those observed for native NP1, purified as reported previously (*44*). The spectra were identical (*46*). Similar procedures have been used for obtaining purified recombinant NP2, NP3, and NP4.

II. Spectroscopic Characterization of the Nitrophorins

A. Optical Spectroscopy of Nitrophorins in the Absence and Presence of NO and Other Ligands

The optical spectra of nitrophorins in the absence of added ligands show Soret band maxima at 403–404 nm. On binding NO, the Soret band shifts to 419–420 nm in NP1 and NP4, and 421–423 nm in NP2 and NP3 (*46, 48–50*). Example spectra are shown in Fig. 5a. The direction of the Soret band shift identifies the oxidation state of the heme iron as being Fe(III) (*51, 52*). The α and β bands of the NO complexes are located near 570 and 535 nm, respectively. The histamine complexes have Soret maxima at 410–412 nm and broad α,β maxima between 580

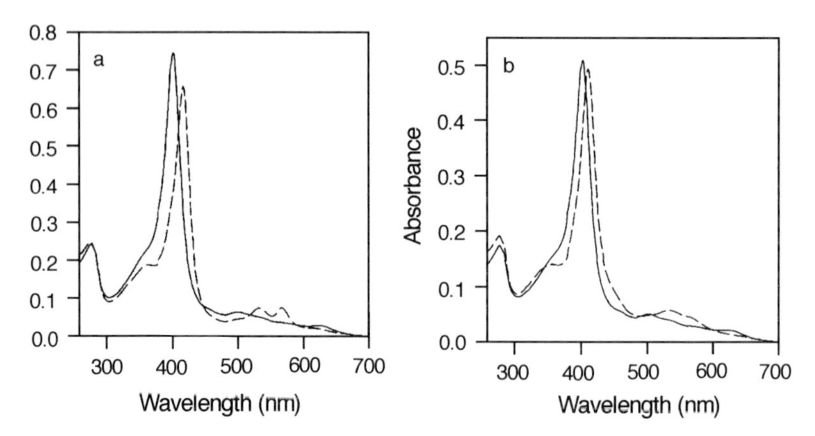

FIG. 5. UV-visible spectra of recombinant *Rhodnius* NP4 (pH 8.0) (a) without ligand (solid line) and NO complex (dashed line); (b) without ligand (solid line) and histamine complex (dashed line). Reproduced with permission from Ref. (*48*).

and 520 nm, as shown in Fig. 5b. Shifts in the Soret band maximum on addition of ligands were used to measure binding or release of NO and histamine in the kinetic and equilibrium studies discussed in Section IV.

The optical spectra of the nitrophorins in the oxidized (Fe(III)) and reduced (Fe(II)) states have been measured by spectroelectrochemical techniques, as shown for NP3 and NP3-NO at pH 7.5 in Figs. 6a and 6b, respectively. As reported previously (*49, 50*), there is only a 3 nm shift in

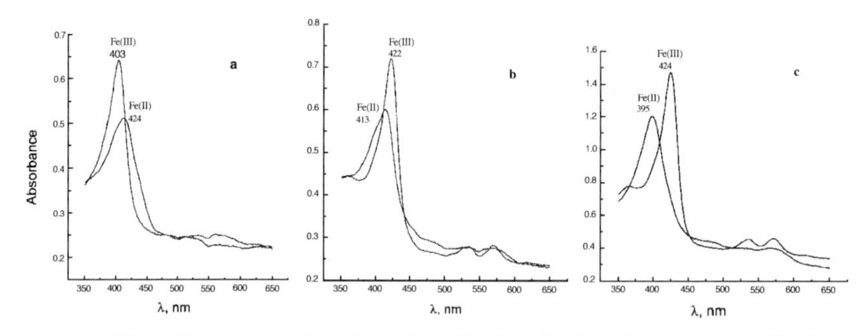

FIG. 6. UV-visible spectra of ~0.05 mM oxidized and reduced recombinant *Rhodnius* NP3 (a) at pH 7.5 without ligand; (b) at pH 7.5 bound to NO; (c) at pH 5.5 bound to NO. In each case, the spectrum of the oxidized nitrophorin is represented by a solid line and the reduced by a dashed line. Spectra were recorded in an optically transparent thin-layer electrochemical cell of approximate window thickness 0.05 mm. To obtain the fully oxidized and reduced spectra, potentials (vs Ag/AgCl) were applied until no change in optical spectrum occurred, of +600 and −400 mV, respectively (a), +200 and −400 mV, respectively (b), and 0 and −280mV, respectively (c).

the position of the Soret band upon reduction of NP^{III}–NO to NP^{II}–NO for NP1 and NP4—from 419 to 416 nm; these Soret band positions do not change over the pH 5.5–7.5 range. In comparison, the band maxima in acetate buffer at pH 5 are slightly different (421 rather than 419 nm for NP1-NO) (*46*). The ferric forms of the nitrophorin–NO complexes have the higher extinction coefficients and sharper Soret and α,β bands, the situation usually reserved for a diamagnetic Fe(II) heme center. This is one bit of evidence that suggests that the electron configuration of the Fe(III)–NO center might be better described as Fe(II)–NO^+, as is the case for other $\{FeNO\}^6$ centers (*2*).

For NP2 and NP3 at pH 7.5, the shift in Soret band positions of the NO complexes for the two oxidation states is somewhat larger—8–10 nm, from 421–423 to 413 nm for the Fe(III) and Fe(II) complexes, respectively (*50*). However, in contrast to NP1-NO (*49*) and NP4-NO (*50*), at pH 5.5 NP2–NO and NP3–NO show very different spectral shifts upon electrochemical reduction, as shown in Fig. 6c for NP3–NO. The Soret band shifts to 395 nm, and both the wavelength maximum and shape of the Soret band are typical of five-coordinate heme–NO centers, including guanylyl cyclase, upon binding NO (*53, 54*). The reduced forms of both NP2–NO and NP3–NO exhibit similar pH dependence of the absorption spectra, whereas NP1–NO and NP4–NO do not show any pH dependence of their absorption spectra over the pH range 5.5–7.5 (*50*).

The fact that the reduced forms of NP2–NO and NP3–NO lose their protein-provided proximal histidine ligands at pH 5.5 attests to the strong *trans*-directing influence of the NO ligand, as has been observed for a number of other ferroheme protein–NO complexes (*39*). Similar spectra are observed upon exposure of guanylyl cyclase (in the Fe(II) form) to NO (*53, 54*). There is, however, no spectrophotometric evidence for the loss of the proximal ligand of the Fe(III)–NO forms of any of the nitrophorins over the pH range investigated (5.5–7.5), or of the Fe(II)–NO forms of NP1 and NP4 at pH 5.5. That the optical spectra of five-coordinate (OEP)Fe(III)–NO are extremely different from those of the six-coordinate (OEP)Fe(III)–NO(L) complexes, where L is an imidazole or pyridine ligand, with a low-intensity Soret band maximum at 359 nm and poorly separated α,β bands at 515 (shoulder) and 557 nm (*64*), suggests that it would be immediately obvious from the optical spectra (Fig. 5) if His 59 were not bound for any of the nitrophorin–NO complexes in the Fe(III)–NO state. Since there are no pH changes in the optical spectra of the Fe(III)–NO complexes of the nitrophorins, it can be concluded that the proximal histidine remains bound over the pH range 5–8 for all four nitrophorins.

The optical spectral data for the $NP2^{II}$–NO and $NP3^{II}$–NO complexes at pH 5.5 suggest that there may be a difference in Fe–N bond strength of the proximal histidine ligand for these two proteins that is only apparent when the metal is reduced to Fe(II). Thus, the kinetic differences in NO release behavior of the NP1,4 group and the NP2,3 group, discussed later, could in part be related to differences in heme–histidine bond strength. If so, then it must be concluded that a weakened Fe–N bond strength produces a more stable Fe–NO complex, since the K_d values for NP2,3–NO are much smaller than for NP1,4–NO (*50*) (see Section IV).

In comparison to the spectra of the nitric oxide complexes, the optical spectra of the ferric and ferrous forms of NP1–4 in the absence of NO over the pH range of 5.5 to 7.5 show the more typically expected changes in Soret band maximum reported previously for NP1 (*49*). The high-spin NP^{III} species have their Soret band maxima at 402–403 nm, whereas the high-spin NP^{II} have their Soret band maxima at 430 nm at pH 5.5 for NP1 and NP4 and 417–418 nm for NP2 and NP3. In all cases the α,β bands are quite broad and of low extinction coefficient (*50*).

In the presence of histamine, the Soret band maximum of the Fe(III) form of the nitrophorins is at 410–412 nm (*50*), as mentioned earlier. Upon reduction, the band shifts to 424 nm, and from this spectral shift it was possible to measure the reduction potentials of the histamine complexes of NP1–4 at pH 5.5 and 7.5 (*50*), as discussed in Section V. The α,β bands are quite broad and of low extinction coefficient for the Fe(III), but sharper and of higher extinction coefficients for the Fe(II) complexes (*50*). The Soret band maxima in the presence of imidazole are the same for the Fe(III) complexes, but slightly larger and somewhat variable, 425–428 nm for the Fe(II) complexes (*55*). The Soret band maxima in the presence of 4-iodopyrazole, the ligand used to produce one of the heavy atom derivatives for solving the first X-ray crystal structure (*31*), are 405 nm for the Fe(III) and 411 nm for the Fe(II) complexes of NP1 and NP4, and 402 nm for the Fe(III) and 409 nm for the Fe(II) complexes of NP2 and NP3 (*55*). As observed for the histamine complexes (*50*), the α,β bands are quite broad and of low extinction coefficient for the Fe(III), but sharper and of higher extinction coefficient for the Fe(II) complexes (*55*).

B. Vibrational Spectroscopy of Nitrophorin–NO and –CO Complexes

The NO complexes of $NP1^{III}$ and $NP1^{II}$ and the CO complex of $NP1^{II}$ have been investigated by FTIR spectroscopy (*49*). As shown in Fig. 7,

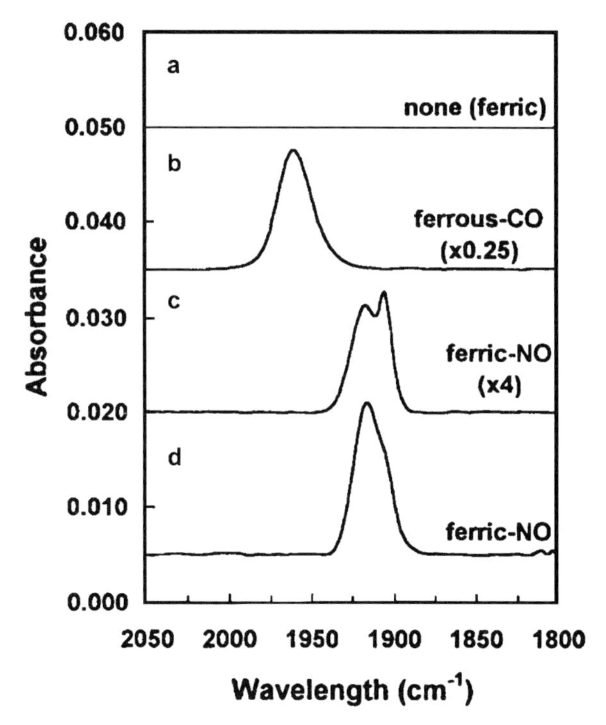

FIG. 7. FTIR spectra of recombinant *Rhodnius* NP1. (a) Ferric NP1 (3.0 mM) exchanged into D_2O (50 mM citrate/NaOD, measured pH 5.6), path length 13 μm. (b) Ferrous-CO derivative (6.7 mM) in 50 mM Tris, 50 mM EDTA, 5 mM $Na_2S_2O_4$, pH 8, path length 56 μm (>90% CO complex). (c) Ferric-NO derivative (3.0 mM) in buffer identical to that of (a), path length 13 μm (73% NO complex). (d) Ferric-NO derivative (11.0 mM) in buffer identical to that of (b) with 1 mM deazaflavin rather than 5 mM $Na_2S_2O_4$, but not illuminated. The path length of the IR cell was 13 μm (95% NO complex). All spectra were recorded at a resolution of 2 cm^{-1} and are averages of 800 scans. Reproduced with permission from Ref. (*49*).

the Fe(III)–NO complex has a somewhat lower N–O stretching frequency than the C–O stretching frequency of the Fe(II)–CO complex (1917 and 1960 cm^{-1}, respectively, for the major bands), indicating that the NO of Fe(III)–NO has a bond order approaching 3, as in Fe(II)–CO, and is thus binding largely as the Fe(II)–NO^+ valence tautomer. In comparison, the 1-electron-reduced Fe(II)–NO complex has the lower frequency of 1611 cm^{-1}, indicating a much lower bond order. Isotopic substitution of ^{15}NO produced the expected drop in frequency of about 28 cm^{-1} for Fe(II)–NO and 37 cm^{-1} for Fe(III)–NO due to the change in the reduced mass (*49*). For both oxidation states, the N–O stretching frequencies and isotopic frequency shifts are typical of those of other heme proteins (*56–61*) and model hemes (*62–65*).

For the Fe(III)–NO and Fe(II)–CO complexes of NP1, there are two components to the X–O stretching band that are separated by 13 and 24 cm^{-1}, respectively. The ratio of the two species depends upon the nature of the buffer (*49*), and has not been investigated further. The other unique feature of the bands shown in Fig. 7 is their widths, which are considerably greater than observed for many heme protein–NO and –CO complexes. At the time of publication of that work (*49*), it was not known that there was a change in conformation of the loop made up of residues 30–39 when NO binds to the Fe(III) form, at least for NP4 (*47*). Therefore, in addition to the much greater exposure of NO in the open conformation, as is the case for the crystal structure of the photoreduced $NP1^{II}$–NO (*49*), the dynamics of closing, and expulsion of water molecules near the bound NO site, discussed in Section III, as determined from crystal structure of the physiologically important oxidation state of $NP4^{III}$–NO (*47*), probably accounts for the observed breadth of the IR stretching bands observed previously (*49*).

C. Mössbauer Spectroscopy of a Nitrophorin–NO Model

Nitric oxide is unique among diatomic molecules in that it can bind to both Fe(III) and Fe(II) centers, including those of hemes and heme proteins. In terms of model hemes, both Fe(III)–NO ($\{FeNO\}^6$) and Fe(II)–NO ($\{FeNO\}^7$) complexes have been characterized spectroscopically and structurally, including those in which there either is or is not another axial ligand present trans to the NO (*2, 65*). Thus, four NO–heme complexes, PFe(III)(NO), PFe(III)(NO)(L), as well as the corresponding Fe(II) analogues, may exist. Of the four oxidation/coordination states possible, only the five-coordinate PFe(III)(NO) center has not been reported in a biological system, although it is probable that additional study will reveal possible roles for this complex as well.

The Fe(II)–NO complexes of porphyrins (*66–68*) and heme proteins (*24, 49, 53, 69–76*) have been studied in detail by EPR spectroscopy, which allows facile differentiation between five-coordinate heme–NO and six-coordinate heme–NO(L) centers. However, only a few reports of the Mössbauer spectra of such complexes have been published (*68, 77–82*), and the only Fe(III)–NO species that have been studied by Mössbauer spectroscopy include the isoelectronic nitroprusside ion, $[Fe(CN)_5(NO)]^{2-}$ (*78*), the five-coordinate complexes $[TPPFe(NO)]^+$ (*68*) and $[OEPFe(NO)]^+$ (*82*), and two reports of the nitro, nitrosyl complexes of iron(III) tetraphenylporphyrins, where the ligand L is NO_2^- (*82, 83*).

Most of these species (except for the five-coordinate Fe(III)–NO complex) have been implicated in the physiological interactions of NO with various heme proteins. For example, EPR spectroscopy shows that the six-coordinate PFe(II)(NO)(L) center is observed for the NO complexes of hemoglobin and myoglobin (*69–74*), cytochrome a_3 (*75, 76*), and cytochromes P450 (*84*), and the five-coordinate PFe(II)(NO) center is formed upon binding NO to the heme of guanylate cyclase (*53, 54*). The EPR-silent six-coordinate PFe(III)(NO)(L) center is observed for the NO complexes of the nitrophorins found in the saliva of the blood-sucking insect *Rhodnius prolixus* (*1, 24, 31, 46–50*).

The Fe(III)–NO complex of NP1 is EPR silent (Fig. 3) because it contains an odd-electron (ferriheme) center bound to the odd-electron diatomic NO (*24*), which creates a $\{FeNO\}^6$ center. The NMR spectrum of NP1 Fe(III)–NO is that of a diamagnetic protein (*85*). However, whether the electron configuration is best described as $Fe(II)–NO^+$ or antiferromagnetically coupled low-spin $Fe(III)–NO^{\cdot}$ is not completely clear, even though the infrared data (*49*) discussed earlier (Fig. 7) are consistent with the former electron configuration. Thus, as a prelude to planned detailed studies of the Mössbauer spectra of the nitrophorins and their NO complexes, we have reported the Mössbauer spectrum of the six-coordinate complex of OEPFe(III)–NO (*86*).

The NO, *N*-methylimidazole complex of 94.5% isotopically enriched octaethylporphyrinatoiron(III) chloride, $[OEP^{57}Fe(NO)(N\text{-}MeIm)]^+Cl^-$, was prepared in dimethylacetamide solution and studied by low-temperature Mössbauer spectroscopy in the presence and absence of a magnetic field. The $[OEPFe(NO)(N\text{-}MeIm)]^+Cl^-$ complex is EPR silent and behaves as a diamagnetic species that exhibits a quadrupole doublet in the absence of a magnetic field. In the presence of a magnetic field, four resolved lines are observed, as shown in Fig. 8, that can be fit with a quadrupole splitting $\Delta E_Q = 1.64$ mm s^{-1}, asymmetry parameter $\eta = 0.4$, isomer shift $\delta = 0.02$ mm s^{-1} and linewidth $\Gamma = 0.3$ mm s^{-1} (*86*). Two electron configurations, Fe(III)–NO (low-spin d^5, strongly antiferromagnetically coupled to NO), or $Fe(II)–NO^+$ (low-spin d^6, purely diamagnetic), are possible. Which is the actual configuration cannot be determined until detailed molecular calculations are carried out.

D. EPR Spectroscopy of the NP1–Histamine Complex

At the time that the first investigation of the histamine complex of native NP1 was reported (*26*), no EPR signal was observed, and it was concluded that the reason for this lack of an EPR signal from an expected low-spin d^5 heme center was that the signal was that of a fast-relaxing "large g_{max}" species whose EPR spectrum was too weak to be observed. However, when recombinant protein became available and

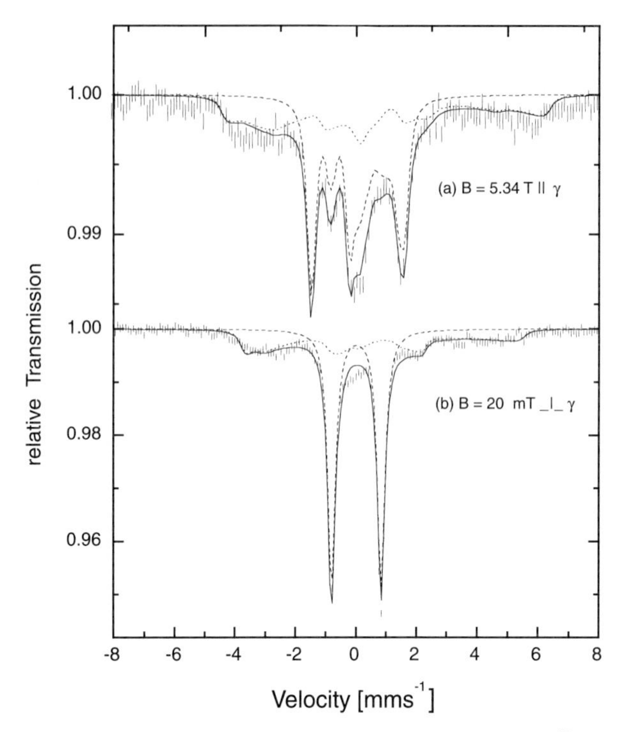

FIG. 8. Mössbauer spectra of a sample containing 9 mM [OEP^{57}FeCl] and 30 mM *N*-methylimidazole in *N,N*-dimethylacetamide, into which NO gas was bubbled for 5 minutes, taken at 4.2 K in a magnetic field of (a) 5.34 T parallel and (b) 20 mT perpendicular to the γ-beam. The dotted line corresponds to the spectrum of the low-spin ferric heme complex [OEP^{57}Fe(NMeIm)$_2$]$^+$Cl$^-$ (39% relative contribution) and the dashed line to the heme–NO complex [OEP^{57}Fe(NMeIm)(NO)]$^+$Cl$^-$ (61% relative contribution). Reproduced with permission from Ref. (*86*).

a more concentrated sample could be investigated, it was found that the EPR signal of the histamine complex was that of a typical rhombic low-spin Fe(III) center with EPR parameters ($g_{zz} = 3.02$, $g_{yy} = 2.25$, and $g_{xx} = 1.46$) (*31*) very similar to those of cytochrome b_5 (*87, 88*). Hence, the sample of native protein investigated initially (*26*) must have been much more dilute than believed at the time, for the signal is, indeed, easy to observe at 4.2 K (*89*), as shown in trace 1 of Fig. 9. The X- and S-band ESEEM spectra of this complex have been reported (*89*) and show that the proton sum combination feature in the X-band (8.706 GHz) ESEEM spectra was well resolved at the low-field g-value (g_{zz}), but unresolved at the high-field g-value (g_{xx}), as shown in Fig. 9, traces

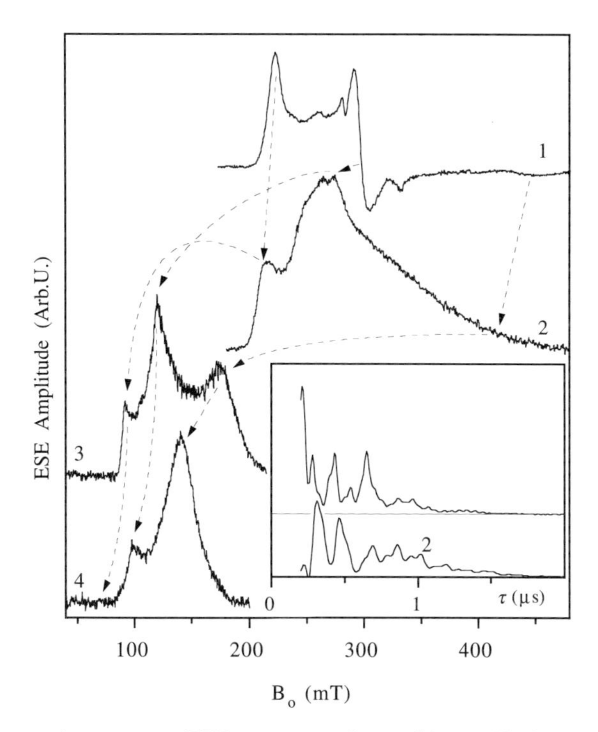

FIG. 9. The continuous wave EPR spectrum of recombinant *Rhodnius prolixus* NP1-histamine at a microwave frequency of 9.338 GHz (trace 1) and field-sweep ESE spectra at 8.706 GHz (trace 2), 3.744 GHz (trace 3), 3.065 GHz (trace 4). Dashed arrows show the changes in magnetic fields corresponding to principal g-values at the different microwave frequencies. Inset trace 1, the primary ESE decay recorded at 8.706 GHz, $B_o = 213$ mT (the low-field turning point in the field-sweep ESE spectrum shown by trace 2 in the main panel). Inset trace 2, the primary ESE decay recorded at 3.065 GHz, $B_o = 140.8$ mT (the high-field turning point in the field-sweep ESE spectrum shown by trace 4 in the main panel). Reproduced with permission from Ref. (*89*).

2,3. Decreasing the microwave frequency to S-band (3.065 GHz) made it possible to obtain well-resolved ESEEM spectra at the high-field (g_{xx}) position (trace 4) and to determine the orientation of g_{xx}. It was found to be oriented in the porphyrin plane 60–90° from the plane of the axial histidine (His 59), rather than at the average orientation of the planes of the two axial ligands (*89*). Thus, the orientation of g_{xx} is consistent with the predictions of counterrotation of the g-tensor with rotation of one of the axial ligands (*90*), indicating that His 59 is much more important in determining the orientation of the g-tensor than is the added histamine ligand.

E. NMR Spectroscopy of Nitrophorins

One of the most important spectroscopic techniques for characterizing heme proteins is proton NMR spectroscopy. In the case of ferriheme proteins, NO bound to Fe(III) hemes, and indeed, the nitrophorins (*85*), produces a diamagnetic complex at ambient temperatures, which should have heme substituent resonances that are buried in the protein proton resonance envelope, whereas the NO-free form of Fe(III) heme proteins is a high-spin ($S = \frac{5}{2}$) complex. Binding of even-electron donor ligands, such as histamine, imidazole, or cyanide, to the NO-free forms of these proteins produces low-spin ($S = \frac{1}{2}$) complexes. The overall goals of detailed NMR spectroscopic studies of the nitrophorins are to assign all of the proton resonances and to determine the three-dimensional structure of the protein in solution, all in order to learn in detail about the dynamics of this beta-barrel heme protein. At the end of this study, it is hoped that an understanding will be achieved of how the beta-barrel nitrophorin structure, the solid-state structures of one or more ligand complexes of NP1, 2, 4 of which have been solved by X-ray crystallography (*31, 32, 47–49*), is stabilized in solution, how heme binding to the protein contributes to the stability and dynamics of the proteins, the points at which the structure and dynamics are most affected by binding of NO or other alternative ligands, and the effect of ligand charge and heme oxidation state on the dynamics of the protein, especially in the vicinity of the ligand-binding pocket. As a first start toward these major goals, the hyperfine-shifted heme and nearby protein proton resonances, which are shifted from their diamagnetic positions by the presence of unpaired electron(s) on the metal, in the two paramagnetic forms of the proteins, have been investigated (*49, 91*).

The unpaired electron(s) on the metal, Fe(III) in the case of the nitrophorins, act as “beacons” that “illuminate” the protons in the vicinity of the metal, by causing shifts (isotropic shifts) of the resonances from those observed in a diamagnetic protein, for example, the NO-bound forms of the nitrophorins. These shifts allow much to be learned about the intimate details of the electron configuration at the iron center. The two contributions to the isotropic shifts are the contact (through bonds) and dipolar or pseudocontact (through space) contributions, and these are discussed in considerable detail elsewhere (*92–94*). For the purposes of this chapter, a detailed physical treatment of these two contributions is not necessary or even desirable, but rather, the differences in the proton NMR spectra of the heme substituents, and what they tell us about the electronic environment at the iron(III) center, the effect of

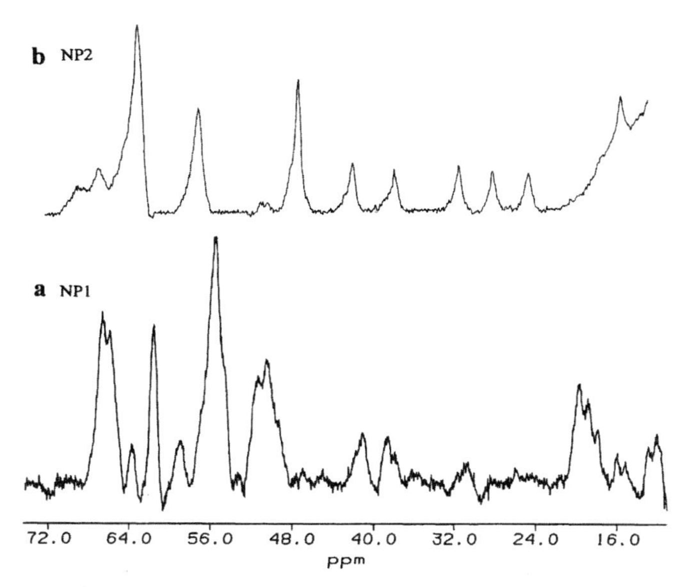

FIG. 10. 1D ^{1}H NMR spectra of the hyperfine-shifted downfield region of (a) 4 mM NP1 and (b) 6 mM NP2 in D_2O at pH* 7 (50 mM phosphate buffer) and 37°C, in the absence of added ligand. Note the well-resolved peaks of NP2, but the broad peaks observed for NP1. The NMR spectrum of NP4 is similar to that of NP1, and that of NP3 is similar to that of NP2. It is concluded that NP1 and NP4 do not have His59 bound to the heme at neutral pH in solution in the absence of a strong field distal-side ligand.

the proximal histidine on the proton NMR shifts, and other features of the protein structure and dynamics, are summarized qualitatively.

1. Proton NMR Spectra of the High-Spin ($S = \frac{5}{2}$) (NO-Free) Forms of the Nitrophorins

The NO-free form of recombinant NP1 at pH 7.0 and $T = 38$°C has resolved heme and hyperfine-shifted protein resonances that extend from 70 to −20 ppm (*91*), the low-field portion of which is shown in Fig. 10a. The pattern of these heme resonances is qualitatively similar to that of aqueous protohemin at neutral pH (*95*), except for the broadening induced by the much longer rotational correlation time of the nearly 20-kDa protein. This pattern is unlike that of any high-spin heme protein reported previously, including those of methemoglobins, metmyoglobins, cytochrome c', and the resting state of horseradish peroxidases (*94, 96–99*). On the other hand, the NMR spectrum of the high-spin form of NP2 is entirely different, as shown in Fig. 10b, with well-resolved, unique resonances of intensities corresponding to one or three protons that are more similar in general appearance to those of previously studied high-spin Fe(III) proteins (*96–99*) especially horseradish peroxidase (*99*) and even cytochrome *c*′ (*98*). The strange appearance of the hyperfine-shifted heme resonances of the NO-free form of recombinant NP1 appears to be indicative of only weak (if any) binding of the proximal histidine to the heme at pH 7 in homogeneous solution (*91*), even though the histidine is clearly bound in the crystalline state (*31, 32, 47–49*), as well as in solution for the ligand-bound low-spin states discussed later. Rather, the spectra are consistent with the heme in NP1 being simply confined in the heme binding pocket, probably by hydrophobic interactions between the heme substituents and the protein side chains making up the pocket, with no direct coordinate bond to the metal at physiological temperatures and pH values, in solution. In contrast, the heme in NP2 is clearly bound to His57 (numbering system based on the deletion of two residues for NP2 and NP3, as compared to NP1 and NP4, Fig. 4) and thus experiences different contact and dipolar shifts at different heme methyl and other substituent positions because of the orientation of the proximal histidine (*91*). Preliminary investigations of the hyperfine-shifted proton resonances of high-spin NP4 and NP3 show that their spectra resemble those of NP1 and NP2, respectively (*85*).

The heme methyl resonances of high-spin NP2 have been assigned using apoprotein samples reconstituted with specifically deuterated protohemins obtained from Professor Kevin M. Smith (*91*). (The samples were prepared by reducing the pH of purified holo-NP2 to 2.7 and extracting the protohemin originally present with 2-butanone, as was done previously with metmyoglobin (*100, 101*), hemoglobin (*101, 102*), and cytochrome b_5 (*103*).) Then, the pH of the apoprotein was raised to near 7 and a specifically deuterated protohemin, dissolved in 0.1 M NaOH, was titrated into the apoprotein solution until the ratio of the UV-visible absorption bands due to the heme Soret and protein residues returned to its original value. It was found that only about 80–90% of the protohemin originally present in the holoprotein could be removed, so these reconstitution experiments always left some residual protonated-methyl protohemin present in the samples.) When 5-CD_3-protohemin is incorporated into apo-NP2, it is observed that the peak at 62 ppm (Fig. 10b) is greatly reduced in intensity, whereas when 5,8-$(CD_3)_2$-protohemin is incorporated into apo-NP2, the peak at 62 ppm is again greatly reduced in intensity, but neither of the other two resolved methyl peaks is changed in intensity, indicating that the 8-methyl resonance is buried under the diamagnetic envelope of protein resonances.

However, another singly deuterated methyl hemin was not available to allow assignment of the 1- and 3-methyl resonances. Therefore, the assignment of these resonances was accomplished by first assigning the low-spin Fe(III) resonances of an appropriate axial ligand complex, and then using chemical exchange between the ligated and unligated forms of the protein to assign the other two heme methyl resonances, as discussed later. From these experiments, it was possible to assign the methyl resonances in Fig. 10b (57173) (*91*).

2. *Proton NMR Spectra of the Low-Spin ($S = \frac{1}{2}$) (Lewis Base–Bound) Forms of the Nitrophorins*

Addition of strong-field ligands such as CN^-, imidazole (ImH), or histamine (Hm) to these high-spin Fe(III) centers of the nitrophorins creates the low-spin Fe(III) state ($S = \frac{1}{2}$), which is characterized by a smaller range of NMR shifts (20–30 ppm) and much sharper resonances than those of the high-spin forms of the same proteins. Both one- and two-dimensional NMR techniques (1D NOE difference spectra, 2D COSY, TOCSY, NOESY, and ROESY spectra) have been extensively utilized to assign the hyperfine-shifted resonances of the heme in the cyanide-bound forms of Fe(III) heme proteins (*104*), where most, but not all, of the heme substituent resonances are found outside the diamagnetic envelope of the protein. In the present work, CN^- has been utilized as an even-electron, diamagnetic substitute for NO for characterizing the nitrophorins from blood-sucking insects in the paramagnetic low-spin Fe(III) state. Imidazole and histamine have also been used extensively to produce low-spin ($S = \frac{1}{2}$) Fe(III) centers for the nitrophorins (*91*), because of the finding that both of these ligands bind to NP1 with higher affinity than does NO (*31, 50*).

a. Lyophilization of Protein Creates Heme Rotational Disorder One of the first observations about the imidazole complex of NP1 was that samples of NP1–ImH, in which the lyophilized protein was simply dissolved in D_2O containing buffer, slightly greater than 1 equiv of imidazole was added, and the NMR spectrum was recorded immediately, had approximately twice as many isotropically shifted heme resonances as did the NMR spectrum of the same sample recorded 12 hours later (*49*). As shown in Fig. 11a, a number of resonances decreased in intensity, such that after 12 hours there were no remaining traces of those resonances, while the others increased in intensity (Fig. 11b). These results were interpreted in terms of the well-known "heme rotational disorder" observed in heme *b* (protohemin)-containing proteins (*103, 105–108*) in

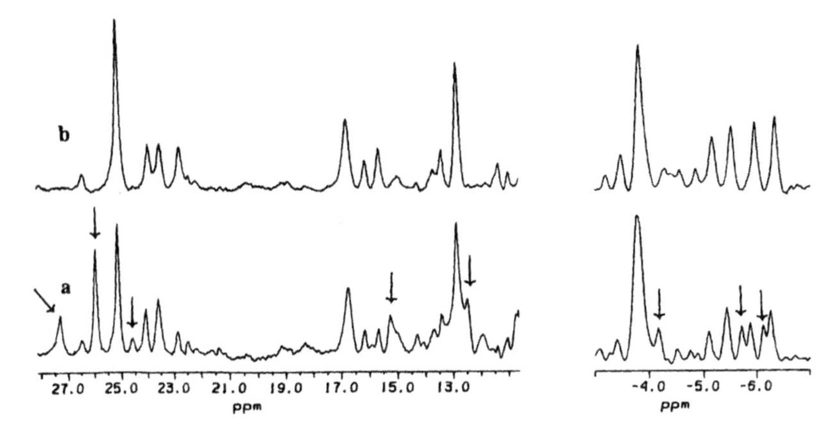

FIG. 11. 1D ^{1}H NMR spectrum of the hyperfine-shifted region of a sample of 4 mM NP1–ImH in D_2O at pH* 7 (50 mM phosphate buffer) and 37°C: (a) freshly prepared sample; (b) same sample 12 hours later. Reproduced with permission from Ref. (*49*).

which the apoprotein has been freshly reconstituted with protohemin, an unsymmetrical molecule. Protohemin can insert itself randomly into the chiral protein pocket in one of two ways that interchange the positions of the vinyl and methyl groups at the 1,4 and 2,3 β-pyrrole positions, hence providing double the number of unique heme substituent resonances for the chiral heme protein (*108*). In most cases, the two rotational isomers are not of equal stability, so that over a period of hours, the heme "turns itself over" to adopt the preferred orientation. (In fact, it is much more likely that the heme leaves and reenters the heme binding cavity many times until it randomly reaches the equilibrium ratio of the more stable to less stable orientation for that particular protein, than that it is able to turn itself over within the heme cavity.) There is probably little functional significance to this heme rotational isomerism, which results in only very small differences in reduction potentials of the two cytochrome b_5 heme rotational isomers (*103*), for example (however, see Ref. *108*), but its presence can certainly complicate the interpretation of the NMR spectral data for heme *b*-containing proteins. For the case of NP1, the existence of heme orientational disorder in a freshly prepared sample suggests that upon lyophilization the heme dissociates from the protein, so that it reinserts randomly when the lyophilized sample is dissolved in D_2O. For NP1–ImH, the ratio of more to less stable heme orientation must be greater than 20 : 1, based upon the undetectably small amount of the "minor" heme orientational isomer of Fig. 11b (*49*).

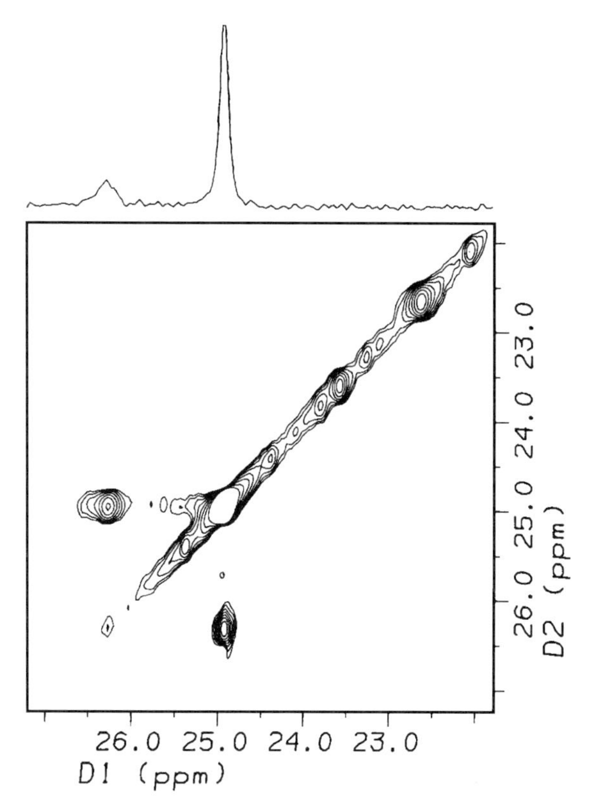

FIG. 12. A portion of the 500 MHz DEFT-NOESY spectrum of NP1–ImH, recorded at 30°C in 50 mM phosphate buffer, pH* 7.0, D_2O. Note the strong EXSY cross peak between the 3-Me signal at 24.9 ppm and a minor 3-Me signal at 26.4 ppm. This is not the other heme orientational isomer, as is evident by comparison of spectrum (a) of Fig. 11. This chemical exchange is also observed for many protein side chains, suggesting fluxionality of some loop, probably the A-B loop that passes close to the heme. In the text, the small peak is defined as species "Y," and the large as species "X."

b. Chemical Exchange Observed for NP1–ImH The most detailed NMR studies of low-spin forms of NP1 and NP2 have been carried out on the imidazole complexes. However, for the imidazole complex of NP1, chemical exchange between at least two sets of resonances of similar chemical shift plagued the attempts to assign the heme resonances. In each case, one set is nearly unobservable in the 1D spectrum, as shown for the small peak near 27 ppm in Fig. 11b, yet gives very strong chemical exchange cross peaks in the DEFT-NOESY/EXSY spectrum, as shown in Fig. 12. This chemical exchange is in the slow-exchange regime in terms of the 1D spectra, as shown in Fig. 11b, but is very readily observed in the NOESY/EXSY spectrum of Fig. 12. It can be further slowed on the NMR time scale by cooling the sample to 12–14°C, but at this temperature the resonances are much broader and the multiple species are still present. The species interconvert too rapidly to involve dissociation, rotation, and reassociation of the hemin, and the pattern of chemical exchange cross peaks is also not consistent with heme rotation. Rather, the multiple resonances must involve some change in conformation of one or more protein side chains, *cis–trans* isomerization about the disulfide bonds of the protein, or, an even more likely possibility, the *cis–trans* isomerism of the two prolines in the A-B loop, made by residues 32–41 (Fig. 4), which is very poorly defined in the NP1 and NP4 structures solved thus far (*46, 48*) (except for the structure of the NO complex of NP4 (*47*), as discussed in Section III) and may thus be quite fluxional. Preliminary analysis of the diagonal and cross peak intensities (Fig. 12) in terms of a two-site chemical exchange process, $X \rightleftarrows Y$, where Y represents the species giving rise to the small peak, yields rate constants of the order of $k_{YX} \approx 70–90\ s^{-1}$ and $k_{XY} \approx 7–10\ s^{-1}$, $K_{eq} \approx 0.1$, at pH 7.0, 30°C (*85*).

No such chemical exchange is observed for NP2–ImH or NP3–ImH, for which WEFT-NOESY spectra have cross peaks due only to through-space nuclear Overhauser effect interactions (*91*). Interestingly, these two proteins have only one proline in the A-B loop (Fig. 4). However, considerable additional NMR investigation needs to be carried out in order to determine whether the chemical exchange process shown in the NOESY spectra of NP1–ImH is indeed due to the dynamics of the A-B loop.

c. Heme Resonance Assignments for the Nitrophorin-Imidazole Complexes As discussed in Section II,E,1 for the high-spin forms of the nitrophorins, the NMR spectra of the low-spin imidazole complexes of the pairs NP1, NP4 and NP2, NP3 also differ significantly, as shown in Fig. 13. In contrast to an earlier report (*109*), deuterated methyl-labeled hemins have confirmed that only *one* heme methyl resonance of NP1–ImH, observed at 25 ppm at 38°C (*91*), Fig. 13a, is resolved outside the diamagnetic envelope, while the other three are buried in the diamagnetic region of the NMR spectrum (−1 to 11 ppm). The peak at 13 ppm that was earlier believed to be a methyl resonance (*109*) is not changed in intensity upon substitution of any of the deuterated hemins (*91*). NP4–ImH also behaves in this manner (Fig. 13b). In contrast, three heme methyl resonances, more closely spaced, at 15.7, 12.4, and 11.3 ppm, are observed for NP2-ImH (Fig. 13c), and a similar pattern

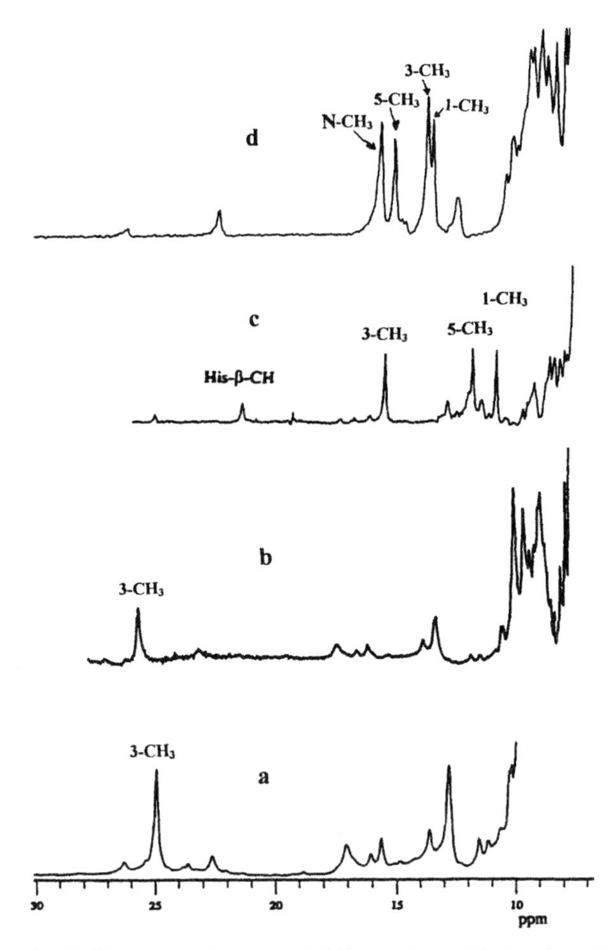

FIG. 13. 1D ^{1}H NMR spectra of the downfield hyperfine-shifted region of (a) NP1–ImH, (b) NP4–ImH, (c) NP2–ImH, and (d) NP2–NMeIm, showing the similarity in chemical shifts of heme resonances for NP1,4–ImH and the difference for NP2–ImH. The spectrum of NP3–ImH is similar to that for NP2–ImH. Spectra recorded in D_2O, pH* 7 (50 mM phosphate buffer), and 37°C.

is observed for NP3–ImH. The NMR spectrum of NP4–ImH is similar to that of NP1–ImH, as shown in Fig. 13b. From the high-resolution crystal structures of NP4 and several of its ligand complexes (*32, 48*), it is known that both heme orientations exist for NP4, in a 60:40 ratio, but this is not apparent from comparison of the NMR spectra of NP1–ImH and NP4–ImH (Fig. 13a,b). For NP1–ImH and NP2–ImH, whose NMR spectra have been investigated in the greatest detail (*91*), only one heme rotational isomer is observed, despite the deceptively open-looking nature of the distal pocket. This is probably because of the tight packing of a number of phenylalanines and tyrosines around the vinyl and methyl substituents of the protohemin in the nitrophorin heme cavities (*46, 48, 49*).

The situation is much more favorable for complete assignment of the heme resonances of NP2–ImH and NP3–ImH than it is for the other two proteins, because of the three resolved heme methyl resonances, as well as the lack of chemical exchange cross peaks in the NOESY/EXSY spectra. Whereas the order of chemical shifts is $3 \gg (5,1,8)$ for NP1,4–ImH, with unknown order of the latter three resonances, which are all buried in the diamagnetic envelope (−3 to 10 ppm), it is $3 > 5 > 1$, with the 8-methyl buried in the diamagnetic envelope, for NP2–ImH (Fig. 13c). Because the NOESY and COSY spectra showed cross-peak connectivities that were consistent with two possible assignments of the heme methyl resonances (*91*), the heme 5-methyl resonance of NP2–ImH was assigned and the 8-methyl resonance was shown not to be resolved outside the diamagnetic envelope of the protein, as described earlier for the high-spin nitrophorins, by using deuterated hemins supplied by Kevin M. Smith. The relative order of the 5- and 1-methyl resonances is always $5 > 1$, because of the substituent effects of the adjacent propionate and vinyl groups, respectively (*109*). Hence, knowing that the 5-methyl is the second heme methyl resonance and that the 8-methyl peak is not resolved outside the diamagnetic envelope fixes the assignment of the 3- and 1-methyl resonances (*91*). Methyl-8 of NP2–ImH has been located at 1.8 ppm from the natural-abundance $^{1}H/^{13}C$ HMQC spectrum (*91*). It was not possible to locate the 5-, 8-, and 1-methyl resonances of NP1–ImH from HMQC spectra of that protein because of the chemical exchange phenomena described in Section II,E,2,b.

The chemical shift of the single resolved heme methyl resonance of NP1–ImH places the orientation of the ligand plane(s) between 125° and 145° from the heme N–N axis connecting pyrrole rings II and IV, since according to the predictions of the effect of ligand plane orientation on the chemical shifts of the heme methyls (*109*), it is only in this range of angles that the other three methyl resonances should lie at less than 10 ppm (Fig. 14). This is indeed within the expected range of angles, as shown by the crystal structures of NP1 and its ligand complexes (*31, 49*), for which the imidazole plane of His 59 in molecule I is oriented 11° counterclockwise of the line connecting the β,δ *meso*-H, or 146° counterclockwise of the porphyrin N–N axis connecting pyrrole rings II and IV, whereas the imidazole plane of the histamine ligand

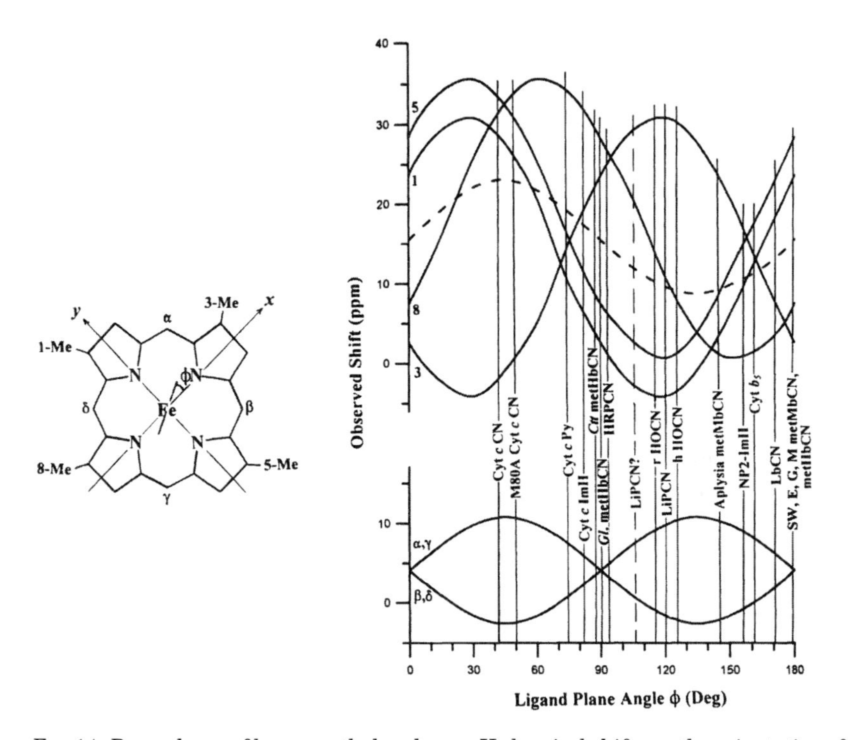

FIG. 14. Dependence of heme methyl and *meso*-H chemical shifts on the orientation of the axial histidine and other planar ligand. (*Left inset*) Heme structure with definition of axes and ligand plane orientation angle, ϕ. For NP1–ImH it is known that His59 makes an angle of 146° with the x-axis, whereas ImH probably makes an angle of 106° with that axis, if it is aligned as is the imidazole ring of histamine in the structure of NP1–Hm (*31*). The best fit of the one resolved heme methyl, Fig. 13(a), 3-Me, is with an angle of 125° to 145°, suggesting that either both ligands are equally important or His59 is more important in determining the orientation of the heme $e(\pi)$ orbital nodal plane. For NP2–ImH, the imidazole ring of His57 makes an angle of +124° with the x-axis; however, the heme methyls match an angle of about +155°, suggesting that in this case, the exogenous imidazole ligand is *at least as* important in determining the orientation of the heme $e(\pi)$ orbital nodal plane as is His57. Modified from Ref. (*109*), with permission.

of NP1–Hm in molecule I is oriented 29° clockwise of the line connecting the β,δ *meso*-H, or 106° counterclockwise of the same porphyrin N–N axis (Fig. 14, left). Thus, either the two ligands contribute equally in determining the orientation of the nodal plane of the $e(\pi)$-orbital that is involved in spin delocalization to the porphyrin ring (average angle 126°), or else His59 plays a more important role than does the added histamine or imidazole ligand. The conclusion reached in the ESEEM investigations of NP1–Hm (*89*), discussed in Section II,D, was that His59 plays a more important role. Because only one heme methyl is resolved, NMR spectroscopy cannot provide supporting information for this conclusion.

In contrast, the difference in the chemical shifts of the three resolved methyl resonances of NP2–ImH suggested that the orientation of the imidazole plane of His57 of NP2 was rotated at least 10° counterclockwise from the position of His59 of NP1 (*31, 49*), an angle of 155° in Fig. 14. On this basis, it was predicted that once the crystal structure of NP2 was solved, the imidazole plane of His 57 would be found to be aligned at about 10° counterclockwise of the line connecting the β,δ *meso*-H. However, when the crystal structure of NP2 was solved, it was shown that although the positions and angles of the histidine ligands of the two proteins are almost identical, the heme orientation of NP2 was reversed as compared to that for NP1, putting His 57's alignment at about 124° counterclockwise of the porphyrin N–N axis connecting pyrrole rings II and IV, or about 10° *clockwise* of the line connecting the β,δ *meso*-H (*110*). And although the structures of neither the histamine nor the imidazole complexes of NP2 have been solved, by analogy to the structure of the histamine complex of NP1 (*31*), the orientation of the imidazole plane of that ligand should be at about 164° counterclockwise of the porphyrin N–N axis connecting pyrrole rings II and IV, or about 30° counterclockwise of the β,δ *meso*-H line.

Hence, the methyl ^{1}H NMR shifts suggest an angle of the nodal plane of the porphyrin orbitals of the unpaired electron of 155°, while the His 57 ligand is positioned at 124° and the added imidazole ligand is probably at about 164°. The average position of the two ligand planes (144°) is closer to the angle indicated by the methyl shifts of Fig. 14 (155°) than it is to the position of His57, but it is smaller than the predicted angle. Thus, it appears that the exogenous ligand is *at least as* important in determining the heme methyl shifts of NP2 than is its proximal histidine ligand, and, in fact, either it is *more* important, or else its angle is greater than expected, based upon the structure of NP1–Hm (*31*), and thus it is equally important in determining the heme methyl shifts as is the His57 ligand plane.

3. Connections between Low- and High-Spin Forms via Saturation Transfer

Although the kinetics of imidazole exchange between NP2 and its low-spin imidazole adduct are very slow and no chemical exchange cross peaks in the NOESY/EXSY spectrum of a mixture of the high-spin and low-spin forms of NP2 are observed, the *N*-methylimidazole complex has faster ligand exchange kinetics that allowed us to connect the assigned heme methyl (and other residue) peaks of the high-spin form of the protein that has no added ligand to those of the low-spin N-MeIm complex (*91*). Based upon the assignment of the 5-methyl resonance of NP2 in its high-spin form (Fig. 10 and discussion), the 5-methyl resonance of the low-spin *N*-methylimidazole complex was assigned by saturation transfer experiments carried out on a sample containing both species (*91*). The 1- and 3-methyl resonances were then assigned based upon the predictions of the angular plot of Fig. 14. The fact that orientation of the *N*-methylimidazole ligand is somewhat different from that of the imidazole ligand is indicated by the even smaller spread of the three resolved methyl resonances (Fig. 13d), the difference in order ($5 > 1,3$ vs $3 > 5 > 1$, respectively), and the fact that the 1- and 3-methyl resonances of the former cross each other at temperatures between 30 and 37°C indicate that the nodal plane of the porphyrin orbital responsible for the methyl contact shifts in NP2–NMeIm has a slightly larger angle (160–162°) than in NP2–ImH (155°).

4. Proton NMR Spectroscopy of the Protein as a Whole

Finally, preliminary investigation of the proton resonances of the diamagnetic part of the NMR spectrum of NP2–ImH indicates that the protein is well behaved for detailed NMR studies of the protein aimed at determining its 3D structure in solution in order to allow study of protein dynamics and protonation/deprotonation equilibria of carboxylates near the heme. Approximately 120 NH–C_αH or related cross peaks are resolved in the DQF-COSY spectrum, which bodes well for the possibility of determining the structure and studying the dynamics of the protein by NMR techniques. Single (^{15}N), double ($^{15}N,^{13}C$), and triple ($^{15}N,^{13}C,^{2}H$) labeling of the protein is currently underway to facilitate these goals.

III. Crystallization and Structure Determination of Nitrophorins

A. Isolation and Crystallization of Native and Recombinant NP1–NP4

The initial crystals of NP1 were obtained with protein painstakingly isolated from *Rhodnius* salivary glands by the Ribeiro laboratory (*44*). This group also cloned all four nitrophorins (*44, 45*), which unfortunately resulted only in inclusion bodies or degraded protein in various expression systems (including *E. coli,* insect, and yeast cells). In order to obtain sufficient protein for the planned experiments, numerous conditions for renaturing the nitrophorins from the *E. coli*–derived inclusion bodies were examined, and eventually conditions were found for obtaining >100 mg of purified, fully active protein per preparation (*31, 46–49*), as detailed in Section I,C,4. This required not only refolding the protein, but also heme insertion and the formation of two correctly linked disulfide bonds. For NP2, several codons early in the coding sequence were altered to be those most commonly used by *E. coli,* which resulted in increased expression. We also showed that the renatured proteins behave identically to the insect-derived material, based on spectroscopic and kinetic analyses (*31, 46–50*). Prior to this, only a few milligrams of protein per year could be obtained from the insects. Now, sufficient amounts of all four nitrophorins can be produced for crystallographic, NMR, and various other spectroscopic analyses.

1. General Information about Rhodnius Nitrophorin Structures

We have completed several structures each of NP1, NP2, and NP4 (*31, 46–49, 110*). These structures reveal the *Rhodnius* nitrophorins to have a fold dominated by an eight-stranded antiparallel beta-barrel, as shown in Fig. 15, and to rely on a remarkable ligand-induced conformational change for NO transport, described later. The structures confirm that the nitrophorins are completely unrelated to the globins, the only other heme-based gas transport proteins whose structures are known. Rather, their fold places them in the lipocalin family, for which several other examples are known (*111–113*). Our initial nitrophorin structure was of NP1 and was determined using standard MIR and

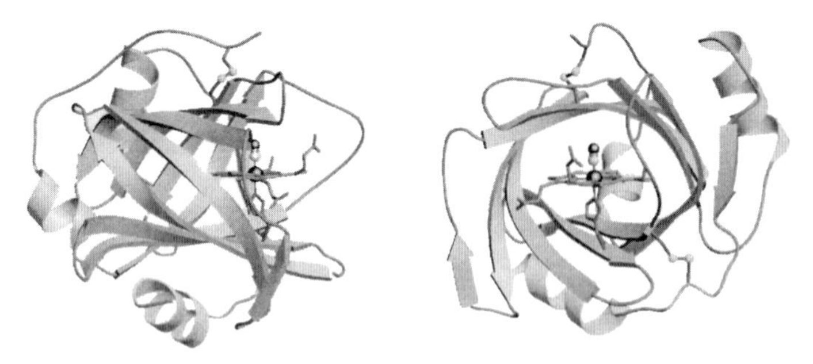

FIG. 15. Ribbon and ball-and-stick diagrams of the NP1–CN structure. The view in (b) is rotated approximately 90° about the vertical axis from the view (a) on the left. The mobile loops are above and next to the heme in this view. Reproduced with permission from Ref. (*31*).

molecular averaging approaches, but with the somewhat unusual heavy atom derivatives of a mercury inserted into a disulfide bond, and a 4-iodopyrazole bound to the distal side of the heme (*31*). Structures of other complexes of NP1, as well as NP4 and NP2, have been determined by molecular replacement.

The nitrophorin heme is inserted into one end of the beta-barrel (the other end is closed) and is ligated on the proximal side to His59, much like the proximal histidine of myoglobin. The distal side is quite open in the absence of NO and does not contain a distal histidine, unlike most myoglobins and hemoglobins. Crystals grown in the presence of $(NH_4)_2HPO_4$ have an NH_3 bound in the distal position (NH_4^+ does not bind), whereas crystals grown in polyethylene glycol (PEG) have a weakly ordered water molecule bound to this site. The distal pocket contains at least four additional water molecules under these conditions. The proximal histidine is oriented in part through a hydrogen bond to a buried water molecule in NP1 and NP4, which is further hydrogen bonded to a buried aspartate, Asp70. It has been suggested that Asp70 is protonated even at pH 7.5 because of its buried position and a second hydrogen bond to one of the heme propionates (Fig. 16), which is not buried and forms a strong hydrogen bond to Lys125 (*31*). Support for this comes from the structure of NP2 (*110*), which has an asparagine at this position that is positioned identically to Asp70 in NP1 and NP4 (sequences, Fig. 4). (However, the NMR data discussed in Section II,E,2,c suggest some difference in proton acceptor strength of Asp70 and Asn68 in the two subgroups of nitrophorins.)

Our electrostatic calculations (using DelPhi (*114*)) suggest that several negatively charged residues near the heme result in an overall negative charge in the heme pocket, which may serve to stabilize the ferric form of the heme (formal charge $= +1$; see Section V,A and Ref. *115* for an example of Glu and Asp mutants stabilizing ferric myoglobin). Another stabilizing factor may be a series of aromatic amino acids that have face-to-heme edge contacts (Y28, F86, and Y105) or edge-to-heme face contacts (Y40 and F68), which may influence the heme through π orbital overlap (*116*). The heme itself is highly "ruffled" (*117, 118*), as is evident in Figs. 16 and 17, with an overall largest out-of-plane deviation of the heme *meso*-carbons of $\sim$0.8 Å, making the nitrophorins the only known proteins with noncovalently linked protohemin to display a highly ruffled heme conformation (*119*). A second unusual feature is that the imidazole plane of His59 is significantly tilted (about 10°) with respect to the normal to the heme plane. Both ruffling and histidine imidazole plane tilt may influence sixth-ligand binding through modifications to the iron *d*-orbitals.

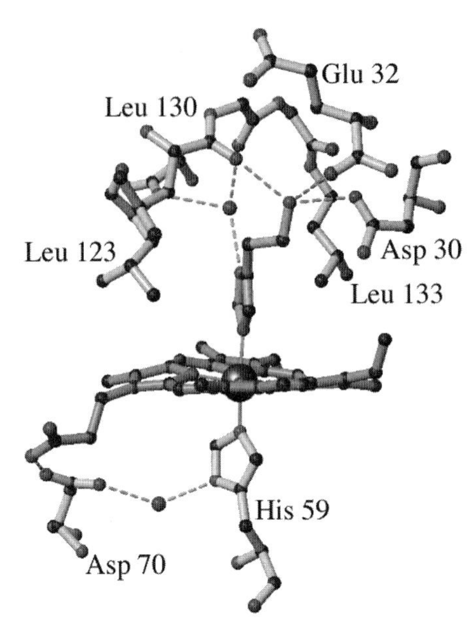

FIG. 16. Histamine binding to the heme of NP1. Shown are hydrogen bonds between histamine and Asp30 (2.7 Å), Glu32 (3.1 Å), Leu130 (2.7 Å), and an ordered water molecule (2.8 Å), which further hydrogen bonds to Thr121 (3.2 Å, not shown), Leu123 (3.0 Å), and Gly131 (2.7 Å). Van der Waals contacts are made to Leu123 (3.7 Å), Leu130 (4.2 Å), and Leu122 (3.7 Å). Also shown are hydrogen bonds between an ordered water molecule and residues His59 (2.7 Å) and Asp70 (2.6 Å), and between Asp70 and a heme propionate (2.5 Å). The other heme propionate has been omitted for clarity. Reproduced with permission from Ref. (*31*).

2. *Comparison of NP4 to Other Lipocalins*

The lipocalin fold was originally discovered in retinol binding protein, the protein required for retinol transport (*120*). Since then, it has become clear that a marvelous variety of biological functions depend on the lipocalin fold. Despite the growing size of the family of well-characterized lipocalins, detection of this fold at the sequence level has proved difficult because the proteins show a low degree of amino acid identity. In fact, as mentioned in Section I, the nitrophorin sequences align as well with the helical globins as they do with the β-barrel-containing lipocalins, and they were not recognized as lipocalins until the structure of NP1 was determined (*31*). Nevertheless, taking a structure-based approach to the alignment of lipocalin sequences, some interesting relationships are detected among some proteins of extremely diverse function.

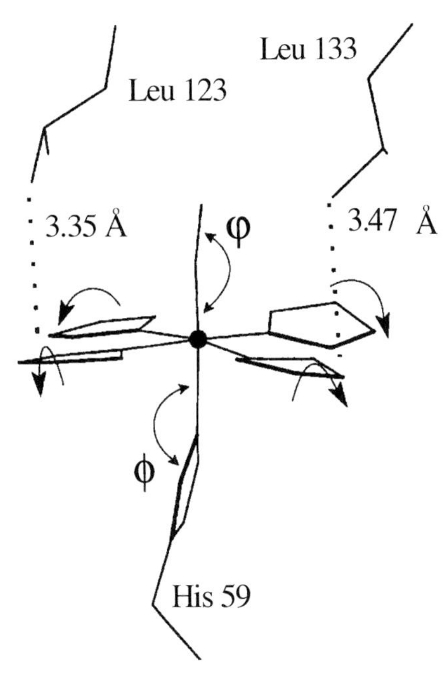

FIG. 17. Structure of the active site of NP4–NO, showing the highly ruffled heme ring and the off-axis binding of the proximal histidine. Only the heme iron and pyrrole rings are included, to emphasize the heme ruffling. Also shown are Leu123 and Leu133, which form close contacts with the heme and may help to induce ruffling. The pyrrole ring rotations are defined to be positive for a clockwise rotation when looking from the pyrrole to the iron. Pyrroles *trans* to one another rotate in opposite directions. The rotation angles range, in absolute value, from 7.5 to 13.6° for NP4–NO, and smaller values (4.0–8.9°) for NP4–NH_3 and NP4–H_2O. The angle between the His59 imdazole ring plane in the proximal bond (ϕ) is defined as the angle between the bond and the (positive) ring normal and is 170° for NP4–NO, 169.9° for NP4–NH_3, and 169.8° for NP4–H_2O.

The structure of NP4 has been superimposed with those of five other lipocalins by first identifying analogous structural regions through inspection, and then determining a least-squares best fit between pairs of structures. The five structures were those of butterfly bilin binding protein (BBP) from *Pierus brassicae* (*113*), moth insecticyanin (INS) from *Manduca sexta* (*112*), human retinol binding protein (RBP) (*120*), bovine β-lactoglobulin (LAC) (*121*), and rat epidydimal retinoic acid binding protein (ERB) (*122*). These proteins are typical lipocalins and contain the same secondary structural elements as the nitrophorins. From these superpositions, an amino acid sequence alignment was derived that was used to determine the true degree of amino acid identity of nitrophorins with other lipocalins and to determine structurally equivalent positions within the group (*48*). The most similar structures were INS and BBP (41% identical, rms deviation in the core atoms of 1.2 Å), while BBP and ERB have very different sequences (12% identical), yet have rms deviation in the core atoms of 1.3 Å. Likewise, NP4 has most sequence homology with the insect-derived proteins BBP (16%) and INS (15%), but the core region aligns better with human-derived RBP (1.3 Å as compared to 1.7 and 1.8 Å), with which it has only 8% sequence identity.

Nevertheless, other features suggest that the insect proteins are more closely related to each other than to the other lipocalins examined. Comparison of disulfide bonding patterns among the proteins shows that indeed, the three insect-derived proteins have the same pattern, whereas the mammalian proteins have different patterns of disulfide bonds. NP4 contains two disulfide bonds, one connecting Cys2 at the N terminus with Cys122 on strand G of the β barrel, and the other connecting Cys41 on strand B with Cys171 at the C-terminus, as shown diagrammatically in Fig. 18. The positions of these cysteines and the pattern of disulfide bonding are identical among NP4, NP1, NP2, and the insect lipocalins BBP and INS. In contrast, the mammalian lipocalins contain from one to three disulfide bonds, and different residues are involved. All of the mammalian proteins examined have a cysteine residue near position 65 on strand D that is not present in insect lipocalins and is disulfide bonded with a cysteine residue near the C-terminus of the

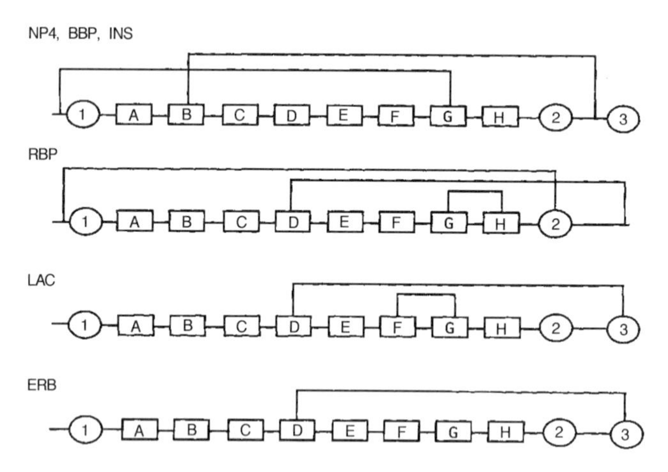

FIG. 18. Disulfide bonding patterns in the aligned lipocalins. Individual strands of the β sheet are shown as rectangles, helices as ovals, and the disulfide bonds as connecting lines. Helix 3 does not occur in BBP and INS. Reproduced with permission from Ref. (*48*).

protein. Thus, disulfide binding patterns appear to be a better predictor of evolutionary relatedness among the lipocalins than overall rms deviations in the protein structure.

3. Comparison of the Ligand-Binding Pocket of NP4 with BBP and INS

Both insecticyanin (INS) and biliverdin binding protein (BBP) bind the breakdown product of heme, biliverdin. These yellow pigments are formed in mammals by the enzyme heme oxygenase (HO), in which the protein HO binds hemin, undergoes 1-electron reduction from Fe(III) to Fe(II), then binds molecular oxygen, and then undergoes another 1-electron reduction to produce a peroxy-bound Fe(III) center (*123*). However, rather than undergoing O–O bond homolysis, as is the case in the cytochromes P450 (*124*) and peroxidases (*124, 125*), the bound peroxide of heme oxygenase reacts electrophilically with the heme itself, at one of the *meso*-carbon positions (*123*). For the mammalian enzymes, the attack is at the α-*meso* carbon. The final products of this reaction are carbon monoxide (CO), free ferric ion, and α-biliverdin (*123*). In contrast, the biliverdin isomer present in INS and BBP is γ-biliverdin (*112, 113*). Despite this difference, the question was raised as to whether the original tetrapyrrole molecule bound in the β-barrels of these two proteins might have been heme, rather than biliverdin (*48*). Hence, the structures of INS and BBP were inspected carefully for the presence of a histidine in a similar position to the His59 of NP4. Indeed, BBP has a similarly positioned histidine, His61 (*113*). However, INS has phenylalanine at this position, but inspection of the distal pocket of INS reveals that His 131 is positioned with its side chain projecting toward the heme iron in such a way that it would be possible for this histidine to bind to the metal, if some rearrangements of side chains occurred (*48*). Preliminary results with NP1 and NP4 show that incubation with hydrogen peroxide (*126*) and NADPH-cytochrome P450 reductase (the redox partner of mammalian heme oxygenase (*127*)) leads to degradation of heme in both proteins. These structural observations and biochemical data suggest that biliverdin production and binding may have served as an evolutionary preadaptation for the nitric oxide binding function of the nitrophorins (*48*).

B. Ligand Binding: NO Induces Complete Distal Pocket Burial, while Histamine Stabilizes an Open Conformation

Structures of nitrophorin complexes with ammonia, histamine, cyanide, and nitric oxide have been obtained (*31, 46–49, 110*). These structures reveal a remarkable feature: the nitrophorins have a broadly open distal pocket in the absence of NO (*31, 46, 48, 110*), as well as for NP1 in the presence of NO, but reduced to Fe(II) (*49*). However, upon NO binding to the physiologically relevant Fe(III) form of NP4, significant changes in the region of the entrance to the distal pocket take place that result in the burial of the bound ligand (*47*), as discussed in Section III,B,1.

The highest-resolution structural data for the nitrophorins are from a small crystal of the NP4–NH_3 complex, pH 7.5, that was flash-frozen and examined at the European Synchrotron Radiation Facility, leading to a model at 1.15 Å nominal resolution (*128*). These data are superb, allowing for fewer restraints during refinement (using SHELX (*129*)), anisotropic temperature factor refinement, hydrogen atom inclusion, multiple atomic position refinement, and atomic resolution in electron density maps. The heme is extremely well defined, and it is clear that it is discretely disordered in the crystal, with ~60% of the heme molecules "right-side-up" and ~40% "upside-down," such that all four possible vinyl positions (two for up and two for down) are partially occupied, and is significantly ruffled. Most other heme proteins appear to have identical properties in the two orientations, and therefore no functional consequences are expected, much like the case of cytochrome b_5 (*103*).

These results with regard to heme disorder differ from those observed in solution by NMR spectroscopy, where one heme orientation is at least highly favored (Section II,E,2). The difference in conditions (high salt, PEG) may be involved in the fact that different ratios of the two heme orientations are observed in the crystalline state than in solution. The difference in heme orientation ratios of 1.5:1 and 10:1, for example, represents a ΔG difference of only 4.9 kJ/mol.

The ruffled heme appears to result from Ile123 and Ile133 pushing down from the distal pocket side (3.4 and 3.5 Å), and from His59, the proximal histidine, pushing up from below (3.4 and 3.5 Å), Fig. 17. Despite the ultrahigh resolution of the structure, residues 31–37 are sufficiently disordered that side-chain positions cannot be modeled.

1. *Structures of the Physiologically Relevant NP4^{III}–NO Complex*

A crystal of NP4–NO was prepared by soaking an NP4 crystal grown in PEG at pH 5.6 in an anaerobic solution containing NO. As discussed later, NO release at low pH is quite slow, and a solution of NP4 prepared under similar conditions was stable for several days, even when exposed to air. Release of NO under these conditions was quickly achieved by blowing argon over the solution, indicating that ferric heme was present (*49*). The NP4–NO crystal was flash-frozen and data measured at the University of Arizona to 1.4 Å resolution, leading to a map with clear electron density for the NO (*47*). It was found that NO binding to the ferric heme induces an extensive conformational change that results in complete burial of the ligand. Closure involves the expulsion of three solvent molecules, the burial of Asp 30, the formation of an extensive hydrogen bonding network, and the packing of hydrophobic groups around the NO molecule, resulting in a filled distal pocket (*47*). Residues 125 to 132 shift toward the distal pocket, allowing Leu130 to pack directly against the NO molecule, as shown in Figs. 19 and 20. Residues 31–37

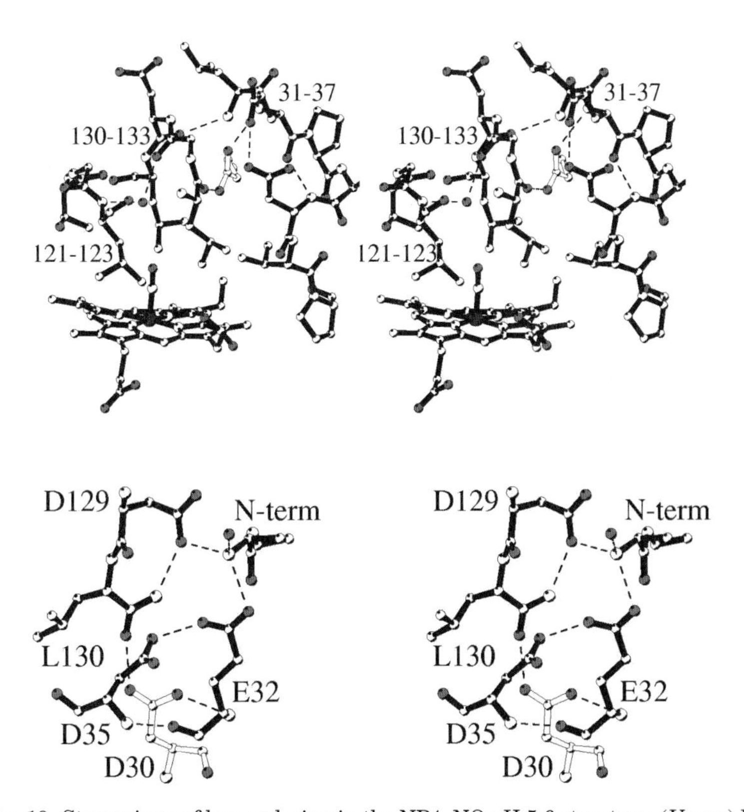

FIG. 19. Stereoviews of loop ordering in the NP4–NO pH 5.6 structure. (*Upper*) Loops 130–133 and 31–37 order over the distal pocket. Asp30 (open bonds) forms two hydrogen bonds and becomes buried in the process. L123, L130, L133, and V36 pack around the NO. (*Lower*) D30 (open bonds), E32, D35, D129, and the N-terminus cluster together in the closed conformer. Reproduced with permission from Ref. (*47*).

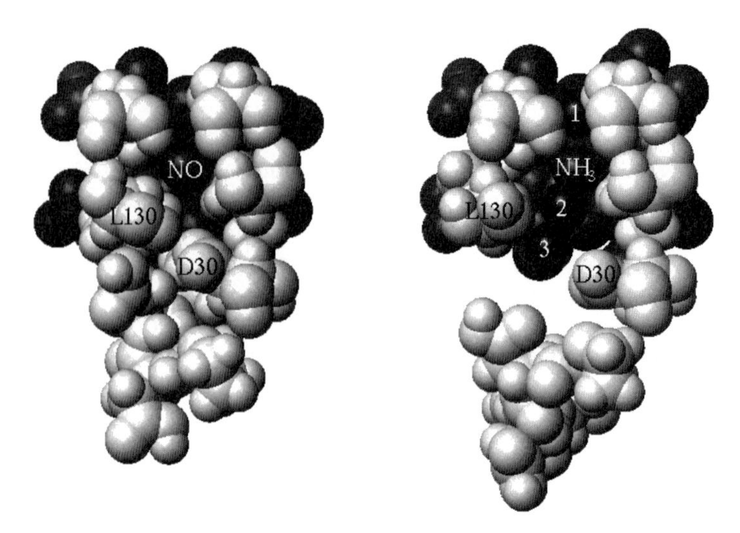

FIG. 20. Space-filling views of distal pockets in NP4–NO (left) and NP4–NH_3 (right) structures. View is from above the distal pocket, with heme, solvent, and ligands shown in black. Waters 1–3 (right) are expelled as L130 and loop 31–37 pack against NO (left). Reproduced with permission from Ref. (*47*).

order over the distal pocket entrance, leading to Val36 packing against Leu130 and Leu133, further burying the bound NO. The new positions for these loops are stabilized through an extensive hydrogen-bonding network formed by Asp30, Glu32, Asp35, Asp129, and the N-terminus (*47*). It is hypothesized that Asp30 and Asp35 are protonated in this arrangement (Fig. 19). The driving force for the conformational change is not yet clear, but it possibly results from the packing of NO with hydrophobic groups (NO is 70 times more soluble in *n*-hexane than in water (*130, 131*)), and by the release of three water molecules from the now more hydrophobic distal pocket (Fig. 20). The key residues involved in this ordering are conserved among the four nitrophorins, suggesting that all four undergo this ligand-induced transition.

The NO in this structure appears to be bound equally in two orientations, a component that is "linear" (Fe–N distance of 1.6 Å and Fe–N–O angle of 170°, Fig. 17), and a second component that is "bent" (Fe–N $\approx$ 2.0 Å, Fe–N–O $\approx$ 110°). To account for this, the NO molecule has been refined as a mixture of both orientations. The roughly linear orientation is indicative of a ferric (Fe^{III}) NO complex (*2, 65*), indicating that the NP4–NO structure represents the first ferric heme–NO complex for any protein. However, the bent orientation is similar in geometry to a ferrous heme–NO complex (*2, 65*). The bend directs the NO toward a small hydrophobic cavity at the back of the distal pocket that is occupied by a solvent molecule in the NP4–NH_3 structure. It is not yet clear if the multiple orientations are due to multiple oxidation states for the heme iron (as could occur by photoreduction), or if the bent conformation results from a steric conflict with Leu130, which is a very close 3.2 Å to the linear NO, but a more reasonable 3.9 Å to the bent NO in the current model. The Leu130 terminal side-chain atoms are less well ordered than the rest of the residue, possibly because of these contacts with NO. That two orientations for the ferric-NO complex exist in solution is supported by the occurrence of two release rates (discussed in Section IV) and two NO stretching frequencies for NP1–NO (measured by FTIR (*49*), Section II,B).

The NP4–NO structure has been examined under several other conditions to further explore the NO-induced conformational change. The closed conformer also exists for NP4–NO at pH 7.5 and at room temperature, but does not occur for the NP4–H_2O or NP4–CN^- complexes at low pH. That cyanide, which is isosteric with NO binding but carries a negative charge, fails to induce the conformational change may support the hypothesis that hydrophobicity drives closure of the binding pocket, although a formal positive charge must be present on the metal or delocalized over iron and the ligand in the formally Fe(III)–NO complex. Thus, NO once again proves to have unusual capabilities for binding and signaling in heme proteins.

2. *Structure of the Photoreduced NP1^{II}–NO Complex*

Binding of NO to NP1 was also examined at room temperature, pH 7.5, which yielded an open conformation and a completely ferrous-NO complex, apparently due to photoreduction (*49*). The increased photoreduction may be due to the more open binding pocket, which may in turn be due to unfavorable crystal contacts and an extra N-terminal amino acid. This is because the recombinant NP1 N-terminus has the initial methionine still present, which is not present in the insect-derived protein, where the N-terminus is formed by cleavage of the translocation signal sequence (*44*). However, for NP4, mass spectrometry, N-terminal sequencing, and the crystal structures all indicate that the N-terminal methionine has been cleaved during synthesis (*47*). The NP4 N-terminus hydrogen bonds to Glu 32 and Asp 129 (Fig. 19), but the importance of these contacts is not yet clear, since similar binding and release kinetics occur for recombinant NP1 and NP4 (see Section IV).

It is interesting that the NP1 ferrous-NO complex has the proximal histidine in place, unlike the heme of soluble guanylyl cyclase, where the proximal histidine is thought to detach from the heme on NO binding, leading to a conformational change and increased catalytic activity (*132–134*). Similar results have been reported for the myoglobin and hemoglobin nitrosyl complexes (*135, 136*). However, the affinity for the added "proximal" imidazole is greatly decreased in the nitrosyl complex of the mutant myoglobin from which the proximal histidine has been mutated to glycine (*137*), and, as discussed in Section II,A and in Section V, NP2–NO and NP3–NO lose their proximal histidine ligands at pH values of 7.0 and lower, when reduced to the Fe(II)–NO forms.

3. Structure of the NP1–Histamine Complex

Histamine binding has only been examined so far in NP1, where the ligand is found to be sandwiched between Leu123 and Leu133, occupying the NO binding site and forming four hydrogen bonds to the protein, as shown in Fig. 16 (*31*). The amino group displaces the water molecule that hydrogen bonds to Asp30 in the aqua complex, leading to an arrangement that stabilizes the open conformation of the A-B loop. The extensive histamine–protein contacts lead to tight binding (K_d = 5–15 nM at pH 8, Section IV) and underscore the specific role of this ligand in NO release.

C. Structure of NP2

NP2 and NP3 have diverged considerably from NP1 and NP4, and several potentially critical residues have been modified (Fig. 4), resulting in considerably tighter NO binding for NP2 (discussed in Section IV). NP2 binds NO ~300-fold tighter than NP1/NP4 and, in addition, contains a third antihemostatic activity, that of Factor X inhibition in the coagulation cascade (Fig. 1). The structure of NP2 at 2.0 Å reveals the protein to have the same general fold as NP1 and NP4 (*110*). The proximal pocket is nearly identical to NP1/NP4, but the heme has settled into a single orientation in the heme pocket, the reverse of that of the major form of NP4. Asp70 is an Asn in NP2, but lies in the identical position (*110*). The distal pocket has greater changes. First, two lysines and a propionate are removed from the entrance, leading to a more open binding pocket. Lys125 is changed to Glu, and the loop containing it and Lys128 rotates away from the binding pocket, causing a heme propionate to move from the distal to the proximal side of the heme (*110*). Second, at the back of the binding pocket, residue 121 is changed from Thr to Ile, resulting in increased hydrophobicity.

Away from the binding pocket, another interesting feature occurs. Invariant residues Lys14 and Tyr82 occupy entirely new positions in

NP2. In particular, Tyr82 flips from outside the protein in NP1/NP4, where it contacts bulk solvent, to inside the protein in NP2, where it hydrogen bonds to Glu55 (*110*). Thus, modeling the NP2 structure based on those of NP1 and NP4, and, as is normally done, keeping the invariant residues structurally conserved, would lead to errors. Structures of protein families such as the nitrophorins therefore provide key information for the future improvement of structure prediction and structural genomics.

D. Structures of NP4 at Multiple pH

Release of NO by the *Rhodnius* insect is assisted by a change in pH from ~5 in the insect salivary gland to ~7.5 in the host tissue (*24*). Binding of NO is about 10-fold tighter at low pH for each of the nitrophorins than it is at neutral pH, with a transition pK_a of 6.5 for the whole-gland homogenate (*24*). On comparison of the room temperature NP4 pH 7.5 and pH 5.6 structures, the only changes found were in the protein interior, relatively far from the ligand binding site. Three buried water molecules that were hydrogen bonded to Glu55, Tyr17, Tyr105, and Ser72 were lost, and nearby amino acids collapsed inward to fill the resulting gap, most notably Phe107 (*138*). The water molecules reappeared in a crystal grown at pH 5.6 and moved to pH 7.5 before data measurement. Oddly, the opposite occurred when the pH 7.5 NP4 crystals were flash-frozen: The buried water molecules were apparently expelled during freezing. Whether these apparent electrostatic changes in the protein interior could alter NO affinity will be explored in experiments currently underway. Glu 55 mutants of NP4 have been prepared for study of the NO binding and release kinetics (see Section IV), and it is hoped that the crystal structures of these mutants can be solved in the near future.

IV. Kinetics and Thermodynamics of Ligand Binding

In order to understand how the nitrophorins function as NO transport proteins and vasodilators, an understanding of the nitrophorin ligand release kinetics and ligand affinities is required, in addition to the structures of the relevant complexes. Several approaches for obtaining these data have been used, all of which rely on changes in the Soret absorption that occur with changes in heme ligation, as was shown in Fig. 5. First, simple difference spectral measurements have been used to obtain ligand dissociation constants (K_d). Second, stopped-flow

TABLE I

EQUILIBRIUM DISSOCIATION CONSTANTS, K_d (μM) FOR NP–NO AND NP–HISTAMINE COMPLEXES[a] (*50*)

	NO (pH 5.0)	NO (pH 8.0)	Histamine (pH 8.0)
NP1	0.12 ± 0.02	0.85 ± 0.06	0.011 ± 0.008
NP2	0.05 ± 0.003	0.54 ± 0.02	0.010 ± 0.001
NP3	—	0.02 ± 0.004	0.009 ± 0.001
NP4	—	—	0.018 ± 0.003

[a] Temperature = 25°C.

spectrophotometry has been used to measure second-order NO association rates (k_{on}), where a solution containing NO at a defined concentration is placed in one syringe and rapidly mixed with nitrophorin protein from the other syringe, while monitoring the spectral change. Third, laser photolysis in the submillisecond time frame is being used as another method for obtaining k_{on}, by monitoring reassociation of NO from bulk solution after NO has been flashed off the protein by absorbing light from a laser pulse. The two methods for measuring k_{on} are in good agreement. Fourth, NO displacement by histamine in a stopped flow apparatus or spectrophotometer has been used to measure k_{off} values. Approaches 1–4 have been completed for all four nitrophorins, NP1–NP4 (*50*). Nano- and picosecond time resolution studies of geminate recombination of NO with the iron of NP4 are under way in collaboration with Professors John Olson and Quentin Gibson at Rice University.

A. EQUILIBRIUM DISSOCIATION CONSTANTS, K_d

The four nitrophorins can be divided into two groups based not only upon sequence homology (Fig. 4), but also on the basis of their rates of NO binding and release. NP2 and NP3 bind NO more tightly, giving smaller values of the equilibrium dissociation constants, K_d, as well as larger second-order association rate constants and smaller dissociation rate constants, than NP1 and NP4.

The pH dependence of K_d could be due to changes in A-B loop disorder rates, perhaps the chemical exchange phenomenon observed for NP1–ImH (Section II,E,2,b), or to changes in ligand bond strength. The change in K_d lies in the off-rates (Tables I–III) consistent with the loop disorder model. Plots of K_d vs pH display an excellent fit with the equation for a titration curve (Fig. 21), indicating that the transition

TABLE II

ASSOCIATION REACTION RATE CONSTANTS FOR NO BINDING TO NP1–4 (*50*)

	pH 5.0			pH 8.0		
	$k_1(\mu M^{-1}s^{-1})$	k_{-1} (s^{-1})	k_2 (s^{-1})	$k_1(\mu M^{-1}s^{-1})$	k_{-1} (s^{-1})	k_2 (s^{-1})
NP1[a]	1.5	5.1	—	1.5	4.3	—
NP4[a]	2.1	3.4	—	2.3	6.3	—
NP2[a]	22	52	51	33	32	42
NP2[b]	7.3	26.5	29.0	6.8	27.4	25.2
NP3[b]	7.0	15.9	23.3	6.7	30.2	33.0

[a] 25°C.
[b] 12°C.

from high to low affinity is governed by a single titratable group, with a pK_a between 6.5 and 7, depending on the protein in question. That the transition is not due to increased bond strength is also suggested by the measured nitrophorin reduction potentials, which are largely pH independent (discussed in Section V). However, the only pH dependent structural changes found so far concern Glu55 (*138*), which is distant from the A-B loop. A recently prepared E55Q mutant of NP4 does not show this pH dependence (*139*), indicating that Glu55 is indeed the residue responsible for the pK_a of 6.5–7 (Fig. 21). Mutants of Asp30 (D30A, D30N), part of the A-B loop, show faster off-rates at all pH values, but the pH dependence of the K_ds (Fig. 21) is still

TABLE III

FAST AND SLOW PHASE DISSOCIATION RATE CONSTANTS (k_{off}, s^{-1}) FOR NO, MEASURED BY HISTAMINE DISPLACEMENT OF NO AT pH 5.0 AND 8.0[a] (*50*)

	pH 5.0			pH 8.0		
	k_{off} (s^{-1})			k_{off} (s^{-1})		
	Fast	Slow	% Fast[b]	Fast	Slow	% Fast[b]
NP1	0.20[c]	0.02[d]	86	2.2[c]	0.6[d]	50
NP4	0.14[c]	0.01[d]	22	2.6[c]	0.6[d]	39
NP2	0.05[c]	0.006[d]	30	0.12[c]	0.01[d]	96
NP3	0.05[c]	0.005[d]	23	0.08[c]	0.01[d]	80

[a] Temperature = 25°C.
[b] Percentage of the total amplitude of the exponential as fast phase.
[c] Rate constant for fast phase.
[d] Rate constant for slow phase.

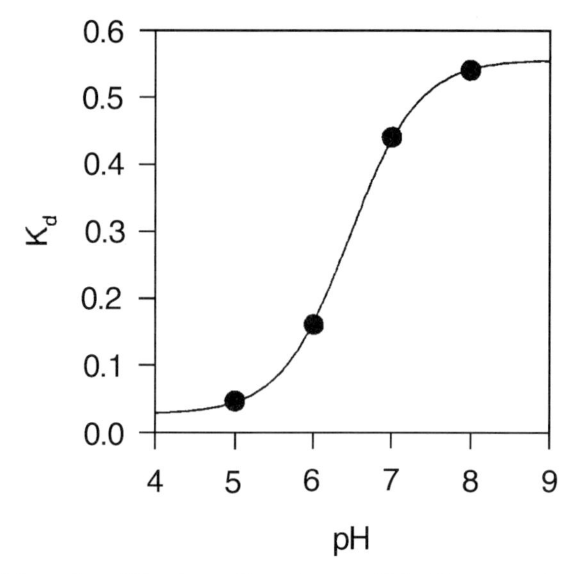

FIG. 21. pH dependent NO release. Plot of K_d vs pH for NP4, fit to the equation for a simple titration curve. The transition from low to high affinity corresponds to the titration of a single ionizable group with a pK_a of ~6.5. Reproduced with permission from Ref. (*50*).

observed (*140*). Apparently there is something favorable about the Asp30 hydrogen bonds, even though they become buried in the NO-bound complex (Figs. 19, 20). Further study of these mutant proteins, including solution of their structures, will be required in order to fully understand the role of Asp30 and to elucidate the effects of mutation of Glu55.

B. ASSOCIATION RATE CONSTANTS, k_1

Although the association kinetics of NO binding to NP1 and NP4 appear to indicate simple, one-step reactions, the kinetics of NO binding and release are biphasic for NP2 and NP3 (*50*). The biphasic association kinetics were analyzed as a two-step reaction in which the first step is represented by the fast phase:

$$\mathrm{NP} + \mathrm{L} \underset{k_{-1}}{\overset{k_1}{\rightleftharpoons}} \mathrm{NP^{*}\text{–}L} \underset{k_{-2}}{\overset{k_2}{\rightleftharpoons}} \mathrm{NP\text{–}L}. \tag{1}$$

The bimolecular association rate constant k_1 and the monomolecular reverse rate constant k_{-1} were obtained from the fast reaction phase and determined using the simple expression from the first step of Eq. (1) alone;

$$k_{obs} = k_1[L] + k_{-1}, \quad (2)$$

where k_{obs} in this case would be the observed pseudo-first-order rate constant for the fast phase of the reaction (*50*). The first-order rate constant k_2 was determined by fitting Eq. (3) with the assumption that k_{-2} is small relative to k_2 and k_{-1}:

$$k_s = k_1k_2[L]/(k_1[L] + k_{-1} + k_2). \quad (3)$$

In Eq. (3), k_s is the observed first-order rate constant for the slow phase of the binding reaction. Values of k_1, k_{-1}, and k_2 for the four nitrophorins, measured at pH 5.0 and 8.0, are summarized in Table II (*50*).

Consistent with the X-ray crystal structures of NP1, NP2, and NP4 discussed in Section III, the association rate constants ($k_1 \approx 1 \times 10^6$ to 3×10^7 $M^{-1}s^{-1}$) for the nitrophorins suggest binding with a heme iron having a coordinated water molecule that is not stabilized by hydrogen bonding with the protein. The association rates for NP1 and NP4 are faster (*50*) than those seen for sperm whale metmyoglobin (*141*) and iNOS (*142*), but 10-fold slower than that for elephant metmyoglobin (*141*). For NP2 and NP3, k_1 is larger than for NP1 and NP4, and about the same as for elephant metmyoglobin (*50, 141*). The faster association rates for the nitrophorins are likely due to the open binding pocket found in the absence of NO, where only a weakly ordered water molecule must be replaced, as has been suggested for ligand binding to elephant myoglobin and several of its distal pocket mutants (*143*). Consistent with this is the more open binding pocket found in NP2 (*110*), which has the fastest nitrophorin k_1 (*50*). For NP2, both propionates lie on the proximal side of the heme, and the loop containing residue 125 is pulled away from the distal pocket. That changes in both bimolecular and geminate rates of association can be achieved through changes in the distal pocket has been demonstrated for myoglobin, although the effect of individual changes can be difficult to predict (*144*). Biphasic association kinetics are seen with NP2 and NP3 that indicate a fast NO association followed by a slower stabilization of the complex, of the order of 51 and 42 s^{-1} for NP2 at 25°C (*50*). The slow phase is not observed in NP1 and NP4, but it is probable that a similar, yet faster, stabilization step occurs. It should be noted that the rate process observed by NMR spectroscopy for NP1–ImH (Section II,E,2,b, Fig. 12) has first-order rate constants $k_{YX} \sim 70$–90 s^{-1} and $k_{XY} \sim 7$–10 s^{-1}. Further investigation is required to determine whether the rate process observed by NMR spectroscopy is related to that observed for NO binding.

C. Dissociation Rate Constants, k_{off}

Histamine binds to the nitrophorins at rates that are roughly similar to the NO association rates at pH 8.0. The NO dissociation rate constants were measured by displacement of NO with histamine, according to the reaction

$$\text{NP–NO} + \text{Hm} \underset{k_1}{\overset{k_{off}}{\rightleftharpoons}} \text{NP} + \text{NO} + \text{Hm} \underset{k_{-1H}}{\overset{k_{1H}}{\rightleftharpoons}} \text{NP–Hm} + \text{NO} \tag{4}$$

(Hm = histamine), which results in a shift in the Soret maximum from near 420 nm to near 413 nm. In conditions of large histamine excess, histamine binds rapidly relative to NO and with high affinity (*50*). The first-order rate constant for NO release was estimated in a displacement reaction using

$$k_{off\text{-}obs} = k_{off}/\{1 + (k_1[\text{NO}]/k_{1H}[\text{Hm}])\}, \tag{5}$$

in which $k_{off\text{-}obs}$ is the observed first-order rate constant for the change in absorbance at 423 nm, k_1 is the bimolecular rate constant for NO binding, k_{1H} is the bimolecular rate constant for histamine binding, and k_{off} is the overall dissociation rate constant. In the two-step mechanism described in Eq. (1), the rate constant k_{-2} can be estimated from k_{off} using Eq. (6):

$$k_{-2} = k_{off}(k_{-1} + k_2)/(k_{-1} - k_{off}). \tag{6}$$

The NO dissociation rate constants are summarized in Table III (*50*) and are smaller than those seen with NO–metmyoglobin complexes. NP2 and NP3 ($k_{off} \approx 0.1\ s^{-1}$) release NO approximately 10 times more slowly than NP1 and NP4 ($k_{off} \approx 2$–$3\ s^{-1}$) at pH 8.0, and the NO release rate for all nitrophorins decreases as the pH is lowered to 5.0. The NO release curves cannot be fit with a single exponential, indicating two off rates at each pH, as previously noted for both recombinant and insect-derived NP1 (*46*), and which has also been recently reported for recombinant NP2 (*145*). The biphasic kinetics suggest the presence of slowly interconverting conformations. The values obtained are pH and protein dependent, ranging from 2.6 to 0.05 s^{-1} (Table III) (*50*), values that are considerably slower than found for sperm whale metmyoglobin

(*141*), elephant metmyoglobin (which does not have a distal histidine) (*141*), or iNOS (*142*).

One possible explanation for the slow nitrophorin off-rates is the burial of the distal pocket that occurs on NO binding (Figs. 19 and 20). For this to be the case, loop disordering must be slow and rate limiting for NO release. Alternatively, the NO–heme bond strength could be greater in the nitrophorins than in metmyoglobin or iNOS. Experiments, including fast kinetic measurements, are planned to distinguish between these two possibilities. In our initial nanosecond kinetic studies (in collaboration with Dr. Olson) no geminate recombination was found at either pH 8 or pH 5. However, the quantum yield at pH 5 was low, suggesting that either faster (picosecond) recombination occurs at this pH, or a stronger bond exists. For biphasic NO release to occur, two forms of the protein that do not rapidly interchange must exist. Again, a possible explanation for this is that two forms of the closed protein exist, giving rise to a faster and a slower escape path. Alternatively, the two NO positions found in the NP4–NO structure may be responsible for the two off-rates. Other possibilities can also be envisioned, and considerably more kinetics data must be obtained before the explanation of the biphasic NO release rates is complete.

We have also obtained the ferrous (Fe^{II}) NO equilibrium dissociation constants for the nitrophorins by combining the electrochemical (Table IV) and ferric equilibrium dissociation constant data (Table I), as discussed in Section V. As expected, binding of NO by the ferrous proteins is $\sim 10^7$-fold tighter than by the ferric proteins. Similar results are obtained with the globins (reviewed in Ref. *141*). Strikingly, iNOS apparently does not display significantly tighter binding to the reduced heme (*142*). The NOS family of proteins differs from the globins and nitrophorins in that the proximal ligand is cysteine, the heme is domed, and substrates and cofactors bind in the distal pocket (*146–148*). The NOS proteins also apparently suffer from product inhibition due to NO binding to either the ferric or ferrous form of the heme, although binding is not particularly strong (*142, 149, 150*).

V. Reduction Potentials of Nitrophorins in the Absence and Presence of NO, Histamine, and Other Ligands

A. Resistance of the Nitrophorins to Autoreduction by NO

It is generally true that Fe(III)–NO species are readily autoreduced to Fe(II)–NO in the presence of excess NO. This is certainly true not only of

TABLE IV

REDUCTION POTENTIALS MEASURED FOR MYOGLOBIN AND THE FOUR NITROPHORINS, AND THE DERIVED LIGAND DISSOCIATION CONSTANTS OF THE NITROPHORINS[a] (49, 50, 55)

Heme protein	pH	E° (mV vs NHE)	Slope of Nernst Plot (mV)	K_d^{III} (μM)	K_d^{II}
Mb	5.5	28 ± 1	61 ± 1		
	7.5	0 ± 2	61 ± 1		
NP1	5.5	−274 ± 2	60 ± 1		
	7.5	−303 ± 4	60 ± 2		
NP1–NO	5.5	+154 ± 5	62 ± 7	~0.1	~6 fM
	7.5	+127 ± 4	61 ± 11	~1.0	~59 fM
NP1–Hm	5.5	−339 ± 2	60 ± 1	—[b]	K_d^{III}/K_d^{II} 12.3
	7.5	−403 ± 1	59 ± 2	~0.02	~1 μM
NP1–ImH	5.5	−421 ± 2	73 ± 4	8	2.7 mM
	7.5	−425 ± 18	59 ± 3	24	2.7 mM
NP1–4IPzH	5.5	−161 ± 2	60 ± 2	27	33 nM
	7.5	−206 ± 1	62 ± 2	48	1.1 μM
NP4	5.5	−259 ± 2	57 ± 2		
	7.5	−278 ± 4	58 ± 3		
NP4–NO	5.5	+94 ± 5	62 ± 7	~0.04	~47 fM
	7.5	—[c]	—[c]	~0.6	—
NP4–Hm	5.5	−393 ± 2	59 ± 2	—[b]	K_d^{III}/K_d^{II} 179
	7.5	−404 ± 1	65 ± 3	~0.01	~1.3 μM
NP4–ImH	5.5	−398 ± 3	60 ± 4	4	0.86 mM
	7.5	−409 ± 5	60 ± 8	12	1.9 mM
NP4–4IPzH	5.5	−148 ± 2	59 ± 2	25	330 nM
	7.5	−198 ± 3	60 ± 4	57	2.6 μM
NP2	5.5	−287 ± 5	59 ± 7		
	7.5	−310 ± 5	62 ± 5		
NP2–NO	5.5	+49 ± 3	60 ± 3	—[b]	K_d^{III}/K_d^{II} 2.3×10^{-6}
	7.5	+8 ± 3	61 ± 2	~0.02	~91 fM
NP2–Hm	5.5	−410 ± 3	58 ± 5	—[b]	K_d^{III}/K_d^{II} 116
	7.5	−452 ± 4	61 ± 4	~0.01	2.4 μM
NP2–ImH	5.5	−418 ± 5	62 ± 12	7	1.1 mM
	7.5	−434 ± 2	59 ± 3	23	2.8 mM
NP2–4IPzH	5.5	−279 ± 6	61 ± 11	37	27 μM
	7.5	−313 ± 5	58 ± 6	22	25 μM
NP3	5.5	−321 ± 5	54 ± 7		
	7.5	−335 ± 3	57 ± 7		
NP3–NO	5.5	+33 ± 4	59 ± 1	—[b]	K_d^{III}/K_d^{II} 1.1×10^{-6}
	7.5	+3 ± 5	66 ± 6	—[b]	K_d^{III}/K_d^{II} 2.1×10^{-6}
NP3–Hm	5.5	−353 ± 8	60 ± 6	—[b]	K_d^{III}/K_d^{II} 3.4
	7.5	−445 ± 3	59 ± 8	~0.02	1.4 μM
NP3–ImH	5.5				
	7.5				
NP3–4IPzH	5.5				
	7.5				

[a] Temperature = 27 ± 1°C.
[b] Too small to measure.
[c] Could not be measured because of facile dissociation of NO.

model hemes (*62, 64*), but also of the NO complexes of hemoglobin and myoglobin (*49, 52, 151*), as well as many other heme proteins. However, the nitrophorins from the saliva of *R. prolixus* are *not* readily autoreduced (*24, 49*) and thus represent the first examples of stable Fe(III)–NO centers that are in need of full characterization. In order to determine quantitatively the relative tendencies for the ferriheme–NO centers of myoglobin and the nitrophorins to be reduced to the ferroheme–NO states, detailed electrochemical investigations have been carried out. First, cyclic voltammetric methods were investigated because they could be carried out more rapidly than spectroelectrochemical titrations. However, as has been reported previously for metmyoglobin (*152*), both this protein and the nitrophorins produced cyclic voltammetric waves that were at best quasi-reversible, and usually irreversible in shape (*153*), probably because of the dissociation of the axially bound water molecule of the high-spin Fe(III) oxidized form upon reduction to Fe(II), and slow reassociation upon reoxidation. Hence, spectroelectrochemical titrations were carried out utilizing very thin electrochemical cells with optically-transparent thin layer gold mesh electrodes (*49, 50, 55, 87, 103, 154–157*). Potentiometric and optical spectral data consisting of absorbance vs applied potential were obtained, as shown in Fig. 22. This method has the further advantage that because optical spectra are obtained, it is usually possible to identify the species present, as in the case of the base-off complexes of $NP2^{II}$–NO and $NP3^{II}$–NO at pH 5.5 discussed later, and to detect nonisosbestic behavior that indicates the presence of more than two absorbing species in solution.

A typical set of optical spectra for the spectroelectrochemical titration of a nitrophorin–NO complex, NP1–NO at pH 5.5, is shown in Fig. 22. These experiments are carried out by setting the potential at some value $E^{o\prime}$ and waiting until the optical spectrum does not change with time, then recording the optical spectrum, followed by setting a new potential $E^{o\prime}$ and repeating the process, etc. As is evident, clear isosbestic behavior is observed, indicating a smooth reduction of the chromophore as the applied potential is lowered, with only two species present. The insert shows the plot, based upon the Nernst equation,

$$E^{o\prime} = E^{o} + 2.303(RT/nF)\log_{10}([\mathrm{Ox}]/[\mathrm{Red}]) \quad (7)$$

where $E^{o\prime}$ is the applied potential, E^{o} is the reduction potential determined from these data, and [Ox] and [Red] are the concentrations of the Fe(III) and Fe(II) states, respectively, in the presence of either no ligand or a high enough concentration of ligand to ensure full complexation

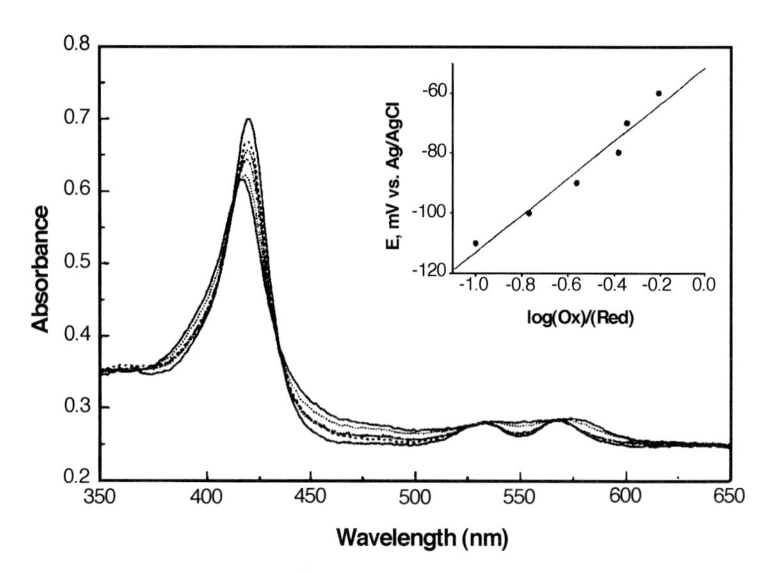

FIG. 22. UV-visible spectra of recombinant *Rhodnius* NP1–NO at pH 5.5 as a function of applied potential. In order of decreasing Soret band heights: +200, −60, −70, −80, −90, −100, −110, and −400 mV vs Ag/AgCl, respectively (add 205 mV for potential vs NHE). (*Inset*) Nernst plot of the spectroelectrochemical data. Reprinted with permission from Ref. (*49*).

of both oxidation states. [Ox] and [Red] are calculated from the optical spectra using Beer's law. E^{o} is determined as the intercept of the plot of log([Ox]/[Red]) vs applied potential $E^{\text{o}\prime}$. The slope of this plot is $2.303RT/nF = 59$ mV at 25°C.

For NP2–NO and NP3–NO at pH 5.5, very different spectral changes upon electrochemical reduction are seen, as were shown in Fig. 6c for NP3–NO. The Soret band shifts to 395 nm, and both the wavelength maximum and shape of the Soret band are typical of five-coordinate heme–NO centers, including guanylyl cyclase upon binding NO (*133, 134*). The reduction of NP2–NO was therefore investigated at pH 5.5, 6.5, 7.0, 7.5, and 8.0. At pH 7.5 and 8.0 the spectral changes are typical of those shown in Fig. 6a, indicating that the proximal histidine (His59) is bound at pH 7.5 and above (*50*). At intermediate pH values, 6.5 and 7.0, nonisosbestic behavior, typical for the presence of three absorbing species in the solution, is observed, whereas at pH 5.5 the spectra again show isosbestic behavior, with the Soret band positions shown in Fig. 6c. It thus appears that the proximal histidine has a pK_a of about 6.5–6.8 for the reduced form of NP2–NO, and the bond to the iron is lost upon binding NO below this pH. Both NP2–NO and NP3–NO exhibit similar pH dependence of the absorption spectra, whereas NP1–NO and NP4–NO do not show any pH dependence of their absorption spectra over the pH range 5.5–7.5 (*50*). In spite of this loss of the proximal histidine ligand at pH 5.5, the change in reduction potential as a function of pH is similar for NP2–NO and NP3–NO to that observed for NP1–NO and NP4–NO.

Using the spectral data of Fig. 22, and similar data obtained for the nitrophorins in the absence of NO and in the presence of histamine, imidazole, or 4-iodopyrazole, Nernst plots such as that shown in the insert of Fig. 22 were constructed, and the midpoint potentials of the nitrophorins and their NO and histamine complexes were calculated. The results are summarized in Table IV, where they are compared to those obtained earlier for NP1 (*49, 50, 55*). All potentials are expressed vs NHE (+205 mV with respect to the Ag/AgCl electrode used in the spectroelectrochemical titrations and the Nernst plot shown in the insert of Fig. 22). It can be seen that the reduction potentials of all four nitrophorins in the absence of NO or histamine are within 20–40 mV of each other. The reduction potentials of their NO complexes, however, differ significantly from each other. For example, the reduction potential of NP4–NO is about 350 mV more positive than that of NP4 in the absence of NO, as compared to a 430 mV shift for NP1 upon binding NO, and the positive shifts for NP2–NO and NP3–NO are somewhat smaller (318 and 336 mV, respectively, at pH 7.5) (*49, 50*). These differences relate to the ratios of the dissociation constants for the two oxidation states, as discussed later.

The reduction potentials of the nitrophorins differ from that of metmyoglobin in the absence of NO by about 278–335 mV (*49, 50*), depending on pH (Table IV), with the nitrophorins having much less tendency to be reduced. However, the reduction potentials differ slightly, with that of NP4 being 15–25 mV more positive than that of NP1, that of NP2 being 7–13 mV more negative than that of NP1, and that of NP3 being 25–36 mV more negative than that of NP2 (*49, 50*). The 278–335 mV difference corresponds to a ΔG° difference of 27–33 kJ/mol for the reduction half-reactions of the two classes of proteins, with metmyoglobin having the more negative ΔG°, and hence the greater tendency to be reduced. The same relative ease of reduction obviously applies to the NO complexes of the two proteins, although the difference in ΔG° cannot be determined because it is not possible to measure the reduction potential of metmyoglobin–NO (*49*).

As discussed in detail previously (*49, 50*), the difference in reduction potential of all four of the nitrophorins and metmyoglobin in the absence of NO cannot be a result of a difference in ligation, since each of these proteins has a proximal histidine ligand. Rather, the 278–335 mV more negative reduction potentials is most likely due to the difference in buried charges near the heme in the two types of proteins (*49, 50*). Myoglobin has an uncharged distal pocket (*158*), and when charges are introduced by mutation of valine E-11 to glutamate or aspartate, the reduction potential is lowered by nearly 200 mV (*115*). Unlike myoglobin (*158*), the nitrophorins have a number of potentially charged residues in or near the heme pocket, as shown in Fig. 16. For NP1 and NP4, these include Asp30, which is near the heme and in the histamine complex of NP1 forms a hydrogen bond to one of the histamine amino protons (*31*) (Fig. 16); Glu32, which is near the surface of the protein on the distal side of the heme; and Glu55 and Asp70 on the proximal side. Asp70 makes a hydrogen bond to an ordered water molecule that in turn is hydrogen-bonded to the proximal histidine ring $N_\delta H$ and is most likely uncharged (*31*). Most buried of all is Glu55, which is at the back of the heme pocket (*31*). The pH-dependent structural changes observed in the region of Glu55 (Section III,D) indicate that this residue may have a pK_a of approximately 6.5, and recent kinetics results for mutants of this residue confirm this (*139*). However, this pK_a makes it unlikely that Glu 55 has a major effect on the reduction potential of the heme, since the potential changes only slightly over the pH range 5.5 to 7.5 (Table IV). Asp30 and Glu55 are conserved among the four nitrophorins, and Glu32 is conservatively replaced by Asp, while Asp70 is nonconservatively replaced by Asn in NP2 and NP3 (Fig. 4). Hence, there are a number of candidates for charged residues that may be responsible for the −280 to −310 mV shift in the reduction potential of the nitrophorins as compared to metmyoglobin. Investigation of the electrochemistry of site-directed mutants, now in progress, will be required to determine the importance of each.

In answer to the originally posed question of why the nitrophorins are stable to autoreduction by excess NO for long periods of time, while metmyoglobin is not, the approximately −300 mV shift in reduction potential in the absence of added ligands (Table IV) is certainly part of the reason, but this alone probably does not account for the large difference in the rate of autoreduction of the two types of proteins. Rather, it is likely this factor is combined with the negative charges near the binding site of the NO, since it has been shown that for autoreduction of metMb–NO, the mechanism involves attack of hydroxide ion on the NO^+ form of the bound ligand (leaving behind Fe(II)), leading to dissociation of nitrite ion and the proton, followed by rapid binding of a second molecule of NO to the Fe(II) center (*159*). Negatively charged residues near the binding site of NO would discourage hydroxide attack.

B. Dependence of Reduction Potential on pH

The similarity in pH dependence of the reduction potentials of myoglobin and the nitrophorins, 10–15 mV per pH unit, suggests that the nitrophorins are, like myoglobin (*160, 161*), between acidic and basic pK_as of heme carboxylates (*162*), bound water (*163*), and histidine imidazole ring NH deprotonation (*164–166*) over the pH range 5.5–7.5, and that this general medium effect is sufficient to explain the small observed pH dependence of the reduction potential of all of the nitrophorins. The NO, histamine, imidazole, and 4-iodopyrazole complexes, however, show somewhat varying pH dependences of their reduction potentials (*49, 50, 55*), but the pH dependences for a given ligand are within experimental error of 10–20 mV/pH unit. This includes the pH dependence of the reduction potentials for the NO complexes of NP2 and NP3, which involve loss of the proximal histidine below pH 7, as discussed in Section V,A. The pH dependence is slightly larger for NP2–NO (21 mV/pH unit) than for NP1–NO (14 mV/pH unit), but that for NP3–NO (15 mV/pH unit) is quite similar to that for NP1–NO, which remains six-coordinate upon reduction, so the differences are within the error limits of the measurements and do not appear to suggest major consequences in terms of redox stability or NO dissociation constant (see next section) because of the loss of the proximal histidine at lower pH.

C. Reduction Potentials of Lewis Base Complexes and the Calculation of Dissociation Constants for the Fe(II) Complexes

In contrast to the NO complexes of the nitrophorins, their histamine and imidazole complexes have much more negative reduction potentials than the nitrophorins in the absence of added ligands by 100–150 mV at pH 7.5 (*50, 55*), whereas the 4-iodopyrazole complexes have less negative or nearly the same reduction potentials (*55*) as the ligand-free nitrophorins. The more negative potentials for the imidazole-based ligands are indicative of the fact that the Fe(III) histamine and imidazole complexes are more stable (have smaller dissociation constants, K_d^{III}) than their corresponding Fe(II) complexes (*50, 55*). In contrast, the large positive shifts of the reduction potentials of the nitrophorin–NO complexes (*49, 50*) indicate that the Fe(II)–NO complexes are much more stable than their corresponding Fe(III)–NO forms. The shift of the Fe^{III}/Fe^{II} reduction potential when a ligand L is bound to the iron is a measure of the ratio of the Fe–L dissociation constants for the two oxidation sates, since the Nernst equation ((Eq. (7)) can be rewritten as

$$E_c^o = E^o + (RT/nF)\ln(K_d^{III}/K_d^{II}) \quad (8)$$

where E_c^o is the measured potential for the nitrophorin fully complexed to the ligand L, E^o is the measured potential for the nitrophorin in the absence of L, and K_d^{III} and K_d^{II} are the dissociation constants for ligand L from the Fe(III) and Fe(II) states, respectively (*49, 50, 55, 167*). This ratio of dissociation constants, K_d^{III}/K_d^{II}, is equal to the ratio of the concentrations of oxidized to reduced species, Eq. (7), in the presence of a large excess of ligand. From Eq. (8), the ratio of the dissociation constants for the two oxidation states is thus $\exp[-(E_c^o - E^o)F/RT]$. As can be seen in Table IV, the K_d values for the Fe(II)–NO complexes are thus calculated to be 4.5×10^{-6} to 5.9×10^{-8} times smaller than those for the respective Fe(II)–NO complexes, with NP1 showing the largest decrease in K_d when the metal is reduced. The ratio K_d^{III}/K_d^{II} either remains constant or decreases by about a factor of 2 as the pH is increased, which, combined with the fact that K_d^{III} increases with pH (see Section IV), means that K_d^{II} increases with pH about the same amount or less than does K_d^{III}. Combined with the K_d^{III} values of Table I, and as reported previously for NP1 (*49*), the Fe(II)–NO dissociation constants for the nitrophorins are smaller by more than a factor of 10^{-5}–10^{-7} than those for Fe(III)–NO, and for NP1 are a factor of 10 smaller than those of most other heme proteins (*39*).

In contrast, for the histamine and imidazole complexes, it is K_d^{III} that is smaller than K_d^{II}, and the Fe(III)–histamine complexes are 48 and 132 times more stable than the Fe(II) complexes for NP1 and NP4, respectively, at pH 7.5, with somewhat larger difference in stability of the two oxidation states for NP2 (244 times), and less difference for NP3 (71 times) (*50*). In the same way, but to a greater extent, the Fe(III)–imidazole complexes are 113 and 158 times more stable than the Fe(II)–ImH complexes for NP1 and NP4, respectively, at pH 7.5, with similar differences in stability of the two oxidation states (122 times) for NP2 (*55*). Thus, the Fe(III) oxidation state of the nitrophorins, which is stabilized by the negative charges near the heme, favors the dissociation of NO and the binding of histamine, whereas the Fe(II) oxidation state would do the opposite. Clearly, the nitrophorins have been engineered to be stabilized in the Fe(III) form in order to perform their roles of releasing NO and binding histamine.

In comparing histamine to imidazole as ligands to the nitrophorins, at pH 7.5 (the approximate pH of the tissues of the victim), the histamine complex is 1200 times more stable for NP1 and NP4, and about

2300 times more stable for NP2 (*55*). (The imidazole and 4-iodopyrazole binding constants have not yet been measured for NP3.) These comparisons point out the importance of the protonated amino side chain of histamine in helping to stabilize its complexes with the nitrophorins. Comparing the two ligands having no protonated amino side chain, the imidazole complex is more stable than the 4-iodopyrazole complex by a ratio of 2 for NP1 and 4.75 for NP4, but is equally stable for NP2. However, in contrast to the imidazole and histamine complexes, it is the Fe(II) complexes that are more stable than the Fe(III), by factors of 44 and 22 for NP1 and NP4, respectively, but the stability order switches for NP2, where the Fe(III) complex is slightly more stable than the Fe(II) (ratio of 1.1) (*55*) (Table II). The crystal structure of NP1–4-iodopyrazole shows the iodine atom to be twofold disordered (*55*), suggesting that hydrogen-bonding of the N–H group of the 4-iodopyrazole is not important in stabilizing the complex.

VI. Summary and Future Directions

The nitrophorins from the salivary glands of *Rhodnius prolixus* are novel heme proteins that are bound to NO at the low pH of the glands and release NO when they are injected into the victim, by dilution and pH increase. These proteins are able to release NO because the heme is stabilized as Fe(III), which has NO dissociation constants of 0.02 to 0.1 μM, rather than the more common Fe(II) state, which has NO dissociation constants of 6–90 fM at pH 7.5. Negative charges near the NO binding pocket are likely responsible for this redox stabilization of Fe(III)–NO. The structure of at least one form of each of three of the four nitrophorins has been determined by X-ray crystallography, and it is found that the structures are those of novel beta-barrel proteins called lipocalins, which have the heme near the mouth of the open end of the barrel. Upon NO binding, a conformational change takes place that involves the ordering of the A-B loop, the N-terminus, and several other residues to expel water molecules in the distal pocket and close the NO inside. Infrared spectroscopic investigations have provided support for the Fe(II)–NO^+ electron configuration, whereas Mössbauer, EPR and NMR spectroscopies cannot differentiate between this electron configuration and that of Fe(III)–$NO^{\cdot}$. The proton NMR resonances of the heme substituents have been assigned for both the high-spin form of NP2 and also for its imidazole and *N*-methylimidazole complexes by a combination of isotopic labeling, 2D NMR, and 1D saturation transfer techniques. It is found that the pattern of heme methyl shifts suggests that the orientation of the nodal plane of the heme in NP1–ImH is controlled mainly by the protein-provided ligand, His59, while it is the exogenous imidazole ligand that controls the orientation of the nodal plane of the heme in NP2–ImH. Pulsed EPR (ESEEM) studies of NP1–histamine also support the conclusion that His59 is most important in determining the orientation of the in-plane g-values, g_{xx} and g_{yy}. Spectroelectrochemical investigations of the Fe(III)/Fe(II) reduction show that the nitrophorin hemes are much more difficult to reduce than is the heme of metmyoglobin. Reduction potentials of the NO, histamine, imidazole, and 4-iodopyrazole complexes of the nitrophorins have been utilized to determine the ratio of the ligand dissociation constants in the two oxidation states, K_d^{III}/K_d^{II}. It is found that the NO complexes of the nitrophorins in the oxidized state are $\sim 10^7$ times less stable than those of the reduced state, making the Fe(III) state nitrophorins excellent as nitric oxide *carriers*. On the other hand, the histamine complexes of the nitrophorins in the oxidized state are ~ 100 times more stable than those of the reduced state, and 50–100 times more stable than the NO complexes of the oxidized state, making the Fe(III) state excellent for binding the histamine released by the mast cells of the victim in response to the bite, and even in displacing the NO. The results of these outcomes for the insect are summarized in Fig. 23. The kinetics and equilibria of NO binding and release have been investigated for the

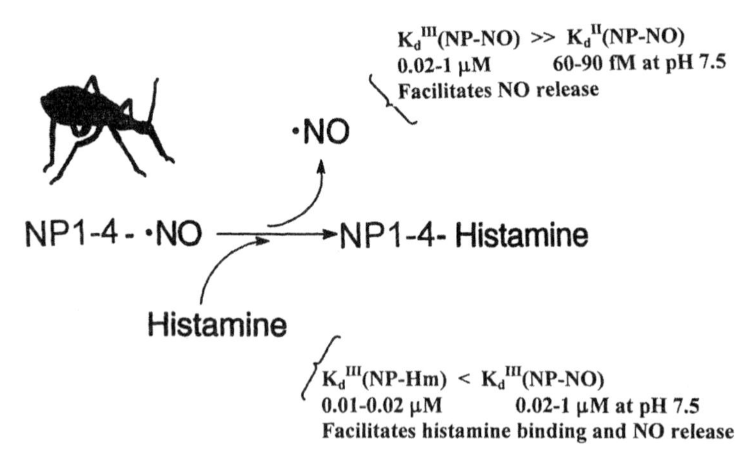

FIG. 23. The dissociation constants and redox stability of the NO and histamine complexes of the nitrophorins from the saliva of *Rhodnius prolixus*, and how they aid the insect in obtaining a sufficient blood meal. Modified from Ref. (*31*).

Fe(III) forms of the nitrophorins. It is found that both association and dissociation kinetics are biphasic, and that dissociation constants K_d vary as expected for a pH titration curve, indicating a pK_a of 6.5–7 for the four nitrophorins. Preliminary investigations of Glu55 mutants of NP4 identify this residue as the one responsible for this pH dependence. Investigation of the kinetics, electrochemistry, structures, NMR spectra and other spectroscopies of additional site-directed mutants, to be prepared in the near future, will be required in order to better understand the basis of the conformational change in the A-B loop and other residues highlighted in Figs. 18 and 20. It may be anticipated that additional fascinating details concerning the mechanism of action of these novel NO- and histamine-binding heme proteins will be uncovered in the process.

Acknowledgments

The financial support of the National Institutes of Health, Grant HL54826, and the assistance in figure preparation by Dr. Tatjana Kh. Shokhireva and Mary Flores are gratefully acknowledged.

References

1. Walker, F. A.; Ribeiro, J. M. C.; Montfort, W. R. In "Metal Ions in Biological Systems"; Vol. 36 "Interrelations between Free Radicals and Metal Ions in Life Processes"; Sigel, H.; Sigel A., Eds.; Marcel Dekker: New York, 1999; pp. 619–661.
2. Enemark, J. H.; Feltham, R. D. *Coord. Chem. Rev.* **1974,** *13,* 339.
3. Bredt, D. S.; Snyder, S. H. *Annu. Rev. Biochem.* **1994,** *63,* 175.
4. Schmidt, H. H.; Walter, U. *Cell* **1994,** *78,* 919.
5. *Nobelförsamlingen*, Karolinska Institutet press release, October 12, 1998.
6. Cornwall, J. W.; Patton, W. S. *Indian J. Med. Res.* **1914,** *2,* 569.
7. Law, J.; Ribiero, J. M. C.; Wells, M. *Annu. Rev. Biochem.* **1992,** *61,* 87.
8. Noeske-Jungblutt, C.; Kratzchmar, J.; Haendler, B.; Alagon, A.; Possani, L.; Verhallen, P.; Schleuning, W.-D. *J. Biol. Chem.* **1994,** *269,* 5050.
9. Grevelink, S. A.; Youssef, D. E.; Loscalzo, J.; Lerner, E. A. *Proc. Natl. Acad. Sci. USA* **1993,** *90,* 9155.
10. Higgs, G. A.; Vane, J. R.; Hart, R. J.; Porter, C.; Wilson, R. G. *Bull. Ent. Res.* **1976,** *66,* 665.
11. Kemp, D. H.; Hales, J. R.; Schleger, A. V.; Fawcett, A. A. *Experientia* **1983,** *39,* 725.
12. Shemesh, M.; Hadani, A.; Shklar, A.; Shore, L. S.; Meleguir, F. *Bull. Ent. Res.* **1979,** *69,* 381.
13. Ribiero, J. M. C.; Evans, P. M.; MacSwain, J. L.; Sauer, *J. Exp. Parasitol.* **1992,** *74,* 112.
14. Dickinson, R. G.; O'Hagan, J. E.; Shotz, M.; Binnington, K. C.; Hegarty, M. P. *Aust. J. Exp. Biol. Med. Sci.* **1976,** *54,* 475.

15. Moro, O.; Lerner, E. A. *J. Biol. Chem.* **1997,** *272,* 966.
16. Ribeiro, J. M. C.; Vachereau, A.; Modi, G. B.; Tesh, R. B. *Science* **1989,** *243,* 212.
17. Lerner, E. A.; Ribeiro, J. M. C.; Nelson, R. J.; Lerner, M. R. *J. Biol. Chem.* **1991,** *266,* 11234.
18. Lerner, E. A.; Shoemaker, C. B. *J. Biol. Chem.* **1992,** *267,* 1062.
19. Cupp, M. S.; Ribiero, J. M. C.; Cupp, E. W. *Am. J. Trop. Med. Hyg.* **1994,** *50,* 241.
20. Champagne, D.; Ribiero, J. M. C. *Proc. Natl. Acad. Sci. USA* **1994,** *91,* 138.
21. Ribeiro, J. M. C.; Nussenzveig, R. H. *J. Exp. Biol.* **1993,** *179,* 273.
22. Ribeiro, J. M. C.; Garcia, E. S. *J. Exp. Biol.* **1981,** *94,* 219.
23. Ribeiro, J. M. C.; Gonzales, R.; Marinotti, O. *Br. J. Pharmacol.* **1990,** *101,* 932.
24. Ribeiro, J. M. C.; Hazzard, J. M. H.; Nussenzveig, R. H.; Champagne, D. E.; Walker, F. A. *Science* **1993,** *260,* 539.
25. Ribeiro, J. M. C. *J. Insect Physiol.* **1982,** *28,* 69.
26. Ribeiro, J. M. C.; Walker, F. A. *J. Exp. Med.* **1994,** *180,* 2251.
27. Lavoipierre, M. M. J.; Kickerson, G.; Gordon, R. M. *Ann. Trop. Med. Parasit.* **1959,** *53,* 235.
28. Francischetti, I. M. B.; Ribiero, J. M. C.; Champagne, D.; Andersen, J. F. *J. Biol. Chem.* **2000,** *275,* 12639.
29. Hellman, K.; Hawkins, R. I. *Nature* **1965,** *207,* 265.
30. Ribeiro, J. M. C.; Schneider, M; Guimaraes, J. A. *Biochem. J.* **1995,** *308,* 243.
31. Weichsel, A.; Andersen, J. F.; Champagne, D. E.; Walker, F. A.; Montfort, W. R. *Nature Struct. Biol.* **1998,** *5,* 304.
32. Montfort, W. R.; Weichsel, A.; Andersen, J. F. "Nitrophorins and Related Antihemostatic Lipocalins from *Rhodnius prolixus* and Other Blood-Sucking Arthropods." *Biochim. Biophys. Acta.* **2000,** in press.
33. Valenzuela, J. G.; Walker, F. A.; Ribeiro, J. M. C. *J. Exp. Med.* **1995,** *198,* 1519.
34. Valenzuela, J. G.; Ribeiro, J. M. C. *J. Exp. Biol.* **1998,** *201,* 2659.
35. Ribeiro, J. M. C.; Nussenzveig, R. H. *J. Exp. Biol.* **1993,** *179,* 273.
36. Ribeiro, J. M. C.; Nussenzveig, R. H. *FEBS Lett.* **1993,** *330,* 165.
37. Nussenzveig, R. H.; Bentley, D. L.; Ribeiro, J. M. C. *J. Exp. Biol.* **1995,** *198,* 1093.
38. Yuda, M.; Hirai, M.; Miura, K.; Matsumura, H.; Ando, K.; Chinzei, Y. *Eur. J. Biochem.* **1996,** *242,* 807.
39. Traylor, T. G.; Sharma, V. S. *Biochemistry* **1992,** *31,* 2847.
40. Kirchhoff, L. V. *N. Engl. J. Med.* **1993,** *329,* 639.
41. Ferreira, L. F.; Britto, C.; Cardoso, M. A.; Fernandes, O.; Reinhard, K.; Araujo, A. *Acta Tropica* **2000,** *75,* 79.
42. Reinhard, K. Personal communication. See Web site http://www.unl.edu/Reinhard/
43. Wayland, B. B.; Olson, L. W. *J. Am. Chem. Soc.* **1974,** *96,* 6037.
44. Champagne, D. E.; Nussenzveig, R. H.; Ribeiro, J. M. C. *J. Biol. Chem.* **1995,** *270,* 8691.
45. Champagne, D. E.; Ribeiro, J. M. C. Unpublished work.
46. Andersen, J. F.; Champagne, D. E.; Weichsel, A.; Ribeiro, J. M. C.; Balfour, C. A.; Dress, V.; Montfort, W. R. *Biochemistry* **1997,** *36,* 4423.
47. Weichsel, A.; Andersen, J. F.; Roberts, S. A.; Montfort, W. R. *Nature Struct. Biol.* **2000,** *7,* 551.
48. Andersen, J. F.; Weichsel, A.; Balfour, C. A.; Champagne, D. E.; Montfort, W. R. *Structure* **1998,** *6,* 1315.
49. Ding, X. D.; Weichsel, A.; Andersen, J. F.; Shokhireva, T. K.; Balfour, C.; Pierik, A. J.; Averill, B. A.; Montfort, W. R.; Walker, F. A. *J. Am. Chem. Soc.* **1999,** *121,* 128.
50. Andersen, J. F.; Ding, X. D.; Balfour, C.; Shokhireva, T. Kh.; Champagne, D. E.; Walker, F. A.; Montfort, W. R. *Biochemistry* **2000,** *39,* in press.

51. Romberg, R. W.; Kassner, R. J. *Biochemistry* **1979,** *18,* 5387.
52. Addison, A. W.; Stephanos, J. J. *Biochemistry* **1986,** *25,* 4104.
53. Zhao, Y.; Hoganson, C.; Babcock, G. T.; Marletta, M. A. *Biochemistry* **1998,** *37,* 12458.
54. Tomita, T.; Ogura, T.; Tsuyama, S.; Imai, Y.; Kitagawa, T. *Biochemistry* **1997,** *36,* 10155.
55. Ding, X. D.; Shokhireva, T. Kh.; Weichsel, A.; Montfort, W. R.; Walker, F. A. To be submitted.
56. Maxwell, J. C.; Caughey, W. S. *Biochemistry* **1976,** *15,* 388.
57. Sampath, V.; Zhao, X.-J.; Caughey, W. S. *Biochem. Biophys. Res. Commun.* **1994,** *198,* 281.
58. Zhao, X. J.; Sampath, V.; Caughey, W. S. *Biochem. Biophys. Res. Commun.* **1994,** *204,* 537.
59. Wang, Y.; Averill, B. A. *J. Am. Chem. Soc.* **1996,** *118,* 3972.
60. Obayashi, E.; Tsukamoto, K.; Adachi, S.; Takahasni, S.; Nomura, M.; Iizuka, T.; Shoun, H.; Shiro, Y. *J. Am. Chem. Soc.* **1997,** *119,* 7807.
61. Miller, L. M.; Pedraza, A. J.; Chance, M. R. *Biochemistry* **1997,** *36,* 12199.
62. Scheidt, W. R.; Lee, Y. J.; Hatano, K. *J. Am. Chem. Soc.* **1984,** *106,* 3191.
63. Ellison, M. K.; Scheidt, W. R. *J. Am. Chem. Soc.* **1997,** *119,* 7404 and references therein.
64. Ellison, M. K.; Scheidt, W. R. *J. Am. Chem. Soc.* **1999,** *121,* 5210.
65. Scheidt, W. R.; Ellison, M. K. *Acc. Chem. Res.* **1999,** *32,* 350.
66. Wayland, B. B.; Olson, L. W. *J. Am. Chem. Soc.* **1974,** *96,* 6037.
67. Yoshimura, T. *J. Inorg. Biochem.* **1983,** *18,* 263.
68. Nasri, H.; Ellison, M. K.; Chen, S., Huynh, B. H., Scheidt, W. R. *J. Am. Chem. Soc.* **1997,** *119,* 6274.
69. Kon, H. *J. Biol. Chem.* **1968,** *243,* 4350.
70. Rein, H.; Ristau, O.; Scheler, W. *FEBS Lett.* **1972,** *24,* 24.
71. Brunori, M.; Falcioni, G.; Rotillo, G. *Proc. Nat. Acad. Sci. USA* **1974,** *71,* 2470.
72. Szabo, A.; Perutz, M. F. *Biochemistry* **1976,** *14,* 4427.
73. Nagai, K.; Hori, H.; Morimoto, H.; Hayashi, A.; Taketa, F. *Biochemistry* **1979,** *18,* 1304.
74. Magliozzo, R. S.; McCracken, J.; Peisach, J. *Biochemistry* **1987,** *26,* 7923.
75. Stevens, T. H.; Chan, S. I. *J. Biol. Chem.* **1981,** *256,* 1069.
76. LoBrutto, R.; Wei, Y.-H.; Mascarenhas, R.; Scholes, C. P.; King, T. E. *J. Biol. Chem.* **1983,** *258,* 7437.
77. Lang, G.; Marshall, W. *Proc. Phys. Soc.* **1966,** *87,* 3.
78. Oosterhuis, W. T.; Lang, G. *J. Chem. Phys.* **1969,** *50,* 4381.
79. Christner, J. A.; Münck, E.; Nanick, P. A.; Siegel, L. M. *J. Biol. Chem.* **1983,** *258,* 11147.
80. Liu, M.-C.; Huynh, B.-H.; Payne, W. J.; Peck, Jr., H.. D.; Dervartanian, D. V.; LeGall, J. *Eur. J. Biochem.* **1987,** *169,* 253.
81. Pfannes, H. D.; Bemski, G.; Wajnberg, E.; Rocha, H.; Bill, E.; Winkler, H.; Trautwein, A. X. *Hyperfine Int.* **1994,** *91,* 797.
82. Ellison, M. K.; Schulz, C. E.; Scheidt, W. R. *Inorg. Chem.* **1999,** *38,* 100.
83. Settin, M. F.; Fanning, J. C. *Inorg. Chem.* **1988,** *27,* 1431.
84. Dawson, J. H.; Andersson, L. A.; Sono, M. *J. Biol. Chem.* **1983,** *258,* 13637.
85. Shokhireva, T. Kh.; Walker, F. A. Unpublished results.
86. Schünemann, V.; Benda, R.; Trautwein, A. X.; Walker, F. A. *Israel J. Chem.* **2000,** in press.
87. Rivera, M. Barillas-Mury, C.; Christensen, K. A.; Little, J. W.; Wells, M. A.; Walker, F. A. *Biochemistry* **1992,** *31,* 12233, and references therein.

88. Guzov, V. I.; Houston, H. L.; Murataliev, M. B.; Walker, F. A.; Feyereisen, R. *J. Biol. Chem.* **1996,** *271,* 26637, and references therein.
89. Astashkin, A. V.; Raitsimring, A. M.; Walker, F. A. *Chem. Phys. Lett.* **1999,** *306,* 9.
90. Shokhirev, N. V.; Walker, F. A. *J. Am. Chem. Soc.* **1998,** *120,* 981.
91. Shokhireva, T. Kh.; Zhao, D.; Jacobsen, N. E.; Andersen, J. F.; Weichsel, A.; Balfour, C.; Montfort, W. R.; Walker, F. A. To be submitted.
92. Walker, F. A.; Simonis, U. In "Biological Magnetic Resonance, Vol. 12: NMR of Paramagnetic Molecules"; Berliner, L. J.; Reuben, J., Eds.; Plenum Press: New York, 1993; pp. 133–274.
93. Walker, F. A. In "The Porphyrin Handbook"; Kadish, K. M.; Smith, K. M.; Guilard, R., Eds.; Academic Press: San Diego; 2000, Chapter 36, Vol. 5; pp. 81–183.
94. La Mar, G. N.; Satterlee, J. D.; De Ropp, J. S. In "The Porphyrin Handbook"; Kadish, K. M.; Smith, K. M.; Guilard, R., Eds.; Academic Press: San Diego; 2000, Chapter 37, Vol. 5; pp. 185–298.
95. Minniear, A. B. M.S. Thesis, University of Arizona, 1998.
96. Krishnamoorthi, R.; La Mar, G.N.; Mizukami, H.; Romero, A. *J. Biol. Chem.* **1984,** *259,* 265.
97. Rajarathanam, K.; La Mar, G. N.; Chiu, M. L.; Sligar, S. G.; Singh, J. P.; Smith, K. M. *J. Am. Chem. Soc.* **1991,** *113,* 7886.
98. La Mar, G. N.; Jackson, J. T.; Dugad, L. B.; Cusanovich, M. A.; Bartsch, R. G. *J. Biol. Chem.* **1990,** *265,* 16173.
99. deRopp, J. S.; Mandal, P.; Brauer, S. L.; La Mar, G. N. *J. Am. Chem. Soc.* **1997,** *119,* 4732.
100. Hoffman, B. M.; Petering, D. H. *Proc. Natl. Acad. Sci. USA* **1970,** *67,* 637.
101. Hoffman, B. M. In "The Porphyrins"; Dolphin, D., Ed.; Academic Press: New York, 1979; Vol. 7, pp. 403–472.
102. Bowen, J. H.; Shokhirev, N. V.; Raitsimring, A. M.; Buttlaire, D. H.; Walker F. A. *J. Phys. Chem. B* **1997,** *101,* 8683.
103. Walker, F. A.; Emrick, D.; Rivera, J. E.; Hanquet, B. J.; Buttlaire, D. H. *J. Am. Chem. Soc.* **1988,** *110,* 6234.
104. La Mar, G. N.; de Ropp, J. S. NMR methodology for paramagnetic proteins, in "Biological Magnetic Resonance, Vol. 12: NMR of Paramagnetic Molecules"; Berliner, L. J. and Reuben, J., Eds.; Plenum Press: New York, 1993; pp. 1–78.
105. La Mar, G. N.; Smith, K. M.; Gersonde, K.; Sick, H.; Overkamp, M. *J. Biol. Chem.* **1980,** *255,* 60.
106. La Mar, G. N.; de Ropp, J. S.; Smith, K. M.; Langry, K. C. *J. Biol. Chem.* **1981,** *256,* 237.
107. La Mar, G. N.; Davis, N. L.; Parish, D. W.; Smith, K. M. *J. Mol. Biol.* **1983,** *168,* 887.
108. Reference *94,* p. 218.
109. Shokhirev, N. V.; Walker, F. A. *J. Biol. Inorg. Chem.* **1998,** *3,* 581.
110. Andersen, J. F.; Montfort, W. R. *J. Biol. Chem.* **2000,** *275,* in press.
111. Flower, D. R. *Biochem. J.* **1996,** *318,* 1.
112. Holden, H. M.; Rypniewski, W. R.; Law, J. H.; Rayment, I. *EMBO J.* **1987,** *6,* 1565.
113. Huber, R.; Schneider, M.; Mayr, I.; Muller, R.; Deutzmann, R.; Suter, F.; Zuber, H.; Falk, H.; Kayser, H. *J. Mol. Biol.* **1987,** *198,* 499.
114. Honig, B.; Nicholls, A. *Science* **1995,** *268,* 1144.
115. Varadarajan, R.; Zewert, T. E.; Gray, H. B.; Boxer, S. G. *Science* **1989,** *243,* 69.
116. Hunter, C. A.; Singh, J.; Thornton, J. M. *J. Mol. Biol.* **1991,** *218,* 837.
117. Jentzen, W.; Song, X.-Z.; Shelnutt, J. A. *J. Phys. Chem. B* **1997,** *101,* 1684.
118. Jentzen, W.; Ma, J. B.; Shelnutt, J. A. *Biophys. J.* **1998,** *74,* 753.

119. Shelnutt, J. A. Personal communication.
120. Newcomer, M. E.; Jones, T. A.; Aqvist, J.; Sundelin, J.; Eriksson, U.; Rask, L.; Peterson, P. A. *EMBO J.* **1984,** *3,* 1451.
121. Brownlow, S.; Morais Cabral, J. H.; Cooper, R.; Flower, D. R.; Yewdall, S. J.; Polikarpov, I.; North, A. C.; Sawyer, L. *Structure* **1997,** *5,* 481.
122. Newcomer, M. E. *Structure* **1993,** *1,* 7.
123. Ortiz de Montellano, P. R. *Acc. Chem. Res.* **1998,** *31,* 543.
124. Ortiz de Montellano, P. R., Ed. "Cytochrome P450: Structure, Mechanism and Biochemistry"; Plenum Press: New York, 1995.
125. Sono, M.; Dawson, J. H. In "Encyclopedia of Inorganic Chemistry"; King, R. B., Ed.; Wiley & Sons, Ltd.: Chichester, 1994; Vol. 4; pp. 1661–1682.
126. Ribeiro, J. M. C. Unpublished results.
127. Andersen, J. F. Unpublished results.
128. Roberts, S. A.; Weichsel, A.; Montfort, W. R. To be submitted.
129. Sheldrick, G. W.; Schneider, T. R. *Methods Enzymol.* **1997,** *277,* 319.
130. Shaw, A. W.; Vosper, A. J. *J. Chem. Soc., Faraday Trans.* **1976,** 1239.
131. Kerwin, J. F., Jr.; Lancaster, J. R., Jr.; Feldman, P. L. *J. Med. Chem.* **1995,** *38,* 4343.
132. Dierks, E. A.; Hu, S.; Vogel, K. M.; Yu, A. E.; Spiro, T. G.; Burstyn, J. N. *J. Am. Chem. Soc.* **1997,** *119,* 7316.
133. Zhao, Y.; Hoganson, C.; Babcock, G. T.; Marletta, M. A. *Biochemistry* **1998,** *37,* 12458.
134. Tomita, T.; Ogura, T.; Tsuyama, S.; Imai, Y.; Kitagawa, T. *Biochemistry* **1997,** *36,* 10155.
135. Brucker, E. A.; Olson, J. S.; Ikeda-Saito, M.; George, N.; Phillips, J. *Proteins: Struct., Funct., Genet.* **1998,** *30,* 352, and references therein.
136. Chan, N.-L.; Rogers, P. H.; Arnone, A. *Biochemistry* **1998,** *37,* 16459, and references therein.
137. Decatur, S. M.; Franzen, S.; DePillis, G. D.; Dyer, R. B.; Woodruff, W. H.; Boxer, S. G. *Biochemistry* **1996,** *35,* 4939.
138. Weichsel, A.; Andersen, J. F.; Roberts, S. A.; Montfort, W. R., unpublished work.
139. Andersen, J. F.; Korsgaard, K.; Shepley, D.; Montfort, W. R. Unpublished results.
140. Andersen, J. F.; Korsgaard, K.; Shepley, D.; Yurek, B.; Montfort, W. R. Unpublished results.
141. Sharma, V. S.; Traylor, T. G.; Gardiner, R. *Biochemistry* **1987,** *26,* 3837.
142. Abu-Soud, H. M.; Wu, C.; Ghosh, D. K.; Stuehr, D. J. *Biochemistry* **1998,** *37,* 3777.
143. Rohlfs, R. J.; Mathews, A. J.; Carver, T. E.; Olson, J. S.; Springer, B. A.; Egeberg, K. D.; Sligar, S. G. *J. Biol. Chem.* **1990,** *265,* 3168.
144. Carver, T. E.; Rohlfs, R. J.; Olson, J. S.; Gibson, Q. H.; Blackmore, R. S.; Springer, B. A.; Sliger, S. G. *J. Biol. Chem.* **1990,** *265,* 20007.
145. Kaneko, Y.; Yuda, M.; Iio, T.; Murase, T.; Chinzei, Y. *Biochim. Biophys. Acta* **1999,** *1431,* 492.
146. Fischmann, T. O.; Hruza, A.; Niu, X. D.; Fossetta, J. D.; Lunn, C. A.; Dolphin, E.; Prongay, A. J.; Reichert, P.; Lundell, D. J.; Narula, S. K.; Weber, P. C. *Nature Struct. Biol.* **1999,** *6,* 233.
147. Raman, C. S.; Li, H.; Martasek, P.; Kral, V.; Masters, B. S.; Poulos, T. L. *Cell* **1998,** *95,* 939.
148. Crane, B. R.; Arvai, A. S.; Ghosh, D. K.; Wu, C.; Getzoff, E. D.; Stuehr, D. J.; Tainer, J. A. *Science* **1998,** *279,* 2121.
149. Hurshman, A. R.; Marletta, M. A. *Biochemistry* **1995,** *34,* 5627.
150. Abu-Soud, H. M.; Wang, J.; Rousseau, D. L.; Fukuto, J. M.; Ignarro, L. J.; Stuehr, D. J. *J. Biol. Chem.* **1995,** *270,* 22997.

151. Hoshino, M.; Maeda, M.; Konishi, R.; Seki, H.; Ford, P. C. *J. Am. Chem. Soc.* **1996,** *118,* 5702.
152. King, B. C.; Hawkridge, F. M.; Hoffman, B. M. *J. Am. Chem. Soc.* **1992,** *114,* 10603.
153. Houston, H. L. M.S. Thesis, University of Arizona, 1995.
154. Hawkridge, F. M.; Kuwana, T. *Anal. Chem.* **1973,** *45,* 1021.
155. Bowen, E. F.; Hawkridge, F. M. *J. Electroanal. Chem. Interfacial Electrochem.* **1981,** *125,* 367.
156. Crutchley, R. J.; Ellis, W. R., Jr.; Gray, H. B. *J. Am. Chem. Soc.* **1985,** *107,* 5002.
157. Hildebrand, D. P.; Tang, H.; Luo, Y.; Hunter, C. L.; Smith, M.; Brayer, G. D.; Mauk, A. G. *J. Am. Chem. Soc.* **1996,** *118,* 12909.
158. Takano, T. *J. Mol. Biol.* **1977,** *110,* 569.
159. Hoshino, M.; Maeda, M.; Konishi, R.; Seki, H.; Ford, P. C. *J. Am. Chem. Soc.* **1996,** *118,* 5702.
160. Taylor, J. F.; Morgan, V. E. *J. Biol. Chem.* **1942,** *144,* 15.
161. Brunori, M.; Saggese, U.; Rotilio, G. C.; Antonini, E.; Wyman, J. *Biochemistry* **1971,** *10,* 1604.
162. Mauk, A. G.; Moore, G. R. *J. Biol. Inorg. Chem.* **1997,** *2,* 119.
163. Brunori, M.; Amiconi, G.; Antonini, E.; Wyman, J.; Zito, R.; Fanelli, A. R. *Biochim. Biophys. Acta* **1968,** *154,* 315.
164. Walba, H.; Isensee, R. W. *J. Org. Chem.* **1956,** *21,* 702.
165. George, P.; Hanania, G. I. H.; Irvine, D. H.; Abu-issa, I. *J. Chem. Soc.* **1964,** 5689.
166. Yagil, G. *Tetrahedron* **1967,** *23,* 2855.
167. Nesset, M. J. M.; Shokhirev, N. V.; Enemark, P. D.; Jacobson, S. E.; Walker F. A. *Inorg. Chem.* **1996,** *35,* 5188, and references therein.

HEME OXYGENASE STRUCTURE AND MECHANISM

PAUL R. ORTIZ DE MONTELLANO* and ANGELA WILKS†

*Department of Pharmaceutical Chemistry, School of Pharmacy, University of California, San Francisco, California 94143-0446, and †Department of Pharmaceutical Sciences, School of Pharmacy, University of Maryland, Baltimore, Maryland 21201-1180

I. Introduction

The action of heme oxygenase is graphically (but perhaps unsuspectingly) familiar to anyone who has observed the gradual discoloration

0898-8838/01 $35.00

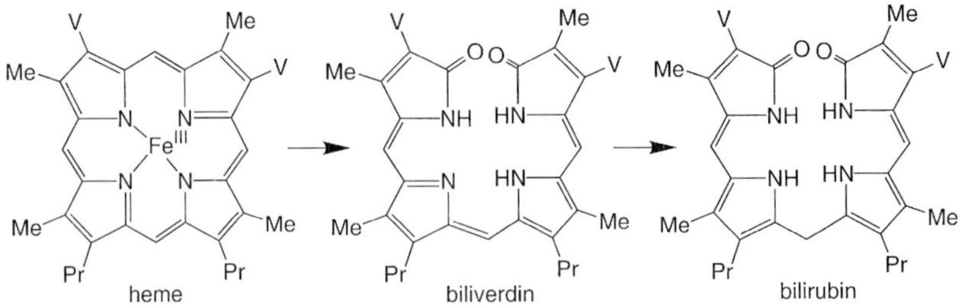

FIG. 1. Metabolism of heme via biliverdin to bilirubin. The abbreviations for the substituents throughout the figures are V, vinyl; Pr, propionic acid.

of a bruise from its initial "black and blue" to green and then yellow. The initial dark color is due to the heme-Fe from the hemoglobin being released into the damaged tissue by ruptured blood vessels. This heme (i.e., heme-Fe) is oxidized by heme oxygenase to biliverdin, which conveys the green tint, and the biliverdin is subsequently reduced to bilirubin, which is responsible for the yellow color (Fig. 1). Heme and biliverdin are highly lipophilic and are relatively difficult to eliminate, but bilirubin is readily excreted in the bile after conjugation with glucuronic acid (*1*).

Heme oxygenase is highly unusual in that it uses heme as both its substrate and its prosthetic group. It is also mechanistically and functionally distinct from the other classes of well-known hemoproteins, including the cytochromes P450, peroxidases, catalases, nitric oxide synthases, prostaglandin synthases, thromboxane synthases, and prostacyclin synthases. Nevertheless, the reactions catalyzed by heme oxygenase are part of the same heme-supported reactivity manifold that underlies the catalytic action of all hemoproteins, and elucidation of its mechanism can be expected to shed considerable light on the function not only of heme oxygenase but of all hemoproteins.

Heme oxygenase catalyzes the oxidation of heme to biliverdin, CO, and free iron in a reaction that requires molecular oxygen and NADPH (*2*). The transfer of electrons from NADPH to heme oxygenase is mediated by cytochrome P450 reductase, the same flavoprotein that transfers electrons to the cytochrome P450 enzymes (*3, 4*). The native enzyme is membrane bound and regiospecifically oxidizes heme at the α-*meso* position (Fig. 2) (*5, 6*). Mass spectrometric analysis of the origin of the two oxygen atoms incorporated into biliverdin and the oxygen incorporated into the CO shows that all three oxygens derive from O_2 (*7*). Indeed, the process requires three distinct molecules of O_2 (*2, 8*): one

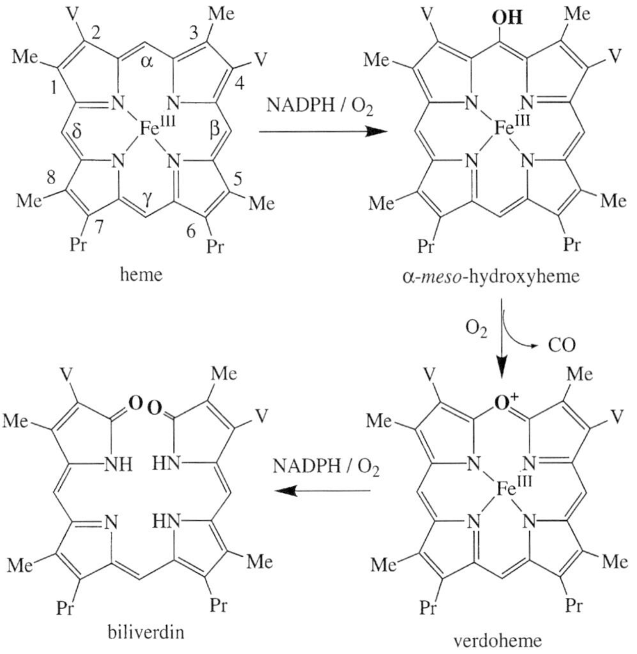

FIG. 2. The distinct steps in the oxidation of heme to biliverdin catalyzed by heme oxygenase. The substituted carbons of the porphyrin ring are labeled, as are the *meso*-positions. The oxygens introduced in the catalytic process are shown in bold type.

for the formation of α-*meso*-hydroxyheme, a second for the conversion of α-*meso*-hydroxyheme to verdoheme, and the third for conversion of verdoheme to biliverdin (Fig. 2).

The existence of two heme oxygenase isoforms, denoted HO-1 and HO-2, is now well established (*9–11*), and a third form (HO-3) with low catalytic activity has been reported (*12*). HO-1 is a heat shock protein also known as HSP32 and is the form that has been primarily examined because it is induced by a variety of agents, including heme, metals, hormones, oxidizing agents, and drugs (*13*, *14*). HO-2, which was discovered more recently, is primarily localized in the brain and testes and is resistant to induction by exogenous factors (*9*). The only inducers of HO-2 that have been identified to date are the adrenal glucocorticoids (*15*). Within the primary sequence of HO-2 are two Cys–Pro pairs flanked by a hydrophobic residue, in this case a Phe, which are the conserved core of the recently identified heme regulatory motif (HRM) (*16*). The HRM represents a new class of heme binding module that binds

1 10 20 30 40
HO-1 MERPQPDSMPPQDLSEALKEATKEV**H**TQAENAEFMRNFQKG
HO-2 MSAEVETSEGVDESEKKNSGALEKENQMRMADLSELLKEGTKEA**H**DRAENTQFVKDFLKG

50 60 70 80 90 100
HO-1 QVTRDGFKLVMASLYHIYVALEEEIERNKESPVFAPVYFPEELHRKAALEQDLAFWYGPR
HO-2 NIKKELFKLATTALYFTYSALEEEMERNKDHPAFAPLYFPMELHRKEALTKDMEYFFGEN

110 120 130 140 150 160
HO-1 WQEVIPYTPAMQRYVKRLHEVGRT**EPELLVAHAYTRYLGDLSGGQVLKK**IAQKALDLPSS
HO-2 WEEQVQCPKAAQKYVERIHYIGQN**EPELLVAHAYTRYMGDLSGGQVLKK**VAQRALKLPST

170 180 190 200 210 220
HO-1 GEGLAFFTFPNIASATKFKQLYRSRMNSLEMTPAVRQRVIEEAKTAFLLNIQLFEELQEL
HO-2 GEGTQFYLFENVDNAQQFKQLYRARMNALDLNMKTKERIVEEANKAFEYNMQIFNELDQA

230 240 250 260 270
HO-1 LT---HDTKDQSPSRAPGLRQRASNKVQDSAPVETPRGKPPL-NTRSQ----APLLRW
HO-2 GSTLARETLEDGFPVHDGKGDMRKCPFYAAEQDKGALEGSSCPFRTAMAVLRKPSLQF

280
HO-1 VLTLSFLVATVAVGLYAM
HO-2 ILAAGVALAAGLLAWYYM

FIG. 3. Alignment of the primary sequences of HO-1 and HO-2. The most highly conserved region is shown in bold, the proximal histidines are shown in bold and are underlined, and the C-terminal membrane anchors are underlined. The residues are numbered for HO-1.

heme transiently and reversibly, allowing changes in heme concentration to be sensed (*17*). However, the role in HO-2 of these noncatalytic heme binding domains remains unknown.

The molecular masses of human HO-1, HO-2, and HO-3 are 33, 36, and 32 kDa, respectively. Comparison of their protein sequences shows that the human HO-1 and HO-2 isoforms are 42% identical, a value somewhat lower than, for example, the 88% identity between the rat and human HO-2 isoforms (*18*). The amino acid sequence of HO-3 differs from that of HO-1 and HO-2 but bears a more striking similarity to that of HO-2 (90%) (*12*). Regions of high sequence conservation are found among all three proteins, most notably in the region encompassed by residues that correspond to residues 125 to 149 in HO-1 (Fig. 3) (*18, 19*). High sequence conservation is also observed in the region that corresponds to residues 11–40 in HO-1 and 30–59 in HO-2. A histidine located in this latter region is involved in the catalytic function of the enzyme (see below). Although less information is available on HO-2, it appears that its catalytic mechanism is similar to that of HO-1. The amino acid sequence of HO-3 has one striking difference from that of the other isoforms in that 21 amino acids that are relatively well conserved in HO-2 (residues 86–107) and HO-1 (residues 67–88) are not present in

HO-3. The catalytic activity of HO-3 is substantially lower than that of either HO-1 or HO-2, giving rise to the hypothesis that it may function as a regulatory protein (*12*).

II. Biological Function of Heme Oxygenase

Heme oxygenase is the only enzyme in mammals known to catalyze the physiological degradation of heme. Alternative mechanisms of heme degradation that lead to nonbiliverdin products have been proposed (*20, 21*) and several systems that generate H_2O_2 and therefore degrade heme have been reported (*22–24*), but the evidence for a physiological alternative heme catabolic pathway remains inconclusive. Heme catabolism both eliminates free heme, a lipophilic potentially toxic oxidizing species, and provides a mechanism for the recovery and reutilization of the iron atom. This is important because only 1–3% of the iron utilized daily in the synthesis of red blood cells is obtained from the diet. The rest of the daily iron requirement is met by recycling of the iron in the body, most of which is present in hemoglobin, myoglobin, and other heme proteins (*25*). The critical role of heme oxygenase in mammalian iron reutilization has been confirmed with mice in which HO-1 was knocked out by genetic methods (*26*). The mice developed an anemia associated with low serum iron levels but, paradoxically, accumulated toxic hepatic and renal iron concentrations.

The heme oxygenase reaction produces biliverdin, which is subsequently reduced to bilirubin by biliverdin reductase. Bilirubin is a powerful antioxidant that may play an important role in minimizing intracellular oxidative damage (*27*). Bilirubin, however, becomes a neurotoxic product when its concentration rises to high levels. The toxic effects of bilirubin can be significant when its conjugation with glucuronide and subsequent excretion are impaired, as in neonatal jaundice and in individuals with Crigler–Najjar syndrome (*13, 28*). Bilirubin, like many chemicals, is thus beneficial at low levels but toxic at high levels.

The third product of heme catabolism is CO. Following theoretical and experimental arguments for a possible role of carbon monoxide as a neural messenger (*29–31*), a great deal of activity has been focused on this potentially exciting new role for the heme oxygenases. The roles attributed to CO by a rapidly growing body of literature include its functions as a neural messenger in memory and learning, as a factor in neuroendocrine regulation, as an endogenous modulator of vascular tone, and as a protective agent in hypoxia and endotoxic shock. However, the role of CO in these processes remains controversial for two primary reasons. First, much of the experimental evidence is not unambiguous because it rests on studies using tin and other metalloporphyrins as inhibitors of heme oxygenase. These metalloporphyrins can inhibit other heme proteins and interact with guanylyl cyclase (*32*); the receptor that mediates the action of nitric oxide and possibly CO. Second, studies of the activation of guanylyl cyclase show that CO is a much poorer activating ligand than NO, (*33, 34*); which puts into question the proposed role of guanylyl cyclase as the receptor for CO. However, the data against the involvement of CO are also inconclusive. The activation of guanylyl cyclase by CO has been found to decrease as the receptor is purified from crude tissue (*34*). It is therefore possible that a factor that activates the receptor toward CO is lost in the purification. The possibility of an activator molecule gains support from the demonstration that 1-benzyl-3-(5′-hydroxymethyl-2′-furyl)indazole, a small organic molecule, sensitizes guanylyl cyclase toward the action of CO and makes the receptor as sensitive to CO as it is to NO (*35*). The possibility also exists that there are multiple guanylyl cyclase receptors and that the CO-sensitive one has not been isolated. Finally, some of the actions of CO may not depend on guanylyl cyclase, as proposed, for example, by Coceani *et al.* (*36*). Thus, the question of a role for CO in mammalian physiology remains open, although sufficient evidence is now available for the possible involvement of CO in diverse regulatory processes to be considered seriously (*37*).

III. Heme Oxygenase Model Systems

Mammalian heme oxygenases are membrane bound and are relatively difficult to purify and study. This made investigation of the enzyme structure and mechanism difficult and, until recombinant methods provided a soluble form of the enzyme, studies of the reaction mechanism were primarily carried out with model coupled oxidation systems. Coupled oxidation refers to the finding that the aerobic reactions of myoglobin, hemoglobin, and other hemoproteins with ascorbic acid, hydrazine, or other reducing agents partially convert their prosthetic heme group to biliverdin (*38*). The same reaction is observed when an aqueous pyridine solution of heme is allowed to react with hydrazine or ascorbic acid (*39, 40*). Indeed, the HO-1-bound heme can be converted to biliverdin by ascorbic acid in the absence of NADPH and cytochrome P450 reductase (*41*).

Prior to the purification of heme oxygenase in sufficient quantities to carry out *in vitro* studies, the coupled oxidation of myoglobin was believed to be the most accurate model for the biological process. In early $^{18}O_2$ studies it was shown for both the biological and chemical reactions that the oxygen atoms incorporated into the final biliverdin product were derived from two separate oxygen molecules, reinforcing the validity of coupled oxidation as a model for biological heme cleavage (*42*). This finding led to the hypothesis that the oxygen-bridged verdoheme (Fig. 2) might not be an intermediate on the pathway to biliverdin, as this step would appear to be a hydrolysis. However, in a subsequent study it was shown that verdoheme from the coupled oxidation of myoglobin could be converted to biliverdin via a nonhydrolytic mechanism (*43*). The overall reaction required three molecules of oxygen and three pairs of reducing equivalents which again was in reasonable stoichiometric agreement with the enzymatic reaction. The assignment of an Fe^{II} oxidation state to the verdoheme was also in agreement with preliminary data emerging on the enzymatic conversion of heme to biliverdin. Identification of an intermediate occurring after *meso*-hydroxylation and prior to biliverdin formation, which retained high affinity for carbon monoxide and formed a species that spectroscopically resembled Fe^{II} CO-verdoheme, suggested that verdoheme was, indeed, an intermediate in the enzymatic reaction (*44*).

The step from α-*meso*-hydroxyheme to verdoheme was studied in greater detail by Sano and colleagues (*45*). In this study apomyoglobin was reconstituted anaerobically with α-*meso*-hydroxyheme, which was believed to be the first intermediate formed in the degradation of heme to biliverdin. Physical characterization of the reconstituted protein by UV/visible and EPR spectroscopic techniques showed absorption maxima at 410, 590, and 640 nm and an EPR signal at $g = 6.3$ attributed to a high-spin Fe^{III} species. Reaction with 1 equiv of oxygen resulted in a new intermediate with maxima at 408, 540sh, 660, and 704 nm. Formation of this intermediate resulted in the extrusion of CO and concomitantly the EPR signal at $g = 6.3$ disappeared and was replaced by a radical signal at $g = 2.0012$. The intermediate formed was concluded to be Fe^{II} verdoheme or its one-electron π-neutral radical form. Since oxygen could not be replaced by hydrogen peroxide, and no ascorbate was utilized, a reductant was not necessary for the conversion of α-*meso*-hydroxyheme to verdoheme. These results were later confirmed in studies of the heme–heme oxygenase complex (see Section X).

Although the coupled oxidation reaction closely resembles the enzymatic reaction, it may not be identical. One clear difference is that heme free in solution is oxidized at all four *meso* positions (*40*), heme in myoglobin only at the α-*meso* position (*6, 46*); and heme in hemoglobin at the α- and β-positions (*6, 46*). Prior to the observation that hemin in solution could be converted to all four isomers, it was suggested that the regiospecificity of the coupled oxidation reaction was due to the inherent electronic properties of the heme (*6*). Subsequent identification of all four isomers from the coupled oxidation of hemin in pyridine contradicted the theory that the regiospecificity was in fact determined by the heme (*39, 40*). It was then proposed that areas of hydrophobicity within the protein cavity influenced the reactivity of the four *meso* positions (*46*). A later model based on the crystal structure of myoglobin, and utilizing an interactive computer display model to explore the relative accessibility of all four methene bridges, proposed that the reaction within the protein was controlled solely by steric effects (*47*).

Although steric effects may account for the regiospecificity observed in the coupled oxidation of hemoglobin and myoglobin, there is evidence that the reaction regiospecificity of heme oxygenase may be sensitive to electronic as well as steric effects (see Section IX,E). Coupled oxidation studies with perfluoroalkyl-substituted porphyrins have shown that substituents that alter the electronic properties of the chromophore can control the oxidation regiospecificity (*48*). The regiospecificity was attributed to the electron withdrawing ability of the CF_3 group, with cleavage occurring at the *meso* position with the highest electron density. These results support those obtained with heme oxygenase in which reconstitution of the protein with *meso*-methyl hemes (electron donating) results in exclusive oxidation of the methyl-substituted position and with α-*meso*-formyl heme (electron withdrawing) exclusively at non-formyl-substituted sites (see Section IX,E) (*49, 50*).

More recently, the coupled oxidation of sperm whale myoglobin mutants in which the distal His64 was relocated through the mutations Leu29His/His64Leu, Phe43His/His64Leu, and His64Leu/Ile107His increased the ratio of the γ-biliverdin isomer to 97, 44, and 22%, respectively. In contrast simple replacement of His64 with Leu gave no change in regiospecificity (*51*). The authors suggested that introduction of a charged histidine at these positions might introduce a biased polarity that enhances γ-specificity, and concluded that both hydrogen bonding and polarity are critical for regiospecificity. Previous coupled oxidation studies on the myoglobin mutant Val67Ala/Val68Ser showed no change in regiospecificity but a substantial increase in the efficiency of the reaction, which was again attributed to the increased polarity of the pocket (*52*). The suggestion that the polarity of the active site may aid in directing the oxygen to the α-*meso* edge again indicates that steric hindrance alone may not determine regiospecificity.

The differences between the myoglobin model and heme oxygenase that account for the increased efficiency of heme cleavage by the latter protein may be explained by the more polar nature of the distal face of heme oxygenase, as evidenced by the three-dimensional structure (see Section V) (*53*). The emerging evidence on the oxidative cleavage of heme in both the biological and coupled oxidation reactions indicates that structural, and possibly also electronic, factors contribute to the regiospecificity of the reaction. The critical factors in the catalytic efficiency of the heme oxygenase reaction when compared to the coupled oxidation systems are most likely a combination of hydrogen bonding interactions, the polarity of the heme binding pocket, and ligand discrimination in which oxygen is bound in preference to carbon monoxide (see Section VII).

IV. Heme Oxygenase: The Protein

A. Protein Expression

Heme oxygenase HO-1 has been purified from a variety of sources, including bovine and pig spleen and rat liver (*9, 54–56*), but the purification is not trivial and the protein is obtained in relatively low yields. The difficulties in obtaining the purified protein in a tractable state impeded elucidation of its structure and mechanism. Efforts to alleviate these problems by heterologously expressing full-length rat HO-1 in *Escherichia coli* were frustrated by the fact that the protein can be expressed in functional form but in unacceptably low yields (*57, 58*). A significant advance was provided by the independent finding in two laboratories that rat HO-1 truncated to remove the membrane binding domain can be expressed in *E. coli* in excellent yields (*58–60*). The recombinant proteins thus obtained are soluble, interact normally with cytochrome P450 reductase, and are catalytically active. The proteins expressed in the two laboratories differ in that one laboratory deleted 23 amino acids from the carboxy-terminus of the recombinant rat and human proteins (*58*), whereas the other deleted 26 amino acids from the rat protein with the concomitant introduction of two mutations (Ser262Arg, Ser263Leu) (*59*). The difference of three amino acids in the truncated terminus, and the two mutations introduced into the shorter protein, perhaps account for the report that the shorter protein is partially active, in contrast to the observation that the longer protein is fully active.

Heme oxygenase HO-2 has been partially purified from rat testes and the full-length protein has been expressed in an active form in *E. coli* (*11, 61–63*). Truncation of the protein at the C-terminus provides a soluble 33-kDa protein and improves the expression yields, but the yields and specific activities of the proteins are significantly lower than those of recombinant HO-1 (*64*). Mechanistic and structural studies on HO-2 have consequently been more limited.

B. Proximal Heme Ligand

The first question that must be addressed with regard to the mechanism of any hemoprotein is the nature of the two axial ligands in the heme–protein complex. Catalytic hemoproteins, in contrast to hemoproteins that function as electron carriers, have a fixed ligand on the proximal side of the heme and no ligand, or a displaceable ligand, on the distal side. The three common proximal ligands are a histidine nitrogen (e.g., myeloperoxidase) (*65*), a cysteine thiolate (e.g., cytochrome P450) (*66*), and, less frequently, a tyrosine phenolate (catalase) (*67*). The proximal ligand in the heme oxygenases has been unambiguously shown to be a histidine. The resemblance between the absorption spectra of heme oxygenase and myoglobin, including an Fe^{II}–CO complex with a λ_{max} at ~418 nm, provided initial evidence for proximal histidine ligation (Fig. 4) (*55, 56*). The involvement of a histidine as the proximal ligand was independently confirmed in two laboratories by resonance Raman spectroscopy (*68–70*). In both studies, the Fe^{III} porphyrin skeletal vibrations were consistent with a six-coordinate, high-spin protein

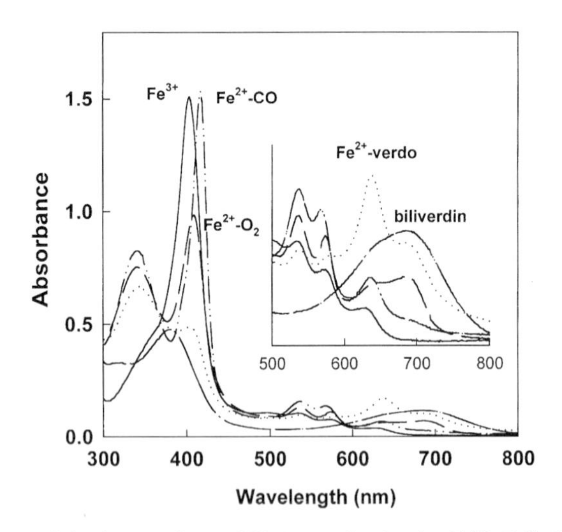

Fig. 4. Spectra of the human heme:HO-1 complex in the Fe^{III}, Fe^{II}–CO, and Fe^{II}–O_2 states and after conversion to the Fe^{II} verdoheme and biliverdin complexes.

with a histidine-ligated heme. Furthermore, a band at 216 cm^{-1} characteristic of a $\nu_{Fe\text{-}His}$ vibration is observed in the spectrum of the Fe^{II} deoxy enzyme. The position of this $\nu_{Fe\text{-}His}$ band in different histidine-ligated hemoproteins ranges from 221 cm^{-1} for myoglobin, in which the histidine ligand is weakly hydrogen bonded (*71–73*), to ~233 cm^{-1} in horseradish peroxidase, in which the imidazole is strongly hydrogen bonded, and ~246 cm^{-1} in a cytochrome *c* peroxidase mutant in which the imidazole is fully deprotonated (*74, 75*). The correlation between the position of the Fe–His vibration and hydrogen bonding of the proximal imidazole suggests that the proximal histidine ligand in heme oxygenase is not significantly hydrogen bonded or deprotonated. This may be relevant to the enzyme mechanism because the ability of the proximal imidazole to assist in cleavage of the dioxygen bond of a peroxo species coordinated to the distal side of the iron is thought to increase as the electron density on the imidazole is increased by partial or complete deprotonation. To the extent that assistance from the proximal ligand is important for cleavage of the distal dioxygen bond, the reaction is attenuated in heme oxygenase by the lack of hydrogen bonding interactions with the proximal histidine ligand.

The specific identity of the proximal histidine ligand has been resolved by a combination of site specific mutagenesis, spectroscopic, and catalytic studies. In an early paper Ishikawa *et al.* aligned the sequences of the rat, human, mouse, and chicken HO-1 enzymes and identified four histidines that were conserved in all four proteins (*59*). Mutation of each of these four histidines to an alanine showed that two of them could be mutated with retention of approximately 40% of the activity, whereas mutation of the other two, His25 and His132, led to essentially complete loss of activity. The interpretation of these results was complicated, however, by the fact that the His132 mutant was expressed in such low amounts that its integrity was questionable. Subsequent mutagenesis studies combined with resonance Raman and EPR characterization of the mutants unambiguously identified His25 as the proximal ligand (*76, 77*). Most telling in this regard is the disappearance of the $\nu_{Fe\text{-}His}$ vibration in the His25Ala HO-1 mutant. As found in the early study by Ishikawa *et al.* (*59*), the His25Ala mutant is completely inactive. The possibility that both the spectroscopic changes and the loss of activity are due to structural changes due to mutation of a histidine other than the proximal ligand is excluded by the demonstration that addition of exogenous imidazole leads to parallel recovery of both the $\nu_{Fe\text{-}His}$ (actually, $\nu_{Fe\text{-}Im}$) band in the resonance Raman spectrum and the catalytic activity of the enzyme (*78*). The $\nu_{Fe\text{-}Im}$ band in this instance is at 228 cm^{-1} because the mass of the exogenous imidazole differs from that of a histidine residue. These results not only confirm the role of His25 as the iron ligand, but show that tethering and subtle orientation of the imidazole ligand are not critical for the catalytic action of heme oxygenase.

Mutation of the proximal His25 in human HO-1 to a cysteine or tyrosine, as found for the His25Ala mutation, yields protein with no heme oxygenase activity (*79*). A resonance Raman spectroscopic analysis indicates that the iron in these proteins is five-coordinate high spin and thus lacks the usual distal water ligand. Resonance Raman isotopic studies confirm that the cysteine and tyrosine are coordinated to the iron in the Fe^{III} state of the corresponding mutants, but both ligands dissociate from the iron when it is reduced to the Fe^{II} state. As a result, the Fe^{II}–O_2 complexes obtained upon the subsequent binding of oxygen are destabilized and autooxidize rather than leading to productive enzyme turnover. Clearly, the histidine ligand plays an important role in the mammalian heme oxygenases. Even though His25 can be functionally replaced by exogenous imidazole when it is mutated to an alanine, it cannot be mutated to other common hemoprotein iron ligands with retention of activity. The inability of a cysteine or tyrosine to substitute for the histidine stems, at least in part, from structural factors that disfavor stable coordination of these ligands to the iron.

An independent approach has provided evidence that His25 is the proximal iron ligand. Incubation of the reduced HO-1:heme complex with $CBrCl_3$ under anaerobic conditions results in covalent binding of the heme to the protein (*80*). Covalent attachment involves the vinyl group of the heme because covalent binding does not occur when mesoheme is used as the substrate. Tryptic digestion of the labeled protein combined with Edman sequencing and mass spectrometric analysis of the peptide identify His25 as the alkylated residue. The site of alkylation can be altered by mutations on the distal side of the heme pocket, but the alkylated residues could not be unambiguously identified in the resulting peptides. These alkylation studies parallel earlier studies of the alkylation of the proximal histidine by the heme in the anaerobic reaction of hemoglobin with $CBrCl_3$ (*81*). Incidentally, these studies provide the first evidence that heme oxygenase, like other hemoproteins, can catalyze the reductive activation of halocarbons to reactive radical species (*80*).

Although more limited, sufficient spectroscopic data is available on the HO-2 isoform to confirm that the proximal ligand is also an imidazole. A 28-kDa catalytically active fragment of human HO-2 obtained by tryptic digestion of the enzyme expressed in *E. coli* was shown by optical, resonance Raman, and EPR studies to have a neutral (unionized) histidine proximal iron ligand (*62*). Mutation of His151 to an alanine in rat HO-2 expressed in *E. coli* yielded inactive protein (*63*), but it is not possible to unambiguously interpret this finding because the protein was not purified or spectroscopically characterized. His151 corresponds to His132 in the rat and human enzymes and is therefore unlikely to be the proximal ligand in the HO-2 isoform. Expression of a truncated soluble human HO-2 has allowed the role of His152 (corresponding to His151 in rat) as well as His45 to be readdressed. In contrast to the report in which rat His151 was reported to be inactive, the purified His152 Ala mutant retained the spectroscopic and enzymatic properties of the wild-type protein. Mutation of His45 to an alanine resulted in a protein that had lost all activity and exhibited a five-coordinate Fe^{II}–NO EPR spectrum. The authors concluded from the lack of activity, the five-coordinate ligation, and the conserved alignment with His25 in HO-1 that His45 is likely to be proximal ligand to the heme.

C. Distal Heme Ligand

The distal iron ligand in HO-1, as suggested by the close similarity in the absorption spectra of HO-1 and myoglobin, is a water molecule. Evidence for this is provided by the observation that increasing the pH causes transitions in the absorption, resonance Raman, and EPR spectra with a pK_a of approximately 8.0, similar to those observed with myoglobin (*68–70*). The transition, as in myoglobin, is assigned to deprotonation of the distal water ligand (Fig. 5). The initial step of heme oxygenase catalysis therefore resembles the conversion of metmyoglobin to oxymyoglobin, in that reduction of the iron is coupled to loss of the distal water ligand and the binding of oxygen.

Mutation of His132 to an alanine, glycine, or serine in the truncated human heme oxygenase, in accord with identification of His25 as the proximal ligand, does not alter the $\nu_{Fe\text{-}His}$ band in the resonance Raman spectrum of the Fe^{II} deoxy protein (*82*). The mutations result in the formation of two separate fractions when the protein is expressed in

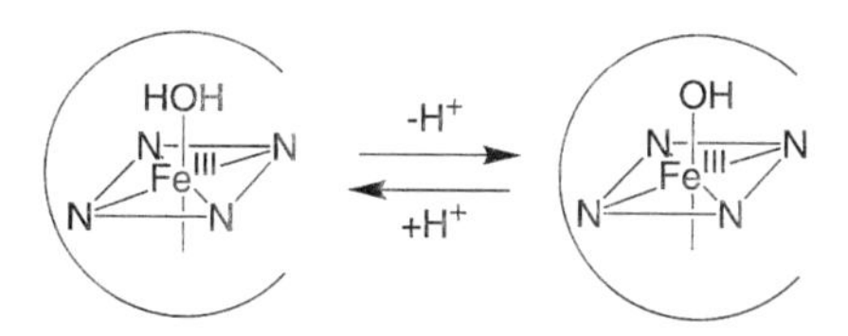

Fig. 5. Scheme showing the deprotonation of the distal water ligand observed with HO-1 at approximately pH 8.

DH5α *E. coli* cells. In one fraction the distal water ligand is lost, much in the same way as the water ligand is lost in myoglobin when His64 is mutated to a valine or leucine (*83–85*), and in the other the water ligand is retained. However, expression of the protein in BL21 *E. coli* cells leads exclusively to formation of the protein retaining the distal water ligand (*86*). These results indicate that His132 is not required for binding of the distal water molecule. As the His132 mutant obtained in the BL21 cells was fully active, it is also clear that His132 is not involved in the catalytic reaction. These results concur with those obtained independently when His132 was mutated to an alanine, glycine, or serine in the truncated rat protein (*87*). The rat mutants expressed in *E. coli* were obtained as inclusion bodies that were refolded to the active proteins. In this instance, the three proteins had spectra identical to those of the wild type, and the His132Ala mutant had a pH-dependent transition similar to that of the wild type. These results indicate, again, that His132 is not required for either retention of the water ligand or catalytic activity. The earlier report that mutation of His132 inactivates the rat enzyme (*59*) is superseded by the more recent results because the quality of the protein in the earlier work was not known. A similar reservation exists concerning the finding that mutation of His151 in the rat HO-2 protein causes loss of catalytic activity (*63*). As already noted, the purified human HO-2 His152Ala mutant (which corresponds to His151 in the rat sequence) has spectroscopic and enzymatic properties identical to those of the wild-type protein (*64, 69*). The absence of a distal histidine that stabilizes the water coordinated to the iron distinguishes the heme oxygenases from the myoglobins, in which mutation of the distal His64 to some but not all alternative residues results in dissociation of the water ligand (*83–85*).

D. Active Site Environment

^{1}H NMR studies of truncated rat heme oxygenase indicate that the heme group binds within the active site in two orientations that differ by a 180° rotation about the α,γ-axis (Fig. 2) (*88*). Both binding orientations leave the α-*meso* carbon in the same place in the active site, so the binding heterogeneity does not affect the α-*meso*-regiospecificity of the enzyme. The two binding orientations are present to about the same extent with heme, but differences in the ratio of the two orientations when heme and iron deuteroporphyrin IX are bound indicate that the ratio is sensitive to the nature of the substituents at the 2 and 4 positions of the heme group. This first NMR study also identified fragments of amino acid side chains in contact with the heme but not the actual amino acids. More importantly, the contact shift patterns for the heme protons determined by isotopic labeling and 2D NMR mapping suggested an unusual heme electronic structure characterized by large differences in delocalized spin density in the two positions within a single pyrrole ring, rather than the more common finding of large differences between adjacent pyrrole rings. This distribution of spin density suggested a direct electronic perturbation of the heme by the protein matrix such as might be produced by the presence of an anionic group close to the α-*meso*-carbon. The presence of an anionic group in the vicinity of the heme is consistent with the finding that the characteristic electronic asymmetry is pH dependent and is lost as the pH is decreased. It is unclear whether this unusual distribution of spin density contributes to the regiospecificity of the enzyme.

A more recent 2D ^{1}H NMR investigation of the human heme–HO-1 complex was undertaken to characterize more fully the nature of the substrate binding pocket (*86*). The NMR data revealed a cluster of at least nine interacting aromatic residues, including three Phe and an additional six Tyr or Phe that form part of the substrate binding site. Although no sequence-specific determination of the individual residues could be made, it was clear from the data that the heme pocket of heme oxygenase is more open than that of the globins or peroxidases. The heme was suggested to be bound with pyrrole rings I (1-methyl, 2-vinyl) and parts of pyrrole rings II (3-methyl) and IV (8-methyl) buried toward the interior of the protein, whereas pyrrole ring III (5-methyl, 6-propionate) and parts of pyrrole ring II (4-vinyl) and IV (7-propionate) are probably exposed to solvent. The proximal side of the heme is readily available to the medium, as evidenced by the fact that both the peptide NH and ring $N_\delta H$ of the proximal His25 are in rapid exchange with the solvent. These assignments have not proven to be consistent with the crystal structure (see later discussion). It is possible that this inconsistency stems from the presence of one predominant heme orientation in solution and another in the crystal structure that differ by a 180° rotation about the α,γ axis of the heme (*vide supra*).

Two labile protons close enough to the iron to influence ligand reactivity were identified in the NMR study: a labile proton originating from a His residue too far away to hydrogen bond to the bound ligand, but close enough that it might influence the stereoselectivity, and a second labile proton from a Tyr or His residue in van der Waals contact with the bound ligand. The previous site-directed mutagenesis studies in which mutation of His132 to an alanine gave a protein with identical spectroscopic and enzymatic properties to that of the wild type, as well as NMR evidence that the labile proton is retained in the same His132

Ala mutant, rule out His132 as the direct hydrogen bonding residue. The fact that mutagenesis of His84 and His119, the two conserved histidines other than His25 and His132, does not result in loss of activity suggested that a tyrosine residue might function as a distal base. Tyrosine as a hydrogen bonding distal residue has been reported in a number of trematode myoglobins (*71, 72*). However, mutation in human HO-1 of three of the conserved Tyr residues (i.e., Tyr134, Tyr137, and Tyr182) to Phe causes neither loss of activity nor loss of the distal water ligand (unpublished results). Interestingly, Tyr134 and Tyr137, like His132, lie within the conserved 24 amino acid region believed to form part of the distal pocket of the heme binding site. These mutagenesis results are consistent with the recently determined X-ray structure of human HO-1, which shows that the distal water ligand is not directly coordinated to a histidine or a tyrosine or, indeed, to any other obvious active site residue (see later discussion).

V. Human HO-1 Crystal Structure

Truncated forms of human and rat heme oxygenase-1 have been successfully crystallized (*53, 89*), and the structure of the human isoform has been determined (Fig. 6) (*53*). The full-length human heme oxygenase is a 288-residue protein, but the protein that was crystallized

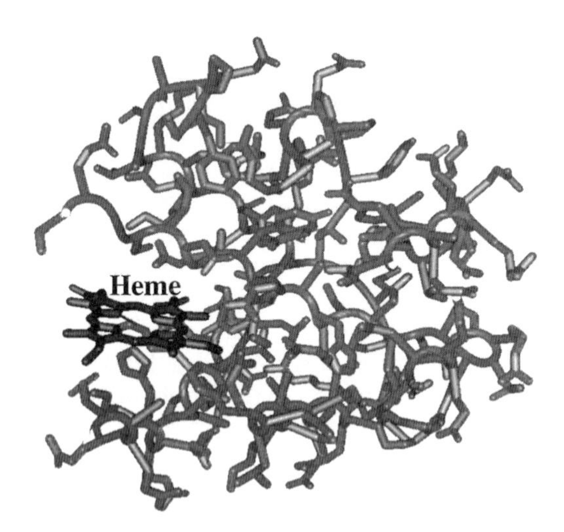

FIG. 6. Structure of truncated human HO-1 showing the location of the heme binding site (*53*). The heme propionic acid groups point away from the viewer, so that the δ-meso edge faces the exterior of the protein. The iron atom in the heme is not shown for clarity.

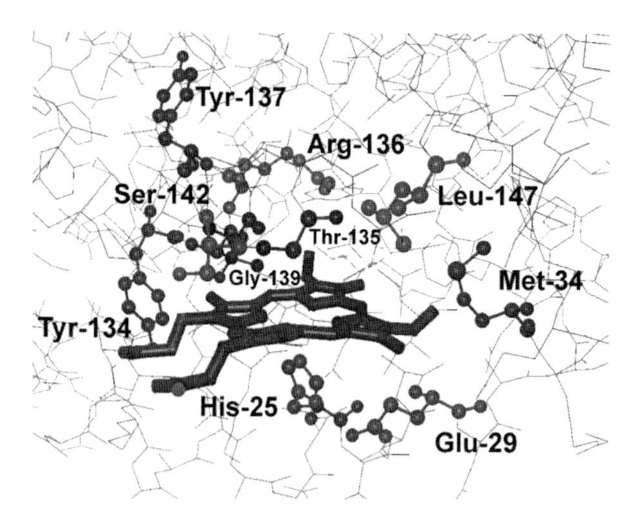

FIG. 7. Structure of the active site of truncated human HO-1 showing residues that are in the active site cavity (*53*).

retained only the first 233 amino acid residues. The 55 amino acid deletion consisted of the 23 C-terminal residues that make up the lipophilic membrane binding domain and an additional 32 C-terminal amino acids. Despite this truncation, the catalytic activity and regiospecificity of the protein were very similar to those of both the native enzyme and the wild-type protein missing only the membrane-binding polypeptide (*90*, *91*). In addition to the truncation, the first nine and the last 10 amino acid residues in the protein were not ordered in the electron maps and therefore could not be placed in the structure. It should be kept in mind, therefore, that the available crystal structure is missing a significant segment of the protein. This segment is not critical for catalytic activity, but could play a role in protein–protein interactions or other subtle aspects of heme oxygenase function.

The heme oxygenase core represents a novel protein fold that consists primarily of α-helices (*53*). As in most hemoproteins, the heme is sandwiched between two helices with the propionic acid carboxyl groups exposed at the molecular surface. As expected from the spectroscopic and mutagenesis data, His25 is the proximate ligand to the heme iron atom (Fig. 7). It is part of the proximal helix that also provides a number of residues that contact the heme group. The helix on the distal ligand-binding side is kinked about 50° directly above the heme because of the flexibility introduced by the three glycines in the highly conserved sequence Gly^{139}-Asp-Leu-Ser-Gly-Gly^{144}. Two of the glycines, Gly139 and

Gly143, are in direct contact with the heme group. The heme is also in contact with the distal helix residues Tyr134, Thr135, Arg136, Ser142, and Leu147 (Fig. 7). As indicated by the spectroscopic and mutagenesis studies, His132 is not appropriately located to stabilize the distal water ligand or to function as an acid–base catalyst in the heme oxygenase reaction. The distal water ligand implicated by the spectroscopic data is located in the crystal structure at a distance of approximately 1.8 Å from the iron, but no polar side chains are found close enough to this water molecule to hydrogen bond with it.

Of the four heme edges, only the δ-*meso* edge is exposed at the surface of the protein (Fig. 6). The other three edges, including the α-*meso* edge that is oxidized by the enzyme, are buried in the interior of the protein. The α-*meso* heme edge abuts against a hydrophobic wall consisting of Phe214, Met34, and Phe37, but a large water-filled cavity extends from the heme site into the protein interior directly above this heme edge. Interactions of the propionate carboxyl groups with Lys179, Arg183, Lys22, and Tyr134 appear to be important in binding the heme and orienting it correctly within the active site (*53*).

Two different conformations of heme oxygenase are found in the asymmetric unit cell. The two molecules differ significantly in the orientation of the distal helix, a region of the molecule that also exhibits relatively high thermal factors (*53*). The flexibility of the helix implied by these results is consistent with the presence of several conserved glycines in the sequence and suggests that motion of the helix opens and closes the heme crevice to allow binding of the heme and dissociation of the product. In the closed structure, the distal helix lies across the top of the heme within 4 Å of its surface, sterically obstructing access of any iron-bound oxidizing species of the β-, γ-, and δ-*meso* carbon atoms. This steric restriction is partially relaxed in the more open structure.

VI. Interaction with Cytochrome P450 Reductase

The crystal structure provides some insight into the interaction of heme oxygenase with cytochrome P450 reductase, its obligatory electron donor partner. The oxidation–reduction potential of −65 mV for the heme in heme oxygenase is appropriate for direct electron transfer to it from cytochrome P450 reductase (*79*), which has potential values of −270 mV for the 1*e*- to 2*e*- state, and −290 mV for the 2*e*- to 3*e*- state (*92*). The crystal structure shows that the surface of the protein surrounding the exposed heme edge is electropositive (*53*). It therefore provides an attractive surface for the docking of cytochrome P450 reductase, which has a predominantly negatively charged surface (*93*). Docking of the reductase to heme oxygenase through these electrostatically complementary surfaces would be consistent with the electrostatic complementarity observed in the binding of the FMN and heme domains in the crystal structure of a P450BM-3 complex (*94*). Binding of the reductase to heme oxygenase in the vicinity of the exposed heme edge offers the shortest route for the delivery of electrons to the heme iron atom, although other protein binding sites with short paths to the iron are available. If heme oxygenase and cytochrome P450 reductase dock as suggested by the crystal structure, the docking requirements may not be highly stringent. This inference derives from the finding that a protein constructed by fusing heme oxygenase to cytochrome P450 reductase exhibits high heme oxygenase activity (*60*). The heme oxygenase and cytochrome P450 reductase domains were incorporated into this fusion protein after deletion of their respective membrane binding peptides. Ionic strength and dilution studies indicated that electrons are transferred intramolecularly in the fusion protein, implying that the reductase domain was able to dock and efficiently transfer electrons to the heme oxygenase domain. It is possible that the fusion protein fortuitously allows normal docking of the flavoprotein domain to the heme domain in the fusion protein, or that an alternative electron mechanism comes into play. However, it is more likely that electron transfer occurs normally and that docking of the two domains is suboptimal because of steric constraints imposed by the covalent link between them, in which case the high activity of the fusion protein suggests that the docking interaction is forgiving.

VII. Gaseous Ligands

The binding of oxygen, carbon monoxide, and nitric oxide to heme oxygenase is of physiological relevance, the first because it is a key substrate, the second because it is a product of the enzyme that can act as an endogenous inhibitor, and the third because it is a potential physiological inhibitor that could inversely couple the synthesis of carbon monoxide with that of nitric oxide. The oxygen affinities of the Fe^{II} HO-1 and HO-2 heme complexes are 30- to 90-fold higher than those of the mammalian myoglobins (*91*). These high affinities are due to slower dissociation of the oxygen from the heme oxygenase than myoglobin sites. The affinities for the binding of CO to the Fe^{II} complexes are less than sixfold higher than those for the binding of oxygen, in contrast to the much higher affinity for CO and O_2 of the globins. The heme oxygenases thus discriminate much more effectively against the binding of CO than the myoglobins or hemoglobins, for which the K_{co}/K_{o2} ratios are ~40 and ~200, respectively (*95*). The crystal structure of truncated human HO-1 suggests two mechanisms for this higher discrimination against CO (*53*). Although there is no distal histidine or strongly polar residue able to hydrogen bond to the iron-coordinated ligand, the distal site is fairly polar due to residues such as Asp140, Arg136, and Asn210 that are near pyrrole ring B of the heme. The ligand may also be hydrogen bonded because of ordering of active site water molecules when the ligand is bound. The observation of a solvent deuterium isotope effect on the EPR spectrum of the oxy-cobalt–HO-1 complex provides evidence that the distal dioxygen ligand is, indeed, hydrogen bonded (*96*). A steric effect on the binding of CO is unlikely because resonance Raman indicates that the Fe–CO bond, if anything, is more linear in heme oxygenase than in the myoglobins (*69*). A second mechanism for diminishing the binding of CO formed in the active site may involve the large hydrophobic chamber located above the α-*meso* edge of the heme. The CO that is formed may escape into this chamber rather than into the more polar heme pocket, decreasing its effective inhibitory concentration.

The binding of nitric oxide to Fe^{II} rat heme oxygenase-1 has been demonstrated by absorption, resonance Raman, and EPR spectroscopy (*68*). EPR studies have similarly demonstrated the formation of an Fe^{II}–NO complex with human heme oxygenase-2 (*62*). In both instances, the EPR superhyperfine coupling of the ^{15}N-labeled iron-bound NO with the nitrogen of the proximal histidine was used as part of the evidence that the proximal ligand was a histidine. The Fe–NO bond appears to be similar to that in the corresponding myoglobin complex. NO has been reported to inhibit the catalytic activity of heme oxygenase in endothelial cell preparations, and EPR studies suggested that this inhibition correlated with formation of the nitric oxide complex of free heme in microsomes (*97*). It has also been reported, however, that nitric oxide inhibits rat heme oxygenase-2 but not rat heme oxygenase-1 or mutants of heme oxygenase-2 in which the cysteine residues of the noncatalytic heme binding motifs have been mutated to alanines (*98*). These latter results led to the proposal that the inhibition was due exclusively to an interaction of nitric oxide with the hemes in these heme binding motifs. However, the inhibition data are not yet sufficiently infinitive to unambiguously determine the effect of nitric oxide on heme oxygenase function. Further work is necessary to determine whether nitric oxide is a physiologically relevant inhibitor of the heme oxygenases.

TABLE I

HEME OXYGENASE SUBSTRATE SPECIFICITY[a]

Name	2	4	Relative rate	Ref.
proto IX	**V**	**V**	**100**	*100*
meso IX	Et	Et	83	*100*
deutero IX	H	H	48	*100*
hemato IX	CHOHMe	CHOHMe	86	*100*
	COMe	COMe	45	*100*
	CHO	CHO	15	*99, 101*
	CHO		17	*99, 101*
		CHO	34	*99, 101*
	$CH{=}CH(CH_2)_2Me$	$CH{=}CH(CH_2)_2Me$	45	*100*
	$CH_2(CH_2)_3Me$	$CH_2(CH_2)_3Me$	28	*100*
	$CHOH(CH_2)_3Me$	$CHOH(CH_2)_3Me$	4	*100*
	$CO(CH_2)_3Me$	$CO(CH_2)_3Me$	37	*100*
	$CH{=}CHCHMe_2$	$CH{=}CHCHMe_2$	50	*100*
	$CH_2CH_2CHMe_2$	$CH_2CH_2CHMe_2$	39	*100*
	$CHOHCH_2CHMe_2$	$CHOHCH_2CHMe_2$	48	*100*
	$COCH_2CHMe_2$	$COCH_2CHMe_2$	27	*100*
	$CH{=}CH(CH_2)_9Me$	$CH{=}CH(CH_2)_9Me$	45	*100*
	$CH_2(CH_2)_{10}Me$	$CH_2(CH_2)_{10}Me$	36	*100*
	$CHOH(CH_2)_{10}Me$	$CHOH(CH_2)_{10}Me$	0	*100*
	$CO(CH_2)_{10}Me$	$CO(CH_2)_{10}Me$	45	*100*

[a]Only the substituents at positions 1 and 4 of the protoporphyrin IX skeleton vary. The other substituents are the same as in protoporphyrin IX: methyl at positions 1, 3, 5, and 8, and propionic acid at 6 and 7. V stands for $-CH{=}CH_2$.

VIII. Substrate Specificity

The enzyme is highly specific for the heme propionic acid side chains at positions 6 and 7, but is much less discriminating with respect to substituents at positions 1, 2, 3, and 4 (Tables I and II; see Fig. 2 for heme numbering scheme) (*99–101*). Exchanging the positions of the methyl and propionic acid substituents at positions 5, 6, 7, and 8 yields inactive substrates, as does decreasing the propionic acid chain length by one carbon. Some activity is retained, however, when the propionic acid side-chain length is increased by one carbon. In contrast, the substituents that are acceptable at one or more of positions 1–4 include CH_2CH_3, CH_2CH_2OH, $CH_2CH_2CO_2H$, $CHOHCH_3$, $CH(OH)CH_2OH$, CHO, $COCH_3$, $CH{=}CH(CH_2)_2CH_3$, $CH_2(CH_2)_3CH_3$, CHOH $(CH_2)_3CH_3$, $CO(CH_2)_3CH_3$, $CH_2CH_2CH(CH_3)_2$, $COCH_2CH(CH_3)_2$, $CHOHCH_2CH(CH_3)_2$, $CH{=}CH(CH_2)_9CH_3$, $CH_2(CH_2)_{10}CH_3$, and $CO(CH_2)_{10}CH_3$ (Tables I and II). Thus, both polar residues and long hydrocarbon chains

TABLE II

HEME OXYGENASE SUBSTRATE SPECIFICITY[a]

1	2	3	4	5	6	7	8	Relative rate	Ref.
Me	**V**	**Me**	**V**	**Me**	**Pr**	**Pr**	**Me**	100	***100***
				Pr	Me	Me	Pr	0	*101*
						Me	Pr	0	*101*
				Pr	Me			0	*101*
	CH_2CH_2OH		CH_2CH_2OH	Me		Me	Pr	0	*100*
				Pr	Me	Me		0	*101*
					C_4–CO_2H	C_4–CO_2H		~50	*101*
					CH_2CO_2H	CH_2CO_2H		0	*101*
V	Me	V	Me					121	*100*
		V	Me					147	*100*
	Me		Me					80	*99, 101*
			Me					118	*99, 101*
	Pr		Pr					0	*100*
Pr	Me		Pr					0	*100*
CH_2CH_2OH	Me		CH_2CH_2OH					175	*100*
	Pr		Pr					10	*100*
			Pr					~50	*99*

[a]The substituents at all 8 positions of the porphyrin ring are indicated. If no substituent is shown, it is the same as in protoporphyrin IX, the first compound. V stands for $—CH{=}CH_2$, Pr for $—CH_2CH_2CO_2H$, and C4-CO_2H for $—CH_2CH_2CH_2CO_2H$.

at positions 1–4 are compatible with substrate binding and catalysis. These results suggest that the heme propionic acid side chains are involved in specific hydrogen bonding or ionic interactions with active site residues and are therefore important for substrate binding, in accord with the HO-1 crystal structure (*53*). The heme orientational heterogeneity revealed by the NMR studies is compatible with this requirement because the propionic acid side chains simply exchange positions in the two heme binding orientations. In contrast, the undemanding specificity for the substituents at positions 1–4 suggests that the active site is relatively open in the region of the active site occupied by the northern (α-*meso*) edge of the heme group (Figs. 2 and 7). An active site structure of this type would allow the catabolism of more complicated heme groups such as those that might arise from cytochrome *c*, including dicysteinyl hematoheme and the heme undecapeptide obtained by digestion of cytochrome *c* in which the heme is crosslinked via the 2 and 4 positions to the two cysteines in the peptide VQKCAQCHTVE (*102*). The crystal structure of human heme oxygenase shows that the active site has a fairly open water channel in the vicinity of the α-*meso*-position, although it does not appear to be large enough to accommodate attachment to the porphyrin of a moiety as large as an undecapeptide.

IX. The First Stage: α-*meso*-Hydroxylation

A. Formation of the Activated Oxygen Species

The first step in the catalytic turnover of heme oxygenase subsequent to the binding of heme is reduction of the heme to the Fe^{II} state by NADPH–cytochrome P450 reductase. This step is readily observed in the presence of CO because of formation of the spectroscopically distinguishable Fe^{II}–CO complex with an absorption maximum at 418 nm (Fig. 4) (*41, 55*). Once the iron is reduced, oxygen is bound to give the Fe^{II}–O_2 complex with an absorption maximum at 410 nm (Fig. 4) (*4*). The spectroscopic resemblance of the heme oxygenase and myoglobin complexes suggests that the properties of the Fe^{II}–O_2 complex may be generally similar to those of oxymyoglobin. However, an abnormal oxygen isotope shift pattern in the resonance Raman spectrum of the Fe^{II}–O_2 HO-1 complex has led to the proposal that the iron–oxygen–oxygen bond is bent, bringing the terminal oxygen of the complex closer to the heme periphery (*103*). The authors proposed that tilting of the oxygen ligand is due to steric interactions with active site residues, although these are not evident in the subsequently determined crystal structure (*53*). Differences in the EPR spectrum of the oxy-cobalt–HO-1 complex in D_2O versus H_2O suggest that the metal-bound dioxygen molecule is hydrogen bonded (*96*), although the identity of the hydrogen bond donor is again not clear from the crystal structure of the protein (*53*). Support for some sort of interaction of distal ligands with the protein is provided by ^1H NMR studies of the rat enzyme, which suggest that the cyano ligand is tilted out of a line perpendicular to the heme plane because of interactions with protein residues (*88*).

B. Substitution of H_2O_2 for O_2 and Reducing Equivalents

The Fe^{II}–O_2 complex can be observed as an intermediate in the catalytic process (Fig. 4), but must be reduced further for heme oxidation to occur (*4*). A two-electron reduction of molecular oxygen produces a species formally equivalent in oxidation state to H_2O_2. We therefore examined the possibility that H_2O_2 might be a viable substitute for molecular oxygen and reducing equivalents in supporting the catalytic turnover of heme oxygenase. Indeed, incubation of the HO-1:heme complex with 1 equiv of H_2O_2, as judged by the observation of a decrease in the Soret absorption coupled with an increase in the absorbance at ~680 nm, results in rapid oxidation of the heme to give the HO-1:verdoheme complex (*58*). Extraction of the prosthetic group from the protein in the presence of pyridine gave the characteristic peaks at 400, 504, 536, and 680 nm of the verdoheme complex. Finally, addition of cytochrome P450 reductase and NADPH to the verdoheme complex formed with 1 equiv of H_2O_2 produces the expected biliverdin IXα (*58*). H_2O_2 is thus equivalent to a molecule of oxygen and two electrons in supporting the oxidation of the heme group as far as verdoheme, but H_2O_2 does not readily support the conversion of verdoheme to biliverdin. Later studies confirmed that H_2O_2 similarly supports the HO-2-catalyzed conversion of heme to verdoheme (*62*). Thus, the oxidation of the heme group involves an oxidizing species equivalent to Fe^{III} heme plus H_2O_2. Anaerobic reaction of the HO-1:heme complex with 1 equiv of H_2O_2 produces a distinct intermediate that is stable in the absence of oxygen (*104*). This intermediate is identical by absorption and resonance Raman spectroscopy to the complex formed anaerobically between synthetic α-*meso*-hydroxyheme and HO-1 (*105*). The fact that the α-*meso*-hydroxyheme:HO-1 complex formed by reaction of the heme complex with H_2O_2 (*104*) or by reconstitution of the apoenzyme with synthetic material (*105–107*) is a competent precursor of verdoheme definitively establishes that α-*meso*-hydroxyheme is a key intermediate in the normal reaction pathway of heme catabolism. The earlier conversion of α-*meso*-hydroxyheme to biliverdin in model systems supports this conclusion (*45, 108–110*) and is consistent with the finding that the oxygen in the carbon monoxide that is eventually released derives from molecular oxygen (*7*).

In one study, it has been reported that radiolytic reduction of the Fe^{II}–O_2 heme oxygenase complex at 77 K produces the Fe^{III}–hydroperoxy (Fe^{III}–OOH) complex with spectroscopic properties similar to those reported previously for the corresponding complex of hemoglobin β-chains (*111*). The Fe^{III}–OOH heme oxygenase complex exhibits a rhombic EPR signal at $g = 2.37, 2.18$, and 1.93. ENDOR spectra identify an exchangeable proton of the hydroperoxide with a hyperfine coupling $A(g_1)$ of approximately 10.5 MHz. When this Fe^{III}–OOH complex is annealed at 200 K it is transformed into Fe^{III} α-*meso*-hydroxyheme. This study confirms that the Fe^{III}–OOH species proposed to be formed with H_2O_2 is indeed formed when the Fe^{II}–O_2 complex is reduced by one electron. Furthermore, it solidifies the evidence that the reactive species in heme oxygenase is the Fe^{III}–OOH complex.

C. Reaction with Peracids and Alkylhydroperoxides

The nature of the species involved in hydroxylation of the heme group is further defined by studies with *meta*-chloroperbenzoic acid (*58*). Reaction with this oxidizing agent produces an intermediate with the spectroscopic properties of a ferryl ($Fe^{IV}{=}O$) species analogous to that of a peroxidase compound II intermediate (*112*). The ferryl intermediate accounts for only one of the two oxidizing equivalents of H_2O_2. The second oxidizing equivalent appears to be used to oxidize the protein because freeze–quench experiments show that a transient EPR detectable radical analogous to that obtained in the reaction of myoglobin with H_2O_2 is simultaneously formed (*113–115*). The spectroscopically observed ferryl species reverts to the starting Fe^{III} state when it is reduced by reaction with ascorbic acid or phenol (*58*). If guaiacol (*ortho*-methoxyphenol) is used instead of phenol or ascorbic acid, the telltale color change due to peroxidation of the phenolic substrate is observed. Although this peroxidase activity is of some interest, the key finding is that the ferryl intermediate is not converted to verdoheme. Furthermore, addition of 1 equiv of H_2O_2 to the enzyme after reaction with 1 equiv of *meta*-chloroperbenzoic acid also does not produce verdoheme, which indicates that the ferryl intermediate actually protects the heme from the reaction with H_2O_2 that produces verdoheme. Clearly, the compound II–like ferryl species is not an intermediate in the normal reaction catalyzed by heme oxygenase! This finding is important because a ferryl moiety is the activated oxidizing species responsible for the reactions catalyzed by most hemoproteins, including the P450 monooxygenases and hemoprotein peroxidases.

The reactions of the HO-1:heme complex with *tert*-butylhydroperoxide, cumene hydroperoxide, and ethylhydroperoxide have also been examined (*58, 116*). Reaction with the first two peroxides follows a course similar to the reaction with *meta*-chloroperbenzoic acid in that a ferryl species appears to be formed, although the reaction is accompanied by some degradation of the heme to nonbiliverdin products. The reaction with ethylhydroperoxide is the most interesting (*116*). Although it also gives rise to a ferryl species, reduction of this intermediate with ascorbic acid to prevent oxidative degradation of the prosthetic group followed by its isolation and HPLC purification shows that the heme group has been partially modified to give iron α-*meso*-ethoxyprotoporphyrin IXα (Fig. 8). The structure of this modified heme was firmly established by absorption and NMR spectroscopy and mass spectrometry. The formation of α-*meso*-ethoxyheme exactly parallels the formation of α-*meso*-hydroxyheme, the first step of the normal

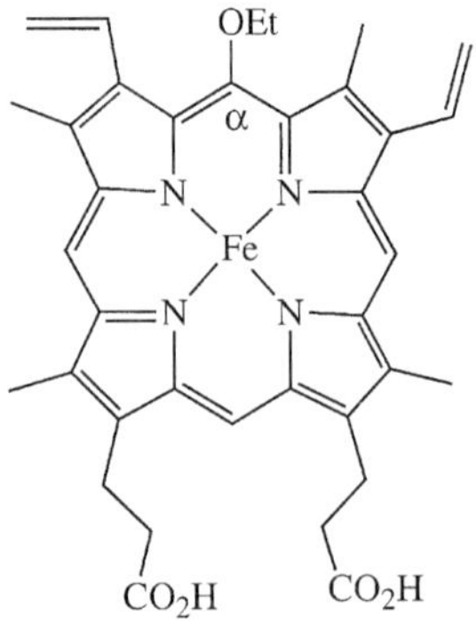

FIG. 8. Structure of the product obtained in the HO-1-catalyzed oxidation of heme supported by ethylhydroperoxide.

reaction. The α-*meso*-ethoxy product is stable toward the downstream reactions of the catalytic process because they depend on deprotonation of the hydroxyl group in the normal α-*meso*-hydroxy intermediate (see later discussion). The ethylhydroperoxide reaction specifically rules out a mechanism in which the terminal oxygen of an Fe^{III}–peroxo anion (Fe^{III}–OO^-) adds as a nucleophile to the porphyrin ring because the terminal oxygen in the ethylhydroperoxide complex is blocked. In view of the evidence that the ferryl species is also not involved, the ethylhydroperoxide results implicate an electrophilic addition of the terminal oxygen of the Fe^{III}–OOH complex to the aromatic ring (Fig. 9), a reaction that could be insensitive to the presence of a peroxide ethyl substituent.

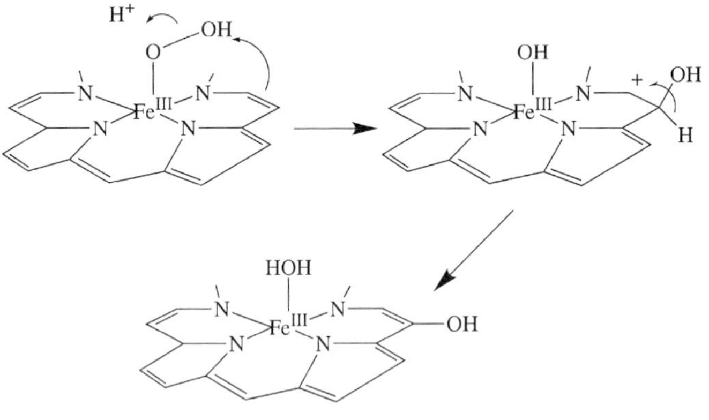

FIG. 9. Electrophilic oxidation of the porphyrin ring by the Fe^{III}–OOH complex formed in the catalytic turnover of heme oxygenase. The heme group is shown in a truncated form.

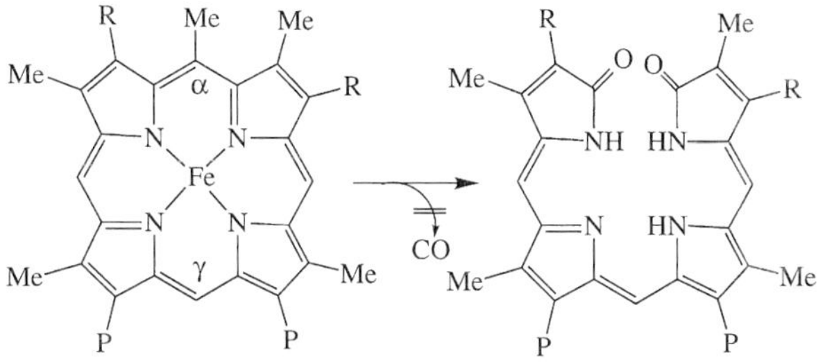

FIG. 10. α-*meso*-Methyl-substituted heme groups and their oxidation to biliverdin products. The substituent R in the heme structure is a methyl in the symmetric porphyrin and an ethyl in mesoheme.

D. THE CATALYTIC OXIDATION OF α-*meso*-METHYL HEMES

Given that α-*meso*-hydroxylation is the first committed step in the conversion of heme to biliverdin, it should be possible to block, or divert, the reaction by placing a substituent at the α-*meso* position. For these studies we first used a symmetric iron porphyrin in which all the substituents were methyl groups except for the 6- and 7-propionic acid groups. The parent iron porphyrin and a derivative with an α-*meso*-methyl substituent were used because their symmetry would facilitate identification of any unusual porphyrin-derived products (Fig. 10). As expected from the substrate specificity of the enzyme, the *meso*-unsubstituted iron porphyrin was oxidized normally to the biliverdin product and CO. To our amazement, the α-*meso*-methyl derivative was also oxidized, albeit without the formation of CO, to exactly the same biliverdin product (*49*)! Similar results were subsequently obtained with α-*meso*-methylmesoheme (Fig. 10) (*50*). The failure to form CO rules out the possibility that the methyl group was oxidatively removed and the protein then underwent normal α-*meso*-hydroxylation. Thus, the mechanism involved in hydroxylation and extrusion of the α-*meso*-carbon as CO can be diverted by α-*meso*-substitution to yield the same biliverdin product, but with the formation of a product other than CO. Efforts to isolate and identify the α-*meso*-carbon containing product, including studies with a ^{13}C-labeled *meso*-methyl group, have excluded acetaldehyde and acetic acid but have not successfully identified the fragment (*117*). It is possible that the fragment eliminated in the oxidation of the α-*meso*-methyl substrate is volatile or is a highly reactive species that binds covalently to the protein.

E. Regiochemistry of Heme Oxidation

The finding that α-*meso*-methyl hemes are oxidized to biliverdin products led us to examine, initially as control experiments, the oxidation of the other three *meso*-methylmesoheme regioisomers (*50*). The required β-, γ-, and δ-*meso*-methylmesohemes were synthesized and their regiochemistry was assigned by ^1H NMR studies (*118*). Given the high specificity of HO-1 for oxidation of the α-*meso*-carbon, we expected that the enzyme would oxidize the β-, γ-, and δ-*meso*-methyl substituted mesohemes at the unsubstituted α-position to give the three methyl-substituted mesobiliverdin IXα isomers. Contrary to expectation, the γ-*meso*-methylmesoheme isomer, as shown by HPLC and mass spectrometric analysis of the product, was oxidized exclusively at the γ-*meso* position to give unsubstituted mesobiliverdin IXγ. The δ-*meso*-methyl isomer was oxidized both at the δ-*meso*-position, yielding mesobiliverdin IXδ, and at an unsubstituted (presumably α) position to give a methyl-substituted mesobiliverdin IX isomer. Finally, the β-*meso*-methyl derivative was a very poor substrate, even though its α-*meso* position was not blocked.

It has been postulated, in part from modeling studies based on the crystal structures of sperm whale myoglobin and human hemoglobin (*47, 119*), that the oxidation regiochemistry is determined by steric orientation of the iron-bound dioxygen ligand. This steric "steering" determines the extent to which the terminal oxygen is located above each of the *meso* positions. Myoglobin and hemoglobin were used as models for the heme oxygenase reaction because coupled oxidation of their heme groups produces low yields of biliverdin isomers (*6, 120*). In sperm whale myoglobin, coupled oxidation of the heme occurs exclusively at the α-*meso* position, but in human hemoglobin the oxidation occurs at both the α- and β-*meso* positions (*47, 121*). The modeling studies were consistent with these experimental product distributions. However, the finding that the regiochemistry of heme oxidation by heme oxygenase is drastically altered by *meso*-methyl substitution is not easily reconciled with the predominant role of steric steering in governing the oxidation regiospecificity that is suggested by both the globin coupled oxidation studies and the crystal structure of heme oxygenase (*53*). It is difficult to rationalize inversion of the reaction regiochemistry to favor the γ- over the α-position when a methyl group is placed at the γ-position by a straightforward steric mechanism. Similar difficulties exist in sterically rationalizing oxidation of the δ-*meso* position when a methyl is at that position. A mechanism is therefore needed that reconciles the oxidation of the *meso*-methyl-substituted hemes with the strong evidence for a dominant role of steric effects in controlling the reaction regiochemistry.

F. Electronic Effects on the Reaction Regiochemistry

In addition to steric effects, the oxidation regiochemistry might be sensitive to electronic effects. The changes in heme oxidation regiochemistry due to methyl substitution support such an electronic control mechanism. If the heme oxidation reaction involves electrophilic addition of the oxidizing agent to the *meso*-carbon, as we believe it does, the addition reaction should be facilitated by electron donating groups. Thus, placement of a methyl group on the α-*meso* carbon should reinforce oxidation of that position resulting, as observed, exclusively in α-*meso* oxidation. On the other hand, introduction of the methyl at the γ-*meso* position would increase its reactivity toward the electrophilic oxygen and, if it overcomes whatever mechanism normally channels the reaction to the α-position, could result in oxidation of the γ-*meso* carbon. Methyl substitution at the δ-*meso* position would similarly increase its reactivity toward oxidation, as observed, but the fact that oxidation occurs at both the δ and (probably) α positions indicates that the δ-*meso* methyl is not as effective as the γ-*meso* methyl in overriding the factors that normally favor α-*meso* oxidation.

To further explore the potential electronic role in controlling the oxidation regiochemistry, the effect of an electron withdrawing rather than donating group was tested by synthesizing the four *meso*-formyl mesoheme regioisomers and examining their HO-1 catalyzed oxidation (*122*). The key finding here is that the formyl-substituted carbon is never the site of the oxidation reaction. Thus, α-*meso*-formylmesoheme is oxidized with essentially quantitative formation of CO to a biliverdin shown by mass spectrometry to retain the *meso*-formyl group. The other three regioisomers are similarly oxidized with the formation of CO. The difference between the site of oxidation in the methyl versus formyl substituted series of hemes confirms that the reaction involves electrophilic addition to the porphyrin ring and provides strong evidence that electronic effects can influence the regiochemistry of heme oxidation by HO-1.

If the α-*meso* regioselectivity of hydroxylation is not inherent to the heme group, as indicated by the fact that all four isomers are obtained by coupled heme oxidation (*40*), how is the regiochemistry of heme oxidation influenced by electronic factors? One possibility is that the puckering or ruffling of the porphyrin caused by steric clashes between the *meso* and flanking substituents in the *meso*-substituted heme probes

(*123, 124*) alters the heme oxidation mechanism. This would appear to be an unlikely explanation because the *meso*-methyl and *meso*-formyl substituents gave rise to quite different oxidation regiochemistries, whereas the porphyrin distortions caused by both types of substituents should be similar. Indeed, the different consequences of *meso*-methyl and *meso*-formyl substitution suggest that selective enhancement of the electron density at the *meso* positions is important. As already noted, NMR studies of the rat heme:HO-1 complex indicate that there are large differences in electron density at the two β-carbons in a given pyrrole ring rather than the more conventional pattern of similar densities at the β-carbons of a given pyrrole ring with large differences between the β-carbons of adjacent pyrrole rings (*88*). Electron density patterns similar to that observed for the heme:HO-1 complex have been observed with iron porphyrins bearing electron donating or withdrawing *meso*-substituents, and the patterns have been modeled by theoretical studies (*125, 126*). It is therefore possible, in principle, that an anionic group near the α-*meso* position could induce an electronic asymmetry that favors oxidation of the α-*meso* carbon. However, there is no such anionic group in the immediate vicinity of the heme in the crystal structure of human heme oxygenase (*53*). The apparent electronic control observed with the *meso*-substituted hemes, and its relevance to the normal reaction with unsubstituted heme, thus remain unclear. One possibility is that the *meso*-substituent, or the porphyrin ruffling associated with it, interferes with closing of the distal helix on top of the heme group. This might decrease the steric control that favors α-*meso* oxidation and allow the electronic properties of the *meso* substituents to determine the oxidation regiochemistry. A hint that incomplete closing of the distal helix can make additional *meso*-positions available for oxidation is provided by the structure of the more open of the two proteins in the asymmetric crystal unit, in which the δ-*meso* carbon is also accessible for oxidation. The bottom line, however, appears to be that steric effects are primarily responsible for the observed α-*meso* regiochemistry of the normal HO-1 catalyzed heme oxidation.

X. The Second Stage: α-*meso*-Hydroxyheme to Verdoheme

The conversion of α-*meso*-hydroxyheme to verdoheme is an oxygen-dependent process because the HO-1:α-*meso*-hydroxyheme complex, whether obtained by reconstitution of the apoenzyme with synthetic α-*meso*-hydroxyheme or from oxidation of the heme complex with H_2O_2, is stable under anaerobic conditions (*104, 105*). EPR analysis of the

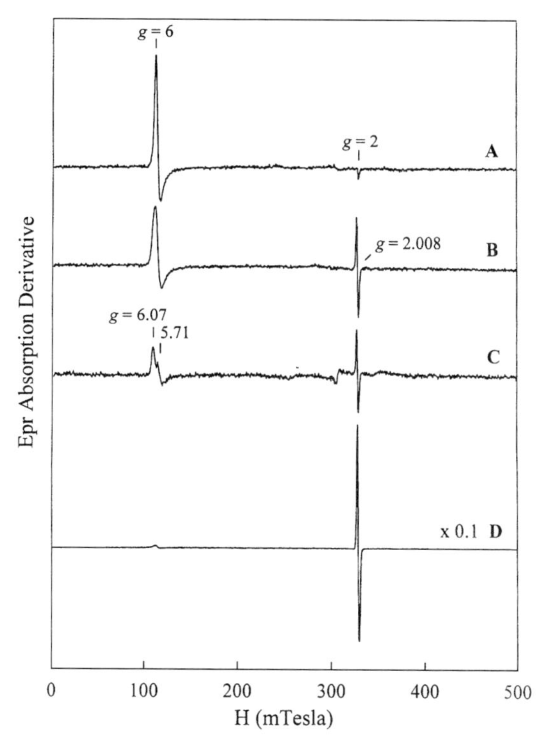

FIG. 11. EPR of the radical in the α-*meso*-hydroxyheme-HO-1 complex: (A) The Fe^{III} heme:HO-1 complex, (B) the α-*meso*-hydroxyheme-HO-1 complex formed anaerobically with 1 equiv of H_2O_2; (C) the α-*meso*-hydroxyheme complex after subtraction of the spectrum of unreacted heme from trace (B); and (D) the spectrum in B after the addition of CO to form the Fe^{II}-CO complex. The spectra are taken from Liu *et al.* (*104*).

anaerobic HO-1:α-*meso*-hydroxyheme complex shows the presence of a rhombic signal at $g = 6.07$ and 5.71 attributable to the Fe^{III} and a signal at $g = 2.008$ due to an organic radical. Addition of CO to the anaerobic system results in virtual disappearance of the g = 5–6 signals and enhancement of the signal at $g = 2.008$ (Fig. 11) (*104*). This finding indicates that an Fe^{III} heme species without an organic radical is in equilibrium with an Fe^{II} heme species with an organic radical (see Fig. 12), and that coordination of CO to the Fe^{II} shifts the resonance equilibrium toward the radical containing species. The formation of a CO complex is supported by a shift in the Soret maximum of the HO-1:α-*meso*-hydroxyheme complex from 405 to 408 nm in the presence of CO. Independent support for this equilibrium is provided by resonance Raman data indicating that deprotonation of the Fe^{III}

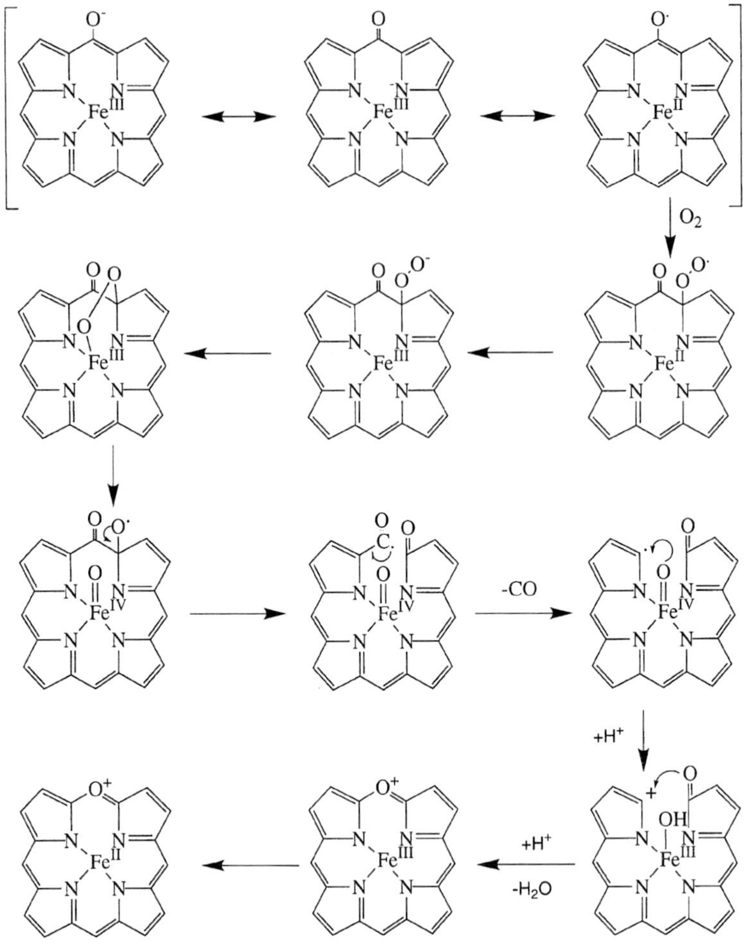

FIG. 12. Proposed mechanism for the HO-1-catalyzed conversion of Fe^{III} α-*meso*-hydroxyheme (shown deprotonated) to Fe^{II} verdoheme. In an equally good variant of this mechanism, the oxygen molecule binds to the Fe^{II} before it binds to the carbon of the porphyrin to give the same peroxo-bridged intermediate.

α-*meso*-hydroxyheme produces an oxophlorin- rather than porphyrin-like structure (*105*). The resonance equilibrium (Fig. 12) was not observed in the EPR by Matera *et al.*, possibly because the EPR conditions saturated the radical signal (*105*). However, they observed a radical signal at $g = 2.004$ when oxygen was added to a solution of HO-1:α-*meso*-hydroxyheme under an anaerobic atmosphere of nitrogen and CO. This signal was attributed to an Fe^{II} heme with an oxygen bound to the porphyrin radical, giving rise to an oxygen radical. Together, these experimental results establish a mechanism in which a deprotonated, free radical form of α-*meso*-hydroxyheme binds molecular oxygen to give a peroxy radical species (Fig. 12).

Exposure to oxygen of an anaerobic solution of the HO-1:α-*meso*-hydroxyheme complex produced from the heme complex with 1 equiv of H_2O_2 results in immediate conversion of the substrate to Fe^{III} verdoheme (*104*). Resonance Raman and absorption spectroscopy indicate that this Fe^{III} verdoheme can be reduced by dithionite to the Fe^{II} verdoheme normally obtained in the presence of reducing agents or NADPH–cytochrome P450 reductase. This basic finding was confirmed by the subsequent demonstration that the reconstituted Fe^{III} α-*meso*-hydroxyheme:HO-1 complex is converted by an approximately stoichiometric amount of oxygen to the Fe^{III} biliverdin complex (*127*). These results indicate that the only requirement for the HO-1-catalyzed conversion of Fe^{III} α-*meso*-hydroxyheme to Fe^{III} biliverdin is one molecule of O_2. These results are consistent with the report that conversion of myoglobin-bound α-*meso*-hydroxyheme to verdoheme does not require exogenous reducing equivalents (*45*). Matera *et al.,* and subsequently Migita *et al.,* have argued that the HO-1-catalyzed conversion of α-*meso*-hydroxyheme to biliverdin requires not only one molecule of oxygen but also one electron (*105, 128*). However, in their earlier experiments they monitored the formation of verdoheme not as the Fe^{III} product but as the Fe^{II}–CO complex. There is no question that the formation of Fe^{II} verdoheme requires one electron—the question is whether the electron must be introduced prior to the formation of Fe^{III} verdoheme. The observation that Fe^{III} verdoheme is formed in good yield from the endogenously formed or reconstituted α-*meso*-hydroxyheme:HO-1 complex in the absence of any reducing equivalents clearly establishes that an electron is not required for this transformation. However, under physiological conditions, in the presence of both cytochrome P450 reductase and NADPH, two possible reaction pathways are possible. In one, Fe^{III} α-*meso*-hydroxyheme may be converted to Fe^{III} verdoheme, followed by verdoheme reduction, whereas in the other pathway an electron may be introduced prior to the extrusion of CO, resulting directly in the formation of Fe^{II} verdoheme. The relative importance of these two pathways will depend on the relative rates of the various steps in the two alternatives, and the endogenous concentrations of NADPH, cytochrome P450 reductase, and oxygen.

The minimal mechanism required for the conversion of α-*meso*-hydroxyheme to verdoheme requires deprotonation of α-*meso*-hydroxyheme to produce an Fe^{II} radical species that binds oxygen to give an Fe^{II}

peroxy radical intermediate (Fig. 12). As already indicated, EPR evidence exists for both of these intermediates. Direct interaction of the peroxy radical with the Fe^{II}, or after internal transfer of an electron to give the Fe^{III}–hydroperoxy species, leads to production of an unstable ferryl alkoxy radical. Formation of the alkoxy radical, which is reactive enough to drive the cleavage of a carbon–carbon bond, provides a trigger for elimination of the α-*meso*-carbon as CO. Elimination of the CO generates an unstable carbon radical that can be oxidized internally to the cation by electron transfer to the ferryl species. Trapping of the cation by the pyrrole carbonyl generated upon extrusion of CO then yields Fe^{III} verdoheme. No direct evidence exists for the individual steps of the mechanism subsequent to binding of oxygen to the porphyrin radical. However, the mechanism is reasonable and is unlikely to differ much, except perhaps for changes in the order of the steps, including the possible uptake of an electron prior to verdoheme formation.

XI. The Third Stage: Verdoheme to Biliverdin

The least information is available on the conversion of verdoheme to biliverdin, the final stage of HO-1-catalyzed heme oxidation. The origin of the oxygen atoms in the final biliverdin product provides important information in this regard. Under an atmosphere of ^{16}O–^{16}O and ^{18}O–^{18}O, the biliverdin produced by HO-1 incorporates one atom each of ^{16}O and ^{18}O (*8*). The two oxygen atoms incorporated into biliverdin thus derive not from one oxygen (*7*), but from different molecules of oxygen. This rules out a hydrolytic mechanism for the conversion of verdoheme to biliverdin, in accord with the finding with both a coupled oxidation system and HO-1 that the reaction requires reducing equivalents and O_2 (*43, 44, 58*). The incorporation of an atom of molecular oxygen into biliverdin starting with verdoheme generated by coupled oxidation has been specifically demonstrated (*43*).

It is of some interest that H_2O_2 does not substitute effectively for O_2 and NADPH–cytochrome P450 reductase in this stage of the transformation (*58*). Indeed, in the presence of excess H_2O_2, the majority of the verdoheme is degraded to nonbiliverdin products, only a small amount of biliverdin being formed. This side reaction may decrease the yield of biliverdin when the reaction is experimentally initiated with H_2O_2 rather than with NADPH–cytochrome P450 reductase (*58*). This side reaction may stem from the fact that the products of the reaction of verdoheme with H_2O_2 terminate with the tetrapyrrole system in too

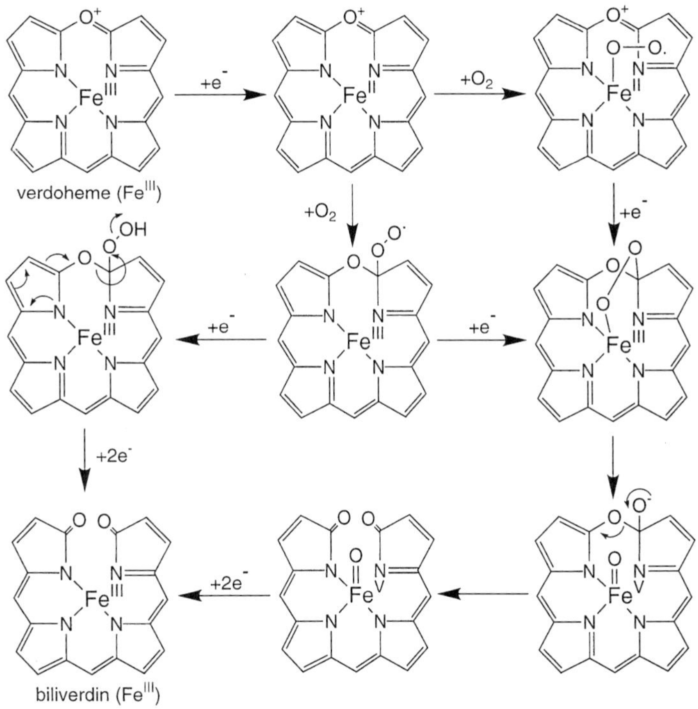

FIG. 13. Hypothetical mechanism(s) for the conversion of verdoheme to biliverdin. Two alternatives are shown that differ in the sequence in which the electrons are introduced and whether the oxygen molecule binds initially to the Fe^{II} or to the porphyrin radical.

high an oxidation state. The formation of biliverdin from Fe^{III} verdoheme requires a two-electron reduction, a reduction that is not possible when H_2O_2 is the only reagent present in the medium. Alternatively, the stepwise reaction of O_2 may permit delivery of the oxygen to the carbon framework, whereas reaction with H_2O_2 may result in a side reaction with the iron to give alternative products.

A schematic but hypothetical mechanism that accommodates the foregoing considerations is shown in Fig. 13. Reduction of the verdoheme iron to the Fe^{II} state, binding of dioxygen, and a second one-electron reduction produce an intermediate formally equivalent to that obtained by reaction of Fe^{III} verdoheme with H_2O_2. A hypothetical sequence of steps involving dioxygen bond cleavage to give either an alkoxy anion or radical species, followed by ring cleavage and a two-electron reduction, produce the desired Fe^{III} biliverdin product. The principal uncertainty in the mechanism is the stage at which the two electrons required to produce biliverdin are introduced. The formation of an $Fe^{V}{=}O$ species is shown, but if such an intermediate is involved, it is likely to be an $Fe^{IV}{=}O$ in conjunction with a porphyrin or protein radical of some kind. Other variations are possible, but most of the elements shown in the mechanism (Fig. 13) are likely to be present in the actual mechanism. In addition, in the presence of an electron source the iron is reduced once more to form Fe^{II} rather than Fe^{III} biliverdin. Initial evidence for this was provided by the finding that coupled oxidation of the heme:heme oxygenase complex produces what appears to be an Fe^{III} biliverdin complex that is not suitable for reduction by biliverdin reductase. The same reaction supported by NADPH–cytochrome P450 reductase produces free biliverdin that is readily reduced to bilirubin (*41*). Direct confirmation of the need to reduce Fe^{III} to Fe^{II} biliverdin to facilitate product release is provided later (*129*).

XII. Kinetics of the Heme Oxygenase Reaction Sequence

The catalytic sequence of heme oxygenase encompasses a minimum of three distinct catalytic reactions separated by the two clearly identified intermediates, α-*meso*-hydroxyheme and verdoheme. The complete process consumes two molecules of oxygen and five electrons provided by NADPH–cytochrome P450 reductase and releases three products: carbon monoxide, biliverdin, and iron (*2*). As might be expected, the kinetics of the reaction sequence are not simple. Analysis of the kinetic sequence is complicated by the fact that heme is poorly water soluble at physiological pH and is usually present in the form of protein complexes, and the fact that a coupled assay involving the conversion of biliverdin to bilirubin by biliverdin reductase is commonly used to monitor the reaction (*55*). In overall terms, the K_m value for heme using the recombinant human protein without the 23 C-terminal amino acids is 3 μM, in close agreement with the values for the rat liver and bovine spleen enzymes (*55, 56, 60*). The V_{max} value for the recombinant, truncated protein is 40 nmol h^{-1} $nmol^{-1}$ (*60*).

A single turnover study of the conversion of the heme–HO-1 complex to free biliverdin has elucidated the relative rates of the catalytic steps (*129*). This transient kinetic study indicates that the conversion of Fe^{III} heme to Fe^{III} verdoheme is biphasic. Electron transfer to the Fe^{III}–heme HO-1 complex occurred at a rate of 0.11 s^{-1} at 4°C and 0.49 s^{-1} at 25°C with a 0.1:1 ratio of NADPH–cytochrome P450 reductase to heme:HO-1 complex. Oxygen binding to the reduced iron was sufficiently rapid under the experimental conditions that the species actually monitored was the Fe^{II}–O_2 complex. Electron transfer to this Fe^{II}–O_2 complex, which occurred at a rate of 0.056 s^{-1} at 4°C and 0.21 s^{-1} at 25°C, led to the formation of α-*meso*-hydroxyheme and its immediate O_2-dependent fragmentation to give the Fe^{3+}–verdoheme HO-1 complex. The HO-1 catalyzed conversion of Fe^{3+}–verdoheme to Fe^{3+}–biliverdin also exhibited biphasic kinetics. Reduction of the Fe^{III} verdoheme complex to the Fe^{II} state occurred at a rate of 0.15 s^{-1} at 4°C and 0.55 s^{-1} at 25°C. Rapid oxygen binding and a second electron transfer to the putative Fe^{II}–dioxy verdoheme intermediate, with rates of 0.025 s^{-1} at 4°C and 0.10 s^{-1} at 25°C, produced the Fe^{3+}–biliverdin complex (*129*). The final conversion of Fe^{3+}–biliverdin to free biliverdin was triphasic. Reduction of the Fe^{3+}–biliverdin complex to the Fe^{II} state occurred at a rate of 0.035 s^{-1} at 4°C and 0.15 s^{-1} at 25°C. The iron was then rapidly released from the Fe^{II} enzyme complex at a rate of 0.19 s^{-1} at 4°C and 0.39 s^{-1} at 25°C, followed by slower dissociation of the biliverdin at a rate of 0.007 s^{-1} at 4°C and 0.03 s^{-1} at 25°C. These results clearly establish that Fe^{II} is released prior to release of the biliverdin. The fact that iron is released in the Fe^{II} rather than relatively insoluble Fe^{III} state may be important, given the central role of heme oxygenase in the recycling of iron and the maintenance of physiological iron homeostasis.

The rate-limiting step in the single-turnover studies of the heme–heme oxygenase complex was release of biliverdin from the enzyme, the final step in the catalytic sequence. However, when biliverdin reductase was present, biliverdin release was accelerated and the rate-limiting step in the single-turnover studies was the conversion of Fe^{2+}–verdoheme to Fe^{3+}–biliverdin. This implies that biliverdin reductase interacts with the biliverdin–heme oxygenase complex in some manner that facilitates either dissociation of the biliverdin or its direct transfer to the reductase. One speculative but attractive possibility is that biliverdin reductase, for which no structure is yet available, interacts with the electropositive surface surrounding the exposed heme edge in the human HO-1:heme complex (*53*) and allosterically promotes release of the product. Two factors must be considered in reaching a conclusion concerning the rate-limiting step under physiological conditions. The role of heme binding, if any, in determining the overall reaction rate is difficult to evaluate because the state in which heme is available under physiological conditions is uncertain. Furthermore, the rates of the electron transfer steps are sensitive to the NADPH–cytochrome P450 reductase:heme oxygenase ratio and thus to the physiological concentrations of the proteins and cosubstrates. Nevertheless, it appears likely that the rate-limiting step in the overall process will be the conversion of Fe^{2+} verdoheme to biliverdin, or the release of biliverdin if

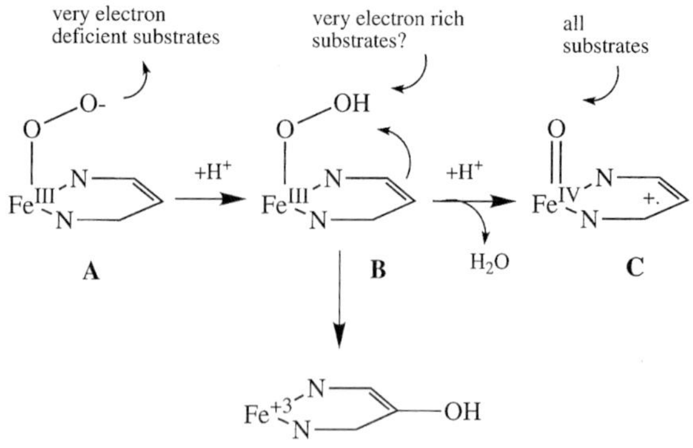

FIG. 14. Manifold of reactive species produced from the reaction of a heme group with oxygen and two reducing equivalents. The rate of conversion of **A** to **B** limits the lifetime (and therefore reactivity) of the Fe^{III} peroxo anion. The rate of formation of the ferryl species **C** via the Fe^{III}–OOH complex **B** competes with the intramolecular hydroxylation reaction to give hydroxyheme. Reactions of the Fe^{III}–hydroperoxy complex **B** with exogenous electrophilic substrates must compete with conversion of the intermediate to both **C** and *meso*-hydroxyheme. The Fe^{III}–OOH complex **B** can also be formed directly with H_2O_2.

the cytochrome P450 reductase to heme oxygenase ratio approaches unity.

XIII. Implications of Electrophilic Heme Oxidation by an Fe^{III}–OOH Intermediate

Studies of the catalytic mechanisms of hemoproteins that employ either oxygen or H_2O_2 as a cosubstrate have provided evidence for two types of oxygen reactivity. The most common oxidative intermediate is the ferryl species (Fig. 14, structure **C**). A ferryl complex coupled with a porphyrin radical cation (e.g., horseradish peroxidase) or a protein radical (e.g., cytochrome *c* peroxidase) is well established as the oxidative intermediate in the peroxidases (*112, 130*). This intermediate undergoes two stepwise electron transfers from substrate molecules, the first of which reduces the porphyrin or protein radical, and the second the ferryl species to the resting Fe^{III} state. A similar but less well characterized ferryl intermediate, formed *in situ* by two-electron reduction of molecular oxygen, is thought to be the primary oxidizing species of cytochrome P450 enzymes (*131*). However, cytochrome P450 enzymes normally transfer the ferryl oxygen to a substrate. As a general observation, the ferryl species is by far the most common intermediate in hemoprotein oxidative catalysis.

A less common reactive species is the Fe^{III} peroxo anion expected from two-electron reduction of O_2 at a hemoprotein iron atom (Fig. 14, structure **A**). Protonation of this intermediate would yield the Fe^{III}–OOH precursor (Fig. 14, structure **B**) of the ferryl species. However, it is now clear that the Fe^{III} peroxo anion can directly react as a nucleophile with highly electrophilic substrates such as aldehydes. Addition of the peroxo anion to the aldehyde, followed by homolytic scission of the dioxygen bond, is now accepted as the mechanism for the carbon–carbon bond cleavage reactions catalyzed by several cytochrome P450 enzymes, including aromatase, lanosterol 14-demethylase, and sterol 17-lyase (*133*). A similar nucleophilic addition of the Fe^{III} peroxo anion to a carbon–nitrogen double bond has been invoked in the mechanism of the nitric oxide synthases (*133*).

The heme oxygenase-catalyzed hydroxylation of heme by an Fe^{III}–OOH species reviewed here is the first well-documented example of a new hemoprotein reactivity. The reactivity of this complex resembles that of the flavin or pteridine hydroperoxides in proteins that utilize these prosthetic groups. With both of these prosthetic groups, the terminal oxygen of what is essentially an alkylhydroperoxide (ROOH) reacts as an electrophile with heteroatoms or aromatic rings. In the same way, the terminal oxygen of the Fe^{III}–OOH intermediate reacts with the porphyrin framework of the heme group (Fig. 14, structure **B**). The rapid reaction of the Fe^{III}–OOH intermediate to give α-*meso*-hydroxyheme illuminates a fundamental aspect of hemoprotein mechanisms. Thus, if an Fe^{III}–OOH intermediate is formed by reductive activation of O_2 or by reaction with H_2O_2, the intermediate must decay to a ferryl species rapidly enough to avoid self-oxidation of the heme group. This conclusion is strengthened by the fact that hydroxylation of the heme group is not unique to heme oxygenase. Myoglobin, hemoglobin, and, indeed, most hemoproteins have been shown to give low to fair yields of biliverdin products when subjected to coupled oxidation (*38, 46*). This argues that the hydroxylation reaction is inherent to the heme Fe^{III}–OOH complex rather than to a specific hemoprotein environment. Heme hydroxylation thus appears to be a self-destructive "clock" reaction of the heme Fe^{III}–OOH species against which the normal catalytic mechanism must compete. This self-destructive ("apoptotic") process built into the Fe^{III}–OOH intermediate suggests that it is unlikely to function effectively as an electrophilic oxidizing agent for exogenous substrates because its reaction with exogenous substrates would have to compete with a favored intramolecular hydroxylation reaction. Such a competition is unlikely to be successful except with highly nucleophilic compounds. In this context it is relevant that the epoxidation of olefins by cytochrome P450 enzymes has been proposed, at least in mutants in which the machinery required to promote oxygen–oxygen bond heterolysis is disabled by mutagenesis, to partially result from reaction of the olefin with an Fe^{III}–OOH intermediate (*134, 135*).

XIV. Heme Degradation in Plants and Bacteria

The role of heme oxygenase in biological systems is not restricted to heme catabolism, but also plays a significant biological role in the synthesis of the tetrapyrrole containing chromophores of photosynthetic organisms. In the cyanobacterium *Synechocystis* sp. PCC 6803 (*136*), the red algae *Cyanidium caldarium* (*137*) and *Rhodella violacea* (*138*), and the higher plant *Arabidopsis thaliana* (*139*), heme oxygenase plays an important role in the biosynthesis of the tetrapyrrole containing phycobilins and phytochromobilins. In contrast to the mammalian microsomal heme oxygenase, the cyanobacteria and rhodophyte biosynthetic heme oxygenases are soluble proteins (*137, 140*) and their biological activities have been shown to be ferredoxin dependent (*136*).

Although there is very little data on the mechanistic aspects of the biosynthetic enzymes, it is thought that their reaction mechanisms are very similar to those of the mammalian heme oxygenases. Indirect evidence that the mechanism of action of the biosynthetic enzymes is similar to that of the mammalian heme oxygenases was provided by ^{18}O-labeling studies showing that the biosynthesis of phycocyanobilin occurred via a two-molecule mechanism (*141, 142*) comparable to that of the mammalian heme oxygenases (*8*).

More recently, the *hmuO* gene, which codes for a protein with 33% identity and 70% similarity to human HO-1, was identified in the Gram-positive pathogen *Corynebacterium diphtheriae* (*143*). The gene product appeared to play a vital role in the acquisition of iron, an element essential for the organism's survival and pathogenicity. The *hmuO* gene is negatively regulated at the transcriptional level by the iron-dependent DNA binding protein DtxR and is positively regulated by heme (*144*). A two-component signal transduction system involved in the regulation of *hmuO* expression has been identified in *C. diphththeriae* (*145*). A response regulator (*chrA*) and its sensor kinase (*chrS*) were shown to strongly activate transcription in a heme-dependent manner (*145*).

The molecular mechanisms by which Gram-positive bacteria such as *C. diphtheriae* acquire heme are not known. However, based on the similarities of heme transport with siderophore uptake in the Gram-negative pathogens, it is reasonable to postulate that a similar mechanism may operate in Gram-positive bacteria (*146, 147*). Identification of an *hmuO* gene capable of complementing mutants deficient in heme utilization strengthens the evidence for such a specific mechanism of heme transport and utilization.

Initial studies have shown that the product of the *hmuO* gene is a soluble protein with a molecular weight of 25 kDa on SDS-PAGE, which oxidatively cleaves heme to biliverdin with the release of carbon monoxide and iron (*148*). The heme was exclusively oxidized at the α-*meso*-carbon as previously observed for the mammalian heme oxygenases. The purified heme oxygenase (Hmu O) forms a 1 : 1 complex with heme with a K_d of 2.5 μM. The Fe^{III} heme–Hmu O complex has a Soret maximum at 404 nm and a characteristic charge transfer band at 632 nm. Reduction of the heme–Hmu O complex with dithionite under a carbon monoxide atmosphere yields an Fe^{II}–CO spectrum with a Soret at 421 nm and α/β bands at 568 and 538 nm, respectively. Following passage of the Fe^{II}–CO heme–Hmu O complex through Sephadex G-25 the Soret band shifts from 421 to 410 nm and the α/β bands to 570 and 540 nm because of formation of the Fe^{II}–O_2 complex.

A synthetic *hmuO* gene was subsequently utilized to further characterize the Hmu O protein (*149*). The spectrophotometric characteristics of the Fe^{III} heme–Hmu O complex were identical to those described earlier (*148*). Utilizing ascorbic acid as a reductant and a partial atmosphere of carbon monoxide the authors identified verdoheme as an intermediate on the pathway to biliverdin. In addition, the conversion of the reconstituted α-*meso*-hydroxyheme–Hmu O complex to verdoheme, and subsequently to biliverdin, provided evidence that α-*meso*-hydroxyheme is also an intermediate on the pathway of heme degradation by Hmu O. The preliminary evidence suggests that oxidative cleavage of the heme in Hmu O is very similar to that of the mammalian heme oxygenases.

Resonance Raman and EPR characterization of the recombinant Hmu O has identified the proximal ligand as a histidine, most likely that of His20, which aligns with the conserved proximal His25 ligand in HO-1 (*150*). We have confirmed His20 as the proximal ligand in site-directed mutagenesis studies (*151*). Site-directed mutagenesis of residue His20 to an alanine gave a shift in the Soret band with evidence of a high-spin marker at 622 nm. The resonance Raman data confirmed loss of the proximal ligand with the disappearance of the Fe^{II}–His stretching mode at 222 cm^{-1}. Titration of the heme–Hmu O complex with imidazole restored full catalytic activity to the enzyme, and coordination of the imidazole to the heme was confirmed by resonance Raman with the appearance of a new Fe^{II}–His stretching mode at 228 cm^{-1}.

However, in contrast to the human His25Ala HO-1:heme complex, which has no detectable activity in the absence of imidazole (*78*), the His20Ala Hmu O:heme complex in the presence of NADPH and NADPH–cytochrome P450 reductase was found to catalyze the initial *meso*-hydroxylation of the heme (*151*). The product of the reaction was Fe^{II} verdoheme, as judged by the electronic absorption spectrum and the detection of carbon monoxide as a product of the reaction. Hydrolytic conversion of the verdoheme product to biliverdin and subsequent HPLC analysis confirmed that the oxidative cleavage of the porphyrin macrocycle was specific for the α-*meso*-carbon.

The reason why the HmuO His20Ala mutant, in contrast to the equivalent mutants of human HO-1 and HO-2, is capable of carrying out the initial hydroxylation of heme is still not known. The regiospecificity of *meso*-hydroxylation in the Hmu O His20Ala mutant is retained, suggesting that the proximal ligand is not essential in orienting the heme for selective hydroxylation of the α-*meso*-carbon. This agrees with the three-dimensional structure of human HO-1, in which charge interactions with the propionic acid groups have been found to be critical for orienting the heme, and specifically the α-*meso* carbon, for selective hydroxylation (*53*). The apparent differences between the reactivity of the bacterial Hmu O and that of mammalian heme oxygenases may not be intrinsic to the proximal ligand, but to the protein's ability to bind the heme so as to allow interactions with critical constituents on the distal side of the heme. The subtle differences between the active sites of Hmu O and those of the mammalian heme oxygenases are highlighted by work showing that the HS to LS transition occurs at a higher pH value than that reported for the mammalian enzymes (see later discussion).

It is unclear why the His20Ala HmuO:heme complex terminates at the level of the Fe^{II} verdoheme complex and does not proceed to biliverdin. The reaction of the His20Ala HmuO:heme complex with NADPH–cytochrome P450 reductase generates an Fe^{II} verdoheme complex, as evidenced by the band at 650 nm in the electronic absorption spectrum similar to that reported for the Fe^{II} verdoheme–CO complex (*152*). In contrast, five-coordinate Fe^{III} verdoheme complexes have a band at 690 nm (*104*). Magnetic circular dichroism (MCD) studies on the His20Ala Hmu O:heme complex when titrated with imidazole show a spectrum typical of a mix of 6cLS and 6cHS species (unpublished data). The MCD spectrum is similar to that of cytochrome b_5, although the intensity is somewhat weaker.

Studies of the H63M mutant of cytochrome b_5 have shown that the protein is capable of carrying out the coupled oxidation of heme (*153*). The reaction was arrested at verdoheme and did not proceed to give biliverdin. The formation of verdoheme rather than biliverdin was proposed to stem from the formation of a six-coordinate His–Met Fe^{II} verdoheme complex. It has previously been shown in coupled oxidation reactions that the high-affinity binding of ligands to both the fifth and sixth coordination sites inhibits the reaction from verdoheme to biliverdin (*154*). This is consistent with inhibition by CO of the conversion of the Fe^{II} verdoheme to biliverdin, which confirms that only five-coordinate verdoheme complexes are capable of undergoing further oxidation to biliverdin (*155*). It is possible that formation of the Fe^{II} verdoheme complex in the His20Ala Hmu O mutant is due to bis-coordination, with the sixth ligand coming either from a protein residue (histidine or methionine) or from CO released as a consequence of oxidative cleavage.

Again, in contrast with the mammalian heme oxygenases, the heme–Hmu O complex does not react very efficiently with hydrogen peroxide (*149*). One possible explanation for this decreased reactivity is the nature of the distal pocket. The deprotonation of the peroxide to generate the reactive peroxo-intermediate has previously been proposed to be facilitated by an ionizable group in the distal pocket (*68, 70*). However, if the deprotonation of the H_2O_2 is deficient or weaker in Hmu O, formation of the reactive intermediate would be inhibited. Support for such a hypothesis was provided by a pH titration of the heme:HmuO complex, which showed a transition to the alkaline form at a pK_a of 9 (*150*). However, the absorption and resonance Raman spectra indicated that a substantial fraction of the high spin heme–Hmu O complex remained. The mixed spin species observed at alkaline pH is not evident in the heme–HO-1 complexes, which are predominantly in the low spin state (*68, 70*). Again, the mixed spin state is thought to be due to the weaker deprotonation efficiency of the distal residue, which therefore forms a weaker hydrogen bond to the bound H_2O. In the three-dimensional X-ray structure of human HO-1 there is no polar side chain analogous to the distal histidine previously observed in the globins and peroxidases that directly stabilizes bound ligands: i.e., there is no polar or basic residue within hydrogen bonding distance of the bound H_2O (*53*). The only functions close enough to be involved in such an interaction are the carbonyl group of Gly139 and the nitrogen of Gly143. Although there is no specific ionizable group on the distal side of the heme, the enzyme has an active site that is considerably more polar than that of the globins. The asymmetry of the polar groups, most of which are located in the vicinity of the heme β-pyrrole ring, may account for the hydrogen-bonding interactions through their effect on polarization and orientation of the internal solvent molecules. If so, subtle changes in the active site of Hmu O that reduce the active site polarity may account for the differences observed in the reactivity of the protein. The specific differences observed between the bacterial Hmu O and the mammalian HO-1 await further structural characterization of the Hmu O protein. The recently reported preliminary crystallization and diffraction analysis of recombinant Hmu O should in the future provide an explanation for the observed differences (*156*).

Acknowledgments

The work of the San Francisco group described in this article has only been possible because of the hard work and insights of Angela Wilks, Justin Torpey, and Yi Liu, and fruitful collaborations with the groups of Tom Poulos, Tom Loehr, Gerd LaMar, and Kevin Smith. The work in San Francisco and the preparation of this review were supported by National Institutes of Health grant DK30297.

References

1. Schmid, R.; McDonagh, A. F. "The Porphyrins", Vol. VI; Dolphin, D.; ed.; Academic Press: New York, 1979; p. 257.
2. Tenhunen, R.; Marver, H. S.; Schmid, R. *J. Biol. Chem.* **1969,** *244,* 6388.
3. Schacter, B. A.; Nelson, E. B.; Marver, H. S.; Masters, B. S. S. *J. Biol. Chem.* **1972,** *247,* 3601.
4. Yoshida, T.; Noguchi, M.; Kikuchi, G. *J. Biol. Chem.* **1980,** *255,* 4418.
5. Tenhunen, R.; Marver, H. S.; Schmid, R. *Proc. Natl. Acad. Sci. USA* **1968,** *61,* 748.
6. O'Carra, P. "Porphyrins and Metalloporphyrins"; Smith, K. M., ed.; Elsevier: Amsterdam, 1975; p. 123.
7. Tenhunen, R.; Marver, H.; Pimstone, N. R.; Trager, W. F.; Cooper, D. Y.; Schmid, R. *Biochemistry* **1972,** *11,* 1716.
8. Docherty, J. C.; Schacter, B. A.; Firneisz, G. D.; Brown, S. B. *J. Biol. Chem.* **1984,** *259,* 13066.
9. Maines, M. D.; Trakshel, G. M.; Kutty, R. K. *J. Biol. Chem.* **1986,** *261,* 411.
10. Shibahara, S.; Müller, R. M.; Taguchi, H.; Yoshida, T. *Proc. Natl. Acad. Sci. USA* **1985,** *240,* 7865.
11. Rotenberg, M. O.; Maines, M. D. *J. Biol. Chem.* **1990,** *265,* 7501.
12. McCoubrey, W. K.; Huang, T. J.; Maines, M. D. *Eur. J. Biochem.* **1997,** *247,* 725.
13. Maines, M. D. "Heme Oxygenase: Clinical Applications and Functions"; CRC Press: Boca Raton, FL, 1992; p. 145.
14. Choi, A. M. K.; Alam, *J. Am. J. Respir. Cell Mol. Biol.* **1996,** *15,* 9.
15. Weber, C. M.; Eke, B. C.; Maines, M. D. *J. Neurochem* **1994,** *63,* 953–62.
16. McCoubrey, W. K., Jr.; Huang, T. J.; Maines, M. D. *J. Biol. Chem.* **1997,** *272,* 12568.
17. Zhang, L.; Guarente, L. *EMBO J.* **1995,** *14,* 313–20.

18. McCoubrey, W. K.; Ewing, J. F.; Maines, M. D. *Arch. Biochem. Biophys.* **1992,** *295,* 13.
19. Rotenberg, M. O.; Maines, M. D. *Arch. Biochem. Biophys.* **1991,** *290,* 336.
20. Bissel, D. M.; Guzelian, P. S. *J. Clin. Invest.* **1980,** *65,* 1135.
21. Ostrow, J. D.; Kapitulnik, J. In "Bile Pigments and Jaundice"; Ostrow, J. D., ed.; Marcel Dekker: New York, 1986; p. 421.
22. Masters, B. S. S.; Schacter, B. A. *Ann. Clin. Res.* **1976,** *8,* 18.
23. Guengerich, F. P. *Biochemistry* **1978,** *17,* 3633.
24. Kutty, R. K.; Maines, M. D. *Biochem. J.* **1987,** *246,* 467.
25. Bothwell, T. H.; Charlton, R. W.; Motulsky, A. G. In "The Metabolic and Molecular Bases of Inherited Disease"; Scriver, C. R.; Beaudet, A. L.; Sly, W. S.; Valle, D., eds.; McGraw-Hill: New York, 1995; p. 2237.
26. Poss, K. D.; Tonegawa, S. *Proc. Natl. Acad. Sci. USA* **1997,** *94,* 10919.
27. Stocker, R.; Yamamoto, Y.; McDonagh, A. F.; Glazer, A. N.; Ames, B. N. *Science* **1987,** *235,* 1043.
28. Criggler, J. F.; Najjar, V. A. *Pediatrics* **1952,** *10,* 169.
29. Marks, G. S.; Brien, J. F.; Nakatsu, K.; McLaughlin, B. E. *Trends Pharmacol. Sci.* **1991,** *12,* 185.
30. Verma, A.; Hirsch, D. J.; Glatt, C. E.; Ronnett, G. V.; Snyder, S. H. *Science* **1993,** *259,* 381.
31. Stevens, C. F.; Wang, Y. *Nature* **1993,** *364,* 147.
32. Grundemar, L.; Ny, L. *Trends Pharmacol Sci.* **1997,** *18,* 193.
33. Burstyn, J. N.; Yu, A. E.; Dierks, E. A.; Hawkins, B. K.; Dawson, J. H. *Biochemistry* **1995,** *34,* 5896.
34. Stone, J. R.; Marletta, M. A. *Biochemistry* **1994,** *33,* 5636.
35. Friebe, A.; Schultz, G.; Koesling, D. *EMBO J.* **1996,** *16,* 6863.
36. Coceani, F.; Kelsey, L.; Seidlitz, E. *Brit. J. Pharmacol.* **1996,** *118,* 1689.
37. Maines, M. D. *Annu. Rev. Pharmacol. Toxicol.* **1997,** *37,* 517.
38. Lemberg, R. Rev. *Pure Appl. Chem.* **1956,** *6,* 1.
39. Lemberg, R. *Biochem. J.* **1935,** *29,* 1322.
40. Petryka, Z.; Nicholson, D. C.; Gray, C. H. *Nature* **1962,** *194,* 1047.
41. Yoshida, T.; Kikuchi, G., *J. Biol. Chem.* **1978,** *253,* 4230.
42. King, R. F.; Brown, S. B. *Biochem J.* **1978,** *174,* 103.
43. Saito, S.; Itano, H. A. *Proc. Natl. Acad. Sci. USA* **1982,** *79,* 1393.
44. Yoshida, T.; Noguchi, M. *J. Biochem.* **1984,** *96,* 563.
45. Sano, S.; Sano, T.; Morishima, I.; Shiro, Y.; Maeda, Y. *Proc. Natl. Acad. Sci. USA* **1986,** *83,* 531.
46. O'Carra, P.; Colleran, E. *FEBS Lett.* **1969,** *5,* 295.
47. Brown, S. B.; Chabot, A. A.; Enderby, E. A.; North, A. C. T. *Nature* **1981,** *289,* 93.
48. Crusats, J.; Suzuki, A.; Mizutani, T.; Ogoshi, H. *J. Org. Chem.* **1998,** *63,* 602.
49. Torpey, J.; Lee, D. A.; Smith, K. M.; Ortiz de Montellano, P. R. *J. Am. Chem. Soc.* **1996,** *118,* 9172.
50. Torpey, J.; Ortiz de Montellano, P. R. *J. Biol. Chem.* **1996,** *271,* 26067.
51. Murakami, T.; Morishima, I.; Matsui, T.; Ozaki, S.; Hara, I.; Yang, H.-J.; Watanabe, Y. *J. Am. Chem. Soc.* **1999,** *121,* 2007.
52. Hildebrand, D. P.; Tang, H.-l.; Luo, Y.; Hunter, C. L.; Smith, M.; Brayer, G. D.; Mauk, A. G. *J. Am. Chem. Soc.* **1996,** *118,* 12909.
53. Schuller, D. J.; Wilks, A.; Ortiz de Montellano, P. R.; Poulos, T. L. *Nature Struct. Biol.* **1999,** *6,* 860.
54. Yoshida, T.; Kikuchi, G., *J. Biol. Chem.* **1978,** *253,* 4224.

55. Yoshida, T.; Kikuchi, G., *J. Biol. Chem.* **1979,** *254,* 4487.
56. Yoshinaga, T.; Sassa, S.; Kappas, A. *J. Biol. Chem.* **1982,** *257,* 7778.
57. Ishikawa, K.; Sato, M.; Yoshida, T., *Eur. J. Biochem.* **1991,** *202,* 161.
58. Wilks, A.; Ortiz de Montellano, P. R. *J. Biol. Chem.* **1993,** *268,* 22357.
59. Ishikawa, K.; Sato, M.; Ito, M.; Yoshida, T. *Biochem. Biophys. Res. Commun.* **1992,** *182,* 981.
60. Wilks, A.; Black, S. M.; Miller, W. L.; Ortiz de Montellano, P. R. *Biochemistry* **1995,** *34,* 4421.
61. Trakshel, G. M.; Kutty, R. K.; Maines, M. D. *J. Biol. Chem.* **1986,** *261,* 11131.
62. Ishikawa, K.; Takeuchi, N.; Takahashi, S.; Matera, K. M.; Sato, M.; Shibahara, S.; Rousseau, D. L.; Ikeda-Saito, M.; Yoshida, T. *J. Biol. Chem.* **1995,** *270,* 6345.
63. McCoubrey, W. K.; Maines, M. D.; Cooklis, M. A. *Arch. Biochem. Biophys.* **1993,** *302,* 402.
64. Ishikawa, K.; Matera, K. M.; Zhou, H.; Fujii, H.; Sato, M.; Yoshimura, T.; Ikeda-Saito, M.; Yoshida, T. *J. Biol. Chem.* **1998,** *273,* 4317.
65. Zeng, J.; Fenna, R. E. *J. Mol. Biol.* **1992,** *226,* 185.
66. Poulos, T. L.; Finzel, B. C.; Howard, A. J. *J. Mol. Biol.* **1987,** *195,* 687.
67. Fita, I.; Rossman, M. G. *J. Mol. Biol.* **1985,** *185,* 21.
68. Sun, J.; Wilks, A.; Ortiz de Montellano, P. R.; Loehr, T. M., *Biochemistry* **1993,** *32,* 14151.
69. Takahashi, S.; Wang, J.; Rousseau, D. L.; Ishikawa, K.; Yoshida, T.; Host. J. R.; Ikeda-Saito, M. *J. Biol. Chem.* **1994,** *269,* 1010.
70. Takahashi, S.; Wang, J.; Rousseau, D. L.; Ishikawa, K.; Yoshida, T.; Takeuchi, N.; Ikeda-Saito, M., *Biochemistry* **1994,** *33,* 5531.
71. Kincaid, J.; Stein, P.; Spiro, T. G. *Proc. Natl. Acad. Sci. USA* **1979,** *76,* 549; 4156 (errata).
72. Kitagawa, T., Nagai, K., Tsubaki, M. *FEBS Lett.* **1979,** *104,* 376.
73. Choi, S.; Spiro, T. G. *J. Am. Chem. Soc.* **1983,** *105,* 3683.
74. Teraoka, J.; Kitagawa, T. *J. Biol. Chem.* **1981,** *256,* 3969.
75. Smulevich, G.; Mauro, J. M.; Fishel, L. A.; English, A. M.; Kraut, J.; Spiro, T. G. *Biochemistry,* **1988,** *27,* 5477.
76. Sun, J.; Loehr, T. M.; Wilks, A.; Ortiz de Montellano, P. R. *Biochemistry* **1994,** *33,* 13734.
77. Ito-Maki, M.; Ishikawa, K.; Matera, K. M.; Sato, M.; Ikeda-Saito, M.; Yoshida, T. *Arch. Biochem. Biophys.* **1995,** *317,* 253.
78. Wilks, A.; Sun, J.; Loehr, T. M.; Ortiz de Montellano, P. R. *Biochemistry* **1995,** *117,* 2925.
79. Liu, Y.; Moënne-Loccoz, P.; Hildebrand, D. P.; Wilks, A.; Loehr, T. M.; Mauk, A. G.; Ortiz de Montellano, P. R. *Biochemistry* **1999,** *38,* 3733.
80. Wilks, A.; Medzihradszky, K. F.; Ortiz de Montellano, P. R. *Biochemistry* **1998,** *37,* 2889.
81. Osawa, Y.; Martin, B. M.; Griffin, P. R.; Yates, J. R., Shabanowitz, J.; Hunt, D. F.; Murphy, A. C.; Chen, L., Cotter, R. J.; Pohl, L. R. *J. Biol. Chem.* **1990,** *265,* 10340.
82. Wilks, A.; Ortiz de Montellano, P. R.; Sun, J.; Loehr, T. M. *Biochemistry* **1996,** *35,* 930.
83. Quillin, M. L.; Arduini, R. M.; Olson, J. S.; Phillips, G. N. *J. Mol. Biol.* **1993,** *234,* 140.
84. La Mar, G. N.; Dalichow, F.; Zhao, X., Dou, Y.; Ikeda-Saito, M.; Chiu, M. L.; Sligar, S. G. *J. Biol. Chem.* **1994,** *269,* 29629.

85. Morikis, D.; Champion, P. M.; Springer, B. A.; Egeberg, K. D.; Sligar, S. G. *J. Biol. Chem.* **1990,** *265,* 12143.
86. Gorst, C. M.; Wilks, A.; Yeh, D. C.; Ortiz de Montellano, P. R.; La Mar, G. N. (1998) *J. Am. Chem. Soc.* **1998,** *120,* 8875.
87. Matera, K. M.; Zhou, H.; Migita, C. T.; Hobert, S. E.; Ishikawa, K.; Katakura, K.; Maeshma, H.; Yoshida, T.; Ikeda-Saito, M. *Biochemistry* **1997,** *36,* 4909.
88. Hernández, G.; Wilks, A.; Paolesse, R.; Smith, K. M.; Ortiz de Montellano, P. R.; La Mar, G. N. *Biochemistry* **1994,** *33,* 6631.
89. Omata, Y.; Asada, S.; Sakamoto, H.; Fukuyama, K.; Noguchi, M. *Acta Cryst.* **1998,** *D54,* 1017.
90. Schuller, D. J.; Wilks, A.; Ortiz de Montellano, P. R.; Poulos, T. L. *Protein Sci.* **1998,** *7,* 1836.
91. Migita, C. T.; Matera, K. M.; Ikeda-Saito, M.; Olson, J. S.; Fujii, H.; Yoshimura, T.; Zhou, H.; Yoshida, T. *J. Biol. Chem.* **1998,** *273,* 945.
92. Vermillion, J. L.; Coon, M. J. *J. Biol. Chem.* **1978,** *153,* 2694.
93. Wang, M.; Roberts, D. L.; Paschke, R.; Shea, T. M.; Masters, B. S. S.; Kim, J.-J. P. *Proc. Natl. Acad. Sci. USA* **1997,** *94,* 8411.
94. Sevrioukova, I. F.; Li, H.; Zhang, H.; Peterson, J. A.; Poulos, T. L. *Proc. Natl. Acad. Sci. USA* **1999,** *96,* 1863.
95. Antonioni, E.; Brunori, M. "Hemoglobins and Myoglobins and Their Reactions with Ligands"; North Holland Publishing Co.: Amsterdam, 1971.
96. Fujii, H.; Dou, Y.; Zhou, H.; Yoshida, T.; Ikeda-Saito, M. *J. Am. Chem. Soc.* **1998,** *120,* 8251.
97. Juckett, M.; Zheng, Y.; Yuan, H.; Pastor, T.; Antholine, W.; Weber, M.; Vercellotti, G. *J. Biol. Chem.* **1998,** *273,* 23388.
98. Ding, Y.; McCoubrey, W. K.; Maines, M. D. *Eur. J. Biochem.* **1999,** *264,* 854.
99. Frydman, R. B., Frydman, B. *Acct. Chem. Res.* **1987,** *20,* 250.
100. Frydman, R. B.; Tomaro, M. L.; Buldain, G.; Awruch, J.; Diaz, L.; Frydman, B. *Biochemistry* **1981,** *20,* 5177.
101. Tomaro, M. L., Frydman, S. B., Frydman, B., Pandey, R. K., Smith, K. M. *Biochim. Biophys. Acta* **1984,** *791,* 342.
102. Kutty, R. K.; Maines, M. D. *J. Biol. Chem.* **1982,** *257,* 9944.
103. Takahashi, S.; Ishikawa, K.; Takeuchi, N.; Ikeda-Saito, M.; Yoshida, T.; Rousseau, D. L. *J. Am. Chem. Soc.* **1995,** *117,* 6002.
104. Liu, Y.; Moënne-Loccoz, P.; Loehr, T. M.; Ortiz de Montellano, P. R. *J. Biol. Chem.* **1997,** *272,* 6909.
105. Matera, K. M.; Takahashi, S.; Fujii, H.; Zhou, H.; Ishikawa, K.; Yoshimura, T.; Rousseau, D. L.; Yoshida, T.; Ikeda-Saito, M. *J. Biol. Chem.* **1996,** *271,* 6618.
106. Yoshida, T.; Noguchi, M.; Kikuchi, G.; Sano, S. *J. Biochem. (Tokyo)* **1981,** *90,* 125.
107. Yoshinaga, T.; Sudo, Y.; Sano, S. *Biochem. J.* **1990,** *270,* 659.
108. Jackson, A. H.; Kenner, G. W.; Smith, K. M. *J. Chem. Soc. Sect. C Org. Chem.* **1968,** 302.
109. Morishima, I.; Fujii, H.; Shiro, Y.; Sano, S. *Inorg. Chem.* **1995,** *34,* 1528.
110. Bonnett, R.: Chaney, B. D. *J. Chem. Soc. Perkin Trans. I* **1987,** 1063.
111. Davydov, R. M.; Yoshida, T.; Ikeda-Saito, M.; Hoffman, B. R. *J. Amer. Chem. Soc.* **1999,** *121,* 10656.
112. Ortiz de Montellano, P. R. *Annu. Rev. Pharmacol. Toxicol.* **1992,** *32,* 89.
113. Gibson, J. F.; Ingram, D. J. E.; Nicholls, P. *Nature* **1958,** *181,* 1398.
114. King, N. K.; Winfield, M. E. *J. Biol. Chem.* **1963,** *238,* 1520.
115. Yonetani, T.; Schleyer, H. *J. Biol. Chem.* **1967,** *242,* 1974.

116. Wilks, A.; Torpey, J.; Ortiz de Montellano, P. R. *J. Biol. Chem.* **1994,** *269,* 29553.
117. Torpey, J., "Mechanistic Studies of Heme Oxygenase", Ph.D. thesis, University of California, San Francisco, 1997.
118. Torpey, J. W.; Ortiz de Montellano, P. R. *J. Org. Chem.* **1995,** *60,* 2195.
119. Brown, S. B. *Biochem. J.* **1976,** *159,* 23.
120. Jackson, A. H. In "Iron in Biochemistry and Medicine"; Jacobs, A.; Worwood, M., eds.; Academic Press: New York, 1974; p. 145.
121. Docherty, J. C.; Brown, S. B. *Biochem. J.* **1984,** *222,* 401.
122. Torpey, J.; Ortiz de Montellano, P. R. *J. Biol. Chem.* **1997,** *272,* 22008.
123. Ema, T.; Senge, M. O.; Nelson, N. Y.; Ogoshi, H.; Smith, K. M. *Angew. Chem. Int. Ed.* **1994,** *33,* 1879.
124. Senge, M. O.; Ema, T.; Smith, K. M. *J. Chem. Soc., Chem. Commun.* **1995,** 733.
125. Tan, H.; Simonis, U.; Shokhirev, N. V.; Walker, F. A. *J. Am. Chem. Soc.* **1994,** *116,* 5784.
126. Shokhirev, N. V.; Walker, F. A. *J. Biol. Inorg. Chem.* **1998,** *3,* 581.
127. Sakamoto, H.; Omata, Y.; Palmer, G.; Noguchi, M. *J. Biol. Chem.* **1999,** *274,* 18196.
128. Migita, C. T.; Fujii, H.; Matera, K. M.; Takahashi, S.; Zhou, H.; Yoshida, T. *Biochim. Biophys. Acta* **1999,** *1432,* 203.
129. Liu, Y.; Ortiz de Montellano, P. R. *J. Biol. Chem.* **2000,** *275,* 5297.
130. Fülop, V.; Phizackerley, R. P.; Soltis, S. M.; Clifton, I. J.; Wakatsuki, S.; Erman, J.; Hadju, J.; Edwards, S. L. *Structure* **1994,** *2,* 201.
131. Ortiz de Montellano, P. R. In "Cytochrome P450: Structure, Mechanism, and Biochemistry"; Ortiz de Montellano, P. R., ed.; Plenum Press: New York, 1995; p. 245.
132. Lee-Robichaud, P.; Shyadehi, A. Z.; Wright, J. N., Akhtar, M. E.; Akhtar, M. *Biochemistry* **1995,** *34,* 14104.
133. Marletta, M. A. *J. Biol. Chem.* **1993,** *268,* 12231.
134. Vaz, A. D. N.; Pernecky, S. J.; Raner, G. M.; Coon, M. J. *Proc. Natl. Acad. Sci. USA* **1996,** *93,* 4644.
135. Vaz, A. D. N.; McGinnity, D. F.; Coon, M. J. *Proc. Natl. Acad. Sci. USA* **1998,** *95,* 3555.
136. Cornejo, J.; Willows, R. D.; Beale, S. I. *Plant J.* **1998,** *15,* 99.
137. Beale, S. I.; Cornejo, J. *Arch Biochem Biophys* **1984,** *235,* 371.
138. Richaud, C.; Zabulon, G. *Proc. Natl. Acad. Sci. USA* **1997,** *94,* 11736.
139. Davis, S. J.; Kurepa, J.; Vierstra, R. D. *Proc. Natl. Acad. Sci. USA* **1999,** *96,* 6541.
140. Cornejo, J.; Beale, S. I. *Photosynthesis Res.* **1997,** *51,* 223.
141. Troxler, R. F.; Brown, A. S.; Brown, S. B. *J. Biol. Chem.* **1979,** *254,* 3411.
142. Brown, S. B.; Holroyd, A. J.; Troxler, R. F. *Biochem J.* **1980,** *190,* 445.
143. Schmitt, M. P. *J. Bacteriol.* **1997,** *179,* 838.
144. Schmitt, M. P. *Infect. Immun.* **1997,** *65,* 4634.
145. Schmitt, M. P. *J. Bacteriol.* **1999,** *181,* 5330.
146. Mills, M.; Payne, S. M. *J. Bacteriol.* **1995,** **177,** 3004.
147. Stojiljkovic, I.; Hantke, K. *EMBO J.* **1992,** *11,* 4359.
148. Wilks, A.; Schmitt, M. P. *J. Biol. Chem.* **1998,** *273,* 837.
149. Chu, G. C.; Katakura, K.; Zhang, X.; Yoshida, T.; Ikeda-Saito, M. *J. Biol. Chem.* **1999,** *274,* 21319-25.
150. Chu, G. C; Tomita, T.; Sonnichsen, F. D.; Yoshida, T.; Ikeda-Saito, M. *J. Biol. Chem.* **1999,** *274,* 24490.
151. Wilks, A.; Moënne-Loccoz, P. *J. Biol. Chem.* **2000,** *275,* 11686.

152. Lagarias, J. C. *Biochim. Biophys. Acta* **1982,** *717,* 12.
153. Rodriguez, J. C.; Rivera, M. *Biochemistry* **1998,** *37,* 13082.
154. Saito, S.; Itano, H. A. *Proc. Natl. Acad. Sci. USA* **1982,** *79,* 1393.
155. Sano, S.; Sano, T.; Morishima, I.; Shiro, Y.; Maeda, Y. *Proc. Natl. Acad. Sci. USA* **1986,** *83,* 531.
156. Chu, G. C.; Park, S. Y.; Shiro, Y.; Yoshida, T.; Ikeda-Saito, M. *J. Struct. Biol.* **1999,** *126,* 171.

ADVANCES IN INORGANIC CHEMISTRY, VOL. 51

DE NOVO DESIGN AND SYNTHESIS OF HEME PROTEINS

BRIAN R. GIBNEY and P. LESLIE DUTTON

The Johnson Research Foundation, Department of Biochemistry and Biophysics, School of Medicine, University of Pennsylvania, Philadelphia, Pennsylvania 19104

I. Introduction

In its purest and most challenging form, *de novo* design involves the construction of a protein, intended to fold into a precisely defined 3-dimensional structure, with a sequence that is not directly related to that of any natural protein. (*1*) With permission, from the *Annual Review of Biochemistry,* Vol. 68, 1999, by Annual Reviews www.AnnualReviews.org.

0898-8838/01 $35.00

Since *de novo* design is a constructive approach, it allows precise tests of fundamental hypotheses about protein structure/function relationships (*2*). Additionally, rapidly advancing chemical synthesis methodologies (*3–5*) allow chemists to explore regions of sequence space (*6*) unpopulated by natural proteins in the hope of designing proteins with novel functions (*7*). *De novo* design has proven an invaluable tool in deciphering the basic protein folding tenets of secondary structure, protein stability, and conformational specificity (*8–12*). Born out of studies over the past 20 years, *de novo* protein design has progressed from the design of stably folded monomeric helices in solution to aggregated structures with well-defined hydrophobic cores. Currently, protein design is at the stage where the design of small, that is, less than 150 amino acids, stably folded protein scaffolds of predictable structure is becoming routine (*1*). Notable successes include the emerging X-ray and NMR structures of *de novo* designed protein scaffolds, including helix–turn–helix hairpins (*13*), three-stranded coiled-coils (*14, 15*), three-α-helix bundles (*16*), four-α-helix bundles (*17–20*), beta sheet proteins (*21*), and mixed α/β architectures (*22, 23*), as well as the design of highly stable and robust designed proteins as models of hyperthermophilic enzymes (*24–26*).

At its current stage of development, *de novo* protein design has evolved beyond the design of simple protein scaffolds to include the incorporation of transition metal ion binding sites (*27–29*) with greater biological import to impart their catalytic (*30*), structural (*31, 32*), regulatory (*33*), and electron transfer properties (*34–36*). These designed metalloproteins provide biologically relevant ligands within a protein scaffold as a bioinorganic model complex (*37, 38*) for the investigation of fundamental chemical issues of metalloprotein assembly, structure, and function under physiological conditions. As such, these novel metalloproteins provide a more elaborate ligand scaffold to the inorganic coordination chemist for the design of metalloenzyme mimetics.

Given that hemes were one of the earliest recognized inorganic biological cofactors having been structurally elucidated by Hans Fischer, it is only fitting that they were one of the initial *de novo* design metalloprotein targets. These novel designed heme proteins attempt to access the diverse range of metabolic, regulatory, and structural roles that hemes play in a multitude of biological systems. As an inherently constructive approach, the *de novo* design of heme proteins complements site-directed mutagenesis studies of heme proteins (*39–41*), as well as structure-based redesign efforts to modify natural heme proteins (*42, 43*) and the total chemical synthesis (*44–46*) of natural metalloproteins. However, the use of *de novo* design protein scaffolds coupled with automated solid-phase peptide synthesis (SPPS) (*3, 47*) offers the advantage that the choice of scaffold, sequence or liganding amino acids is not strictly limited by biological constraints.

The focus of this review is on *de novo* designed proteins that bind a metalloporphyrin, regardless of encapsulated metal or peripheral porphyrin architecture. The designed heme proteins are classified as to whether their protein architecture contains a stable hydrophobic core in the absence of heme (protein systems) or whether heme binding induces folding and/or association of the peptide scaffold (peptide systems). Simple systems, containing single or multiple copies of one type of metalloporphyrin, are reviewed before delving into the complex systems containing different types of hemes and hemes with other biochemical cofactors. The chemical, physical, and spectroscopic properties of the designed heme proteins are compared to each other and to natural systems to evaluate the success of the design concepts. The redox chemistry of designed heme proteins is a fundamental facet of their biological chemistry that is directly compared to that of natural heme proteins and enzymes. Functional aspects, including the induction of protein secondary structure, dioxygen reactivity, and proton coupled electron transfer, are discussed in light of recent papers that offer insight into the details of peptide–porphyrin interactions. The materials science applications of *de novo* designed heme proteins are also reviewed. Finally, this overview of the collection of *de novo* designed heme proteins is presented so as to assess the current status of this emerging field and areas in need of further research.

II. Natural Protein Engineering

Heme proteins are ubiquitous in biological systems, underscoring their functional importance, which varies from biological energy conversion (*48*) to programmed cell death (*49, 50*). As with all proteins, the global fold of a protein lies at a free energy minimum and is fully encoded within the amino acid sequence (*51*). The primary structure folds into secondary structure elements such as α-helices and β-sheets, which in turn form tertiary and quaternary structures. The bulk of the driving force for the collapse of an amino acid chain into a singular global protein fold is provided by the hydrophobic effect, with lesser contributions from hydrogen bonding interactions and electrostatic effects (*1*). In this section we survey three primary functions of heme proteins (biological electron transfer, reversible dioxygen binding, and dioxygen activation) to illustrate how the protein ligand greatly influences the chemical properties of the bound heme. In fact, the multifunctional utility observed for the heme cofactor is not restricted only to hemes; it is also becoming recognized in other inorganic cofactors such as the iron–sulfur proteins (*52*). As such, this may represent how natural systems utilize to greatest advantage a limited collection of biochemical cofactors and prosthetic groups, both inorganic and organic, with a restricted number of global protein folds in a combinatorial manner.

A. Electron Transfer

Cytochromes serve as electron donors and electron acceptors in biological electron transfer chains, and with >75,000 members (*53*) they provide the bulk of natural heme proteins in biology. Cytochromes may be fixed into place within an extended electron transfer chain, such as the membrane-bound b_L and b_H of the cytochrome bc_1 complex, or may be soluble and act as mobile electron carriers between proteins, for example, cytochrome *c* (*54*). In either role, the cytochrome may be classified by the peripheral architecture of the porphyrin macrocycle. Figure 1 shows the dominant heme types in biological systems, which are hemes *a*, *b*, *c*, and *d*, with cytochomes *b* and *c* being most prevalent. The self-association of a protein with heme via two axial ligands is a

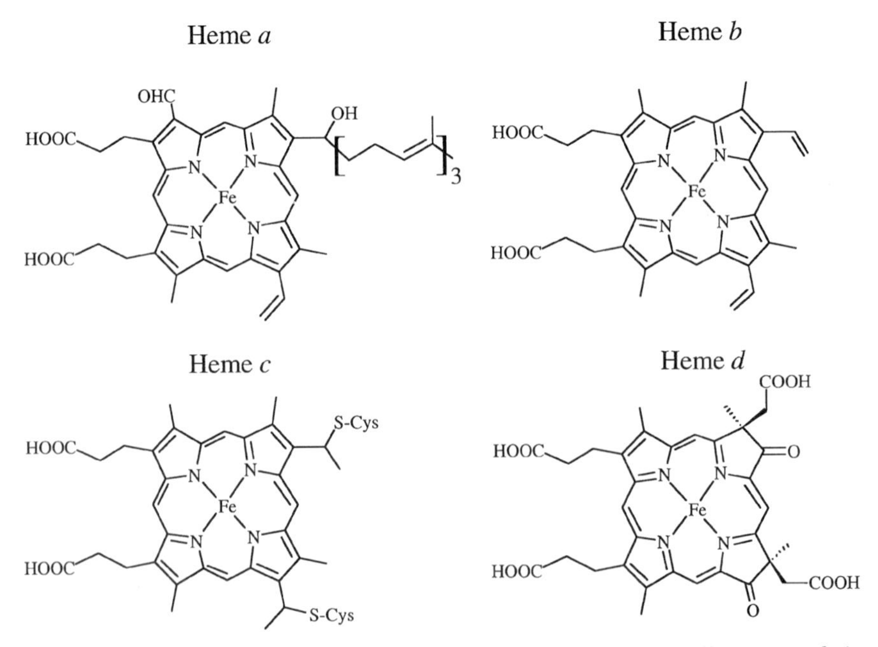

Fig. 1. Chemical structures of the predominant biological hemes illustrating their respective peripheral architectures.

common theme used in the construction of hemes *a*, *b*, and *d*. The incorporation of iron(protoporphyrin IX) signifies a *b*-type heme, such as the bis-histidine ligated cytochrome b_5 (*55*). Heme *a*, a derivative of heme *b* that contains a formyl group and a farnesyl side chain on the porphyrin periphery, is utilized only in terminal oxidases, such as cytochrome *c* oxidase, and is the site of dioxygen reduction to water in aerobic respiration. Cytochrome *d* of cytochrome cd_1 (*56*), utilized both in respiration and in denitrification, contains peripheral ketone groups. The covalent attachment of the heme moiety via cysteine residues indicates a *c*-type heme, such as cytochrome *c*, which also includes cytochrome *f* of the cytochrome b_6f complex from plants (*57*). The covalent attachment of the heme moiety to cytochromes *c* is affected by the highly conserved -CXXCH- sequence, which not only affixes the heme macrocycle but also provides an axial ligand, the histidine. Both of these biologically relevant heme-protein construction methods, self-assembly and covalent linkages, have been employed in *de novo* heme protein design.

As with any metalloprotein, the chemical and physical properties of the metal ion in cytochromes are determined by the both the primary and secondary coordination spheres (*58–60*). The primary coordination sphere has two components, the heme macrocycle and the axial ligands, which directly affect the bound metal ion. The pyrrole nitrogen donors of the heme macrocycle that are influenced by the substitutents on the heme periphery establish the base heme properties. These properties are directly modulated by the number and type of axial ligands derived from the protein amino acids. Typical heme proteins utilize histidine, methionine, tyrosinate, and cysteinate ligands to affect five or six coordination at the metal center.

The heme microenvironment is further established beyond its primary coordination sphere where the low dielectric protein interior allows a multitude of factors to indirectly adjust the properties of the bound heme. The simple act of inserting the heme into a low dielectric environment destabilizes the formally charged ferric state altering the heme electrochemistry (*61–63*). Hydrogen bonding between the primary coordination sphere and second shell amino acids plays critical roles in the push–pull mechanisms (*64*) of heme peroxidases (*65*) and oxygenases (*66*). Steric hinderance with tightly packed hydrophobic cores may result in distortions of the porphyrin ring system, which has been proposed to affect the spectroscopy (*67*) and electrochemistry (*68*) of *c*-type cytochromes. The local electrostatic field established by coulombic charge interactions transmitted within low dielectric hydrophobic cores has been calculated from the heme protein structure and generally found to agree with the observed trends in heme reduction potentials (*69, 70*).

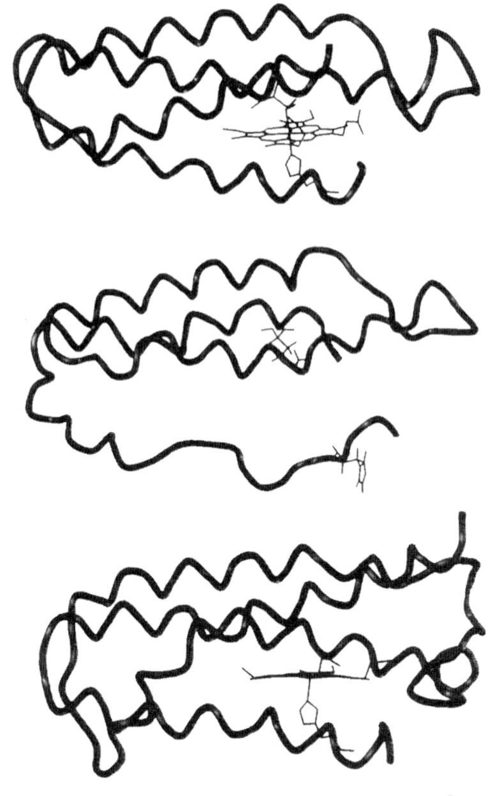

FIG. 2. Comparison of the X-ray structures of (top) ferric cytochrome b_{562} (middle) apo cytochrome b_{562}, and (bottom) ferric cytochrome c'.

Since the primary structure of a peptide determines the global fold of any protein, the amino acid sequence of a heme protein not only provides the ligands, but also establishes the heme environmental factors such as solvent and ion accessibility and local dielectric. The prevalent secondary structure element found in heme protein architectures is the α-helix; however, it should be noted that β-sheet heme proteins are also known, such as the nitrophorin from *Rhodnius prolixus* (*71*) and flavocytochrome cellobiose dehydrogenase from *Phanerochaete chrysosporium* (*72*). However, for the purpose of this review, we focus on the structures of cytochromes b_{562} (*73*) and c' (*74*) shown in Fig. 2, which are four-α-helix bundle protein architectures and lend themselves as resource structures for the development of *de novo* designs.

Figure 2 shows the X-ray structure of Fe(III) cytochrome b_{562} at 1.4 Å resolution (*73, 75, 76*). The unit cell comprises two 106 amino acid four-helix-bundle monomers with 83% helical content. The helices are each composed of 20–23 amino acids and pack inclined 20° to one another into a left-handed twist. The heme binding residues, M7 and H102, are found on the N-terminal and C-terminal helices, respectively, and the heme propionates are oriented toward the solvent. An interesting structural detail at the heme binding site is high-level interactions between aromatic residues and the heme macrocycle. The phenylalanines (F61 and F65) form the top of a clamp with the tyrosines (Y101 and Y105) at the base with the heme macrocycle in between.

The apocytochrome b_{562} solution NMR structure is shown for comparison (*77*). Apocytochrome b_{562} maintains the general overall topology observed in the holo structure, but with significant loss of secondary structure in the C-terminal helix that provides the histidine ligand. Rearrangements of the interhelix hydrophobic packing lead to exposure of the heme binding pocket to solvent. These structural differences between the apo and holo forms are not unique to cytochrome b_{562}. The binding of heme to apoproteins invariably induces formation of the final structure, which is partially to completely disordered in the absence of the heme macrocycle (*78, 79*). The binding of heme not only determines the final structure, it is also involved in the folding process with adventitious ligand binding being a predominant misfolding event in cytochrome *c* (*80, 81*).

Figure 2 also shows the X-ray structure of the Fe(III) cytochrome c' monomer as determined to 1.67 Å resolution (*74*). The structure consists of a pair of identical 128 amino acid four helix bundle subunits containing 73% helical content. The helical regions are connected by well-structured turn regions including a very short one between helices C and D. The high-spin, five-coordinate histidine ligated heme is covalently linked to helix D, the C-terminal helix, and is nestled in a group of aromatic amino acids. The heme propionates are partially buried within the interior of the protein with one hydrogen-bonded to a conserved arginine on helix A. Finally, the similarities (*82*) between cytochrome b_{562} and cytochrome c' have been exploited for the conversion of cytochrome b_{562} into a *c*-type cytochrome by the judicious choice of cysteine mutants (*83–85*).

B. Dioxygen Transport

The first three-dimensional structure of a protein determined was the dioxygen transporter myoglobin isolated from sperm whale in the ferric form (*86, 87*). A 153 amino acid globular protein, myoglobin contains eight helical regions (A to H) with a single heme *b* bound between helices

E and F. The heme is coordinated to a single histidine residue, H8 of helix F, via the N_ε nitrogen with the sixth coordination site open for reversible dioxygen binding, which binds in a bent geometry (*88*). The prototype globin fold, myoglobin shares some stuctural similarity to the subunits (2α and 2β) of the tetrameric hemoglobin (*89*). However, structural studies of a hemoglobin, FixL from *Bradyrhizobium japonicum*, which serves as a dioxygen sensor in the nitrogen fixing bacteria demonstrate a novel protein fold. In the FixL structure, the five-coordinate heme *b* is bound to H200 on the central α-helix of a mixed α/β architecture. The distal face of the heme macrocycle is protected from solvent by a five-strand antiparallel beta sheet. The structure of FixL clearly demonstrating that reversible dioxygen binding is plausable in nonglobin protein folds (*90, 91*).

C. Dioxygen Activation

Cytochromes P450 are a class of heme monooxygenases that catalyze the hydroxylation of aromatic and aliphatic substrates by molecular oxygen (*92, 93*). These heme proteins serve numerous critical biochemical roles, including drug metabolism and steroid biosynthesis. The structure of the soluble bacterial cytochrome P450cam with bound camphor substrate consists of 414 amino acids in a mixed α/β protein scaffold composed of 13 helical segments, helixA–helixL, connected by loop regions and strands of β-sheet (*64, 65*). The thiolate ligated heme resides buried in the protein interior, 8 Å from the protein surface, between the L-helix and the I-helix with the heme propionates in a cluster of hydrogen bonding and charged residues. The mechanism of dioxygen activation by cytochromes P450 has been explored using time-resolved X-ray crystallography, which may provide insight into the functional heme protein design (*94*).

The four-electron/four-proton reduction of dioxygen to water is carried out by the terminal oxidases of the respiratory chain with the concomitant pumping of protons across the membrane for use in ATP synthesis (*95*). The bacterial (*96*) and mammalian (*97*) cytochrome *c* oxidases are membrane bound heme proteins containing four redox centers: the dinuclear Cu_A center, heme *a*, heme a_3, and Cu_B. The Cu_A center resides in the intermembrane space bound to subunit II close to the proposed site of cytochrome *c* docking. The hemes, heme *a* and heme a_3, are equidistant from Cu_A and are buried within the membrane spanning helices. The Cu_B site lies above the heme a_3 macrocycle forming the site of dioxygen binding and reduction.

III. *De Novo* Heme Proteins

The goal of the *de novo* heme protein designer is to understand the various factors that specify a singular global fold that suitably orients amino acid ligands for heme ligation. Additionally, the heme protein designer may wish to modulate the properties of the bound heme via environmental factors such as solvent accessibility and local dielectric that must be encoded in the primary structure. Because of the prevalence of α-helical heme proteins (*98, 99*) and the well-established rules of *de novo* helix design (*1, 2, 100–102*), protein designers have utilized helical scaffolds for *de novo* heme protein design. The full complement of design strategies developed for scaffold design are being utilized for heme protein design. For the purposes of this review, we differentiate the designs into two groups based on the status of the hydrophobic core in the absence of heme: heme–peptide systems and heme–protein systems. Heme–peptide systems are those designs that are mostly unfolded in the apo-state, and metal ligation induces the formation of secondary structure. These heme–peptide systems are generally smaller than the corresponding heme–protein systems, which possess stably folded hydrophobic cores prior to heme incorporation. Just as observed in natural heme proteins, the incorporation of heme into either type of system generally causes structural reorganization, since the metal ligation and hydrophobic interactions of the heme with the protein are intimately coupled to protein folding (*78, 81, 103, 104*).

A. Peptide Systems

The coupling of short alanine based peptides based on the work of Baldwin and co-workers (*105–107*) to hemes provides several related heme–peptide systems whose minimal nature allow for detailed examination of the individual factors controlling heme protein function. Benson's initial report of the synthesis of a heme–peptide system relied on the covalent attachment of lysine side chain amino groups in a 13 amino acid peptide to the propionic acids groups of iron(III) mesoporphryin IX (*108*). Figure 3 shows the resulting di-α-helical heme construct, termed a peptide-sandwiched mesoheme or PSM, and the mono-α-helical construct that were utilized to study the induction of helical structure by the porphyrin moiety and the factors contributing to induced heme CD signals. Because of the diastereomeric nature of the initial PSMs, the synthesis of iron(III) mesoporphyrin II, a C_2 symmetric heme, and its incorporation into PSMs has been described (*109*).

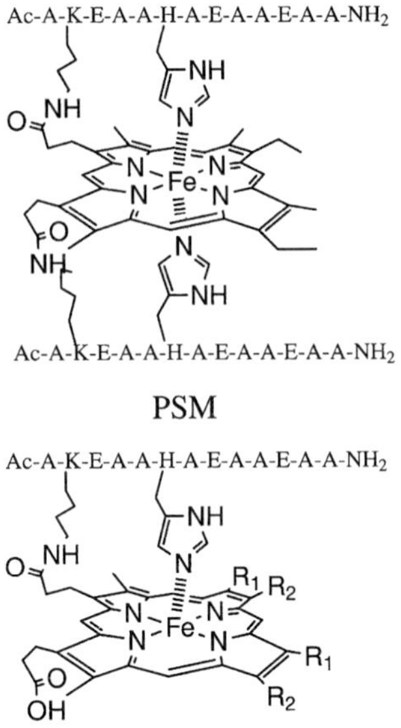

Ac-A-K-E-A-A-H-A-E-A-A-E-A-A-NH_2

Ac-A-K-E-A-A-H-A-E-A-A-E-A-A-NH_2

PSM

Ac-A-K-E-A-A-H-A-E-A-A-E-A-A-NH_2

Monopeptide-PSM: R^1 = Me; R^2 = Et

FIG. 3. Chemical structures of (*top*) the peptide sandwiched mesoheme (PSM) and (*bottom*) its monopeptide analog. Reprinted with permission from Ref. (*108*). Copyright 1995 American Chemical Society.

The mono-α-helical construct is reminiscent of the microperoxidases, *c*-heme peptide conjugates derived from cytochrome *c* (*110–112*), and provides the first examples of the design of five-coordinate hemes in *de novo* heme design. A second synthetic strategy for the synthesis of five-coordinate hemes is to affix both ends of the peptide to the porphyrin periphery. Geier and Sasaki utilized thioether linkages to bind a 14 amino acid peptide to a modified free base tetraphenylporphyrin (*113*). In an elegant approach, Karpishin and co-workers utilized Cu(II) binding as the sites of peptide attachment to a Zn(II)–porphyrin (*114*).

In a covalent architecture related to the PSMs, Pavone and co-workers have synthesized covalent heme–peptide conjugates, called mimochromes, utilizing the free base of deuteroporphyrin (3,7,12,17-tetramethylporphyrin-2,18-dipropionic acid) and a nine amino acid peptide based on the β-chain F helix of human deoxyhemoglobin (*115*). The resulting mimochrome architecture possess the advantage that the synthetic method allows for the insertion of various transition metal ions; to date these include Fe(III)/Fe(II) and Co(III) (*116, 117*). Figure 4 shows the NMR-derived solution structures of the resolved Λ and Δ isomers of the diamagnetic, exchange inert Co(III)–mimochromes (*118*).

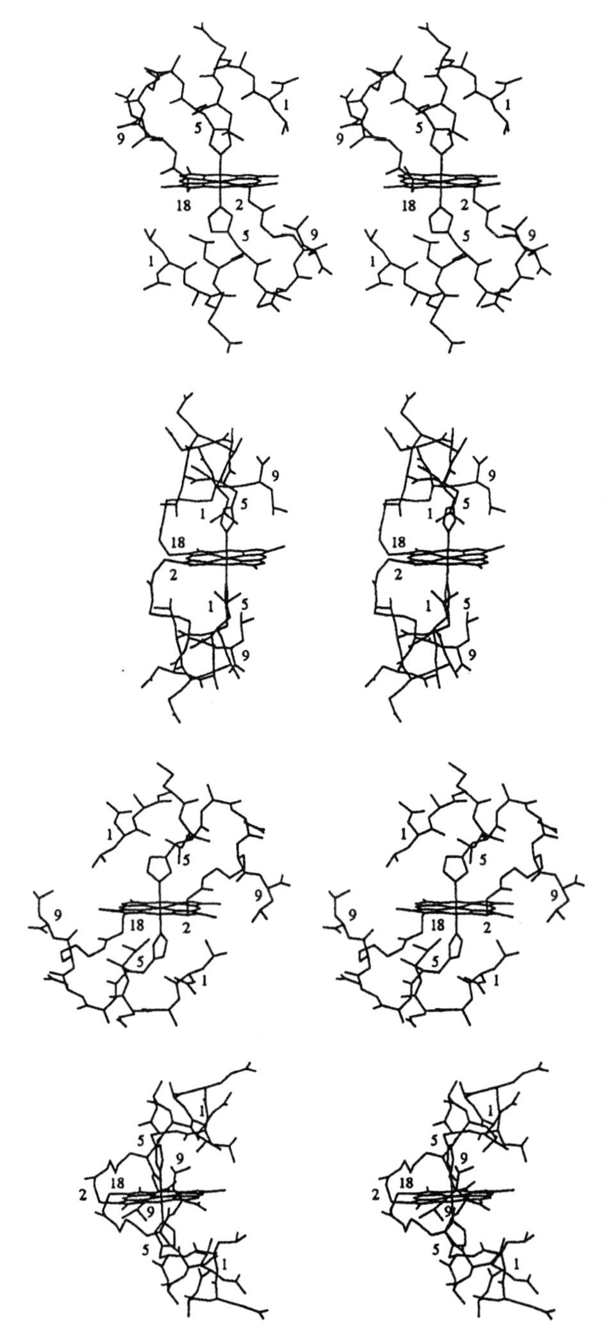

FIG. 4. Stereoscopic views of the (top pair) Δ and (bottom pair) Λ isomers of Co(III) mimochrome I complex structure as determined by NMR spectroscopy. Reprinted with permission from Ref. (*118*); copyright 1997 Wiley-VCH.

TABLE I

COMPARISON OF DESIGNED HEME PROTEIN PROPERTIES[a]

Peptide	Ligation motif	K_d	E_m (mV vs NHE)	Ref.
Heme peptide systems				
PSM	His/His	CA	ND	*108*
monopeptide-PSM	His/-	CA	ND	*159*
porphyrin template ***4***	free base	CA	NA	*113*
Assembly **I**	Zn (II)	CA	NA	*114*
Mimochrome I	His/His	CA	ND	*115*
His-2α	His/His	mesoheme 3.5 μM	 ND	*119*
H2α-17	His/His	mesoheme 91 nM	 ND	*121*
H2α (17)-L6	His/His	mesoheme 178 nM		*123*
H2α (17)-L4	His/His	mesoheme 137 nM	 ND	*123*
cH2α (17)-L4	His/His	mesoheme 137 nM	 ND	*123*
H2α (17)-L4S	His/His	mesoheme 120 nM	 ND	*123*
cH2α (17)-L4S	His/His	mesoheme 100 nM	 ND	*123*
H2α-17-FF	His/His	mesoheme 21 nM	 ND	*120*
H2α-17-LV	His/His	mesoheme 333 nM	 ND	*120*
AA-A	His/His	[a]0.23 mM^{-2}	Fe (CP-I) −223	*125*
FF-Aib	His/His	[a]47 mM^{-2}	Fe (CP-I) −304	*125*
SS-Aib	His/His	[a]0.25 mM^{-2}	Fe (CP-I) −218 mV	*125*
Heme protein systems				
$VAVH_{25}$(S-S)	His/His	700 nM	−170	*126*
retro-(S-S)	His/His	10 nM	−220	*126*
$[H10H24]_2$	His/His	200 nM, 3 μM	−80, −130 −180, −230	*127*
$[H10A24]_2$	His/His	heme *b* 15 nM, 800 nM heme *a* 0.1 nM, 5 nM mesoheme ND	heme B −170, −265 heme A −45 mesoheme −196	*127, 148, 160*
$[A10H24]_2$	His/His	800 nM, 20 μM	−75, −205	*127*

TABLE I (*Continued*)

Peptide	Ligation motif	K_d	E_m (mV vs NHE)	Ref.
$[H11A24]_2$	His/His	9.5 μM	−166	*168*
$[\alpha\text{-}\ell\text{-}\alpha]_2$	His/His	0.1 nM, 50 nM	−110, −190	*135*
		800 nM, 10 μM	−235, −270	
$[\alpha\text{-}\ell\text{-}\alpha\text{-}SS\text{-}\alpha\text{-}\ell\text{-}\alpha]$	His/His	0.1 nM, 50 nM	−110, −190	*135*
		800 nM, 10 μM	−235, −270	
$[H{:}\alpha\text{-}\ell\text{-}\alpha]_2$	His/His	35 nM, 135 nM	−105, −220	*149*
$[H{:}Fl\text{-}\alpha\text{-}\ell\text{-}\alpha]_2$	His/His	130 nM, 1.5 μM	−153	*149*
Protein 86	His/Met	740 nM	ND	*137*
Protein F	His/His	12 μM	ND	*137*
Protein G	His/His	2 μM	ND	*137*
DG-1	His/His	80 nM	ND	*139*
TASP systems				
MOP1	His/His	<1 μM	−110, −170	*143*
Ru-MOP2	His/His	<1 μM	−170 mV	*150*
Ru-MOP3	His/His	<1 μM	−164 mV	*150*

CA, covalently attached; ND, not determined; NA, not applicable.

[a] Expressed as a binding constant between peptide and metalloporphyrin forming a 2:1 complex.

These NMR structures reveal that the Δ isomer conforms to the design hypothesis, whereas the less regular peptide conformation of the Λ isomer forms an alternate structure.

The use of disulfide linked di-α-helical peptides for the self-assembly of a heme–peptide model compounds has also been explored by Benson *et al.* (*109*). Conceptually analogous to the larger heme-protein systems utilized by Dutton and co-workers, to be detailed later, the incorporation of C_4 symmetric Co(III)–porphyrins, based on coproporphyrin and octaethylporphyrin, resulted in helical induction comparable to that observed in the covalent PSM systems.

Mihara and co-workers (*119–124*) have prepared a variety of platforms for *de novo* heme protein design, including utilizing a series of disulfide linked di-α-helical peptides of variable length. The initial designs contained helical regions of 14, 17, or 21 amino acids with heme binding histidines at helical position 9 and homodimerized by C-terminal β-Ala–Cys linkers. Consistent with protein design principles, the longer peptides were more helical in solution (*123*). The identity of the loop regions was altered in later designs by the incorporation of GGGC loops on one or both ends of the helical regions with the bis-disulfide linked di-α-helical peptide having the highest helical content and greatest affinity for heme, K_d values of ≈100 nM as given in Table I (*123*). Incorporation of heme into some of these di-α-helical peptides results in assembly of tetrameric structures (eight helix bundles), illustrating the ability of heme ligation to alter the preferred peptide aggregation state (*121*).

Finally, Huffman *et al.* have described a peptide–heme system that utilizes monomeric peptides containing a single histidine residue to ligate Fe(III)–coproporphyrin in aqueous solution (*125*). The solution spectroscopy of the peptide–heme coordination compounds is similar to bis-histidine ligated Fe(III)–coproporphyrin. Although this system bears the largest entropic barrier to complex formation, significant peptide–heme binding constants lead to stable complexes that have been electrochemically characterized (*vide infra*).

These minimalistic peptide scaffolds potentially provide a biologically relevant laboratory in which to explore the details of heme–peptide interactions and, with development, perhaps approach the observed range of natural heme protein function. These heme–peptide systems are more complex than typical small molecule bioinorganic porphyrin model compounds, and yet are seemingly not as enigmatic as even the smallest natural heme proteins. Thus, in the continuum of heme protein model complexes these heme–peptide systems lie closer to, but certainly not at, the small molecule limit which allows for the effects of single amino acid changes to be directly elucidated.

B. Protein Scaffolds

Designed heme proteins potentially offer more diversity in peptide scaffold compared to designed heme peptide systems because of the presence of stably folded hydrophobic cores. As protein design advances, the number and variety of global protein folds available for heme–protein design will expand. In practice, however, the majority of protein designers have so far limited themselves to a small subset of protein architectures, most being based on four-α-helix bundles akin to those originally described by Ho and DeGrado (*100*). The similarity of these four-helix-bundle heme protein scaffolds facilitates comparison of their properties, and the development of novel scaffolds remains an exciting and demanding area of research as the field of protein design continues to mature.

Two reports in 1994 began to develop the concepts of coordination chemistry based self-assembly of heme into designed protein scaffolds. In collaborative work, the laboratories of DeGrado and Dutton provided two related architectures for *de novo* heme protein design based on Ho and DeGrado's α_2 peptide. Choma *et al.* (*126*) utilized a series of helix–loop–helix peptides containing one histidine residue dimerized via N-terminal cysteine residues to form monomeric four-α-helix bundles with a single bis-histidine heme binding site. All of the designs, including α_2(S-S), which contained no potentially ligating histidine residues, bound heme tightly enough to elute with the four-α-helix bundles in size exclusion chromatography, which provided a dissociation constant (K_d value) estimate of 100 μM. Indeed, it was surprising that the introduction of the histidine residues into α_2(S-S) to produce H25 did not yield a bis-histidine ligated heme as evidenced by optical spectroscopy. The inclusion of histidine residues and introduction of a hydrophobic cavity in $VAVH_{25}$ enhanced the heme affinity to 700 nM, given in Table I, with spectroscopic and electrochemical properties typical of bis-histidine ligated *b*-type cytochromes. The affinity for heme was further improved to 10 nM in retro(S-S), shown in Fig. 5, which was partially unfolded in the absence of heme.

In a contemporaneous report, Robertson *et al.* (*127*) introduced multiheme synthetic proteins using a helix–disulfide–helix monomer architecture containing one or two bis-histidine binding sites between the parallel helices as shown in Fig. 6. These (α-SS-α) architectures spontaneously self-associate in solution to yield four-α-helix bundles, $(\alpha\text{-SS-}\alpha)_2$, with either two or four heme binding sites named $[H10A24]_2$ and $[H10H24]_2$, respectively. The design of the spacing of the heme binding sites in $(\alpha\text{-SS-}\alpha)_2$ was derived from an analysis of the known sequences of cytochrome bc_1 complexes that contain a pair of bis-histidine ligated hemes whose ligands are separated by 14 amino acids. In addition to the ligands, a phenylalanine (F17) placed directly between the histidines (H10 and H24) and an arginine (R27) placed four amino acids removed from one of the histidines were introduced by analogy to the cytochrome bc_1 respiratory complex. The H10, H10′ sites, closest to the loop regions, bound heme tightly, K_d of 50 nM, with Fe(III) and Fe(II) state optical spectra consistent with bis-histidine ligation. The H24,H24′ sites bound hemes more weakly, K_d of 1–20 μM, with elevated reduction potentials presumably due to destabilization of the formally charged ferric heme by the presence of the positively charged arginine residue at position 27. Furthermore, the 70–100 mV difference between the pair of hemes at H10,H10′ (and also the pair at H24,H24′) due to oxidized heme–oxidized heme electrostatics demonstrated the presence of a low dielectric hydrophobic core in this model system.

Robertson *et al.* defined these elaborate peptide-based inorganic model complexes using a term extracted from art and architecture, *maquette,*

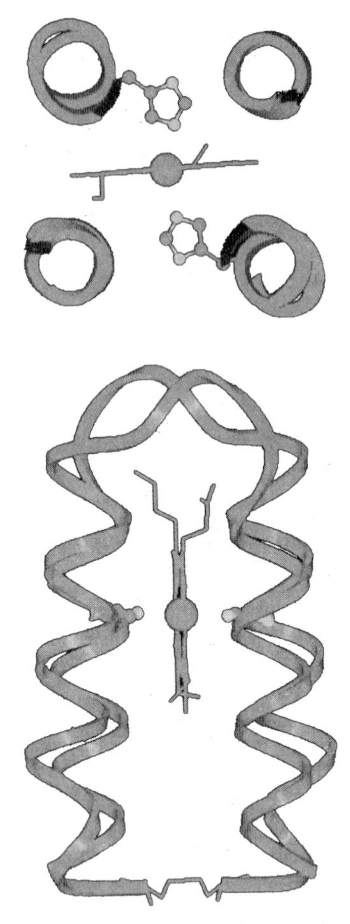

FIG. 5. Top and side view of a retro(S-S) molecular model containing heme. Reprinted with permission from Ref. (*126*). Copyright 1994 American Chemical Society.

a scale model of a sculpture or building. In the field of protein design, a maquette is a functional synthetic protein that is simplified compared to its natural counterpart. As physiologically relevant bioinorganic model complexes, maquettes offer a minimal peptide/protein scaffold in which to constructively study metalloprotein structure–function relationships.

The $(\alpha\text{-SS-}\alpha)_2$ architecture of Robertson *et al.* has shown wide utility as a basis structure in the systematic design of ever more sophisticated metalloprotein maquettes of the photosynthetic (*36*) and respiratory complexes toward the goal of testing the fundamentals of biological

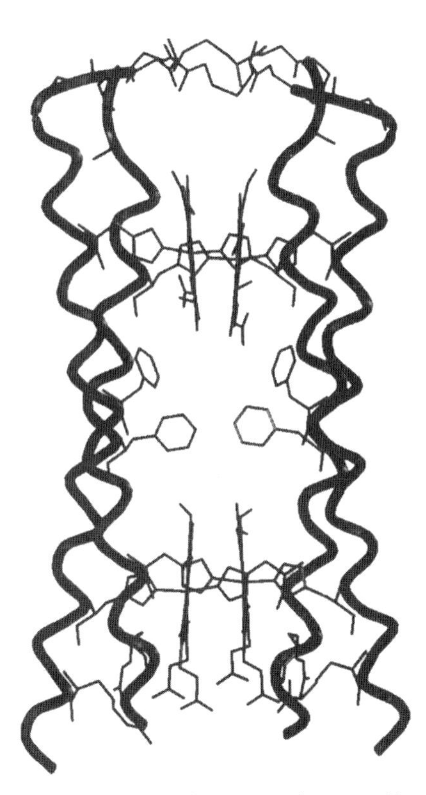

FIG. 6. Molecular model of the H10H24 heme protein maquette with four hemes bound. Reprinted with permission from *Nature* (*127*); copyright 1994, Macmillan Magazines Ltd.

electron transfer (*128–130*). Rabanal *et al.* (*131*) introduced a synthetic method for the high yield construction of asymmetric disulfide bonds that yielded the first $(\alpha\text{-SS-}\alpha')_2$ heme–protein architectures, which contain two distinct helices per monomer. This synthetic protocol has led to the development of a model of the bacterial photosynthetic reaction center containing a free base porphyrin dimer appended to a four-α-helix bundle containing up to four hemes (*132*).

The $(\alpha\text{-SS-}\alpha)_2$ maquette architecture has also found utility in the apo-state for the investigation of the amino acid determinants of conformational specificity in designed proteins. The systematic iterative redesign of by $(\alpha\text{-SS-}\alpha)_2$ Gibney *et al.* (*133*) has demonstrated the robustness of this maquette scaffold as the variety of amino acid modifications did not alter the four helix bundle global fold. In the apo-state, several of these redesigns yielded NMR spectra typical of natural proteins, and Fig. 7 illustrates the NMR derived solution structure of the $(\alpha'\text{-SS-}\alpha')$

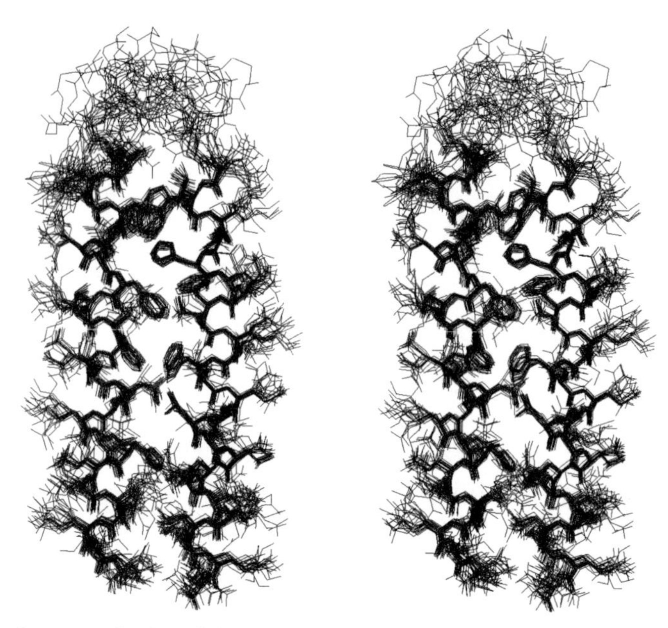

FIG. 7. Stereoscopic view of the apo-maquette monomer, (α'-SS-α'), solution structure. A superposition of the best solution structures derived from the NMR data is shown with the hydrophobic core facing the viewer. Reprinted with permission from Ref. (*20*). Copyright 1999 American Chemical Society.

monomer determined by Skalicky *et al.* (*20*). This study provided the initial structural confirmation of the apo-maquette scaffold designs and demonstrated that the tightly packed hydrophobic core did not contain a cavity for the heme macrocycle. Furthermore, this NMR structure serves as a modeling framework for the structure-based redesign of metalloproteins.

The group of Dutton *et al.* (*134, 135*) has also expanded their heme protein designs beyond the original (α-SS-α)$_2$ system to include the more common helix–loop–helix (α-ℓ-α)$_2$ architecture of *de novo* protein design. The single peptide monomer, (α-ℓ-α), contains two antiparallel helical regions connected by a flexible loop region which self-assemble in solution to yield the four helix bundle. These (α-ℓ-α) monomers can also be linked via an N-terminal disulfide to yield monomeric four helix bundles of the generic (α-ℓ-α-SS-α-ℓ-α) fold similar to those of Choma *et al.* (*126*). The reorganization of the peptide backbone in (α-ℓ-α)$_2$ has been shown not to interfere with the binding or spectroelectrochemical properties of the heme as long as the constellation of amino acids local to the heme binding sites remains invariant. Furthermore, the flexible

FIG. 8. Molecular model of Protein 86 heme complex showing the proposed His–Met ligation sphere. Reprinted from Ref. (*137*); copyright 1997 with permission of Cambridge University Press.

loop region provides a site for the incorporation of other inorganic cofactors, such as [4Fe-4S] clusters (*136*), which are typically not observed within helical bundles in biological systems.

In a novel approach, Rojas *et al.* (*137*) studied the binding of heme to a combinatorial library of four-α-helix bundles designed based simply on the pattern of hydrophobic and hydrophilic amino acids (*138*). The use of histidine and methionine residues as hydrophobic amino acids in the "binary code" strategy for *de novo* protein design provided a series of 30 distinct four-α-helix bundle sequences containing potential ligands with the hydrophobic cores. Despite the *a priori* lack of precise heme binding site design, the presence of multiple histidine and methionine residues provides appreciable heme affinity for fully half of the peptides studied; however, as observed by Choma *et al.* (*126*), the mere presence of ligands does not elicit heme binding. The most fully characterized heme protein of the library, protein 86 shown in Fig. 8, bound heme tightly, K_d of 740 nM, and has resonance Raman spectra consistent with histidine/methionine ligation. However, the presence of four methionine residues and five histidines complicates the precise identification of the liganding residues, which were deduced from molecular models and remain to be confirmed in future detailed experiments.

In an exciting new development, Isogai *et al.* (*139*) introduced the initial design of a globin fold using a computer-aided protein design algorithm. The resultant 153 amino acid protein was designed based on the myoglobin backbone fold, but contained only 26.1% sequence identity to the natural protein. Consistent with the design, the characterization of the recombinately expressed apoprotein, DG1, demonstrated that it was monomeric and had secondary structure contents, a radius of gyration and a global stability similar to those of myoglobin. DG1 bound iron(III) heme with a K_d of 80 nM, but the reduced spectra demonstrated the presence of some bis-histidine ligation to the ferrous iron as well as the designed mono-histidine ligation motif found in myoglobin. Clearly, the data validate the computational algorithm for *de novo* protein design and demonstrate a novel and significant global protein fold for future heme protein design.

C. TASP Scaffolds

Similar to their peptide–heme counterparts, several groups have utilized covalent linkages between porphyrins and peptides to construct heme–protein moieties. Since the covalent linkage restricts the entropy of the system and may template the helical structure, these covalent constructs are termed template assisted synthetic proteins, TASPs. Åkerfeldt *et al.* (*140*) utilized a C_4 symmetric tetrakis(carboxylphenyl)-porphyrin derivative to covalently attach a hydrophobic α-helical sequence, $(LSLBLSL)_3$, where B is α-aminoisobutyric acid, to yield a membrane spanning, voltage independent proton channel. The use of the template porphyrin not only increased the conductance lifetime from <0.2 to 5 ms, it virtually eliminated the voltage dependence observed with the parent peptide alone, demonstrating the considerable structural utility of the template.

A membrane peptide TASP based on the Mn(II) complex of tetrakis(*o*-aminophenyl) porphyrin bound to hydrophobic helices containing a covalently linked flavin was designed by Mihara *et al.* as an artificial electron transfer system (*141*). The Zn(II) complex of this porphyrin template was later elaborated with four identical helices containing histidine residues for heme ligation by Ushiyama *et al.* (*142*). The presence of the covalently attached porphyrin induced helical structure in the peptides and provided for heme binding as designed. However, the binding of hemes caused the aggregation of the heme–protein system into a complex containing three TASPs based on three Zn(II)porphyrins, and nine hemes despite 12 heme binding sites per complex. This report

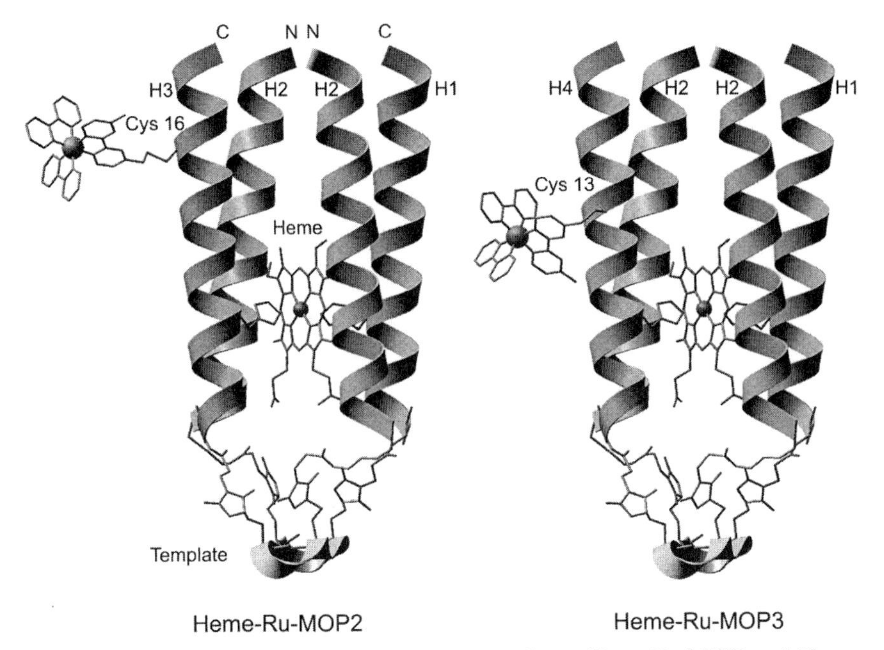

FIG. 9. Molecular models of the TASP heme complexes, Heme-Ru-MOP2 and Heme-Ru-MOP3 (*150*). Copyright 1998 National Academy of Sciences, USA.

clearly demonstrates the complicating influence that heme ligation can exert over peptide secondary, tertiary, and quaternary structure.

Haehnel and Rau made use of a cyclic peptide template to synthesize well-behaved four-α-helix bundle architectures composed of different helices in an antiparallel arrangement (*143*). Using the orthogonal protecting groups available in SPPS, Rau *et al.* (*143*) initially synthesized a four-α-helix bundle designed as a model for the core of the cytochrome b_6f and bc_1 complexes that shows sequence similarity to the maquettes of Robertson *et al.* (*127*). Further elaboration of their synthetic methodology has yielded four-α-helix bundle TAPSs with three different helices for the study of photoinitiated electron transfer reactions as shown in Fig. 9. The recent extension to combinatorial synthesis of heme–protein TASPs provides a wealth of designed heme proteins (*144*).

D. HEMES WITH OTHER COFACTORS

The presence of multiple different inorganic cofactors in a protein is a common theme in natural metalloproteins, best exemplified by the respiratory complexes, which may contain as many as 10 redox active cofactors within the protein matrix (*145*). The design of complex metallopeptides, containing more than one type of biochemical cofactor, represents an area where *de novo* design potentially offers greater flexibility than traditional small molecule coordination complexes if the engineering principles can be elucidated. Several key reports indicate the feasibility of fabricating multicofactor metalloproteins using *de novo* design. Two construction methods are currently available to the multicofactor protein designer: self-assembly and covalent attachment. High-yield self-assembly reactions allow for the incorporation of both cofactors using of selective metal ion ligation motifs, whereas the covalent attachment of one or both of the cofactors to the peptide scaffold obviates the necessity for designed metal ion specificity. The following section outlines two examples of each methodology illustrating the successful design of complex metalloproteins.

Gibney *et al.* (*136*) illustrated the feasibility of combining different bioinorganic cofactors within the same maquette scaffold using the inherently different coordination geometric and ligand preferences of hemes and iron–sulfur clusters. Building on the success of the four heme binding, four-α-helix bundles, a cysteine rich $[4Fe–4S]^{2+/+}$ binding site was designed into the loop region of a helix–loop–helix peptide using insight drawn from natural ferredoxin proteins. The resulting helix–loop–helix peptide self-associated in solution to yield a four-helix bundle capable of binding two $[4Fe–4S]^{2+/+}$ clusters and four hemes, as shown in Fig. 10. In addition to the construction of this complex oxidoreductase maquette, the authors have since utilized the 16 amino acid ferredoxin maquette, the excised loop region, to investigate the minimal protein design principles critical to $[4Fe–4S]^{2+/+}$-peptide assembly and stabilization in aqueous solution (*146, 147*). Thus, the design and construction of complex oxidoreductase maquettes can be achieved using the intrinsic ligand and geometric preferences of the transition metal ions.

Identical ligation motifs may also be used for the binding of different cofactors if the kinetics of cofactor exchange are appropriate. Using a diheme maquette scaffold, Gibney *et al.* (*148*) have constructed a maquette with two different types of heme, heme *b* and heme *a*, as an initial model complex for the cytochrome ba_3 type oxidases. Despite the equivalence of the heme binding sites, related by a twofold symmetry axis, addition of 1 equiv of either heme type followed by addition of 1 equiv of the other heme type results in the exclusive formation of the mixed heme maquette as evidenced by detailed electrochemistry. Any entropic driving force for the rearrangement of the hemes to yield a

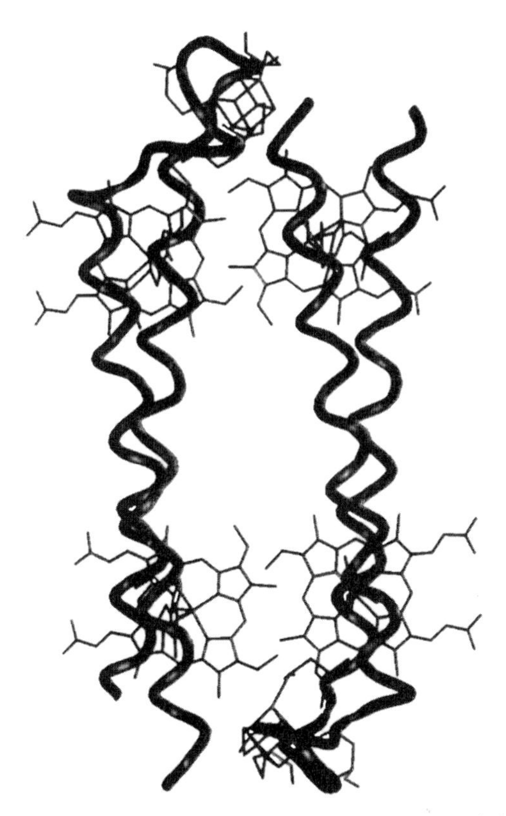

FIG. 10. Molecular model of the ferredoxin–heme maquette containing four hemes and two [4Fe–4S] clusters (*136*). Copyright 1996 National Academy of Sciences, USA.

statistical mixture of heme type combinations in the proteins is effectively combated by the tight binding K_d values and slow dissociation kinetics of the hemes. The simple order of addition can direct the formation of mixed cofactor metalloproteins in cases with slow dissociation kinetics.

In contrast to the use of self-assembly reactions and metal ion coordination preferences to direct the construction of mixed cofactor systems, the use of SPPS or selective chemical ligation allows for the direct covalent attachment of cofactors for the construction of mixed cofactor systems within *de novo* design. Figure 11 shows the flavocytochrome maquette constructed by Dutton and co-workers (*149*) using a flavin moiety covalently attached to a unique cysteine residue inside a four helix bundle with bis-histidine binding sites for heme

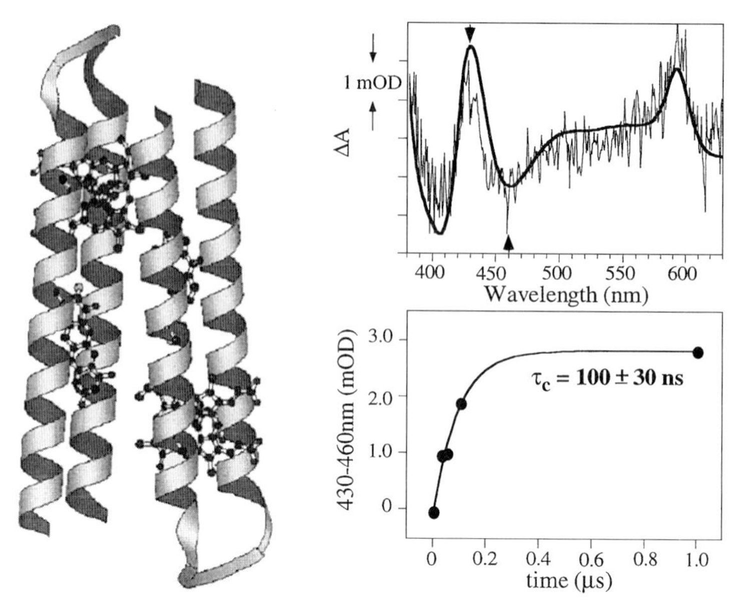

FIG. 11. Molecular model and electron transfer kinetics of the flavocytochrome maquette containing two bis-histidine ligated hemes and two covalently bound flavins (*149*). Copyright 1998 National Academy of Sciences, USA.

incorporation. The direct covalent attachment of the flavin obviated the necessity for the construction of a specific flavin binding domain, thus simplifying the construction of the designed protein. The utility of this type of system for the constructive study of biological electron transfer was illustrated by the photoinduced electron transfer between the flavin and a bound heme of suitable reduction potential, 1-methyl-2-oxomesoheme XIII (E_{m8} of -30 mV vs NHE), on the 100-ns time scale.

In a contemporaneous report, Rau *et al.* (*150*) illustrated the construction of two TASPs, MOP2 and MOP3 shown in Fig. 9, containing pendent ruthenium tris-bipyridyl complexes designed to act as photoinitiatior for electron transfer to the heme bound in the interior of the bundle. Although the electron transfer kinetics could not be extracted because of complicating parallel processes, the data clearly demonstrate that the electron transfer is faster in Heme-Ru-MOP3, which has the shorter modeled ruthenium tris-bipyridyl to heme distance in accordance with electron transfer theory (*129*). These two examples of electron transfer donor-to-heme systems illustrate the benefits of using a constructive protein design approach to begin to test hypotheses about the fundamental factors involved in biological electron transfer.

Although the direct covalent attachment of cofactors obviates the necessity for selective cofactor binding sites or appropriate binding kinetics for the construction of mixed cofactor systems, it is not incompatible with them leaving room in the future for the design of more elaborate multicofactor systems, as will be needed to make models of more complex metalloproteins such as the respiratory complexes, which contain as many as 10 redox cofactors. The design of such systems will undoubtedly provide keen insight into the structure and function of complex metalloproteins that cannot be easily accessed using bioinorganic complexes in isotropic solution, demonstrating a clear advantage of *de novo* metalloprotein design.

IV. Physical/Electrochemical Studies of *de Novo* Designed Heme Proteins

The heme moiety provides *de novo* designed heme proteins with an intrinsic and spectroscopically rich probe. The interaction of the amide bonds of the peptide or protein with the heme macrocycle provides for an induced circular dichroism spectrum indicative of protein–cofactor interactions. The strong optical properties of the heme macrocycle also make it suitable for resonance Raman spectroscopy. Aside from the heme macrocycle, the encapsulated metal ion itself provides a spectroscopic probe into its electronic structure via EPR spectroscopy and electrochemistry. These spectroscopic and electrochemical tools provide a strong quantitative base for the detailed evaluation of the relative successes of *de novo* heme proteins.

A. Induced Circular Dichroism Spectropolarimetry of the Heme Group

Monomeric hemes possess a mirror plane and are hence achiral (*151*). Incorporation of the heme macrocycle into the anisotropic protein matrix distorts the heme environment, inducing a circular dichroism spectrum (*57, 152, 153*). From the design standpoint, the presence of an induced heme CD spectrum qualitatively confirms intimate communication between the heme and the surrounding protein matrix, which indicates the heme is most likely specifically bound. This spectroscopic signature serves as a first indication that the heme resides within the designed protein scaffold and has been used by various groups to confirm the design concept (*108, 115, 122, 142, 154, 155*). Additionally, Gibney *et al.* (*135*) utilized induced heme-CD to characterize the binding of exogenous ligands to the Fe(III) and Fe(II) hemes within the maquette.

B. Resonance Raman Spectroscopy

Resonance raman (rR) spectroscopy of heme proteins is highly sensitive to the structure of the macrocycle and its interactions with the anisotropic protein matrix (*156–158*). Since the method relies on the optical properties of the heme macrocycle rather than the iron oxidation state, it can yield information in both heme oxidation states. Resonance Raman spectroscopy has been utilized by a variety of groups to demonstrate the success of the initial design concept by illustrating that the rR spectrum is consistent with the intended design (*116, 139, 159–161*). In a detailed rR study of the various hemes in the $[H10A24]_2$ maquette scaffold, Kalsbeck *et al.* (*162*) demonstrated that the ketones of the bound heme variant, 1-methyl-2-oxomesoheme XIII, reside in a local environment more hydrophobic than CH_2Cl_2 or DMSO, illustrating the presence of a low dielectric peptide interior. Additionally, the rR of heme in $[H10A24]_2$, containing ferric protoheme III, showed vibrational bands consistent with the existence of one high-spin five-coordinate heme due to steric interactions between the vinyl groups of juxtaposed hemes. Thus, the rR data clearly indicate the presence of a sterically constrained low dielectric hydrophobic core in the maquette scaffold. Resonance Raman spectroscopy has also demonstrated its utility in assaying the combinatorial heme protein library of Rojas *et al.*, where it has identified the initial His/Met ligation motif in *de novo* heme protein design in the reduced state rR of Protein 86 (shown in Fig. 8) (*137*).

C. EPR Spectroscopy

Electron paramagnetic resonance (EPR) spectroscopy is one of the primary tools for elucidating the primary coordination sphere of ferric hemes. Whereas optical and CD spectroscopies investigate the iron ions via the porphyrin macrocycles, EPR spectroscopy directly probes the encapsulated metal ion. Since the ligand field strength of the strong-field porphyrin ligands is not great enough to force low-spin states of Fe(II) and Fe(III), the EPR spectra of Fe(III) hemes are indicative of the axial ligands bound to the metal (*163, 164*). In general, the widely utilized bis-histidine coordination motif of *de novo* heme protein design should yield low-spin Fe(III) with rhombic EPR spectrum *g*-values between 3.8 and 1.0, whereas high-spin hemes are generally five-coordinate displaying axial EPR spectra with $g_{\perp} \approx 6$ and $g_{\parallel} \approx 2.0$.

The vast majority of designed heme–peptide and heme–protein complexes yield rhombic low-spin ferric EPR spectra (typically observed *g*-values 2.95, 2.25. 1.54) consistent with their utilization of bis-histidine coordination of the heme iron (*125, 126, 162, 165*). These low-spin EPR spectra are indicative of bis-histidine ligated hemes with virtually parallel histidine planes (*166*). One detailed report of EPR and ENDOR (electron nuclear double resonance) spectra has demonstrated the parallel orientation of the histidine planes in a TASP-based diheme four-helix bundle, MOP, with a minor subpopulation of highly anisotropic low-spin ferric heme (1% of spins) with twisted or tilted histidine planes (*167*). Axial high-spin spectra (*g*-values of 6.0 and 2.0) consistent with pentacoordinate iron(III) has also been reported sometimes, consistent with the design (*159*) and sometimes consistent with loss of a single histidine ligand to form a peptide-bound five-coordinate heme (*116, 126*) or loss of both histidine ligands leading to heme dissociation from the designed heme binding site (*168*).

D. Electrochemistry

The characteristic reduction potential (E_m) of a cytochrome is a fundamental property critical to its biological function of mediating electron transfer. Since the reduction potential directly reflects the relative stabilization of the Fe(II) and Fe(III) states of the bound heme, the base electrochemical properties of cytochrome class are set by the porphyrin macrocycle and adjusted by the primary coordination sphere of the encapsulated iron ion and the protein environment. As such, the reduction potentials of natural cytochromes, which span a range of nearly a full volt (24 kcal/mol) (*169, 170*), demonstrate exceptional control of the Fe(II)/Fe(III) equilibrium. Thus, the reduction potentials of designed heme proteins potentially provide direct insight into the heme microenvironment as well as the success of their respective designs. However, despite the importance of the reduction potential in comparing designed proteins to each other and to natural cytochromes, many designed heme proteins that purport to be cytochrome models have not been studied electrochemically.

In the smaller peptide systems, only three reports containing heme peptide reduction potentials have been published; however, this limited data set reveals two important aspects of cytochrome design. Figure 12 shows that Huffman *et al.* (*125*) have convincingly illustrated a direct

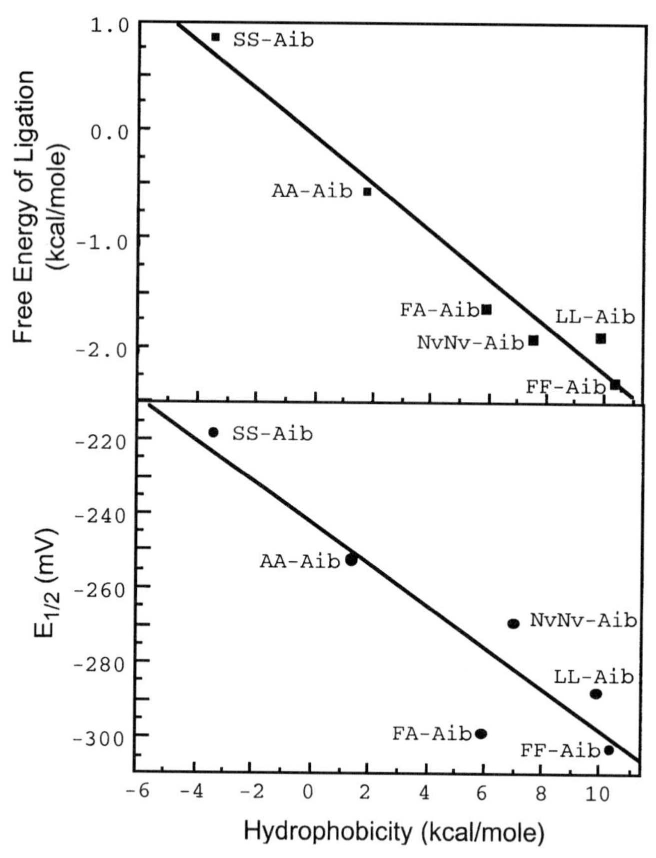

FIG. 12. Correlation between peptide hydrophobicity and both the free energy of ligation and equilibrium reduction potential in Fe(III)–coproporphyrin I–peptide complexes. Reprinted with permission from Ref. (*125*). Copyright 1998 American Chemical Society.

correlation between the reduction potential of the iron ion vs NHE and the relative hydrophobicity of their heme–peptide coordination complexes. This clearly shows that a more hydrophobic peptide has a higher affinity for ferric heme, a feature also observed by Sakamoto *et al.* (*120*), and yields a more negative reduction potential vs NHE. Kennedy *et al.* (*171*) demonstrate using a covalently linked heme in a PSM scaffold that stabilizing the designed peptide secondary structure, using aromatic–porphyrin interactions or an organic cosolvent, results in a lowering of the bis-histidine ligated heme reduction potential. Thus, stabilization of the protein fold, using hydrophobic amino acid–macrocycle interactions or an organic cosolvent, leads to lower heme reduction potentials in heme–peptide systems. Clearly, more data are required before general principles on how peptide systems establish heme reduction potentials can be extracted.

In contrast to the peptide systems, there exists a somewhat more robust literature on the electrochemistry of hemes in the larger protein systems that provides a wealth of insight into how proteins modulate the reduction potential of heme cofactors. A variety of groups have reported the reduction potential of various hemes in four-α-helix bundle architectures, which is beginning to reveal the fundamental factors that control heme reduction potentials.

The first general principle is that the type of metalloporphyrin incorporated into a designed protein or natural cytochrome offers a method to adjust the reduction potential of the heme. Sharp *et al.* (*149*) have demonstrated that a synthetic heme, 1-methyl-2-oxomesoheme XIII, incorporated into a designed heme protein has a reduction potential 90 mV higher than the same protein with heme *b*. Additionally, Gibney *et al.* (*148*) have illustrated that heme *a* has a reduction potential 160 mV higher than heme *b* in the identical protein scaffold. This heme-dependent effect provides protein designers with a predictable modulation of the heme reduction potential in a synthetic protein.

The second general principle that is evident from the literature is that incorporating a heme into a protein with a stably folded hydrophobic core invariably results in an elevation in the reduction potential as compared to the E_m of the bis-imidazole ligated heme in aqueous solution (−235 mV vs NHE). This E_m differential can be as great at 150 mV, representing up to 4 kcal/mol destabilization of the formally charged ferric form within the hydrophobic core. This finding directly contrasts with the result that incorporating a heme into the heme peptide systems results in a lowering of the E_m value relative to bis-imidazole ligated heme in aqueous solution, because of stabilization of the ferric iron by the strong field ligands. This simple comparison clearly delineates the fundamental difference between heme peptide systems and heme protein systems and indicates that principles learned from one type of design system may be obscured or overriden by different factors in other designed heme proteins.

The third principle observed in designed heme protein systems is that local electrostatics can have a significant impact on the measured heme reduction potential values. The initial heme protein maquette work demonstrated the elevation of heme reduction potentials using local charges, based on both amino acids, Arg^{+}, and adjacent oxidized hemes, $[Fe(III)(por)]^{+}$. In the case of heme-amino acid electrostatic interactions, Robertson *et al.* (*127*) concluded that the presence of an arginine four amino acids removed from the histidine ligand, -HEERL-, resulted in a 50 mV increase in E_m due to charge–charge repulsion that destabilizes the formally charged ferric heme. The second type of charge–charge electrostatic interaction is derived from pairs of adjacent oxidized hemes, Fe(III)-to-Fe(III) distance of $\approx$13 Å. This heme–heme interaction of 80–110 mV is transmitted through the low dielectric hydrophobic core of the protein scaffold and demonstrates heme–heme proximity, which has been utilized not only to determine the global topology of the oxidized diheme protein maquettes, but also to cause a massive protein conformational change (*172, 173*).

Local charge compensation of the formally charged Fe(III) heme, as discussed more fully in a later section, demonstrates a significant modulation ($\approx$210 mV) of the heme reduction potential in a designed heme protein. This scaffold-dependent effect has been shown to be additive to the heme-dependent effect of porphyrin peripheral architecture to demonstrate the modulation of a designed heme protein reduction potential by 450 mV using a single maquette scaffold.

These fundamental studies into the heme protein reduction potential provide protein designers not only with greater predictability in designing the reduction potential of designed cytochrome models, but also with guidelines and concepts to decipher the results to redox activity in combinatorial heme protein libraries (*137, 144*). Additionally, good control of heme reduction potentials opens new possibilities; for instance, this will be critical for the construction of air-stable ferrous hemes necessary for dioxygen transport functions or designing catalytic redox functions.

The rich spectroscopy and electrochemistry of the heme moiety yields a wealth of opportunities for the *de novo* heme protein design to evaluate the success of the heme binding site design. Combinations of these spectroscopic and electrochemical methods are elucidating the structure and function of *de novo* heme proteins and illustrating that they serve as excellent bioinorganic model complexes for simple cytochromes.

V. Functional Aspects

A. Helix Induction

The substantial free energy available in the binding of hemes into natural proteins is frequently harnessed to fold or stabilize the final protein structure as observed in the differences in the apo- (*77*) and holo-structures (*73*) of cytochrome b_{562} shown in Fig. 2. Although detailed thermodynamic studies of natural heme proteins are beginning to reveal the energetics of heme binding (*79, 174*), the use of smaller peptide systems that do not possess stably folded hydrophobic cores in the absence of bound heme allows for the detailed characterization of the effect of heme binding on protein folding as evidenced by secondary structure content. These studies done in concert with studies of natural heme proteins may help delineate the forces responsible for metalloprotein stabilization. Sucess will inevitably provide a fundamental understanding of the porphyrin macrocycle–peptide and ligand–metal energetics and will provide useful comparison with the better-developed energetics of protein–protein interactions of *de novo* design (*2*).

Benson and co-workers in a series of manuscripts have investigated the induction of helical structure within their designed peptide-sandwiched mesoheme (PSM) systems. Their initial report (*108*) illustrates the stabilization of the designed helical peptide fold by the covalent attachment of iron(III) mesoporphyrin IX, a variant of heme *b* with the vinyls reduced to ethyl groups. The apo-peptide existed as a random coil in solution, whereas after heme attachment the peptide was 52% helical and showed optical spectra consistent with bis-histidine ligation. The moderate helical content of the PSM could be increased to 97% helix in the presence of organic cosolvents such as trifluoroethanol or isopropanol that stabilize helical structure. Consistent with metal ion ligation-derived stabilization of the peptide fold, loss of the ligands by protonation of the histidine residues resulted in loss of helical content. Further studies have illustrated that other metalloporphyrins, for instance cobalt(III) porphyrins and symmetric iron(III) porphyrins, and other peptide architectures, such as the helix–disulfide–helix, can also be utilized to study helix induction. Appropriately placed aromatic amino acids that form hydrophobic interactions with the heme moiety analogous to those observed in ferric cytochrome b_{562} (*73*) can be used to stabilize the helix conformation (*155*).

The energetics of peptide–porphyrin interactions and peptide ligand–metal binding have also been observed in another self-assembly system constructed by Huffman *et al.* (*125*). Using monomeric helices binding to iron(III) coproporphyrin I, a fourfold symmetric tetracarboxylate porphyrin, these authors demonstrate a correlation between the hydrophobicity of the peptide and the affinity for heme as well as the reduction potential of the encapsulated ferric ion, as shown in Fig. 12. These data clearly demonstrate that heme macrocycle–peptide hydrophobic interactions are important for both the stability of ferric heme proteins and the resultant electrochemistry.

Mihara and co-workers (*120, 123*) have shown in a self-assembly system composed of a helix–disulfide–helix and an iron(III) mesoporphyrin

IX that heme ligation increases not only the helical content of the peptide scaffold but also the aggregation state. Furthermore, the observed increase in helicity, from 54% to 85%, could be reversed by the addition of cyanide ion, which inhibited histidine coordination. Additionally, in the absence of heme, the peptide exists as monomeric helix–disulfide–helix unit that forms a tetramer upon addition of 1 equiv of heme per binding site, demonstrating the influence that the binding of hemes can have on protein structure. This is further demonstrated by a report by Mihara in which a peptide that forms β-sheet aggregates converts to a tetramer of helix–disulfide–helix peptides upon the addition of a single equivalent of heme per binding site (*175*). These studies clearly indicate that iron porphyrin binding to peptides can exert a large influence on peptide secondary structure and aggregation state that, if controlled, can be used to stabilize desired structures.

B. Engineering Principles for Heme Protein Design

One of the unique advantages of the constructive methodology of *de novo* heme protein design is the direct evaluation of the engineering principles and tolerances in heme protein design. Although the literature demonstrates a variety of successful heme protein designs, the limits of this apparent success and the reasons for it are necessary to evaluate the relative impact of hydrophobic interaction and heme ligation energetics to improve future designs. Such studies may prove critical to our understanding of the design of heme proteins with tighter heme affinity, which may allow for the design of heme proteins with vacant coordination sites as models of peroxidases and oxygen transport proteins. Heme affinity clearly illustrates the relative failure of *de novo* heme protein designs, because the best current designs bind heme a factor of 10^5 more weakly than myoglobin (*176*), and some successful designs bind heme more weakly than does bovine serum albumin (*177*), an adventitious heme binding protein.

Growing from the original report by Williamson and Benson (*178*), there is consensus that the presence of aromatic residues one helical turn removed from the heme histidine ligand results in greater stabilization of the peptide architecture due to edge-to-face aromatic-porphyrin hydrophobic interations, which leads to tighter heme binding. This specific case of aromatic-porphyrin hydrophobic interactions, which has recently been demonstrated by H/D exchange rates as well as CD and NMR spectroscopy of the diamagnetic Co(III) PSM analogs (*155*), may be generalizable, as both Sakamoto *et al.* (*171*) and Huffman *et al.* (*125*) demonstrate significant correlations between the relative hydrophobicities of the peptides and heme binding constants, with peptides containing aromatics having the highest relative hydrophobicities and tightest heme binding constants. The hydrophobic interaction of the heme with these peptide systems is evident due to the lack of a stable hydrophobic core and is certainly obscured in the larger designed heme protein systems. Whether these heme macrocycle–aromatic interactions, commonly observed in natural heme proteins, can be utilized to optimize heme binding in larger protein systems where protein–protein hydrophobic interactions may interfere remains to be resolved with future research.

In terms of optimal ligand arrangement, Arnold *et al.* (*159*) and Gibney and Dutton (*168*) have evaluated their initial designs to demonstrate the engineering tolerances of their respective systems. In the peptide sandwich mesoheme system, reducing the spacing between the covalent attachment site, lysine, and the heme ligand, histidine, leads to increased springboard stain in the system (*159*). In two cases helix formation is incompatible with histidine ligation to the iron in the heme, which results in a bis-histidine liganded heme with a random coil peptide conformation rather than a helical protein without bis-histidine heme ligation, illustrating the comparative energetics of ligation to the heme and peptide folding. Additionally, the lack of optimal design in the helical state is evidenced by the increased susceptibility of these PSMs to proton competition, and by the presence of more high-spin Fe(III) in their EPR spectra and UV-vis spectra consistent with the loss of a ligand in the ferrous state to form five-coordinate PSMs. Thus, these PSM studies illustrate the control of protein conformation by the coordination preferences of the transition metal ion.

In a larger protein system, Gibney and Dutton tested the initial heme protein maquette scaffold design by evaluating the heme binding properties of a series of related peptides with histidines in various positions (*168*). Moving the histidines through an entire heptad repeat validated the original design of a heptad **a**–heptad **a** bis-histidine heme binding site that bound heme 600-fold more tightly than any of the other bis-histidine sites studied. However, molecular models indicated that the second tightest heme binding protein in the series, $[H11A24]_2$, which bound a single heme with a 9.5 μM K_d, required significant deformation of the helices to bind the heme. Thus, these two reports illustrate that the energetics of metal–ligand bonding in heme binding may in some cases compensate for nonoptimal designs. Furthermore, these studies demonstrate the rather narrow engineering tolerances of these *de novo* designed protein architectures, where moving an amino acid one position away leads to compromised heme binding.

C. Dioxygen Reactivity

The interaction of natural heme proteins with dioxygen is critical to aerobic life in that it provides for both the capture and transport of dioxygen to tissues as well as the $4e^-/4H^+$ reduction of dioxygen in terminal oxidases requisite for life. Although vital to the survival of aerobic organisms, the design of a dioxygen binding protein represents a daunting challenge to the *de novo* heme protein designer. Recently, small molecule inorganic coordination compounds have been designed that perform both the reversible dioxygen binding and the $4H^+/4e^-$ reduction of dioxygen to water in isotropic solution (*179, 180*); however, designed protein systems are in their infancies with respect to this synthetic challenge, as they both require an open coordination site at the heme for the binding of dioxygen.

The interaction of dioxygen has been observed in several systems, mostly due to autooxidation of ferrous hemes with dioxygen, but only characterized in a few instances. Sakamoto *et al.* (*119*) have illustrated peroxidase-type activity using a helix–disulfide–helix system that binds a single heme as shown in Fig. 13. The initial communication illustrated that the addition of an organic cosolvent, trifluoroethanol, increases the helical content of the peptide, the affinity for heme (1.7 μM K_d at maximal affinity, 15% TFE), and the peroxidase activity (conversion of

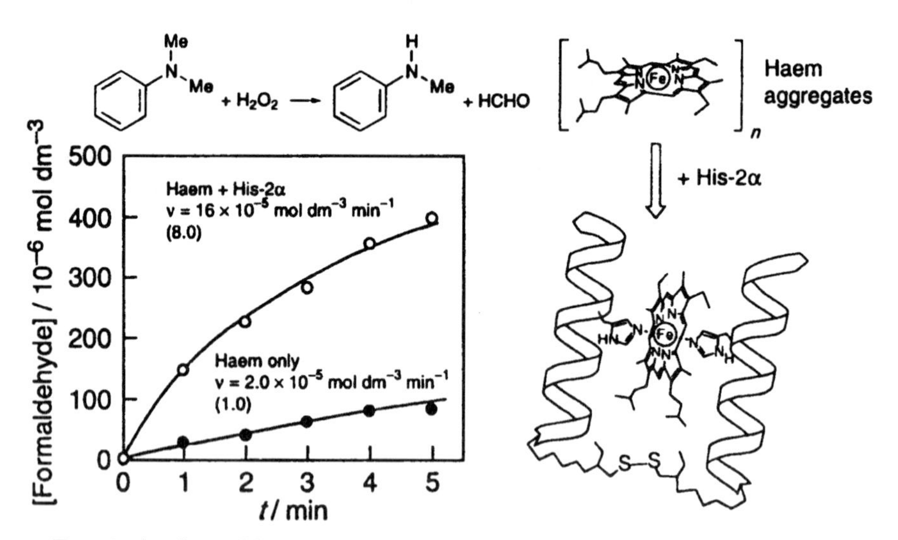

Fig. 13. Accelerated kinetics of formaldehyde formation observed in the presence of His-2α compared the free hemin in 100 mM Tris·HCl buffer (pH 7.4) with 15% (v/v) TFE (*119*). Reproduced with permission of The Royal Society of Chemistry.

N,N-dimethylanaline into formaldehyde), eightfold higher than those of heme in isotropic solution. Furthermore, the reaction was characterized to be rate limited by the interaction of hydrogen peroxide with the heme. In a more recent design series, an inverse correlation was demonstrated between heme affinity and peroxidase-type reaction, conversion of *o*-methoxyphenol into tetraguaiacol. These data, and those of Nastri *et al.* (*116*), illustrate that tightly bound six-coordinated hemes are incompetent for peroxidase activity and indicate that future designs of five-coordinate heme–proteins or heme–peptides, such as the monopeptide PSM analogs synthesized by Arnold *et al.* (*159*), may hold promise for developing designed heme based biocatalysts.

D. Proton Coupled Electron Transfer

The oxidation/reduction of redox cofactors in biological systems is often coupled to proton binding/release either at the cofactor itself or at local amino acid residues, which provides the basic mechanochemical part of a proton pump such as that found in cytochrome *c* oxidase (*95*). Despite a thermodynamic cycle that provides that coupling of protonation of amino acids to the reduction process will result in a 60 mV/pH decrease unit in the reduction potential per proton bound between the pK_a values in the Fe(III) and Fe(II) states, the essential pumping of protons in the respiratory complexes has yet to be localized within their three-dimensional structures.

Using a heme protein maquette with a single heme bound, Shifman *et al.* (*160*) investigated the amino acid side chain responsible for the observed redox–Bohr effect, which resulted in a 210-mV change in the reduction potential between pH 4 and 11 as illustrated by Fig. 14. The stability of the maquette over the wide pH range studied was critical to these studies, as was the robustness of the scaffold, which withstood various alterations without changing aggregation state or ability to bind heme, further illustrating the utility of *de novo* designed heme proteins. The majority of the proton/electron coupling was localized to a single glutamate with a pK_a^{ox} of 4.25 and a pK_a^{red} of 7.0, regardless of the number or type of heme bound, one or two hemes *b* or its dimethyl ester derivative. Studies of glutamate ($R–COO^-$) to glutamine ($R–CONH_2$) variants of this protein localized the majority of the effect to Glu11, one amino acid from the histidine heme ligand H10. The minor effect (60 mV) observed at high pH (>8) was assumed to be localized to lysine residues. Since the proton coupling is a function of the scaffold and not the iron porphyrin ligated as shown by Shifman *et al.* (*181*) and many of the designed heme proteins based on DeGrado's α_2 architecture have

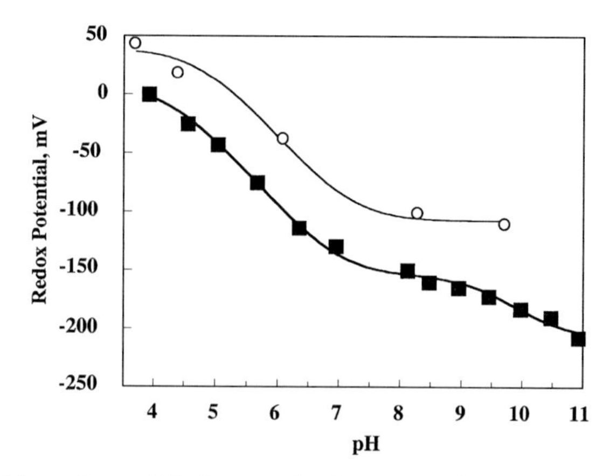

FIG. 14. pH dependence of the heme reduction potential vs NHE in the heme protein maquette, $[H10A24]_2$, demonstrating the 210 mV range of E_m observed between pH 4 and 11. Data for Fe(protoporphyrin IX, filled squares) and Fe(protoporphyrin IX dimethyl ester, open circles) are shown. Reprinted with permission from Ref. (*160*); copyright 1998 American Chemical Society.

a glutamate residue adjacent to the iron ligating histidine, analogous to the H10E11 sequence in the maquette, these other designs may also have similar pH-dependent E_m values, demonstrating the importance of specifying the pH at which the E_m is measured. Additionally, heme protein systems based on Baldwin-type helices may have different E_m/pH behavior, which might be uniform between the various designs. Consistent with the first assertion, a related heme maquette has identical pK_a values despite a lower reduction potential in solution (*182*).

E. MATERIALS SCIENCE APPLICATIONS

The robust scaffolds of designed heme proteins, their redox tunability, and their modular nature are attractive properties for the rational design and construction of biomolecular materials that may include bioelectronic devices such as biosensors and biological memory storage devices. Engineering biomaterials requires a reliable methodology for the attachment of the redox protein to the desired surfaces in an ordered array and retention of the desired chemical and physical properties of the protein. The high fidelity coupling of redox active proteins onto electrode surfaces with high surface coverages in regular arrays has been approached using both self-assembled monolayer (SAM) and Langmuir–Blodgett (LB) methodologies. Although the chemical reactions requisite for SAM construction generally offer control of the heme protein surface orientation, recent research with LB films demonstrates methods to control the orientation of designed heme proteins in the final LB films using the attachment of hydrophobic tails and engineering the external electrostatic interactions. Once constructed, the retention of optical and electronic properties similar to those observed in bulk solution is critical to the eventual applications of these biomaterials. Advances in the rational design of protein–surface interactions coupled with control of the electronic properties of designed heme proteins in solution and translated onto surfaces reveals a fertile area of continuing research toward bioelectronic devices.

1. Langmuir–Blodgett Monolayers

The construction of ordered protein arrays on surfaces using Langmuir–Blodgett techniques has been extensively studied using natural electron transport proteins. The advantage of the LB technique resides in the fact that the proteins can be ordered prior to transfer to substrate using surface pressure and that the transfer process can be repeated to build up multilayer assemblies. In a pair of papers, Dutton and co-workers (*183, 184*) have initiated the use of LB techniques as a feasibility study into the design bioelectronic devices based on *de novo* designed heme proteins. Using a dimeric heme protein maquette scaffold and one modified with a N-terminal palmitoyl chain, Chen *et al.* observed the dissociation of the dimeric protein scaffold into di-α-helical peptide monomers on the air–water interface of the Langmuir trough. Interestingly, these dihelix peptides retained their secondary structure and heme binding ability after transfer to quartz or CaF_2 substrates. Linear dichroism of the hemes showed they were well ordered with a 40° angle between the heme plane and the substrate surface. Furthermore, the palmitoylated heme protein maquette formed a two-dimensionally ordered array when transferred at high surface pressures that exhibited properties of a polarizer, as illustrated in panel D of Fig. 15.

Utilizing both a novel *de novo* designed heme protein maquette architecture and the pattern of external electrostatic charges on the heme protein maquette, Chen *et al.* (*184*) have been able to engineer the transfer of an intact four-helix-bundle scaffold onto an LB film with the desired final orientation. Despite the use of the monomeric (α-ℓ-α-SS-α-ℓ-α) heme protein maquette architecture, the forces at the air–water interface still dissociated the hydrophobic core of the bundle, exposing the interior to the air. Only when the external electrostatic charges were radically redesigned by placing all negatively charged glutamates on one helix with all positively charged lysine residues on the adjacent helix, (α^+-ℓ-α^--SS-α^+-ℓ-α^-), did the protein retain the four-helix-bundle

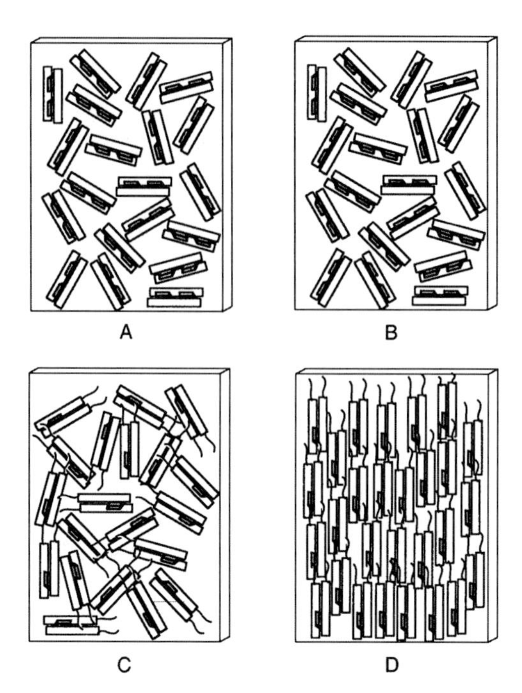

FIG. 15. Representation of the molecular orientations of LB films of heme maquettes. Transfer of the N-terminal palmitoylated maquette at high pressures, panel D, leads to a film with polarizer properties. Reprinted with permission from Ref. (*183*); copyright 1998 American Chemical Society.

structure on the air–water interface; thus, the additional external electrostatic stabilization was able to buttress the structure against opening on the air–water interface. In the apo state, this maquette transferred to the substrate with helices parallel to the substrate, whereas in the holo state this maquette transferred perpendicular to the substrate. Thus, these studies demonstrate the rational engineering of protein films using LB techniques as a prelude to the assembly of bioelectronic devices.

2. *Self-Assembled Monolayers*

The use of self-assembled monolayers of proteins by direct chemical attachment of the peptide to a modified surface obviates some of the protein orientation difficulties observed in the construction of LB monolayers. An initial report of the construction of peptide SAMs on silanized quartz was published in early 1998 (*185*). Pilloud *et al.* have utilized the covalent ligation of protein cysteine thiols to thiol-terminated silanized

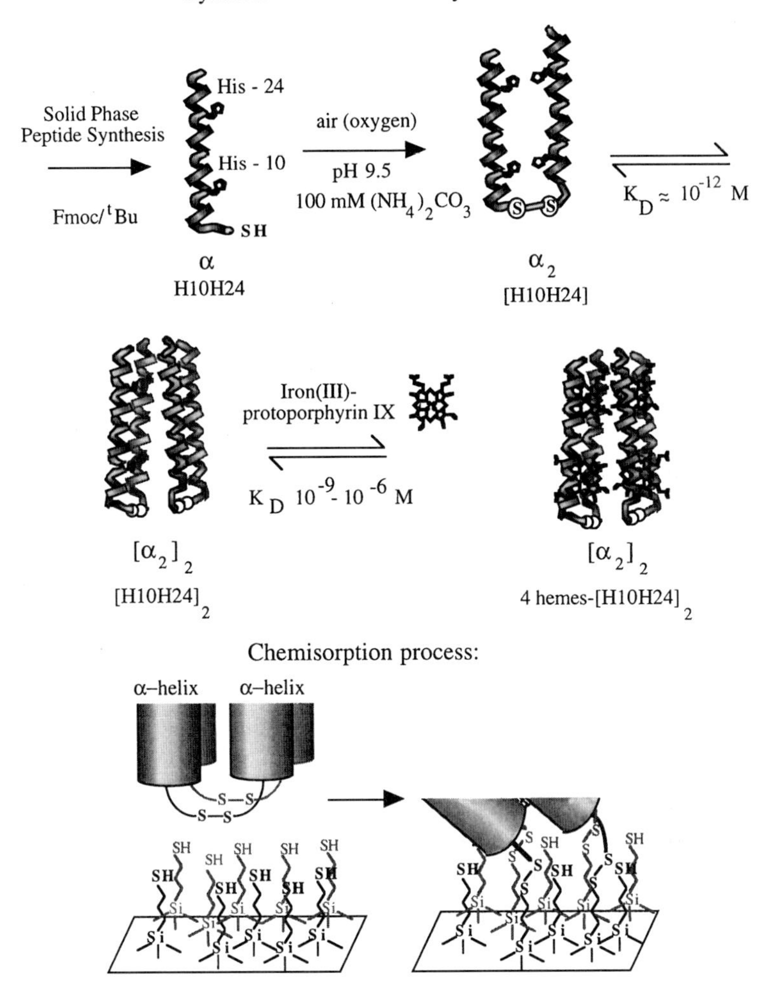

FIG. 16. Schematic representation of the assembly of designed heme protein SAMs on silanized quartz substrates. Designed peptides are synthesized, homodimerized, and self-associate to form four-helix bundles prior to heme incorporation, followed by chemisorption on prepared quartz surfaces. Reprinted with permission from Ref. (*185*); copyright 1998 American Chemical Society.

quartz substrates to assemble self-assembled monolayers (SAM) of heme protein maquettes as shown in Fig. 16. This SAM construction method yields a high surface coverage of active thiols (675 pm/cm^2) that react to give close packed of α-helices (110 pm/cm^2) with a $41 \pm 2^\circ$ tilt of the

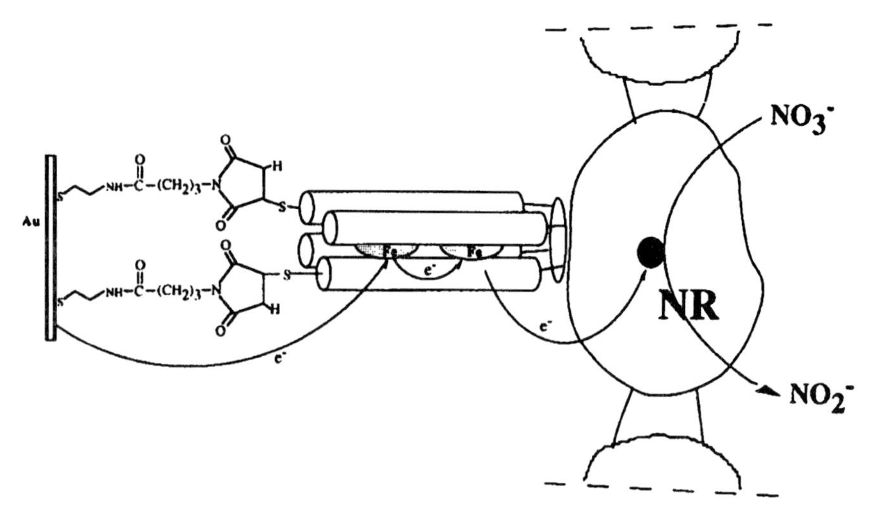

FIG. 17. Bioelectric catalysis on a designed heme protein modified Au electrode. Schematic electron transfer. Reprinted with permission from Ref. (*187*); copyright 1999 American Chemical Society.

heme with respect to the surface as measured using linear dichroism, similar to the LB results of Chen *et al.* Most notably, the secondary structure of the helical peptides is preserved on the surface, the optical spectra of the hemes is strikingly similar to that observed in solution in both Fe(III) and Fe(II) forms, and the heme protein SAMs retain their ability to bind exogenous diatomics, laying the foundation for future biosensor design in a highly reproducible and reversible protein SAM construct.

Later in 1998, Katz *et al.* utilized an *N*-succinimidyl-3-maleimidopropionate coupled to a cysteamine SAM on gold to covalently attach a modified form of MOP1, the template assisted heme-binding four-helix bundle (*186*). In comparison to the SAM of Pilloud *et al.*, which are covalently linked only to the surface, the presence of the cyclic template at one end of the bundle and the surface attachment on the other yields a significantly rigidified system. The MOP1 SAM could be reconstituted with heme, and the electrochemical properties of the hemes were nearly identical to those observed in solution. Coulometric analyses of the heme redox waves indicate a surface coverage of 25 pm/cm^2 (equivalent to 100 pm/cm^2 of helices, as observed by Pilloud *et al.*), close to the theoretical limit of 27 pm/cm^2, indicating a densely packed heme protein monolayer. Chronoamperometry illustrated the differential rates of electron transfer from the gold electrode to the first heme (40 s^{-1})

and the second heme (1 s^{-1}) consistent with electron transfer theory (*129*). This heme protein SAM was utilized as a redox mediator between the gold electrode and nitrite reductase, which was crosslinked on the surface, for the electrocatalytic reduction of nitrite (NO_3^-) to nitrate (NO_2^-) as shown in Fig. 17. Additionally, the heme protein SAM was utilized for the hydrogenation of acetylene dicarboxylic acid at cobalt(II) protoporphyrin IX substituted myoglobin (*187*). These exciting results illustrate the potential of *de novo* designed heme protein SAMs for biocatalysis on designed surfaces.

ACKNOWLEDGMENT

We acknowledge the National Institutes of Health for funding our research endeavors (GM41048 and GM48130 to P.L.D.).

REFERENCES

1. DeGrado, W. F.; Summa, C. M.; Pavone, V.; Nastri, F.; Lombardi, A. *Annu. Rev. Biochem.* **1999,** *68,* 779–819.
2. Bryson, J. W.; Betz, S. F.; Lu, H. S.; Suich, D. J.; Zhou, H. X.; O'Neil, K. T.; DeGrado, W. F. *Science* **1995,** *270,* 935–941.
3. Merrifeld, R. B.; Stewart, J. M. *Nature* **1965,** *207,* 522–523.
4. Dawson, P. E.; Muir, T. W.; Clark-Lewis, I.; Kent, S. B. H. *Science* **1994,** *266,* 776–779.
5. Tam, P.; Lu, Y.-A.; Lui, C.-F.; Shao, J. *Proc. Natl. Acad. Sci. USA* **1995,** *92,* 12485–12489.
6. Smith, M. J. *Nature* **1970,** *225,* 563–564.
7. Cordes, M. H. J.; Davidson, A. R.; Sauer, R. T. *Curr. Opin. Struct. Biol.* **1996,** *6,* 3–10.
8. O' Neil, K.; DeGrado, W. F. *Science* **1990,** *250,* 246–250.
9. Padmanabhan, S.; Marqusee, S.; Ridgeway, T.; Laue, T. M.; Baldwin, R. L. *Nature* **1991,** *344,* 268–270.
10. Kim, C. A.; Berg, J. M. *Nature* **1993,** *362,* 267–270.
11. Minor, P. L.; Kim, P. S. *Nature* **1994,** *371,* 264–267.
12. Hecht, M. H.; Richardson, J. S.; Richardson, D. C.; Ogden, R. C. *Science* **1990,** *249,* 884–891.
13. Fezoui, Y.; Connolly, P. J.; Osterhout, J. J. *Protein Sci.* **1997,** *6,* 1869–1877.
14. Lovejoy, B.; Choe, S.; Cascio, D.; McRorie, D. K.; DeGrado, W. F.; Eisenberg, D. *Science* **1993,** *259,* 1288–1293.
15. Ogihara, N. L.; Weiss, M. S.; DeGrado, W. F.; Eisenberg, D. *Protein Sci.* **1997,** *6,* 80–88.
16. Walsh, S. T. R.; Cheng, H.; Bryson, J. W.; Roder, H.; DeGrado, W. F. *Proc. Natl. Acad. Sci. USA* **1999,** *96,* 5486–5491.
17. Schafmeister, C. E.; Miercke, L. J. W.; Stroud, R. M. *Science* **1993,** *262,* 734–738.
18. Schafmeister, C. E.; LaPorte, S. L.; Miercke, L. J. W.; Stroud, R. M. *Nature Struct. Biol.* **1997,** *4,* 1039–1046.

19. Hill, R. B.; DeGrado, W. F. *J. Am. Chem. Soc.* **1998,** *120,* 1138–1145.
20. Skalicky, J. J.; Gibney, B. R.; Rabanal, F.; Bieber–Urbauer, R. J.; Dutton, P. L.; Wand, A. J. *J. Am. Chem. Soc.* **1999,** *121,* 4941–4951.
21. Kortemme, T.; Ramirez-Alvarado, M.; Serrano, L. *Science* **1998,** *281,* 253–256.
22. Struthers, M. D.; Cheng, R. P.; Imperiali, B. *Science* **1996,** *271,* 342–345.
23. Dahiyat, B. I.; Mayo, S. L. *Science* **1997,** *278,* 82–7.
24. Jiang, X.; Bishop, E. J.; Farid, R. S. *J. Am. Chem. Soc.* **1997,** *119,* 838–839.
25. Malakauskas, S. M.; Mayo, S. L. *Nat. Struct. Biol.* **1998** *5,* 470–475.
26. Jiang, X.; Farid, H.; Pistor, E.; Farid, R. S. *Protein Sci.* **2000,** *9,* 415–428.
27. Lu, Y.; Valentine, J. S. *Curr. Opin. Chem. Biol.* **1997,** *7,* 495–500.
28. Gibney, B. R.; Rabanal, F.; Dutton, P. L. *Curr. Opin. Chem. Biol.* **1998,** *1,* 537–542.
29. Hellinga, H. W. *Fold. Res.* **1998** *3* R1–R8.
30. Pinto, A. L.; Hellinga, H. W.; Caradonna, J. P. *Proc. Natl. Acad. Sci. USA* **1997,** *94,* 5562–5567.
31. Ghadiri, M. R.; Choi, C. *J. Am. Chem. Soc.* **1990,** *112,* 1630–1632.
32. Regan, L.; Clarke, N. D. *Biochemistry* **1990,** *29,* 10878–10883.
33. Dieckmann, G. D.; McRorie, D. K.; Tierney, D. L.; Utschig, L. M.; Singer, C. P.; O'Halloran, T. V.; Penner-Hahn, J. E.; DeGrado, W. F.; Pecoraro, V. L. *J. Am. Chem. Soc.* **1997,** *119,* 6195–6196.
34. Hay, M.; Richards, J. H.; Lu, Y. *Proc. Natl. Acad. Sci. USA* **1996,** *93,* 461–464.
35. Mutz, M. W.; McLendon, G. L.; Wishart, J. F.; Gaillard, E. R.; Corin, A. F. *Proc. Natl. Acad. Sci. USA* **1996,** *93,* 9521–9526.
36. Rabanal, F.; Gibney, B. R.; DeGrado, W. F.; Moser, C. C.; Dutton, P. L. *Inorg. Chim. Acta* **1996,** *243,* 213–218.
37. Ibers, J. A.; Holm, R. H. *Science* **1980,** *209,* 233–235.
38. Karlin, K. D. *Science* **1993,** *261,* 701–708.
39. Varadarajan, R.; Zewert, T. E.; Gray, H. B.; Boxer, S. G. *Science* **1989,** *243,* 69–72.
40. Lloyd, E.; King, B. C.; Hawkridge, F. M.; Mauk, A. G. *Inorg. Chem.* **1998,** *37,* 2888–2892.
41. Rodgers, R. R.; Sligar, S. G. *J. Am. Chem. Soc.* **1991,** *113,* 9419–9421.
42. Yeung, B. K.; Wang, X.; Sigman, J. A.; Petillo, P. A.; Lu, Y. *Chem. Biol.* **1997,** *4,* 215–221.
43. Wilcox, S. K.; Putman, C. D.; Sastry, M.; Blankenship, J.; Chazin, W. J.; McRee, D. E.; Goodin, D. B. *Biochemistry* **1998,** *37,* 16853–16862.
44. Low, D. W.; Hill, M. G. *J. Am. Chem. Soc.* **1998,** *120,* 11536–11537.
45. Meadows, K. A.; Parkes-Loach, P. S.; Kehoe, J. W.; Loach, P. A. *Biochemistry* **1998,** *37,* 3411–3417.
46. Cheng, R. P.; Fisher, S. L.; Imperiali, B. *Science* **1996,** *271,* 342–345.
47. Bodansky, M. "Peptide Synthesis: A Practical Approach"; Springer-Verlag: New York, 1983.
48. Dutton, P. L.; Wilson, D. F.; Lee, C.-P. *Biochemistry* **1970,** *9,* 5077–5082.
49. Yang, J.; Liu, X.; Bhalla, K.; Kim, C. N.; Ibrado, A. M.; Cai, J.; Peng, T. I.; Jones, D. P.; Wang, X. *Science* **1997,** *275,* 1129–1132.
50. Kluck, R. M.; Bossy-Wetzel, E.; Green, D. R.; Newmeyer, D. D. *Science* **1997,** *275,* 1132–1136.
51. Anfinsen, C. B. *Science* **1973,** *181,* 223–230.
52. Beinert, H.; Holm, R. H.; Munck, E. *Science* **1997,** *277,* 653–659.
53. As determined by a search of the European Bioinformatics Institute (http://www.ebi.ac.uk/) protein sequence databases.

54. Lemberg, R.; Barrett, J. "Cytochromes"; Academic Press: New York, 1973, p. 580.
55. Argos, P.; Mathews, F. S. *J. Biol. Chem.* **1975,** *250,* 747–51.
56. Fulop, V.; Moir, J. W.; Ferguson, S. J. *Cell* **1995,** *81,* 369–77.
57. Moore, G. R.; Pettigrew, G. W. "Cytochromes *c*"; Springer-Verlag: New York, 1990.
58. Karlin, S.; Zhu, Z.-Y.; Karlin, K. D. *Proc. Natl. Acad. Sci. USA* **1997,** *94,* 14225–14230.
59. Karlin, S.; Zhu, Z.-Y. *Proc. Natl. Acad. Sci. USA* **1997,** *94,* 14231–14236.
60. Karlin, S.; Zhu, Z.-Y.; Karlin, K. D. *Biochemistry* **1998,** *37,* 17726–17734.
61. Kassner, R. J. *Proc. Natl. Acad. Sci. USA* **1972,** *69,* 2263–2267.
62. Stellwagen, E. *Nature* **1978,** *275,* 73–74.
63. Tezcan, F. A.; Winkler, J. R.; Gray, H. B. *J. Am. Chem. Soc.* **1998,** *120,* 13383–13388.
64. Poulos, T. L. *Adv. Inorg. Biochem.* **1988,** *7,* 1–36.
65. Poulos, T. L.; Finzel, B. C.; Howard, A. J. *J. Mol. Biol.* **1987,** *195,* 687–699.
66. Dawson, J. H. *Science* **1988** *240* 433–439.
67. Blauer, G.; Sreerama, N.; Woody, R. W. *Biochemistry* **1993,** *32,* 6674–6679.
68. Shelnutt, J. A.; Song, X. Z.; Ma, J. G.; Jia, S. L.; Jentzen, W.; Medforth, C. J. *Chem. Soc. Rev.* **1998,** *27,* 31–41.
69. Churg, A. K.; Warshel, A. *Biochemistry* **1986,** *25,* 1675–1681.
70. Gunner, M. R.; Honig, B. *Proc. Natl. Acad. Sci. USA* **1991,** *88,* 9151–9155.
71. Weichsel, A.; Andersen, J. F.; Champagne, D. E.; Walker, F. A.; Montfort, W. R. *Nature Struct. Biol.* **1998,** *5,* 304–309.
72. Hallberg, B. M.; Bergfors, T.; Backbro, K.; Pettersson, G.; Henriksson, G.; Divne, C. *Structure with Folding & Design* **2000,** *8,* 79–88.
73. Lederer, F.; Glatigny, A.; Bethge, P. H.; Bellamy, H. D.; Mathews, F. S. *J. Mol. Biol.* **1981,** *148,* 427–448.
74. Finzel, B. C.; Weber, P. C.; Hardman, K. D.; Salemme, F. R. *J. Mol. Biol.* **1985,** *186,* 627– 643.
75. Itagaki, E.; Hager, L. P. *J. Biol. Chem.* **1966,** *241,* 3687–3695.
76. Barker, P. D.; Nerou, E. P.; Cheesman, M. R.; Thomson, A. J.; de Oliveira, P.; Hill, H. A. O. *Biochemistry* **1996,** *35,* 13618–13626.
77. Feng, Y.; Sligar, S. G.; Wand, A. J. *Nature Struct. Biol.* **1994,** *1,* 30–35.
78. Moore, C. D.; Al-Misky, O. N.; Lecomte, J. T. J. *Biochemistry* **1991,** *30,* 8357–8365.
79. Robinson, C. R.; Liu, Y.; Thomson, J. A.; Sturtevant, J. M.; Sligar, S. G. *Biochemistry* **1997,** *36,* 16141–16146.
80. Brems, D. N.; Stellwagen, E. *J. Biol. Chem.* **1983,** *258,* 3655–3660.
81. Elöve, G. A.; Bhuyan, A. K.; Roder, H. *Biochemistry* **1994,** *33,* 6925–6935.
82. Weber, P. C.; Salemme, F. R.; Matthews, F. S.; Bethge, P. H. *J. Biol. Chem.* **1981,** *256,* 7702–7704.
83. Arnesano, F.; Banci, L.; Bertini, I.; Ciofi-Baffoni, S.; Woodyear, T. D.; Johnson, C. M.; Barker, P. D. *Biochemistry* **2000,** *39,* 1499–1514.
84. Barker, P. D.; Nerou, E. P.; Freund, S. M. V.; Fearnley, I. M. *Biochemistry* **1995,** *34,* 15191–15203.
85. Barker, P. D.; Ferrer, J. C.; Mylrajan, M.; Loehr, T. M.; Feng, R.; Konishi, Y.; Funk, W. D.; MacGillivray, R. T. A.; Mauk, A. G. *Proc. Natl. Acad. Sci. USA* **1993** *90* 6542–6546.
86. Kendrew, J. C.; Bodo, G.; Dintzis, H. M.; Parrish, R. G.; Wyckoff, H. W.; Phillips, D. C. *Nature* **1958,** *181,* 662.
87. Kendrew, J. C.; Dickerson, R. E.; Strandberg, B. E.; Hart, R. G.; Davies, D. R.; Phillips, D. C.; Shore, V. C. *Nature* **1960,** *185,* 422–427.

88. Phillips, S. E. V. *J. Mol. Biol.* **1980,** *142,* 531–554.
89. Perutz, M. F.; Rossmann, M. G.; Cullis, A. F.; Muirhead, H.; Will, G.; North, A. C. T. *Nature* **1960,** *185,* 416–422.
90. Gong, W.; Hao, B.; Mansy, S. S.; Gonzales, J.; Gilles-Gonzales, M. A.; Chan, M. K. *Proc. Natl. Acad. Sci. USA* **1998,** *95,* 15177–15182.
91. Perutz, M. F.; Paoli, M.; Lesk, A. M. *Chem. Biol.* **1999,** *6,* R291–297.
92. Ortiz de Montellano, P. R. "Cytochrome P450: Structure, Mechanism and Biochemistry"; Plenum: New York, 1995; 2nd ed.
93. Sono, M.; Roach, M. P.; Coulter, E. D.; Dawson, J. H. *Chem. Rev.* **1996,** *96,* 2841–2887.
94. Schlichting, I.; Berendzen, J.; Chu, K.; Stock, A. M.; Maves, S. A.; Benson, D. E.; Sweet, R. M.; Ringe, D.; Petsko, G. A.; Sligar, S. G. *Science* **2000,** *287,* 1615–1622.
95. Ferguson-Miller, S.; Babcock, G. T. *Chem. Rev.* **1996,** *96,* 2889–2907.
96. Iwata, S.; Ostermeier, C.; Ludwig, B.; Michel, H. *Nature* **1995,** *376,* 660–669.
97. Tsukihara, T.; Aoyama, H.; Yamashita, E.; Tomizaki, T.; Yamaguchi, H.; Shinzawa-Itoh, K.; Nakashima, R.; Yaono, R.; Yoshikawa, S. *Science* **1995,** *269,* 1069–1074.
98. Kohn, W. D.; Mant, C. T.; Hodges, R. S. *J. Biol. Chem.* **1997,** *272,* 2583–2586.
99. Beasley, J. R.; Hecht, M. H. *J. Biol. Chem.* **1997,** *272,* 2031–2034.
100. Ho, S. P.; DeGrado, W. F. *J. Am. Chem. Soc.* **1987,** *109,* 6751–6758.
101. Munson, M.; O'Brien, R. O.; Sturtevant, J. M.; Regan, L. *Protein Sci.* **1994,** *3,* 2015–2022.
102. Munson, M.; Balasubramanian, S.; Fleming, K. G.; Nagi, A. D.; O'Brien, R. O.; Sturtevant, J. M.; Regan, L. *Protein Sci.* **1996,** *5,* 1584–1593.
103. Englander, S. W.; Sosnick, T. R.; Mayne, L. C.; Shtilerman, M.; Qi, P. X.; Bai, Y. W. *Acc. Chem. Res.* **1998,** *31,* 737–744.
104. Dumont, M. E.; Corin, A. F.; Campbell, G. A. *Biochemistry* **1994,** *33,* 7368–7378.
105. Scholtz, J. M.; Baldwin, R. L. *Annu. Rev. Biophys. Biomol. Struct.* **1992,** *21,* 95–118.
106. Chakrabartty, A.; Baldwin, R. L. *Adv. Protein Chem.* **1995,** *46,* 141–176.
107. Baldwin, R. L. *Nature Struct. Biol.* **1999,** *6,* 814–817.
108. Benson, D. R.; Hart, B. R.; Zhu, X.; Doughty, M. B. *J. Am. Chem. Soc.* **1995,** *117,* 8502–8510.
109. Arnold, P. A.; Shelton, W. R.; Benson, D. R. *J. Am. Chem. Soc.* **1997,** *119,* 3181–3182.
110. Harbury, H. A.; Cronin, J. R.; Fanger, M. W.; Hettinger, T. P.; Murphy, A. J.; Myer, Y. P.; Vinogradov, S. N. *Proc. Natl. Acad. Sci. USA* **1965,** *54,* 1658–1664.
111. Munro, O. Q.; Marques, H. M. *Inorg. Chem.* **1996,** *35,* 3752–3767.
112. Low, D. W.; Winkler, J. R.; Gray, H. B. *J. Am. Chem. Soc.* **1996,** *118,* 117–120.
113. Geier III, G. R.; Sasaki, T. *Tetrahedron Lett.* **1997,** *38,* 3821–3824.
114. Karpishin, T. B.; Vannelli, T. A.; Glover, K. J. *J. Am. Chem. Soc.* **1997,** *119,* 9063–9064.
115. Nastri, F.; Lombardi, A.; Morelli, G.; Maglio, O.; D'Auria, G.; Pedone, C.; Pavone, V. *Chem. Eur. J.* **1997,** *3,* 340–349.
116. Nastri, F.; Lombardi, A.; Morelli, G.; Pedone, C.; Pavone, V.; Chottard, G.; Battioni, P.; Mansuy, D. *JBIC* **1998,** *3,* 671–681.
117. Fattorusso, R.; De Pasquale, C.; Morelli, G.; Pedone, C. *Inorg. Chim. Acta* **1998,** *278,* 76–82.
118. D'Auria, G.; Maglio, O.; Nastri, F.; Lombardi, A.; Morelli, G.; Morelli, G.; Paolillo, L.; Pedone, C.; Pavone, V. *Chem. Eur. J.* **1997,** *3,* 350–362.
119. Sakamoto, S.; Sakurai, S.; Ueno, A.; Mihara, H. *Chem. Commun.* **1997,** *9,* 1221–1222.
120. Sakamoto, S.; Obataya, I.; Ueno, A.; Mihara, H. *J. Chem. Soc. Perkin Trans. 2,* **1999,** 2059–2069.
121. Sakamoto, S.; Ueno, A.; Mihara, H. *Chem. Commun.* **1998,** *10,* 1073–1074.
122. Mihara, H.; Haruta, Y.; Sakamoto, S.; Nishino, N.; Aoyagi, H. *Chem. Lett.* **1996,** 1–2.

123. Sakamoto, S.; Ueno, A.; Mihara, H. *J. Chem. Soc. Perkin Trans.* **1998,** *2,* 2395–2404.
124. Mihara, H.; Tomizaji, K.; Fujimoto, T.; Sakamoto, S.; Aoyagi, H.; Nishino, N. *Chem. Lett.* **1996,** 187–188.
125. Huffman, D. L.; Rosenblatt, M. M.; Suslick, K. S. *J. Am. Chem. Soc.* **1998,** *120,* 6183–6184.
126. Choma, C. T.; Lear, J. T.; Nelson, M. J.; Dutton, P. L.; Robertson, D. E.; DeGrado, W. F. *J. Am. Chem. Soc.* **1994,** *116,* 856–865.
127. Robertson, D. E.; Farid, R. S.; Moser, C. C.; Urbauer, J. L.; Mulholland, S. E.; Pidikiti, R.; Lear, J. D.; Wand, A. J.; DeGrado, W. F.; Dutton, P. L. *Nature* **1994,** *368,* 425–432.
128. Beratan, D. N.; Onuchic, J. N.; Winkler, J. R.; Gray, H. B. *Science* **1992,** *258,* 1740–1741.
129. Page, C. C.; Moser, C. C.; Chen, X.; Dutton, P. L. *Nature* **1999,** *402,* 47–52.
130. Moser, C. C.; Keske, J. M.; Warncke, K.; Farid, R. S.; Dutton, P. L. *Nature* **1992,** *355,* 796–802.
131. Rabanal, F.; DeGrado, W. F.; Dutton, P. L. *Tetrahedron Lett.* **1996,** *37,* 1347–1350.
132. Rabanal, F.; DeGrado, W. F.; Dutton, P. L. *J. Am. Chem. Soc.* **1996,** *118,* 473–474.
133. Gibney, B. R.; Rabanal, F.; Skalicky, J. J.; Wand, A. J.; Dutton, P. L. *J. Am. Chem. Soc.* **1999,** *121,* 4952–4960.
134. Gibney, B. R.; Johansson, J. S.; Rabanal, F.; Skalicky, J. J.; Wand, A. J.; Dutton, P. L. *Biochemistry* **1997,** *36,* 2798–2806.
135. Gibney, B. R.; Rabanal, F.; Reddy, K. S.; Dutton, P. L. *Biochemistry* **1998,** *37,* 4635–4643.
136. Gibney, B. R.; Mulholland, S. E.; Rabanal, F.; Dutton, P. L. *Proc. Natl. Acad. Sci. USA* **1996,** *93,* 15041–15046.
137. Rojas, N. R. L.; Kamtekar, S.; Simons, C. T.; Mclean, J. E.; Vogel, K. M.; Spiro, T. G.; Farid, R. S.; Hecht, M. H. *Protein Sci.* **1997,** *6,* 2512–2524.
138. Kamtekar, S.; Schiffer, J. M.; Xiong, H.; Babik, J. M.; Hecht, M. H. *Science* **1993,** *262,* 1680–1685.
139. Isogai, Y.; Ota, M.; Fujisawa, T.; Izuno, H.; Mukai, M.; Nakamura, H.; Iizuka, T.; Nishikawa, K. *Biochemistry* **1999,** *38,* 7431–7443.
140. Åkerfeldt, K. S.; Kim, R. S.; Camac, D.; Groves, J. T.; Lear, J. D.; DeGrado, W. F. *J. Am. Chem. Soc.* **1992,** *114,* 9656–9657.
141. Mihara, H.; Tomizaki, K.; Fujimoto, T.; Sakamoto, S.; Aoyagi, H.; Nishino, N. *Chem. Lett.* **1996,** 187–188.
142. Ushiyama, M.; Yoshino, A.; Yamamura, T.; Shida, Y.; Arisaka, F. *Bull. Chem. Soc. Jpn.* **1999,** *72,* 1351–1364.
143. Rau, H. K.; Haehnel, W. *J. Am. Chem. Soc.* **1998,** *120,* 468–476.
144. Rau, H. K.; DeJonge, N.; Haehnel, W. *Angew. Chem. Int. Ed.* **2000,** *39,* 250–253.
145. Dutton, P. L.; Moser, C. C.; Sled, V. D.; Daldal, F.; Ohnishi, T. *Biochim. Biophys. Acta* **1998,** *1364,* 245–257.
146. Mulholland, S. E.; Gibney, B. R.; Rabanal, F.; Dutton, P. L. *J. Am. Chem. Soc.* **1998,** *120,* 10296–10302.
147. Mulholland, S. E.; Gibney, B. R.; Rabanal, F.; Dutton, P. L. *Biochemistry* **1999,** *38,* 10442–10448.
148. Gibney, B. R.; Isogai, Y.; Reddy, K. S.; Rabanal, F.; Grosset, A. M.; Moser, C. C.; Dutton, P. L. *Biochemistry,* in press.
149. Sharp, R. E.; Moser, C. C.; Rabanal, F.; Dutton, P. L. *Proc. Natl. Acad. Sci. USA* **1998,** *95,* 10465–10470.
150. Rau, H. R.; DeJonge, N.; Haehnel, W. *Proc. Natl. Acad. Sci. USA* **1998,** *95,* 11526–11531.

151. Myer, Y. P.; Pande, A. In "Circular Dichroism Studies of Heme Proteins"; Dolphin, D., Ed.; Academic Press: New York, 1978; Vol. III, p. 271.
152. Blauer, G.; Sreerama, N.; Woody, R. W. *Biochemistry* **1993,** *32,* 6674–6679.
153. Nakanishi, K.; Berova, N.; Woody, R. W. "Circular Dichroism: Principles and Applications"; VCH: New York, 1994, p. 570.
154. Wang, M. X.; Kennedy, M. L.; Hart, B. R.; Benson, D. R. *Chem. Commun.* **1997,** *9,* 883–884.
155. Lui, D.; Williamson, D. A.; Kennedy, M. L.; Williams, T. D.; Morton, M. M.; Benson, D. R. *J. Am. Chem. Soc.* **1999,** *121,* 11798–11812.
156. Spiro, T. G. "Biological Applications of Raman Spectroscopy"; Wiley: New York, 1987; Vol. 3.
157. Kitagawa, T.; Ozaki, Y. *Struct. Bonding* **1987,** *64,* 71–114.
158. Procyk, A. D.; Bocian, D. F. *Annu. Rev. Phys. Chem.* **1992,** *43,* 465–496.
159. Arnold, P. A.; Benson, D. R.; Brink, D. J.; Hendrich, M. P.; Jas, G. S.; Kennedy, M. L.; Petasis, D. T.; Wang, M. *Inorg. Chem.* **1997,** *36,* 5306–5315.
160. Shifman, J. M.; Moser, C. C.; Kalsbeck, W. A.; Bocian, D. F.; Dutton, P. L. *Biochemistry* **1998,** *37,* 16815–16827.
161. Sharp, R. E.; Diers, J. R.; Bocian, D. F.; Dutton, P. L. *J. Am. Chem. Soc.* **1998,** *120,* 7103–7104.
162. Kalsbeck, W. A.; Robertson, D. E.; Pandey, R. K.; Smith, K. M.; Dutton, P. L.; Bocian, D. F. *Biochemistry* **1996,** 3429–3438.
163. Peisach, J.; Blumberg, W. E.; Adler, A. *Ann. NY Acad. Sci.* **1973,** *206,* 310–327.
164. Taylor, C. P. S. *Biochim. Biophys. Acta* **1977,** *491,* 137–149.
165. Gibney, B. R.; Rabanal, F.; Skalicky, J. J.; Wand, A. J.; Dutton, P. L. *J. Am. Chem. Soc.* **1997,** *119,* 2323–2324.
166. Walker, F. A.; Huynh, B. H.; Scheidt, W. R.; Osvath, S. R. *J. Am. Chem. Soc.* **1986,** *108,* 5288–5297.
167. Fahnenschidt, M.; Rau, H. K.; Bittl, R.; Haehnel, W.; Lubitz, W. *Chem. Eur. J.* **1999,** *5,* 2327–2334.
168. Gibney, B. R.; Dutton, P. L. *Protein Sci.* **1999,** *8,* 1888–1898.
169. Horton, P.; Whitmarsh, J.; Cramer, W. A. *Arch. Biochem. Biophys.* **1976,** *176,* 244–247.
170. Izadi, N.; Haladjian, H. Y.; Goldberg, M. E.; Wandersman, C.; Delepeirre, M.; Lecroisey, A. *Biochemistry* **1997,** *36,* 7050–7057.
171. Kennedy, M. L.; Gibney, B. R.; Dutton, P. L.; Rodgers, K. R.; Benson, D. R. Submitted for publication.
172. Grosset, A. M.; Rabanal, F.; Farid, R. S.; Robertson, D. E.; DeGrado, W. F; Dutton, P. L. In "Peptides: Chemistry, Structure and Biology"; Kaumaya, T. P. and Hodges, R. S. Eds.; Mayflower Scientific, 1996, pp. 573–574.
173. Grosset, A. M.; Gibney, B. R.; Rabanal, F.; Moser, C. C.; Dutton, P. L. Submitted for publication.
174. Feng, Y.; Sligar, S. G. *Biochemistry* **1991,** *30,* 10150–10155.
175. Sakamoto, S.; Obataya, I.; Ueno, A.; Mihara, H. *Chem. Commun.* **1999,** 1111–1112.
176. Hargrove, M. S.; Barrick, D.; Olson, J. S. *Biochemistry* **1996,** *35,* 11293–11299.
177. Beaven, G. H.; Chen, S. H.; D'Albis, A.; Gratzer, W. B. *Eur. J. Biochem.* **1974,** *41,* 539–546.
178. Williamson, D. A.; Benson, D. R. *Chem. Commun.* **1998,** *10,* 961–962.
179. Collman, J. P.; Rapta, M.; Bröring, M.; Raptova, L.; Schwenninger, R.; Boitrel, B.; Fu, L.; L'Her, M. *J. Am. Chem. Soc.* **1999,** *121,* 1387–1388.
180. Collman, J. P.; Fu, L. *Acc. Chem. Res.* **1999,** *32,* 455–463.

181. Shifman, J. M.; Gibney, B. R.; Sharp, R. E.; Dutton, P. L. *Biochemistry,* in press.
182. Gibney, B. R.; Huang, S. S.; Skalicky, J. J.; Fuentes, E. J.; Wand, A. J.; Dutton, P. L. Submitted for publication.
183. Chen, X.; Moser, C. C.; Pilloud, D. L.; Dutton, P. L. *J. Phys. Chem. B* **1998,** *33,* 6425–6432.
184. Chen, X.; Moser, C. C.; Pilloud, D. L.; Gibney, B. R.; Dutton, P. L. *J. Phys. Chem. B* **1999,** *103,* 9029–9037.
185. Pilloud, D. L.; Rabanal, F.; Gibney, B. R.; Farid, R. S.; Moser, C. C.; Dutton, P. L. *J. Phys. Chem. B* **1998,** *102,* 1926–1937.
186. Katz, E.; Heleg-Shabtai, V.; Willner, I.; Rau, H. K.; Haehnel, W. *Angew. Chem. Int. Ed.* **1998,** *37,* 3253–3256.
187. Willner, I.; Heleg-Shabtai, V.; Katz, E.; Rau, H. K.; Haehnel, W. *J. Am. Chem. Soc.* **1999,** *121,* 6455–6468.

SUBJECT INDEX

D

O

CONTENTS OF PREVIOUS VOLUMES

VOLUME 45

VOLUME 46

VOLUME 47

ISBN 0-12-023651-6